Improving comfort in clothing

The Textile Institute and Woodhead Publishing

The Textile Institute is a unique organisation in textiles, clothing and footwear. Incorporated in England by a Royal Charter granted in 1925, the Institute has individual and corporate members in over 90 countries. The aim of the Institute is to facilitate learning, recognise achievement, reward excellence and disseminate information within the global textiles, clothing and footwear industries.

Historically, The Textile Institute has published books of interest to its members and the textile industry. To maintain this policy, the Institute has entered into partnership with Woodhead Publishing Limited to ensure that Institute members and the textile industry continue to have access to high calibre titles on textile science and technology.

Most Woodhead titles on textiles are now published in collaboration with The Textile Institute. Through this arrangement, the Institute provides an Editorial Board which advises Woodhead on appropriate titles for future publication and suggests possible editors and authors for these books. Each book published under this arrangement carries the Institute's logo.

Woodhead books published in collaboration with The Textile Institute are offered to Textile Institute members at a substantial discount. These books, together with those published by The Textile Institute that are still in print, are offered on the Woodhead web site at: www.woodheadpublishing.com. Textile Institute books still in print are also available directly from the Institute's website at: www.textileinstitutebooks.com

A list of Woodhead books on textile science and technology, most of which have been published in collaboration with the Textile Institute, can be found on pages xv–xxi.

Woodhead Publishing Series in Textiles: Number 106

Improving comfort in clothing

Edited by
Guowen Song

WP
WOODHEAD PUBLISHING

Oxford Cambridge Philadelphia New Delhi

Published by Woodhead Publishing Limited in association with The Textile Institute
Woodhead Publishing Limited,
80 High Street, Sawston, Cambridge CB22 3HJ, UK
www.woodheadpublishing.com

Woodhead Publishing, 1518 Walnut Street, Suite 1100, Philadelphia,
PA 19102-3406, USA

Woodhead Publishing India Private Limited, G-2, Vardaan House,
7/28 Ansari Road, Daryaganj, New Delhi – 110002, India
www.woodheadpublishingindia.com

First published 2011, Woodhead Publishing Limited

British Library Cataloguing in Publication Data
A catalogue record for this book is available from the British Library.

ISBN 978-0-08-101432-5 (print)
ISBN 978-0-85709-064-5 (online)
ISSN 2042-0803 Woodhead Publishing Series in Textiles (print)
ISSN 2042-0811 Woodhead Publishing Series in Textiles (online)

The publisher's policy is to use permanent paper from mills that operate a sustainable forestry policy, and which has been manufactured from pulp which is processed using acid-free and elemental chlorine-free practices. Furthermore, the publisher ensures that the text paper and cover board used have met acceptable environmental accreditation standards.

Typeset by RefineCatch Limited, Bungay, Suffolk, UK

Contents

Contributor contact details

(* = main contact)

Editor

Dr Guowen Song
Department of Human Ecology
331 Human Ecology Building
University of Alberta
Edmonton, Alberta
T6G 2N1
Canada
E-mail: guowen.song@ualberta.ca
gwsongsgs@gmail.com

Chapter 1

Dr Asim Kumar Roy Choudhury,* Dr Prabal Kumar Majumdar and Dr Chakradhar Datta
Government College of Engineering and Textile Technology
Serampore-712201
Dt. Hooghly (W.B.)
India
E-mail: akrc2008@yahoo.in

Chapter 2

Professor S. A. Hosseini Ravandi*
Department of Textile Engineering
Isfahan University of Technology
Isfahan
84154
Iran
E-mail: hoseinir@cc.iut.ac.ir
hoseinir2000@gmail.com

Assistant Professor Masoumeh Valizadeh
Faculty of Engineering
Department of Textile Engineering
University of Guilan
Rasht 3756
Iran

Chapter 3

A/Professor John Stanton
Department of Agriculture and Food Western Australia
3 Baron-Hay Court
South Perth Western Australia 6151
Australia
E-mail: jstanton@agric.wa.gov.au

Chapters 4 and 6

Dr Fatma Selcen Kilinc-Balci
National Institute for Occupational Safety and Health
National Personal Protective Technology Laboratory Pittsburgh
626 Cochrans Mill Road
P.O. Box 18070
Pittsburgh, PA 15236
USA
E-mail: fselcen@gmail.com

Chapter 5

Lubos Hes
Technical University of Liberec
Czech Republic

Dr John Williams*
TEAM Research Group
De Montfort University
Leicester
LE1 9BH
UK
E-mail: jtw@dmu.ac.uk

Chapter 7

Chu Po Ho, Professor Jintu Fan,* Professor Edward Newton and Dr Raymond Au
Institute of Textiles and Clothing
The Hong Kong Polytechnic University
Hung Hom
Kowloon
Hong Kong
P.R. China
E-mail: tcfanjt@inet.polyu.edu.hk

Chapter 8

Dr R. S. Rengasamy
Department of Textile Technology
Indian Institute of Technology, Delhi
Hauz Khas
New Delhi – 110016
India
E-mail: rsr60@hotmail.com
rsrengasamy@gmail.com

Chapter 9

Dr Apurba Das* and Professor R. Alagirusamy
Department of Textile Technology
Indian Institute of Technology
Hauz Khas
New Delhi – 110016
India
E-mail: apurba65@gmail.com
apurba_das@hotmail.com

Chapter 10

Dr Penelope Watkins
Research Fellow 3D Design and Technical Fashion
Associate Director Centre for Fashion Science
London College of Fashion
20 John Princes Street
London
W1G 0BJ
UK
E-mail: p.a.watkins@fashion.arts.ac.uk

55 **Innovation and technology of women's intimate apparel**
W. Yu, J. Fan, S. C. Harlock and S. P. Ng

56 **Thermal and moisture transport in fibrous materials**
Edited by N. Pan and P. Gibson

57 **Geosynthetics in civil engineering**
Edited by R. W. Sarsby

58 **Handbook of nonwovens**
Edited by S. Russell

59 **Cotton: Science and technology**
Edited by S. Gordon and Y-L. Hsieh

60 **Ecotextiles**
Edited by M. Miraftab and A. R. Horrocks

61 **Composite forming technologies**
Edited by A. C. Long

62 **Plasma technology for textiles**
Edited by R. Shishoo

63 **Smart textiles for medicine and healthcare**
Edited by L. Van Langenhove

64 **Sizing in clothing**
Edited by S. Ashdown

65 **Shape memory polymers and textiles**
J. Hu

66 **Environmental aspects of textile dyeing**
Edited by R. Christie

67 **Nanofibers and nanotechnology in textiles**
Edited by P. Brown and K. Stevens

68 **Physical properties of textile fibres Fourth edition**
W. E. Morton and J. W. S. Hearle

69 **Advances in apparel production**
Edited by C. Fairhurst

70 **Advances in fire retardant materials**
Edited by A. R. Horrocks and D. Price

71 **Polyesters and polyamides**
Edited by B. L. Deopura, R. Alagirusamy, M. Joshi and B. S. Gupta

72 **Advances in wool technology**
Edited by N. A. G. Johnson and I. Russell

73 **Military textiles**
Edited by E. Wilusz

74 **3D fibrous assemblies: Properties, applications and modelling of three-dimensional textile structures**
J. Hu

75 **Medical and healthcare textiles**
Edited by S. C. Anand, J. F. Kennedy, M. Miraftab and S. Rajendran

76 **Fabric testing**
Edited by J. Hu

77 **Biologically inspired textiles**
Edited by A. Abbott and M. Ellison

78 **Friction in textile materials**
Edited by B. S. Gupta

79 **Textile advances in the automotive industry**
Edited by R. Shishoo

80 **Structure and mechanics of textile fibre assemblies**
Edited by P. Schwartz

81 **Engineering textiles: Integrating the design and manufacture of textile products**
Edited by Y. E. El-Mogahzy

82 **Polyolefin fibres: industrial and medical applications**
Edited by S. C. O. Ugbolue

83 **Smart clothes and wearable technology**
Edited by J. McCann and D. Bryson

84 **Identification of textile fibres**
Edited by M. Houck

85 **Advanced textiles for wound care**
Edited by S. Rajendran

86 **Fatigue failure of textile fibres**
Edited by M. Miraftab

87 **Advances in carpet technology**
Edited by K. Goswami

88 **Handbook of textile fibre structure Volume 1 and Volume 2**
Edited by S. J. Eichhorn, J. W. S. Hearle, M. Jaffe and T. Kikutani

89 **Advances in knitting technology**
Edited by K-F. Au

90 **Smart textile coatings and laminates**
Edited by W. C. Smith

91 **Handbook of tensile properties of textile and technical fibres**
Edited by A. R. Bunsell

92 **Interior textiles: design and developments**
Edited by T. Rowe

93 **Textiles for cold weather apparel**
Edited by J. T. Williams

94 **Modelling and predicting textile behaviour**
Edited by X. Chen

95 **Textiles, polymers and composites for buildings**
Edited by G. Pohl

96 **Engineering apparel fabrics and garments**
J. Fan and L. Hunter

97 **Surface modification of textiles**
Edited by Q. Wei

98 **Sustainable textiles**
Edited by R. S. Blackburn

99 **Advances in textile fibre spinning technology**
Edited by C. A. Lawrence

100 **Handbook of medical textiles**
Edited by V. T. Bartels

101 **Technical textile yarns**
Edited by R. Alagirusamy and A. Das

102 **Applications of nonwovens in technical textiles**
Edited by R. A. Chapman

103 **Colour measurement: principles, advances and industrial applications**
Edited by M. L. Gulrajani

104 **Textiles for civil engineering**
Edited by R. Fangueiro

105 **New product development in textiles**
Edited by B. Mills

106 **Improving comfort in clothing**
Edited by G. Song

107 **Advances in textile biotechnology**
Edited by V. A. Nierstrasz and A. Cavaco-Paulo

108 **Textiles for hygiene**
Edited by B. McCarthy

109 **Nanofunctional textiles**
Edited by Y. Li

110 **Joining textiles**
Edited by I. Jones and G. Stylios

111 **Soft computing in textile engineering**
Edited by A. Majumdar

112 Textile design
Edited by A. Briggs-Goode and K. Townsend

113 Biotextiles as medical implants
Edited by M. King and B. Gupta

114 Textile thermal bioengineering
Edited by Y. Li

115 Woven textile structure
B. K. Behera and P. K. Hari

116 Handbook of textile and industrial dyeing Volume 1: Principles, processes and types of dyes
Edited by M. Clark

117 Handbook of textile and industrial dyeing Volume 2: Applications of dyes
Edited by M. Clark

118 Handbook of natural fibres Volume 1: Types, properties and factors affecting breeding and cultivation
Edited by R. Kozlowski

119 Handbook of natural fibres Volume 2: Processing and applications
Edited by R. Kozlowski

120 Functional textiles for improved performance, protection and health
Edited by N. Pan and G. Sun

121 Computer technology for textiles and apparel
Edited by J. Hu

122 Advances in military textiles and personal equipment
Edited by E. Sparks

123 Specialist yarn, woven and fabric structure: Developments and applications
Edited by R. H. Gong

Preface

Human comfort is complex and subjective, and is influenced psychologically and physiologically by clothing and surrounding environmental conditions. Clothing as a near environment of the human body plays a vital role in achieving human comfort and over the past few decades, extensive and systematic investigations of clothing comfort, function, and ergonomics have been conducted, specifically with protective clothing. The mechanisms and underlying principles associated with human physiological needs, comfort attributes of clothing, and their interaction with a variety of environments have been formalized and established. Methods for the study and evaluation of human comfort and clothing function have also been developed, and findings and discoveries from these studies have led to the development of high performance fibers, novel structures for yarns and fabrics, and new concepts for clothing systems. The development of hollow and profiled fibers, which manage heat and moisture transport in sportswear and cold weather clothing, are excellent examples of new functional fibers. Numerous mathematical models involving the human body, clothing, and environment provide useful tools for identifying key parameters in material design and for predicting clothing performance under extreme environmental conditions. However, there is still much work left to do, particularly for protective clothing. The additional requirements of these garments to provide protection against hazards while simultaneously maintaining an acceptable level of human comfort poses a tremendous challenge. As a result, the performance provided by the individual pieces of protective clothing or the clothing system ensemble has been significantly compromised.

This book presents a holistic and theoretical review of knowledge concerning the physiological theory of human comfort, the role and function of clothing, and the interaction of clothing with a variety of environmental conditions. Included are discussions of the impact of thermal (heat and flame), chemical, biological, radiation, and nuclear (CBRN) hazards on human thermal comfort when wearing protective clothing. The comprehensive reviews integrate the development of theories, textile materials and garments, and testing and evaluation.

There are three parts in the book. Part I introduces the fundamentals of clothing comfort and comfort assessment. Part II discusses the key principles of clothing thermal comfort, moisture management in apparel, sensorial comfort, garment design factors, and clothing movement comfort. Part III presents discussions of comfort and heat stress issues for protective clothing used by firefighters, military soldiers, medical personnel, as well as cold weather clothing and intimate wear.

Part I covers six chapters. Chapter 1 defines the principles of human thermal comfort and how this relates to heat and moisture transfer between the human body, clothing, and environment. Factors affecting thermal and skin sensorial comfort are presented, and comfort properties and heat stresses associated with wearing protective clothing are reviewed. Chapters 2 and 3 are concerned with the properties of fibers and fabrics and their contribution to clothing comfort. The development of functional fibers and the management of heat and moisture from these fiber structures are reviewed. The unique properties of wool fibers that contribute to garment comfort are also discussed. Chapter 4 provides a review of consumer comfort perception. The dimensions of human comfort are prescribed from human physiological perspectives and the specific properties of clothing. Testing and evaluation of clothing properties, and human physiological comfort and prediction using developed models are focused in Chapters 5 and 6. Extensive studies of these clothing properties and the effects on human comfort have led to the development of numerous models for the study of clothing comfort and performance.

Five chapters are included in Part II. Chapters 7 and 8 consider approaches that improve human thermal comfort by examining heat and moisture transport in clothing. The approaches cover topics on textile materials, garment design, and the possible attachment of wearable devices to garment systems. An in-depth review of moisture transport and relevant mechanisms is also provided. Chapters 9 to 11 deal with sensorial and movement comfort that result from the physical interaction between human skin and clothing. The relationship between sensorial comfort and fabric mechanical properties and applied finishes is discussed, as well as the contribution of garment fit, size, and design in the achievement of movement comfort. The underlying principles covered in these chapters imply that clothing comfort is a result of the complex engineering of textile fibers, yarns, fabric structures, and finishes, and the proper fit of garment designs.

Part III covers chapters concerning current issues in protective clothing, sportswear, cold weather clothing and intimate wear. The heat stress produced when wearing protective clothing can significantly decrease work performance and becomes an important issue for health and safety. Chapter 12 provides a detailed review of existing lab methods for evaluating the heat stress and comfort of firefighter and first responder protective clothing. Methods include the measurement of textile material properties, sweating mannequin evaluation, and human subject trials. Research needs in clothing comfort evaluation are also identified. A comprehensive review of protective clothing systems for military use against chemical and biological warfare (CBW) agents is described in Chapter 13. The unique requirements for military use, individual protective equipment

(IPE) development, and performance issues are discussed. In Chapter 14, protective clothing used by hospital personnel is described, focusing on issues involved in achieving a balance between function and comfort. Interference with human activity, performance, and extreme environmental conditions are considered in Chapters 15 and 16 for sportswear and cold weather clothing. Approaches for achieving comfort are emphasized, with a focus on understanding the mechanisms associated with heat and moisture transport in textiles and clothing. Chapter 17 presents a detailed review of intimate wear comfort. Given the proximity to human skin, intimate wear is an important layer contributing to overall clothing comfort.

The challenges posed by the multifunctional requirements of a high level of protection with appropriate physiological burdens of protective clothing have led to the development of new materials and novel clothing systems. There is no doubt that the next generation of textiles will benefit from advanced technology, including nanotechnology, wearable sensors, embedded electronics, and processors. It is my hope that this volume will provide useful knowledge and helpful information on clothing comfort for researchers and engineers in universities, research institutes, and in industry. This volume is the collective effort of many authors, and I wish to extend my sincerest appreciation for their contributions, cooperation, and patience. Special thanks to Kathryn Picking, Beatrice Bertram, Mandy Kingsmill, Cathryn Freear and Francis Dodds at Woodhead Publishing Limited in Cambridge for their patience, persistence, and rapid response in the development of this volume.

Guowen Song
Edmonton, 2010

Part I

Fundamentals of comfort and assessment

1
Factors affecting comfort: human physiology and the role of clothing

A. K. ROY CHOUDHURY, P. K. MAJUMDAR and C. DATTA, Government College of Engineering and Textile Technology, India

Abstract: This chapter discusses the role of body components in maintaining body temperature and the principles of heat transfer to and away from the human body. Various aspects of thermal and skin sensational clothing comfort are explained. Special clothing for protection of the body from external hazards and their comfort properties are reviewed.

Key words: metabolic rate, human heat balance, mean radiation temperature, clothing comfort, clo value, protective clothing.

The human environment must be aesthetically pleasing and must provide light, air and thermal comfort. The benefits of human-friendly atmosphere are:

- increased attention to work resulting in increased productivity, improved quality of products and services with fewer errors
- reduced absenteeism
- lesser number of accidents
- reduced health hazards.

When the comfort condition exists, the mind is alert and the body operates at maximum efficiency. It has been found that maximum productivity occurs under comfortable conditions and that industrial accidents increase at higher and lower temperatures. Postural discomfort due to a cold feeling results in just as many accidents as does mental dullness caused by a too warm environment.

1.1 Definition of comfort

Comfort is a fundamental and universal need of a human being. However, it is very complex and is very difficult to define. According to Fourt and Hollies (1970) comfort involves thermal and non-thermal components and is related to wear situations such as working, non-critical and critical conditions. The physiological responses of the human body to a given combination of clothing and environmental conditions are predictable when the system reaches steady state. According to Slater (1985), comfort is a pleasant state of physiological, psychological, neuro-physiological and physical harmony between a human being and the environment.

He identified the importance of environment to comfort and defined the following three types:

1. physiological comfort is related to the human body's ability to maintain life,
2. psychological comfort to the mind's ability to keep it functioning satisfactorily without external help, and
3. physical comfort to the effect of the external environment on the body.

Although it is difficult to describe comfort positively, discomfort can be easily described in such terms as prickle, itch, hot and cold. According to Hatch (1993), comfort is 'freedom from pain and from discomfort as a neutral state'. The discomfort arises from too hot, too cold, and odorous or stale atmosphere. Comfort conditions are those that do not cause unpleasant sensation of temperature, drafts (unwanted local cooling), humidity or other aspects of the environment. In ideally conditioned space, people should be unaware of noise, heat or air motion. Comfort depends on subjective perceptions of visual, thermal and tactile sensations, psychological processes, body–apparel interaction and external environmental effects (Li, 2001).

1.2 Human physiological aspect of comfort

1.2.1 Physiological interpretation

Physiological comfort is defined as the achievement of thermal equilibrium at normal body temperature with the minimum amount of bodily regulation. The body feels uncomfortable when it has to work too hard to maintain thermal equilibrium. Under the conditions of comfort, the production of heat is equal to the loss of heat without any action necessary by the heat control mechanisms. When the comfort condition exists, the mind is alert and the body operates at maximum efficiency. When the environmental temperature changes, the body tries to acclimatise by different temperature-regulating mechanisms – clothing also helps in acclimatisation.

1.2.2 Physiology and body temperature

Human beings are warm temperature animals and have a normal internal body temperature of 37 °C (98.6 °F) with tolerance of ±0.5 °C under different climatic conditions. Any departure of body temperature from 37 °C causes changes in the rates of heat loss or heat production to bring the body temperature back to 37 °C. This crucial temperature level is called the set point of the various temperature control mechanisms that regulate the body.

Metabolic activity or oxidisation of foods results in the production of heat which can be controlled partially by controlling metabolic rate. However, metabolism during various activities of the body generates heat at varying rates. Hence, the body must reject heat at the proper rate to keep body temperature constant.

The mental state and physical operations done by the body are disturbed if the internal body temperature rises or falls beyond its normal range and serious

Table 1.1 Physiological responses at different body temperatures

Body temperature	Physiological response
43.3 °C (110 °F)	Brain damage, fainting, nausea
37.8 °C (100 °F)	Sweating
37 °C (98.6 °F)	Normal
< 37 °C (98.6 °F)	Shivering and goose bumps
< 32.2 °C (90 °F)	Speechless
26.5 °C (80 °F)	Stiff and deformed body
< 26.5 °C (80 °F)	Irreversible body cooling

physiological disorders or even death may occur if the temperature rises or falls to extreme levels. Often, the human body's own immunological system also causes the rise of body temperature in order to kill infections or viruses.

The physiological reactions of body temperature will largely depend on the geographical location of the human being. The human being is accustomed to live in a certain atmosphere and can tolerate the temperature range existing in the surrounding area throughout the year. The reported physiological responses at various internal body temperatures are given in Table 1.1. When body temperature falls, the respiratory activity, particularly in muscle tissue, automatically increases and generates more heat. The extreme symptom of this form of body control is shivering (essentially rapid muscle contractions). Studies have shown that shivering can result in a five times increase in metabolism. 'Goose bumps' is really an attempt to raise the body hairs which doesn't work too well since most humans are quite hairless.

1.2.3 Role of body components in regulating body temperature

We have separate heat and cold sensors in our body. Heat sensors, located in the hypothalamus, send signals when skin temperature is higher than 37 °C. Cold sensors, located in the skin, send signals when skin temperature is below 37 °C. The higher the temperature difference, the more is the impulse. If impulses from both types of sensors are of the same magnitude, the body feels thermally neutral – if not, one feels cold or warm.

Role of anterior hypothalamus pre-optic area

The blood which circulates to all body tissues is warmed by the heat released within the body, thereby keeping various parts of the human body at the same temperature. The body temperature is a result of the balance between heat production and heat loss and is mostly regulated by a nervous feedback mechanism. An extremely sensitive portion of brain, called the 'hypothalamus', continuously records the temperature of blood and regulates body temperature, using the nervous system's pathways, to a constant set point of around 37 °C (98.6 °F). It is stimulated when there is a minute change in temperature at any part of the body, say while drinking, eating or touching hot or cold materials.

The hypothalamus is the body's thermostat and the large numbers of heat sensitive as well as cold sensitive neurons in the anterior hypothalamic pre-optic area of the hypothalamus are the temperature sensors for controlling body temperature. The hypothalamus triggers heat controlling mechanisms to increase or decrease heat loss by controlling the flow of blood to the skin, which is decreased or increased by constricting or expanding the blood vessels (vasoconstriction or vasodilatation) within the skin. The sensors in the skin send signals to the brain to show the level of heat gain or loss.

Role of peripheral receptors and posterior hypothalamus

The peripheral receptors located in the skin, the deep body temperature receptors in the spinal cord, abdominal viscera and in and around great veins, mainly detect cold temperatures.

The temperature signals generated from the central and peripheral receptors are transmitted to the posterior hypothalamus where both these signals are combined to control heat-producing as well as heat-conserving reactions of the body.

Role of peripheral blood vessels

Blood has very high thermal conductivity. So when blood flows to the skin from the body core, it transmits heat to the skin. By controlling peripheral blood flow to the skin the body is able to:

- increase the temperature of the skin to speed up elimination of body heat
- support sweating.

With increase in body temperature, the blood vessels in the skin dilate (vasodilatation), resulting in more blood transferring to the skin. As a result skin temperature increases, with consequent increase of heat loss and decrease in body temperature.

In a cold environment, the body may lose more heat than it produces. To avoid this higher rate of heat loss, the outer blood vessels are constricted (vasoconstriction), thereby reducing blood flow to the outer surface of the skin and decreasing heat loss and conserving body heat.

The skin surface acts as a layer of insulation between the interior of the body and the environment. This may also happen when a light sweater is put on the body. If the body is still losing too much heat, the control device increases heat production by involuntary muscular activity or shivering. When heat loss is too great, the body tends to bend up and undergo muscular tension, resulting in a strained posture and physical exhaustion if the condition persists for any length of time.

Role of the lungs and respiratory tract

Evaporation of some water from the lungs and respiratory tract causes a minor amount of heat loss from the body.

Role of the heart

While losing a significant quantity of body heat, an increased amount of the blood pumped by the heart goes directly from the heart to the skin and back to the heart, bypassing the brain and other organs. As a result, people experience a feeling of lethargy and mental dullness. In a hot environment, there is increased strain on the heart – it beats more rapidly to pump the blood to the periphery and causes more rapid heat loss.

Role of the autonomic nervous system

When the temperature of the body is increased, the sweat glands in the skin are stimulated resulting in opening of the pores of the sweat glands and passing of body fluid through the pores. When this fluid is evaporated it causes cooling of the body. The evaporation of perspiration is largely responsible for heat loss.

Role of the sympathetic nervous system

The sympathetic nervous system stimulation causes liberation of catecholamine (norepinephrine and epinephrine) hormones which increase the metabolic rate of the many tissues of the body and ultimately result in heat generation. In the liver and muscle these two hormones cause glycogenolysis (production of glucose from glycogen). Sympathetic stimulation causes brown fat burning to generate heat by non-shivering thermogenesis. Premature babies do not have sufficient brown fat and so are more vulnerable to hypothermia (cooling of the body). Vasoconstriction in the peripheral blood vessels is the result of sympathetic stimulation. So prevention of excessive heat loss from the body is the main function as far as the sympathetic nervous system is concerned.

When a person is cold stressed, the skin temperature receptors send signals to the central hypothalamic region, resulting in general sympathetic nervous system stimulation and rapid rise in the level of circulating norepinephrine. This catecholamine surge mediates several important thermal responses:

- It causes lipolysis and re-esterification of brown fat stores to release heat.
- The heart rate rises, delivering more oxygen in order to meet the high metabolic needs of non-shivering thermogenesis.
- Peripheral vasoconstriction diverts blood from the skin towards the organs and drives thermogenesis.

Thus thermogenesis not only leads to warming of the body but also depletes the endogenous substances because of excessive metabolism (Ghai, 2004). The different mechanisms to regulate body temperature are, therefore, closely interrelated.

1.2.4 Acclimatisation

Within a limited range of temperature, the body can acclimatise itself to thermal environmental change. Such limits are not large, especially when the change is abrupt, such as when passing from indoors to outdoors. The slower seasonal changes are accommodated more easily and changes in clothing assist this acclimatisation. Whenever the body cannot adjust itself to the thermal environment, heat stroke (at very high temperature) or frost bite (at very low temperature) to death is inevitable.

When exposed to high temperature, sweat secretion occurs. At first the sweat gland secretes primary secretion whose component is similar to the plasma, except that it does not contain plasma proteins. Sodium chloride (NaCl) is excreted from the body in this mechanism resulting in mild hyponatraemia (blood sodium deficiency). When a person is exposed to hot weather for 4–6 weeks, the constituent of the sweat is modified to prevent excessive hyponatraemia. In this condition the secretion of the aldosterone hormone is increased resulting in increased renal absorption of NaCl by the renin-angiotensin-aldosterone mechanism and decrease in the NaCl concentration in the sweat. Thus hyponatraemia is prevented.

1.3 Energy metabolism and physical work

1.3.1 Definition of metabolism

A human requires energy for growth, regeneration, and operation of the body's organs, such as muscle contraction, blood circulation, and breathing. The process of liberation, transformation and utilisation of energy in the body is known as energy metabolism and to be alive, people must metabolise or oxidise food taken into the body, converting it into electrochemical energy so that they can carry out normal bodily functions.

With every energy conversion (from one form to another) process, there is certain conversion efficiency. For the human body, only about 20% of all the potential energy stored in food is available for useful work. The remaining 80% takes the form of heat as a by-product of the conversion. This results in the continuous generation of heat within the body, which must be rejected by means of sensible heat flow (radiation, convection, or conduction) to the surrounding environment or by evaporating body fluids like sweat. If more food energy is ingested than is needed, it is stored as fat tissue for later use.

In the engineering fields, a machine converts fuel into energy for the purpose of doing work. A similar phenomena happens with the human body – the more active the body, the more fuel that is consumed. The rate of heat production within the body is known as the metabolic rate and includes all of the heat given off by all of the chemical reactions taking place in the body. The metabolic rates are the heat released from the body per unit skin area expressed in met units. A met is the average amount of heat produced by a sedentary man, and any metabolic rate can be expressed in multiples of this standard unit. Met is defined in terms of body surface area as:

$$1 \text{ met} = 18.4 \text{ Btuh/ft}^2 \text{ (of body surface)} = 58.2 \text{ W/m}^2 \text{ (of body surface)} = 50 \text{ kcal/m}^2\cdot\text{hr} \quad [1.1]$$

The body surface for a normal adult is 1.7 m^2. Hence, for an average size man, the met unit corresponds to 1.7 × 58.2 or 100 W (approximately) = 360 Btuh = 90 kcal/hr.

While in the idle level of bodily activity corresponding to the state of rest, energy is continuously drawn by life-sustaining organs such as the heart. It requires minimum energy conversion, and thus a minimum amount of heat is released as a by-product. When the body is engaged in additional mental or physical activity, metabolism increases to provide the necessary energy. At the same time, more heat is generated as a by-product. The food currently being digested or, if necessary, from the fat stored inside the body is used as fuel during that time. Again, when the body loses more heat with consequent dropping of internal body temperature, metabolism increases in an effort to stabilise the temperature even though there is no additional mental or physical activity. All of the additional energy metabolised is then converted into heat. The average activity level for the last hour should be used when evaluating the metabolic rate, due to the body's heat capacity. Some examples of typical metabolic rates are given in Table 1.2 (ASHRAE, 1989).

Table 1.2 Metabolic rates for selected human activities

Activity	Metabolic rate		Activity	Metabolic rate	
	Met	Wm^{-2}		Met	Wm^{-2}
Sleeping	0.7	40.7	Walking (2 mph)	2.0	116.4
Seated and reading	0.9	52.4	Walking (3 mph)	2.6	151.3
Seated and writing	1.0	58.2	Walking (4 mph)	3.8	221.2
Seated and typing/talking	1.2	69.8	Car driving	1.5	87.3
Cooking	1.6	93.1	Motor cycle riding	2.0	116.4
House cleaning/ ironing	2.0–3.4	116.4–197.9	Heavy vehicle driving	3.2	186.2
Shopping	1.4	81.5	Fishing	1.2–2.0	69.8–116.4
			Golf	1.4–2.6	81.5–151.3
			Dancing	2.4–4.4	139.7–256.1
			Wrestling	5.0–7.6	291–442.3

Note: actual metabolic rates depend on the relation between the intensity of the given activity and the individual's peak capacity.

1.3.2 Measurement of metabolic rate

The whole-body metabolic rate can be measured in the following ways:

- Direct calorimetry – the total quantity of heat liberated from the body in a given time is measured to determine the whole-body metabolic rate.
- Indirect calorimetry – the energy equivalent of oxygen is measured. For the average diet, the quantity of energy liberated per litre of oxygen used in the body averages about 4.825 calories. The whole-body metabolic rate can be calculated with a high degree of accuracy from the rate of oxygen utilisation.
- The metabolator – an apparatus which records the rate at which the body uses oxygen.

1.3.3 Factors affecting metabolic rate

There are many factors affecting metabolic rate. The factors are:

- Age: The metabolic rate of a young child is almost double that of an old person due to rapid synthesis of cellular materials and growth of the body. Metabolism peaks at ten years of age and minimum at old age.
- Physical exercise: Strenuous exercise causes most dramatic increase in the metabolic rate.
- Body weight and surface area: Metabolic rate increases with the increase in body surface area.
- Hormones: Thyroxine increases the metabolic rate of the whole body by increasing the rates of activity of almost all chemical reactions. Growth hormone and testosterone increase the metabolic rate by increasing basal metabolic rate.
- Food consumption: After consumption of a meal containing large quantities of protein, cellular chemical processes are stimulated. The metabolic rate starts increasing within one hour of food ingestion. The increased level lasts for about 3–12 hours.
- Sympathetic stimulation: This causes release of norepinephrine and epinephrine which increase the metabolic rates of many tissues of the body.
- Climate: Metabolic rate is lowest between 20–30 °C. It increases in cold environmental conditions if the body is not thermally protected.
- Sleep: During sleep, the metabolic rate decreases 10–15% below normal.
- Fever: The metabolic rate increases with fever.
- Malnutrition: In malnutrition there is paucity of necessary food substances in the cell. The metabolic rate, therefore, decreases up to 20–30%.
- Physiological condition: The metabolic rate is increased by 10% in pregnancy and lactation.
- Amount of clothing: The heavy, protective clothing worn in cold weather may add 10–15% to the rate.

1.3.4 Basal metabolic rate

The basal metabolic rate (BMR) is the metabolic rate of a person measured under basal conditions, i.e. when a person is awake and in absolute physical and mental rest after 12 hours of absolute fasting, and when the environmental temperature is 20–25 °C. As long as the person remains healthy, his/her BMR does not vary more than 5–10% except for the age related change, and 85% of normal people have a BMR within 10% of the mean. BMR increases with the increase in body surface area, so to compare BMR between different people, it is expressed as calories per hour per square metre of body surface area.

1.4 Human heat balance

1.4.1 Means for heat transfer to or from the body

Like all mammals, humans 'burn' food for energy and must discard the excess heat. This is accomplished by latent heat loss through evaporation along with the three modes of sensible heat transfer, namely conduction, convection, and radiation. For health reasons, the heat loss should not be too fast or too slow, and a very narrow range of body temperature must be maintained. The body thermal balance depends on the following body systems:

- cardiovascular system
- skeleto-muscular system
- central nervous system
- pulmonary system
- digestive system
- thermoregulatory mechanism.

'Heat' is a form of energy that flows from a point of higher temperature to another point of lower temperature. This heat is rejected in two forms:

1. sensible heat transfer and
2. latent heat transfer.

Sensible heat transfer is accompanied by the change in temperature. The gain or loss of sensible heat changes the temperature of the material depending on its property called 'specific heat'. Sensible heat depends on the degree of molecular excitation caused by exposure to radiation, chemical reaction, inter-object friction or contact with a hotter object.

Latent heat changes the state of matter from solid to liquid or liquid to gas. The latent heat of fusion and latent heat of vaporisation are the heat needed per unit mass of solid for melting to liquid or per unit mass of liquid for vaporising to gas respectively. During cooling when a gas liquefies or a liquid solidifies latent heat is released.

1.4.2 Mechanism of heat balance

The human body remains in a state of thermal equilibrium with its environment when it loses heat at exactly the same rate as it gains heat. Mathematically the relationship between the heat production and heat loss can be calculated by the heat balance equation (Ogulata, 2007) as follows:

Heat production = Heat loss

or

$$M - W = C_v + C_k + R + E_{sk} + E_{res} + C_{res} \qquad [1.2]$$

where M = metabolic rate (internal heat production, W/m^2)
W = external work (W/m^2)
C_v = heat loss by convection
C_k = heat loss by thermal conduction (W/m^2)
R = heat loss by thermal radiation (W/m^2)
E_{sk} = heat loss by evaporation from the skin (W/m^2)
E_{res} = evaporative of heat loss due to respiration (W/m^2)
C_{res} = sensible heat loss due to respiration (W/m^2).

The metabolic rate (M) is always positive as the body always produces heat. However, it varies with the degree of exertion. When the body's combined heat loss through radiation, conduction, convection, and evaporation is less than the body's rate of heat production, the excess heat must be stored in body tissue. But the body has a limited thermal storage capacity. Therefore, as its interior becomes warmer, the body reacts to correct the situation by increasing blood flow to the skin surface and increasing perspiration. As a result, body heat loss is increased, thereby maintaining the desired body temperature and the heat balance expressed by equation [1.2]. The dependence of radiation, convection and evaporation on various factors is shown in Table 1.3 (Blankenbaker, 1982).

Table 1.3 Dependence of radiation, convection and evaporation on various factors

	Radiation (R)	Convection (C)	Evaporation (E)
Dry-bulb temperature		√	√
Wet-bulb temperature			√
Temperature difference between body and surroundings	√		
Emissivity of surfaces	√		
Velocity of air		√	√
Surface area	√	√	√
Clothing	√	√	√
Ability to sweat			√
Mean body surface temperature		√	

When heat loss is greater than body heat production, a reversal of the above process occurs and, if necessary, shivering. This increased activity raises the metabolic rate.

Radiation

All bodies emit 'thermal radiation' and loss of heat by radiation occurs in the form of infrared waves. A nude person staying in a room at normal room temperature may lose about 60% of the total heat by radiation. Heat loss may occur by radiation to cooler surfaces or heat gain from warmer surfaces and when radiation encounters a mass, three phenomena may occur:

1. radiation continues its journey unaffected or transmitted,
2. it is deflected from its course or reflected, or
3. it may be absorbed.

Usually, the response of radiation to a material is a combination of transmission, reflection, and absorption. The net exchange of radiant heat between two bodies depends on the difference in temperature between the two bodies. The radiation characteristics of a material are determined by its temperature, emissivity (emitting characteristics), absorptivity, reflectivity, and transmissivity.

Radiation is the net exchange of radiant energy between two bodies across an open space. The human body gains or loses radiant heat, for example, when exposed to an open fire, the sun, or a window on a cold winter day. The earth, the sun, a human body, a wall, a window, or a piece of furniture gains or loses heat by radiation with every other body in the direct line of sight with it. The radiant energy cannot go around corners or be affected by air motion. For example, to keep away from uncomfortable heat of direct sunlight, we take shelter under the shade of a tree as the radiant energy coming directly from the sun cannot bend and enter into the shade of the tree. Nearly all radiant exchanges are between solid surfaces as air is a poor absorber of radiant heat.

If the radiating temperatures of the surrounding surfaces are higher or lower than the body temperature, the radiant heat moves towards or away from the body respectively. In a cold room, the warmer body or its clothing transmits radiant heat to all cooler surfaces such as walls, glass, and any other construction within view. When we sit near a cold window, it will drain a large amount of heat away from our body, making the body feel colder. By putting curtains between the cold window and a person the radiant transfer of heat can be blocked in the same way that a person can cut off the radiant energy from the sun by stepping into the shade of a tree.

The rate of radiant transfer depends on the temperature differential, the thermal absorptivity of the surfaces, and the distance between the surfaces. The body gains or loses heat by radiation according to the difference between the body surface (bare skin and clothing) temperature, and the mean radiation temperature

(MRT) of the surrounding surfaces. Since the surrounding surface temperatures may vary widely, the MRT is calculated as a weighted average of the temperatures of all radiating surfaces in direct line of sight of the body. For two dimensional spaces, it may be calculated as follows:

$$MRT = \frac{\sum T\theta}{360} = \frac{T_1\theta_1 + T_2\theta_2 + \cdots + T_n\theta_n}{360} \quad [1.3]$$

where T is surface temperature and θ is the exposure angle of the surface relative to the occupant, in degrees.

For example, in winter when external temperature is 0 °C, the MRT of a person sitting inside a room with one solid exterior wall, one glass exterior wall and two interior partitions (Fig. 1.1) may be calculated as follows:

Let the temperatures of glass exterior, solid exterior and internal partition walls be 5 °C, 15 °C and 20 °C respectively.
Let the angles made by the body with glass, solid and partition walls be 125 °, 75 ° and 160 ° respectively.
The MRT of the occupant = Σ temperature × angle for each type of wall/360
= (125 × 5 + 75 × 15 + 160 × 20)/360 = 13.75 °C

With the same location of the occupant in summer with outside temperature of 35 °C, MRT may be calculated as follows:

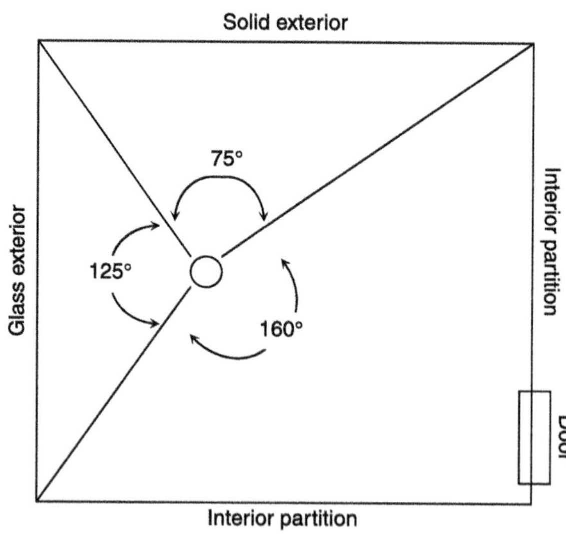

1.1 Calculation of mean radiation temperature.

Let the temperatures of shaded glass exterior, solid exterior and internal partition walls be 30 °C, 27 °C and 25 °C respectively.
The MRT of the occupant = (125 × 30 + 75 × 27 + 160 × 25)/360 = 27.15 °C.

The MRT affects the rate of radiant heat loss from the body or gained by the body from the surface and tends to be close to room air temperature. However, the closeness is affected by the presence of open or uninsulated doors and windows, degree of insulation of the room, presence of hot lights and any other heating medium. The inside surface temperature of a wall will be very close to room air temperature, if insulated.

When MRT is below or above body temperature, the heat will radiate away or towards the body respectively and the value of radiant heat (R) will be positive and negative in respective cases.

A cooled room is comfortable because the body can lose heat by radiation; on the other hand, a hot and humid condition is unpleasant as the body cannot reject excess heat.

The body loses radiant heat according to its surface temperature. For a normally dressed adult in a comfortable situation, the weighted average temperature of the bare skin and clothed surfaces is about 80 °F (27 °C). When air is not flowing (motionless), radiation is the only means for exchange of heat between the body and the environment. Consider a person during the cold season, seated with his or her back near a cool outside wall. Because the radiant heat loss to the cold wall is so high, he or she will feel chilly. As a rule of thumb, if the MRT is 10 °C (5 °F) hotter or colder than comfortable room air conditions, an occupant will feel uncomfortable. Alternate ways to make conditions comfortable are:

- Insulation of the outside wall or hanging an insulating tapestry or wall hanging over the outside wall.
- Changing the position of the desk, moving the person closer to an inside wall. The radiant exchange would then be predominantly influenced by the surface temperature of the inside wall, which would be near the air temperature.
- If the desk cannot be moved, the temperature of the air may be increased by turning up the thermostat. Increasing the air temperature would decrease the convective heat loss from the body. This would balance the heat loss from the body. However, everyone else in the room not sitting near an outside wall will feel too warm.

During the hot season, in a similar situation a person might feel too warm because of the radiant heat the body gains from a warm outside wall or window. In this case, the sensible heat loss from the body could be increased by decreasing the air temperature. This puts one person's body heat loss in balance, but everyone else in the room would be too cool. Thus, not only good, properly operated heating and cooling equipment are important for maintaining comfort, but the building construction itself can also have a strong influence.

Poorly insulated walls and windows should be flagged as comfort problems. Furthermore, the type of occupancy must be borne in mind when analysing the intended comfort conditions.

Convection

Convection is the process of carrying heat stored in a particle of the fluid into another location. Heat loss may occur by convection to cooler surrounding air or heat gain from surrounding warmer air. Air passing over the skin surface not only evaporates moisture, but also transfers sensible heat to or from the body. The faster the rate of air movement, the larger is the temperature difference between the body and surrounding air; and the larger the body surface area, the greater is the rate of heat transfer. When the air temperature is lower than that of the skin (and clothing), the convective heat term (C_v) in equation [1.2] is positive and the body loses heat to the air. If the air is warmer than the skin temperature, the convective heat term (C_v) is negative and the body gains heat from the air. Convection becomes increasingly effective at dissipating heat as air temperature decreases and air movement increases.

Conduction

In this process molecular excitation spreads through a substance or from one substance to another by direct contact. Conduction allows us to lose heat through the soles of the feet or our body when lying or sitting on colder ground. Heat is also lost by dry respiration to cooler air entering the lungs and the warmer air being exhaled, but the amount of heat lost by conduction is usually insignificant. Clothing slows down the rate of conduction and the nature of the clothing also influences the rate of loss (Threlkeld, 1970).

The conduction heat loss or gain occurs through contact of the body with physical objects such as the floor and chairs. If two chairs – one with a metal seat and the other with a fabric seat – have been in a 70 °F (21 °C) room for a period of time, they will both have a temperature of 70 °F (21 °C), but the metal one will feel colder than the one with the woven seat because metal is a good conductor and we sense the rate at which heat is conducted away, not the temperature. Moreover, the metal chair has a smoother surface, which makes a good contact between chair and body, facilitating better conduction. Clothing also plays an important role in conductive heat transfer, insulating us from the warm or cold surface, just as a pot holder protects us from a hot pot.

Evaporation

When the surrounding temperature is higher than that of the skin, the only means by which the body can release heat is the evaporation of perspiration from the skin. When each gram of water evaporates from the body surface, 0.58 calories of heat is lost. Water evaporates insensibly from the skin and the lungs, which causes continual heat loss at a rate of 12–16 calories per hour. The evaporation loss is dependent upon the mass transfer coefficient and the air humidity ratio for a given body surface temperature (Threlkeld, 1970). The heat loss by evaporation is made up of the insensible heat loss by skin diffusion and the heat loss by regulatory sweating. The latent heat loss mechanisms include:

- latent respiration heat loss,
- water diffusion through the skin, and
- evaporation of sweat (skin wetting).

Depending on the temperature of the surrounding objects and air, the human body can either gain or lose heat by radiation, conduction or convection processes of heat transfer. On the other hand, evaporation is exclusively a cooling process. At lower temperature, evaporation usually plays an insignificant role in the body's heat balance. At high temperature, when heat loss by radiation or conduction cannot occur, evaporation becomes the predominant factor for body heat loss.

When surrounding temperatures are comfortable, sensible heat steadily flows from the skin to the surrounding air. The flow rate of this sensible heat depends upon the temperature difference between the skin and air. Depending on the surrounding temperature, humidity, and air velocity, the skin temperature may vary from 4 to 41 °C (40 to 105 °F), even though the internal temperature of the body may remain largely constant.

During the hot summer season, the average surface temperature of an adult staying indoors and wearing comfortable clothing may be approximately 80 °F (27 °C). As the surrounding temperature falls, the skin temperature decreases correspondingly. When the surrounding environment is about 70 °F (21 °C), most people lose sensible heat at a rate which makes them feel comfortable. If the ambient temperature increases and becomes close to the skin temperature, there will be no loss of sensible heat. If the ambient temperature continues to rise, the body cannot lose heat but starts gaining heat from the environment, and the only way of losing heat is by increasing evaporation. When humans are highly active, more metabolic heat is produced with a corresponding increase in evaporative heat losses. A person engaged in strenuous physical work may sweat as much as a quart (¼ gallon) of fluid in an hour.

The evaporation potential of the air determines the rate of evaporation and corresponding heat loss. It depends less on the relative humidity of the surrounding air and largely on the velocity of air. Evaporated moisture is carried away by the passing air from the skin surface. Sufficient heat is taken away from the body by the evaporation of perspiration. The amount of heat lost is equal to the latent heat of vaporisation of the moisture evaporated and the phenomenon is known as latent heat transfer from the body.

Sweating from the skin occurs only when the surrounding temperature is moderately high. However, water evaporation from the respiratory passages and lungs occurs uninterruptedly.

This can be observed when breathing, by exhaling in frosty weather. We generally exhale saturated (100% RH) air, and even at rest, the body requires about 100 Btuh (30 W) of heat to evaporate this moisture from the lungs into the inhaled air. The evaporative heat loss from our lungs and skin plays an important role in disposing of body heat.

The convection and radiant heat loss is greater for lightly clothed subjects. Both decrease with increasing air temperature, while evaporative heat loss increases with increasing air temperature. Heat loss by evaporation is relatively constant below certain air temperatures – approximately 75 °F (24 °C) for the heavily clothed subject and 85 °F (29 °C) for the lightly clothed subject.

1.5 Clothing as near environment

Various environments surround human beings to varying degrees, and among them clothing is the nearest mobile environment. The primary function of clothing is to protect the body against an unsuitable physical environment by forming a layer or layers of barrier. However, clothing serves several functions in human life such as decoration, social status, protection and modesty. Aesthetic clothing according to latest fashion gives the wearer mental comfort and a feeling of looking good, while well-fitting and luxurious dresses enhance the status of the wearer. Clothing can provide a feeling of modesty and also the mental comfort of having the body covered properly as per the standard of the society. At the interface between the human body and its surrounding environment, clothing plays a very important role in determining the subjective perception of comfort status of a wearer. Sometimes it is called a 'second skin'.

Clothing is the aspect of our environment with which we are in closest contact and over which we have the most control and is often used as an extension of one's own body. Clothing is an integral part of human life and to some older adults, becomes the part of their lives over which they can maintain some degree of control.

1.5.1 Perception of comfort

Human perception of clothing comfort is an interaction between physical, physiological and psychological factors with the surrounding environment when wearing a garment. Different aspects of clothing comfort have been studied for many years. Thirty (2003) overviewed apparel comfort issues including the effect of environment, available test methods, fabric handle, moisture and thermal management and psychological comfort.

Comfort is a multidimensional and complex phenomenon. Subjective perception of comfort involves complicated processes in which a large number of stimuli from clothing and external environments communicate to the brain through multi-channels of sensory responses to form subjective perceptions. These perceptions involve a psychological process in which all relevant sensory perceptions are formulated, weighed, combined, and evaluated against past experiences and present desires to form an overall assessment of comfort status.

Pontrelli (1977) developed a Comfort's Gestalt in which the variables influencing comfort status of a wearer were listed and were classified into three groups:

1. physical variables of the environment and the clothing,
2. psycho-physiological parameters of the wearer, and
3. psychological filters of the brain.

The flowchart for the subjective perception of comfort illustrates the process of how the subjective perception of overall comfort is formulated (Fig. 1.2). The physical processes provide the signals or stimuli to the sensory organs of the human body, which will receive them, produce neuro-physiological impulses, send this to the brain, and take action to adjust sweating rate, blood flow, and sometimes heat production by shivering. The brain will process the sensory signals to formulate subjective perception of various individual sensations, and further evaluate and weigh them against past experience and desires, which is influenced by many factors such as physical, environmental, social and cultural surroundings, and state of being.

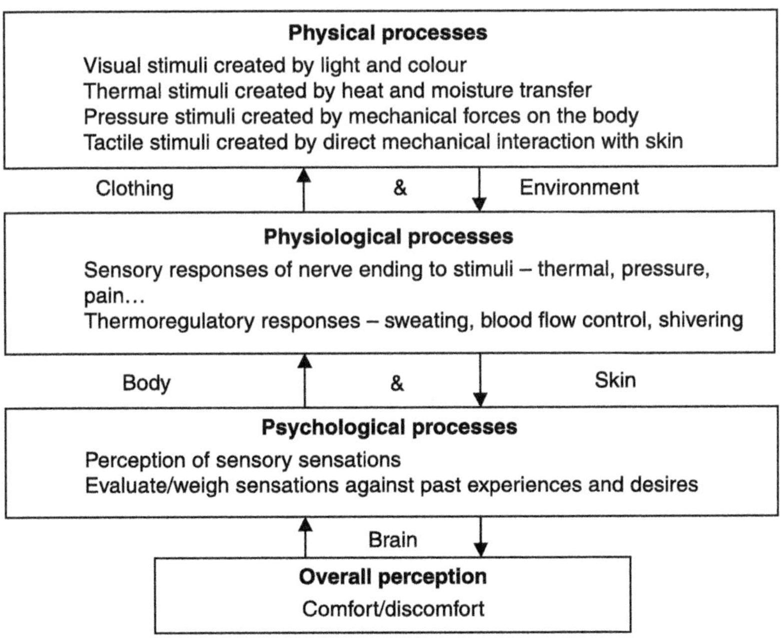

1.2 Subjective perception of comfort.

Bornais (1997) suggested that the thermo-physiological, tactile and psychological stimuli influence the degree of comfort in apparel. For understanding psychological processes, these perceptions are to be measured subjectively. Since no physical instrument can measure thinking or feeling of a wearer (of clothing) objectively, the only way to measure subjective perception is by psychological scaling.

1.5.2 Comfort equation

Human comfort is not just heat balance, but takes into account complex psychological processes. Human judgements or preferences are based on the processing of thermal sensations through several mental processes, but the thermal comforts perceived by different humans are not always the same. Attempts have been made to correlate comfort perceptions with specific physiological processes. Among the various models for quantitative estimation of thermal comfort, the most widely used one was suggested by Fanger (1970). He developed a mathematical model to define the neutral thermal comfort zone of man in different combinations of clothing and at different activity levels. Mean skin temperature and sweat secretion rates were used as physical measures of comfort. He developed a comfort equation (Hui, www.hku.hk) which is as follows:

$$f(M, I_{cl}, V, t_r, t_{db}, P_s) = 0 \qquad [1.4]$$

where M = metabolic rate (met), I_{cl} = cloth index (clo), V = air velocity (m/s), t_r = mean radiant temp. (°C), t_{db} = dry-bulb or ambient temp. (°C), and P_s = water vapour pressure (kPa).

Thermal comfort has been defined as the condition of mind that expresses satisfaction with the thermal environment. Fanger's equation is complex, but it may be transformed to comfort diagrams and can also be used to yield three indices expressing dissatisfaction caused by warm or cool discomfort for the body, namely:

1. Predicted mean vote (PMV)
2. Predicted percentage of people dissatisfied (PPD)
3. Lowest possible percentage dissatisfied (LPPD)

Predicted mean vote

The predicted mean vote (PMV) predicts mean value of the subjective ratings of a group of people in a given environment.

The PMV index is calculated through a complex mathematical function of six major comfort variables. This equation has been empirically developed following an extensive study and monitoring of human beings under varying conditions and a comprehensive statistical analysis of their responses. The PMV relates imbalance between the actual heat flow from the human body in a given environment and the heat flow required for optimum comfort at the specified activity.

$$\text{PMV index} = (0.303e^{-0.036M} + 0.028)*L \qquad [1.5]$$

where M = metabolic rate and L = thermal load defined as the difference between the internal heat production and the heat loss to the actual environment for a person hypothetically kept at comfort values of skin temperature and evaporative heat loss by sweating at the actual activity level. This is similar to the difference of both sides of equation [1.2].

The PMV may be calculated from the available tool on the internet (http://www.atmos.es.mq.edu.au/~rdedear/pmv/) and this method is the basis of International and European Standards (EN ISO, 1984). The PMV index quantifies the degree of discomfort, giving the predicted mean vote of a large group of subjects according to the psychological scale ranging between –3 and + 3.

PMV value	Thermal sensation
– 3	cold
– 2	cool
– 1	slightly cool
0	neutral
+ 1	slightly warm
+ 2	warm
+ 3	hot

Percentage of people dissatisfied

The percentage of people dissatisfied (PPD) with the thermal environment at various conditions has been mathematically related to PMV.

Dissatisfaction is defined as anybody not voting –1, +1 or 0. As per the recommendation of the International Standards Organisation (ISO), PPD of 10% is acceptable and it corresponds to the PMV range of –0.5 to +0.5. Even with PMV equal to zero, about 5% of the people remain dissatisfied. Fanger found that the optimal operative temperature, which satisfies most people at given clothing and activity, ranges between 18 and 22 °C.

Gagge *et al.* (1986) proposed a new index PMV* which simply replaces dry-bulb temperature (t_{db}) in Fanger's comfort equation with MRT. They pointed out that Fanger's PMV is primarily based on heat load; it is not sensitive to changes in relative humidity or vapour pressure, or to the vapour permeability of clothing worn. The new PMV* also respond to heat strain associated with changing humidity of the environment and vapour permeability of clothing.

In the prediction of clothing comfort using a stepwise regression method (Wong and Li, 2006), six regression models with different numbers of fabric properties were generated. However, only the following model showed high correlation:

$$\text{Overall comfort} = -6.44\,(\text{MIU}) + 2.24\,(\text{WC}) + 2.44\,(\text{MWRL}) + 1.15\,(\text{SMD}) + 55.35 \quad [1.6]$$

where MIU is frictional coefficient, WC is compressional energy, MWRL is maximum wetted radius (lower), and SMD is geometric roughness. These physical properties can be represented as the dimensions of moisture, tactile and pressure in relation to clothing comfort. Moisture and thermal sensations can be abstracted into a single sensory factor. The limitation of this model is that the perception of thermal comfort has not been properly incorporated and the number of garments considered for the experiment was very low (only eight).

1.6 Various aspects of clothing comfort

Comfort is related to subjective perception of various sensations. It may be psychological or physiological. Three aspects of clothing comfort are:

1. Thermal comfort – attainment of a comfortable thermal and wetness state; it involves transport of heat and moisture through fabric.
2. Sensorial comfort – the elicitation of various sensations when a textile comes into contact with skin.
3. Body movement comfort – ability of a textile to allow freedom of movement, reduced burden, and body shaping, as required.

External environments (physical, social, and cultural) have great impact on the comfort status of the wearer (Hatch, 1993) and researches have shown that there is a close relationship between moisture and thermal comfort. Moisture comfort and pressure comfort are the most important considerations for denim apparel purchases in both summer and winter. Kamata *et al.* (1988) observed that in fabrics that are less permeable to air, such as twill and denim, the heat transfer coefficient is reduced by the fabric covering.

Ishtiaque (2001) stated that the comfort of athletic apparel depends on optimising interactions between fibre types, spinning, weaving or knitting parameters, fabric density and weight, finishing, fit, and manufacture. Critical functions include thermal retention, UV light resistance, cooling capacity, sweat absorption, rapid drying, antibacterial properties and relaxation without fatigue.

1.6.1 Thermal comfort

Thermal comfort is that condition of mind which expresses satisfaction with the thermal environment (ISO 7730). Human thermal comfort depends on the metabolic rate (internal heat production), the heat loss from the body and the climatic conditions. Clothing modifies the heat loss and moisture loss from the skin surface, so it plays a vital role in the maintenance of heat balance. A clothing system which is suitable for one climate may not be suitable for another as clothing insulation is very important for human thermal comfort (Ogulata, 2007). Good thermal insulation properties are needed in clothing and textiles used in cold climates. The thermal insulation depends on different factors like thickness and number of layers, drape, fibre density, flexibility of layers and adequacy of closures. The thermal insulation value of clothing when it is worn is not just dependent on the insulation value of each individual garment but on the whole outfit as the air gaps between the layers of clothing can add considerably to the total thermal insulation value.

1.6.2 Sensorial comfort

Human skin is the interface between a human body and its environment and contains specialised sensory receptors to detect various external stimuli. The fundamental function of the sensory receptors is to transduce various external stimuli into the standard code by which the nervous systems work. It has been found that the common feature of the transduction is the generation of current flows within the receptor, recorded as a potential change that is proportional to the intensity of the applied stimulus. There are three major stimuli, namely:

1. Mechanical contact with external objects.
2. Temperature changes due to heat flow to or from the body.
3. Damaging traumatic and chemical insults.

Sensorial properties of a fabric depend on the fibre types, the fabric construction (surface structure) and the fabric finishing treatments. Surface properties like friction and roughness, physical properties like tensile, shear, compression and bending and surface coolness or warmness are the important parameters for clothing comfort. A smooth fabric surface has a large contact area with the skin and thus it may feel cool to the skin because a thermal insulative air layer is absent.

Sensorial comfort is a perception of clothing comfort which is the sensory response of nerve endings to external stimuli including thermal, pressure, pain etc. producing neuro-physiological impulses sent to the brain. These sensory signals are processed by the brain to formulate subjective perceptions of sensations and are suitably responded to by adjusting the blood flow, sweating rate or heat production by shivering.

Li (1998) investigated psychological sensory responses to clothing of consumers living in different countries and 26 sensory descriptors were selected. The sensory responses to these descriptors were analysed by oblique principal component cluster analysis. For summer wear and sport-wear, the cluster analysis showed that the 26 sensory descriptors could be classified into four clusters as shown below:

1. Tactile sensations – prickly, tickling, rough, raggy, scratchy, itchy, picky, sticky.

2. Moisture sensations – clammy, damp, wet, sticky, sultry, non-absorbent, clingy.
3. Body fit (pressure) sensations – snug, loose, lightweight, heavy, soft, stiff;
4. Thermal sensations – cold, chilly, cool, warm, hot.

The components of tactile sensations are well defined and do not change much with type of clothing. Some sensations from other clusters (such as heavy, stiff, etc.) become closely associated with this cluster in certain wear conditions. Moisture sensations are also relatively stable and do not change with the type of clothing. However, they interact with thermal sensations (hot and chilly) in sportwear and with tactile sensations in summer wear. The pressure and thermal sensations are not stable – the components are not clearly clustered and change their membership frequently. The pressure sensations interact with tactile and thermal sensations, while thermal sensations interact strongly with moisture sensations. Tactile comfort is associated with the sensations involving direct skin–fabric mechanical interactions. This factor responds largely with the pain receptors in the skin and relates mainly to the surface characteristics of the fabric.

Fabric prickliness

Fabric-evoked prickle has been identified as one of the most irritating discomfort sensations for clothing wear next-to-skin. The degree of discomfort caused by prickle varies from person to person and with the wear situation, and prolonged irritation that evokes the action of scratching the affected area may lead to skin inflammation. Fabric containing wool is unsatisfactory for underwear garments because it causes prickle or skin irritation.

Garnsworthy *et al.* (1985) identified a special type of pain nerve responsible for prickle sensation, which is triggered by a threshold of force of about 0.75 mN. Individual protruding fibre ends from a fabric surface are responsible for triggering the pain nerve endings during contact with the skin.

Fabric itchiness

Like fabric prickle, itch is also found to result from activation of some superficial pain receptors. It has been found that the perception of itchiness in clothing is highly correlated with the perception of prickliness. Li (1988) observed that perception of itchiness is correlated with fibre diameter, fabric thickness at low and high pressures, and fabric surface roughness.

Fabric smoothness

The friction and mechanical interaction between fabric and skin during contact are the key factors determining the perception of roughness, smoothness and scratchiness. These are important tactile sensations determining the comfort performance of next-to-skin wear. A fabric that is perceived to be comfortable at low-humidity conditions may be perceived to be uncomfortable at higher humidity or sweating conditions. The fabric roughness or smoothness is associated with a number of physical properties objectively measured such as surface roughness, friction, prickle, shear and bending stiffness, thickness and aerial density.

Garment fit and pressure comfort

A garment needs to be cut neatly in appearance and should be able to maintain a reserve of comfort for the wearer's dynamic movements. Kirk and Ibrahim (1966) identified three essential components involved in meeting the skin strain requirements – garment fit, garment slip and fabric stretch. 'Garment fit' provides the space allowance for skin strain, which is affected by the ratio of garment size to body size and the nature of garment design. The skin strain is also accommodated by another mechanism called 'garment slip' which is mostly determined by the coefficient of friction between skin and fabric and between different layers of garments. 'Fabric stretch', an important factor in pressure comfort, depends largely on elastic characteristics and elastic recovery properties of fabrics. If a fabric has high friction and stretching resistance, high clothing pressure is likely to be exerted on the body, which could result in discomfort sensations. They also identified that the critical strain areas of the body are the knee, the seat, the back and the elbows. Denton (1970) pointed out that there are four mechanical factors relating to garment comfort namely weight, ease of movement, stretch and ventilation. Ease of movement is largely dependent on garment design and the relative size between body and clothing. Loose fitting allows freedom of movement but may not be desirable in many situations. He also pointed out (Denton, 1971) that the discomfort level of clothing pressure was found to be between 20 and 40 g/cm^2, depending on the individual and the part of the body concerned, which is similar to blood pressure in the capillary blood vessels near the skin surface.

Fabric hand

The concept of fabric hand has long been used in the textile and clothing industries as a description of fabric quality and performance. During wear, clothing continuously comes into contact and interacts dynamically with the skin of the whole body. The fabric hand property is a subjective sensory complex sensation obtained by active manipulation of neural sagaciousness of our hands. A fabric hand or handle depicts the way a fabric feels when it is touched by a human hand and gives an indication of texture of the fabric. Various psychological sensations such as stiffness, softness or hardness, warm or cool, wet or dry are also perceived.

Neuro-physiological researches have shown that the various sensations resulting from the skin–fabric interaction are triggered by three categories of sensory receptors which cover pain, temperature and touch sensations. During fabric–skin contact, the fabric produces pressure and vibration on the skin and stimulates touch receptors. Peirce (1930) was first to describe the relationship of fabric properties and handle. He concluded that fabric stiffness is the key factor in deciding fabric handle. Kawabata and Niwa (1995) separated handle into three levels – mechanical properties, primary handle value and total handle value.

According to ASTM Standard D123 (2003), the following terms are important for describing fabric handle:

Flexibility – ease of bending
Compressibility – ease of squeezing
Extensibility – ease of stretching
Resiliency – ability to recover from deformation
Density – mass/unit volume
Surface contour – divergence of surface from the fabric plane
Surface friction – resistance to slipping
Thermal character – apparent temperature difference between fabric and skin.

The touch may be active or passive, synthetic or analytic. Active touch may be classified into four categories:

1. Gliding touch
2. Sweeping touch
3. Grasping touch
4. Kinematic touch.

1.6.3 Non-sensorial comfort

Non-sensorial comfort deals with physical processes which generate the stimuli like heat transfer by conduction, convection and radiation, moisture transfer by diffusion and evaporation. It also includes mechanical interactions in the form of pressure, friction and dynamic irregular contact. Non-sensorial comfort is not only comprised of thermal and moisture transmission but also includes air permeability, water repellency and water resistance (Das, 2005).

Air permeability

The air permeability of a fabric is the measure of how well it allows the passage of air through it. The passage of air is of importance for a number of fabric end uses such as industrial filters, tents, sail-cloths, parachutes, raincoat materials, shirting, waterproof fabrics and airbags. In outdoor clothing, it is important that air permeability is as low as possible because it should function as a wind protection. A material that is permeable to air is usually permeable to water, in either the vapour or the liquid phase. Thus, the moisture–vapour permeability and the liquid–moisture transmission are normally closely related to air permeability. On the other hand, the thermal resistance of a fabric is strongly dependent on the enclosed still air, and this factor is in turn influenced by the fabric structure.

Water vapour transmission

The human body cools itself by sweat production and evaporation during periods of high activity. The clothing must be able to remove this moisture in order to maintain comfort and reduce the degradation of thermal insulation caused by moisture build-up in a cold environment. Water vapour transmission is essential in determining the breathability of clothing and textiles in outdoor and indoor wear. A breathable textile allows extra heat loss by evaporation of moisture through the clothing layers. If clothing layers are impermeable the moisture is captured between skin and clothing and heat is accumulated in the body. As a consequence, heat and moisture build up, causing discomfort, wet skin and skin abrasion.

Water repellency and water absorption

Water repellency treatment modifies the surface tension properties of fibres or fabrics so that they repel water drops. The treatment may also improve soil repellency. Water resistance is needed in outdoor clothing for protection against rain and is a requirement for furniture and bed-coverings to protect against liquid excretions. On the other hand, water generated at the body surface as perspiration should be removed quickly if comfort is desired. Some textile end uses such as towels, cleaning cloths, diapers and sanitary pads are made of material capable of absorbing water to achieve comfort.

1.7 Comfort variables

Thermal comfort variables are of two types, namely:

1. Two personal variables, controlled by the individual:
   (a) clothing insulation value, termed the 'clo' value, and
   (b) activity level deciding metabolism rate, with units of 'met'.
2. Four environmental variables which represent the environment surrounding the body:
   (a) temperature of the surrounding air (dry-bulb temperature)
   (b) radiant temperature of the surrounding surfaces represented by 'mean radiant temperature' (MRT)
   (c) humidity of the air denoted by 'relative humidity'
   (d) air movement.

The above six variables are considered as primary comfort variables. Non-thermal comfort depends on the following environmental factors:

- odours
- dusts
- acoustics
- lighting.

1.7.1 Factors affecting thermal comfort variables

Air (dry-bulb) temperature

Air temperature affects the rate of heat loss from the body by convection and evaporation. It is perhaps the most important determinant of thermal comfort, since a narrow range of comfortable temperatures can be established almost independently of the other variables. A fairly wide range of temperature, with properly combined relative humidity, MRT, and air flow, can provide comfort. With variation of the above conditions, the surrounding air temperature must be adjusted in order to maintain comfort conditions.

Temperature drifts and ramps are passive and actively controlled gradual temperature changes over time, respectively. People may feel comfortable with temperatures that rise or fall like a ramp over the course of time, even though they would be uncomfortable if some of the temperatures were held constant. Ideal comfort standards call for a change of no more than 1 °F/hr (0.6 °C/hr) during occupancy, provided that the temperature excursion doesn't extend far beyond the specified comfort conditions, and for very long.

Air temperature in an enclosed space generally increases from floor to ceiling. If this variation is sufficiently large, discomfort could result from the temperature being overly warm at the head and/or overly cold at the feet, even though the body as a whole is thermally neutral. Therefore, to prevent local discomfort, the vertical air temperature difference within the occupied zone should not exceed 5 °F (3 °C). The occupied zone within a space is the region normally occupied by people. It is generally considered to be the first 6 feet (1.8 m) above the floor and 2 feet (0.6 m) or more away from walls or fixed air conditioning equipment. The floor temperature should be between 65 and 84 °F (18 and 29 °C) to minimie discomfort for people wearing appropriate indoor footwear.

The hot or cold objects can be quickly identified just by touching, but one may mislead while describing how hot or cold the objects are. The touching sensation depends more on the rate of conduction of heat to or from the body than the actual temperature of the objects. Even if steel and wood are at the same temperature, the former will be felt cooler or hotter, if touched, depending on whether both being cooler or hotter than the body, respectively. This is because steel conducts heat away from or to fingers or other touching body parts very quickly. In other words, the sensors on our skin are poor judges of temperature, but are designed to sense the degree of heat flow.

Effect of humidity

The humidity in air can be measured in various ways.

- Absolute humidity is the weight of water in a unit volume of air (pound/ft^3 or gm/ml).
- The humidity ratio or specific humidity is the weight of water vapour per unit weight of dry air (pound per pound or kg/kg). The retention capacity of moisture by air is a function of the temperature: the warmer the air, the more moisture it can hold. The degree of saturation is the amount of water present in the air relative to the maximum amount it can hold at a given temperature without causing condensation.
- Percentage humidity is the amount of water present in air expressed as percentage of maximum retention capacity. Low percentages indicate relative dryness, and high percentages indicate high moisture. Percentage humidity is often mistakenly called relative humidity. Relative humidity (RH) is the actual vapour pressure of the air–vapour mixture expressed as a percentage of the pressure of saturated water vapour at the same dry-bulb temperature. Percentage and relative humidity are numerically close to each other but are not identical.

Human beings can tolerate more variation of humidity than of temperature. However, high humidity can cause condensation problems on cold surfaces and retards human heat loss by evaporative cooling (sweating and respiration). Air with high moisture content cannot absorb much more from the skin. The drier and warmer the air, the greater is the evaporation rate and consequently the higher the heat loss from the skin. However, low humidity tends to dry the throat and nasal passages and can accumulate static charge, causing discomfort. To minimise problems with static charge, carpets interlaced with a conductive material such as copper or stainless steel yarn is now available commercially.

For people at rest, comfort is maintained over a wide range of humidity conditions. In winter the body comfort is maintained over RH ranging from 20 to 50%. In summer the tolerance range extents up to 60% RH. When the temperature exceeds 75 °F (24 °C), the skin feels sweaty. Nevertheless, some types of industrial applications, such as textile manufacturing, optical lens grinding, and food storage, maintain an RH above 60%, while certain pharmaceutical products, plywood cold pressing, and some other processes require an RH below 20%. Hospitals should maintain RH between 50 and 55% as the level of bacteria propagation is lowest in this range.

Mean radiant temperature

The importance of radiant temperature can be quickly understood when we enter a room with thick cold walls in a prickly heated summer or in a room with heated surfaces during a chilling cold winter.

The MRT for office workers should be in the range of 65° to 80 °F (18° to 27 °C), depending on the clothing worn and the activity to be done by them. In winter, levels of wall, roof, and floor insulation together with window treatments such as double glazing, blinds, and drapes in accordance with good design practice codes should generally result in indoor surface temperatures that are no more than 5 °F (2.8 °C) below the indoor air temperature.

Air movement

The body heat loss or gain by convection and evaporation is significantly affected by the movement of air. Air movement results from free (natural) and forced convection (by fans, etc.) as well as from the occupants' bodily movements. The faster the motion, the greater the rate of heat flow by both convection and evaporation. When ambient temperatures are within acceptable limits, there is no minimum air movement that must be provided for thermal comfort as the natural convection of air over the surface of the body allows for the continuous dissipation of body heat. However, when ambient temperature is high, natural air flow velocity is no longer sufficient and air movement must be increased artificially using fans. Typical human responses to air motion are shown in Table 1.4.

Insufficient air motion promotes stuffiness and variation of air temperatures from floor to ceiling. On the other hand, excessive movement of air causes unpleasant drafts to the room occupants. The exact limits to acceptable air movement in an occupied space are functions of the overall temperature, humidity, and MRT of the room along with the temperature and humidity conditions of the moving air stream. While perspiration is present on the skin, noticeable air movement across the body may be felt as a pleasant cooling breeze. However, the same air movement may be considered a chilly draft when the surrounding surface and room air temperatures are cool. The neck, upper back, and ankles are most sensitive to drafts, particularly when the entering cool air is 3 °F (1.5 °C) or more below normal room temperature.

Table 1.4 Human response to the velocity of air

Air velocity (m/s)	Occupants' response
< 0.05	Stagnation
0.05-0.25	Comfortable
0.25-0.51	Air motion is felt, comfort depends on air temperature and room conditions
0.51-1.02	Strong feel of air motion, acceptable in working place, if intermittent and if air temperature and room conditions are acceptable
≥1.02 (about 2 mph)	Annoyances like blowing of papers, hair, etc.

Above a velocity of 0.15 m/s, every 0.075 m/s increase in air movement is sensed by the body as a 1° temperature drop. Air systems are usually designed for a maximum motion of 0.25 m/s in the occupied zone, but that is typically exceeded at the outlet of air registers. Warm air introduced into a room may cool off before reaching the occupant or the air is intended to cool the occupant. In either case, when the temperature of the air impinging on an occupant is below the ambient temperature, the individual becomes more sensitive to air motion and may complain of drafts. Therefore, careful attention must be given to air distribution as well as velocity.

The warm air used for heating can greatly affect occupant comfort due to convective air motion, and thus the heat source in a room should be placed in the correct position. Air outlet design is determined by the air distribution pattern it is intended to create. Besides the removal of heat and humidity, another function of air motion in alleviating stuffiness is the dispersion of body odours and air contaminants.

1.7.2 Clothing insulation

Thermal comfort is significantly affected by clothing. Clothing, through its insulation properties, is an important modifier of body heat loss and comfort. The insulation properties of clothing are due to the presence of a large number of small air pockets between interlacements of warp and weft yarns preventing air from migrating through the material. Similar insulation may be achieved when several sheets of newspaper are wrapped one over another around our body as there remain layers of air between the sheets. In general, all clothing makes use of this principle of trapped air within the layers of fabric.

Clothing insulation is described in terms of its clo value. Developed in 1941, it was the first real attempt to explain the insulation value of clothing so that people would know how much clothing they might need to stay warmer or cooler in a given temperature environment. The higher the clo number, the more is the insulating value. Clo is the thermal resistance of an assembly of clothing expressed numerically.

$$1 \text{ clo} = 0.88 \text{ ft}^2 \cdot \text{hr} \cdot {}^\circ\text{F/Btu} = 0.155 \text{ m}^2 \cdot {}^\circ\text{C/W} \qquad [1.7]$$

A clo value of 1 is defined as the amount of clothing required by a resting human (in other words, sitting, lying down or standing, but not moving) to be comfortable at a room temperature of 21 °C (approx. 71 °F). One (1) clo corresponds to a person wearing a typical business suit – shirt, undershirt, trousers and suit jacket. Zero (0) clo corresponds to a naked person.

Studies have shown that a clo value of

- 2.5 is needed to keep a sleeping person comfortable at an air temperature of 15 °C (59 °F)
- 4 is required for comfort at 9 °C (48.2 °F)
- 6 is required for comfort at 0 °C (32 °F).

Clo values for a few common articles of clothing are listed in Table 1.5. The total insulation value of a clothing ensemble can be estimated as the sum of the individual garment clo values.

1.7.3 Thermal indices

Thermal sensation can be described as hot, warm, neutral, cool, cold, and a range of classifications in between. However, it depends on four environmental factors described before. There have been numerous attempts to find a single index – integrating some or all of the environmental factors that determine thermal comfort conditions for a given metabolic rate and amount of clothing. The most common of these indices still in use are:

- dry- and wet-bulb temperatures
- operative temperature
- globe thermometer temperature
- new effective temperature.

Dry- and wet-bulb temperatures

These represent air temperature and humidity respectively. The simplest practical index of cold and warmth is the reading obtained with an ordinary dry-bulb thermometer. This long-established gauge is fairly effective in judging comfort for average humidity (40 to 60% RH), especially in cold conditions.

Table 1.5 Clo values of various clothing types

Clothing type	Clo	Clothing type	Clo	Clothing type	Clo
T-shirt	0.09	Light trousers (men)	0.26	Light sweater (men)	0.20
Briefs	0.05	Heavy trousers (men)	0.32	Heavy sweater (men)	0.37
Long underwear	0.10	Light dress (women)	0.22	Sleeveless sweater (women)	0.17
Bra and panties	0.05	Heavy dress (women)	0.70	Long sweater (women)	0.37
Light shirt (men)		Long sleeve blouse (women)		Light jacket (men)	0.22
short sleeve	0.14	light	0.20	Heavy jacket (men)	0.49
long sleeve	0.22	heavy	0.29	Light jacket (women)	0.17
Hat and overcoat	2.00	Light slacks (women)	0.10	Heavy jacket (women)	0.37
Heavy shirt (men)		Heavy slacks (women)	0.44	Socks knee high (men)	0.10
short sleeve	0.25	Sandals	0.02	Socks any length (women)	0.01
long sleeve	0.29	Boots	0.08		

In the hot season, the significance of the dry-bulb temperature is limited as humidity greatly affects the efficiency of control of body temperature by sweating. The wet-bulb temperature represents an improvement over the simple dry-bulb temperature by taking humidity into account.

Operative temperature

Operative temperature is a combined measure of air temperature and MRT. It is the uniform temperature of an imaginary enclosure in which the occupant would exchange the same heat by radiation and convection as in the actual environment. In other words, operative temperature is an average of MRT and dry-bulb temperatures weighted by the respective radiation and convection heat transfer coefficients.

Humid operative temperature is the uniform temperature of an imaginary environment at 100% RH with which the occupant would exchange the same heat by radiation, convection, conductance through clothing, and evaporation as in the actual environment.

Globe thermometer temperature

Globe thermometer temperature is usually used as a simple device for determining MRT. The globe thermometer uses a 6-inch (150-mm) diameter black globe. Here the shown temperature accounts for the effects of radiation and air movement. The equilibrium temperature of the globe is a single temperature index describing the combined physical effect of dry-bulb temperature, air movement, and net radiant heat received from the surrounding surfaces. The globe temperature is an approximate measure of operative temperature.

New effective temperature

Effective temperature cannot be measured by a thermometer. It is an experimentally determined index of the various combinations of dry-bulb temperature, humidity, radiant conditions (MRT), and air movement that induce the same thermal sensation. The new effective temperature (ET*) of a given space is defined as the dry-bulb temperature of a thermo-equivalent environment at 50% RH and a specific uniform radiation condition. The combinations that induce the same feeling of warmth or cold are called thermo-equivalent conditions.

The thermo-equivalent heat exchange is based on clothing at 0.6 clo, still air (40 fpm = 0.2 m/s or less), 1-hour exposure time, and a sedentary activity level (approximately 1 met). Thus, any space has an ET* of 70 °F (21 °C) when it induces a sensation of warmth like that experienced in still air at 70 °F (21 °C), 50% RH, and the proper radiant conditions. ET* is, in general, a reliable indicator of discomfort or dissatisfaction with the thermal environment.

1.7.4 Combined effects

It is difficult to visualise comfort conditions and sometimes an air temperature range is mentioned as a comfortable range. About 100 years ago, Willis Carrier developed the 'Psychrometric Chart' (see Fig. 1.4), which shows combined effects of air temperature and outlines the conditions at which most sedentary or inactive humans are comfortable. Any point located on the chart establishes the temperature (dry bulb) and the amount of water vapour in unit quantity of air. The comfort envelope accounts for three comfort variables: air temperature, mean radiant temperature and relative humidity. The fourth environmental variable, air motion, shifts the envelope laterally. As the air motion increases, the envelope moves to the right and vice-versa. As different amounts of garments are worn in summer and winter seasons the chart shows separate envelopes or boundary for summer comfort and winter comforts. The single temperature at which most humans would be comfortable all-year round is about 74 °F or 23 °C. For human comfort, the amount of humidity is not so critical when the relative humidity is below 55% (exactly below dew point of 62 °F).

The common misconception is that control of air temperature and humidity are primary factors to achieve comfort. The average temperature of the surrounding surfaces or mean radiation temperature (MRT) is as important a comfort variable as air temperature. When a person moves to a warmer window, the body's view angle of the window becomes larger and hence MRT is higher. For maximum comfort, the body prefers that the average temperature of surrounding air and surfaces should be around 74 °F or 23 °C. When air temperature and MRT are equal, radiation and convection heat losses are equal.

The relationship between clothing insulation and room temperature necessary for a neutral thermal sensation, as reported in ASHRAE (The American Society of Heating, Refrigerating and Air-conditioning Engineers Inc) Standard 55 is schematically presented in Fig. 1.3 for occupants at rest and specified air speed (1/3 mph) and humidity (50% RH). Comfortable clothing levels are expressed as a function of operative temperature, which is based on both air and mean radiant temperatures. At air speeds of 0.4 m/s or less and MRT less than 120 °F (50 °C), the operative temperature is approximately the average of the air temperature and the mean radiant temperature and is equal to the adjusted dry-bulb temperature.

There is no combination of conditions that would satisfy all people. The optimum operative temperature represented by the middle solid line in Fig. 1.3, is the temperature that satisfies the greatest number of people with a given amount of clothing and specified activity level. The upper and lower thermal acceptability limits (shown by large dashed and small dashed lines respectively) demarcate a room environment that at least 80% of the occupants would find thermally acceptable. By adjusting clothing as desired, the remaining occupants can satisfy their own comfort requirements.

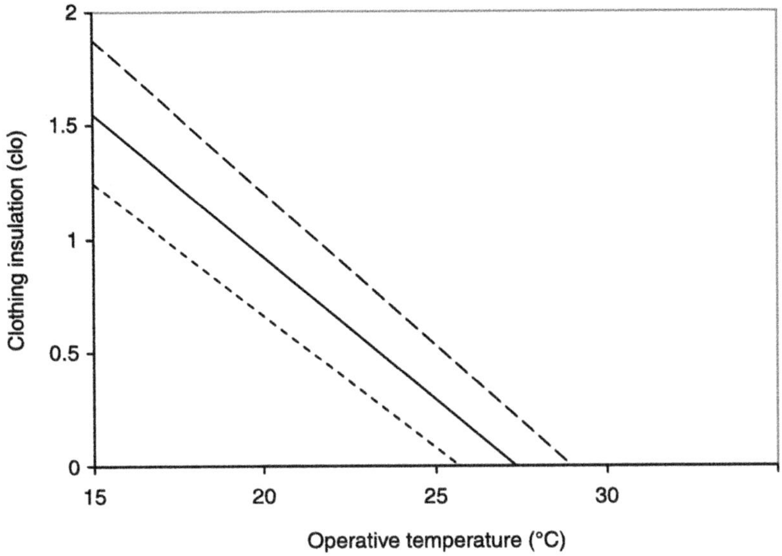

1.3 Clothing insulation (in clo units) necessary at different operating temperatures.

Since the 1970s, energy has become increasingly scarce. Earlier than this, the preferred amount of clothing worn by building occupants was less, and correspondingly the preferred temperatures increased from about 68 °F (20 °C) for winter to the year-round range of 72 ° to 78 °F (22 ° to 25.5 °C). Present conditions, however, make it desirable to minimise energy consumption for providing thermal comfort.

Energy savings can be achieved if the insulation value of clothing worn by people indoors is appropriate to the season and outside weather conditions. During the summer months, suitable clothing in commercial establishments consists of lightweight dresses, lightweight slacks, short-sleeved shirts or blouses, stockings, shoes, underwear, accessories, and sometimes a thin jacket. These ensembles have insulation values ranging from 0.35 to 0.6 clo. The winter season needs thicker, heavier clothing. A typical winter ensemble – including heavy slacks or skirt, long-sleeved shirt or blouse, warm sweater or jacket, and appropriately warm accessories – would have an insulation value ranging from 0.8 to 1.2 clo.

During more moderate seasons, the clothing would likely consist of medium-weight slacks or skirt, long-sleeved shirt or blouse, and so on, having a combined insulation value of 0.6 to 0.8 clo. These seasonal clothing variations of building occupants allow indoor temperature ranges to be higher in the summer than in the winter and yet remain comfortable. In the wintertime, additional clothing lowers the ambient temperature necessary for comfort and for thermal neutrality. Adding 1 clo of insulation permits a reduction in air temperature of approximately 13 °F (7.2 °C) without changing the thermal sensation. At lower temperatures, however, comfort requires a fairly uniform level of clothing insulation over the entire body. For occupancy of more than an hour at rest, the operative temperature should not be less than 65 °F (18 °C).

The insulation of a given clothing ensemble can be estimated by adding up the clo values of the individual items worn and multiplying the sum by 0.82. A rough approximation of the clo value may also be estimated by multiplying the weight in pounds of clothing by 0.15 clo (or each kilogram by 0.35 clo).

The closeness of fit of a garment has a great influence on its insulation value as well as the fabric from which it is constructed. The resistance of a fabric to the movement of heat through it is important in its thermal comfort. Thermal resistance to transfer of heat from the body to the surrounding air is the sum of three parameters:

1. the thermal resistance to transfer heat from the surface of the body,
2. the thermal resistance of the clothing material, and
3. the thermal resistance of the air interlayer.

The entrapped air is the most significant factor in determining thermal insulation. Increase of either 'microlayers' (those between contacting surfaces of the materials) or 'macrolayers' (between non-contacting surfaces) of air enclosed within an assembly can increase thermal insulation. However, the characteristics of fibre, yarns, fabrics and garments also have a major contribution towards thermal comfort.

1.7.5 Comfort under non-steady conditions

The comfort standard ASHRAE 55-1992 discusses 'non-steady state' conditions. When thermal mass is used to achieve comfort conditions, temperature fluctuations will be experienced in the form of 'drifts' or 'ramps' which are acceptable provided the operative temperature is within comfort zone limits. The comfort standard specifies that the rate of change in operative temperature during drift (flow) or ramp (rise) should not exceed 0.6 °C/h (1 °F/h).

Vertical temperature difference

Air temperature generally increases from floor to ceiling. If the difference is too high, there may be local discomfort. Local discomfort may start when the maximum temperature difference between head and feet is above 3 °C (5 °F).

Asymmetrical radiant temperature

Even when the air temperature is within the comfort zone, large temperature differences between surrounding surfaces (e.g. hot or cold windows, walls, ceilings, improper heating panels or direct sunlight) may cause discomfort. Humans are more sensitive to asymmetric radiant temperature between horizontal surfaces (i.e. ceiling and floor) than between vertical surfaces (i.e. walls). For comfort, the recommended temperature difference in opposite directions (asymmetry) should be less that 5 °C (9 °F) in the vertical planes and less than 10 °C (18 °F) in the horizontal planes.

Floor temperature

A too hot or too cold floor may be uncomfortable even for people wearing shoes. For foot thermal comfort, the floor temperature rather than the material for floor covering is most important. The allowable floor temperature should be between 19 °C (66.2 °F) and 29 °C (84.2 °F).

Draft

Draft is unwanted local cooling of the body by air movement. Both air speed and temperature affect draft sensation and Some people are more sensitive to air motion than others. Uncovered skin, especially the head and lower leg regions, is more sensitive to draft.

1.8 Effective temperature and the comfort chart

Effective temperature is an experimentally determined index of the various combinations of dry-bulb temperature, humidity, radiant conditions, and air movement that induce the same thermal sensation. The combinations that generate the same feeling of warmth or cold are called thermo-equivalent conditions.

The effective temperature (ET*) is the dry-bulb temperature of a thermo-equivalent environment at 50% relative humidity (RH) and a specific uniform radiation condition. The thermo-equivalent heat exchange is based on clothing at 0.6 clo (standard indoor office clothing), still air (40 fpm = 0.2 m/s or less), one-hour exposure time, and a sedentary activity level (approximately 1 met). Thus, any space has an ET* of 70 °F (21 °C) when it induces a sensation of warmth like that experienced in still air at 70 °F (21 °C), 50% RH, and the proper radiant conditions. ET* is, in general, a reliable indicator of discomfort or dissatisfaction with the thermal environment.

The psychometric comfort chart, schematically shown in Fig. 1.4, correlates the perception of comfort with the various environmental factors known to influence it. The dry bulb temperature is plotted along the horizontal axis. The right side of the chart shows a dew point scale and the left side a wet-bulb temperature scale indicating guide marks for imaginary lines sloping diagonally down from left to right. The lines curving upward from left to right represent RHs. ET* lines are also drawn. These are the sloping dashed lines that cross the RH lines and are labelled in increments of 5 °F. At any point along any one of

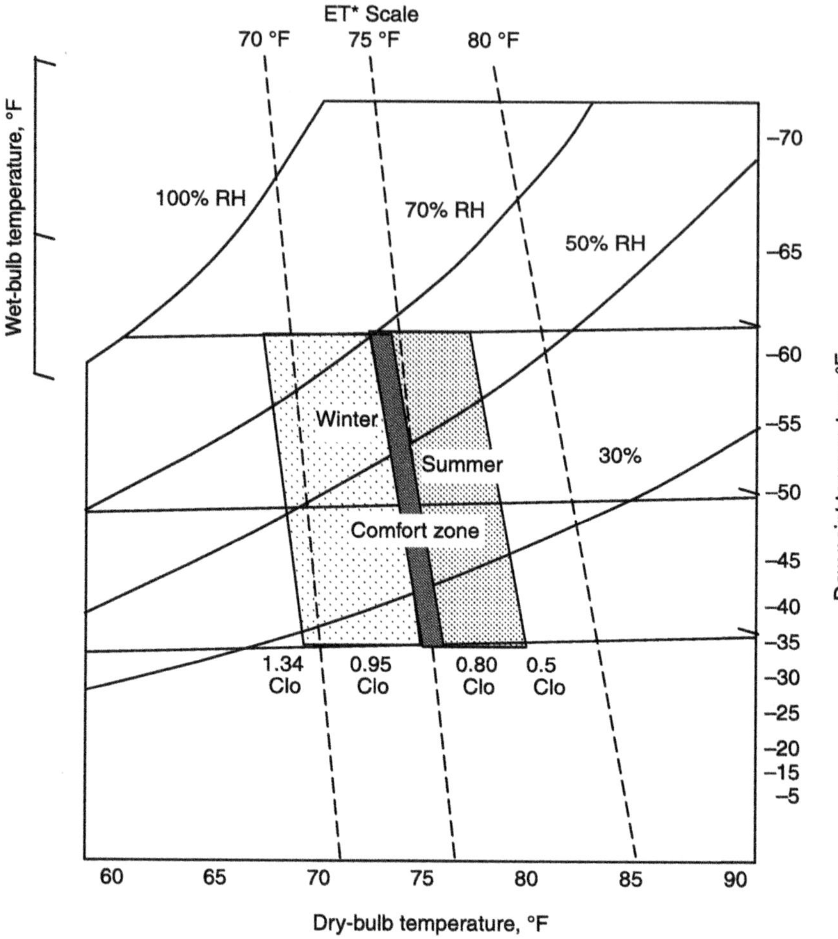

1.4 Psychometric chart for comfort.

these lines, an individual will experience the same thermal sensation and will have the same amount of skin wetness due to regulatory sweating.

The comfort chart in Fig. 1.4 is derived from what is called the psychrometric chart. Two comfort envelopes or zones are defined by the shaded regions on the comfort chart – one for winter and one for summer. The thermal conditions within these envelopes are estimated to be acceptable to 80% of the occupants when wearing the clothing ensemble (clo value) indicated below the comfort zones. To satisfy 90% of the people, the limits of the acceptable comfort zone are sharply reduced to one-third of the above ranges.

The zones overlap in the 73° to 75°F (23° to 24°C) range. In this region, people in summer dress tend to be slightly cool, while those in winter clothing feel a slightly warm sensation.

The right hand corners above and below the comfort zones in the figure represent the 'too warm zone' and the zone with 'static electricity problems' respectively, while the lower left corner represent the 'too cold' zone.

Figure 1.4 generally applies when altitudes range from sea level to 7000 feet (2134 m), MRT is nearly equal to dry bulb air temperature, and air velocity is less than 40 fpm (0.2 m/s). Under these conditions, thermal comfort can be defined in terms of two variables: dry-bulb air temperature and humidity.

Mean radiant temperature is equally important as air temperature in affecting comfort. When air movement in an indoor environment is low, the operative temperature is approximately the average of air temperature and MRT. When the MRT in the occupied zone significantly differs from the air temperature, the operative temperature should be substituted for the dry-bulb temperature scale along the bottom of Fig. 1.4.

The comfort chart is also useful to identify the risk of sedentary heat stress at higher temperatures. The ET* scale is based on a 1-hour exposure, however, data show no significant changes with longer exposures unless the limits of heat stress (ET* greater than 90 °F or 32 °C) are approached. When RH is about 20% and 55%, humidity has only a small effect on thermal comfort. With increase of both temperature and the degree of regulatory sweating, humidity increases discomfort. Near the comfort zone, evaporative heat loss is very low – about 25% of the total heat loss. With increase in temperature, evaporation loss increases and when the ambient temperature is equal to that of skin and clothes, the evaporation accounts for 100% of the heat loss.

The upper and lower humidity limits on the comfort zone of Fig. 1.4 are decided on the basis of respiratory health, mould growth, and other moisture-related phenomena in addition to comfort.

The temperature, radiation, humidity, and air movement necessary for thermal comfort depend upon the occupant's clothing and activity level. As 90% of people's indoor work and leisure time is spent at or near the sedentary activity level, the comfort zones developed by ASHRIE is limited to lightly clothed occupants (0.5 to 0.6 clo) engaged in sedentary activities. The comfort zone shown in Fig. 1.4 strictly applies only to sedentary and slightly active, normally clothed persons at low air velocities, when the MRT is equal to air temperature. For other conditions, the comfort zone must be adjusted accordingly. Comfort can be maintained even at 68 °F (20 °C) wearing a clothing ensemble measuring less than 1.34 clo and moving around for at least 10 minutes out of every half-hour. Comfort conditions may be extended upward to 82 °F (28 °C) with a fan-induced air velocity of 160 fpm (0.8 m/s) in the comfort zones shown in Fig. 1.4, and no minimum air movement is necessary for thermal comfort.

However, the maximum allowable air motion is lower in the winter than in the summer. In the wintertime, the average air movement within the occupied zone should not exceed 30 fpm (0.15 m/s). At temperatures lower than neutral sensation, slight increases in air velocity or irregularity of air movement can cause uncomfortable localised drafts. In the summer, the average air movement in the occupied zone can go as high as 50 fpm (0.25 m/s) under standard temperature and humidity conditions. Above 79 °F (26 °C), the average air motion is to be increased by 30 fpm for each °F (0.275 m/s for each °C) increase of temperature to maintain comfort. The air movement can be increased to a maximum of 160 fpm (0.8 m/s) when loose paper, hair, and other light objects start blowing around.

Sweating increases with increase in steady-state activity. To maintain comfort, clothing must be adjusted, the air motion must be increased, or the operative temperature must be decreased.

The reliable thermal comfort indices are very important for the effective control of the thermal environment without using further energy and equipment. The five variables affecting comfort for a given activity or room function are dry-bulb temperature, humidity, MRT, air movement, and clothing. Sometimes when any of these conditions is out of the comfort range, adjusting one or more of the other conditions will restore comfort with the addition of little or no additional energy.

The most commonly recommended conditions for comfort are:

ET* = 75 °F (24 °C)
Dry-bulb air temperature = MRT
Relative humidity = 40% (20 to 60% range)
Air velocity < 40 fpm (0.2 m/s).

ASHRAE's Standard 55 (Thermal Environmental Conditions for Human Occupancy) recommends conditions that have been found experimentally to be acceptable to at least 80% of the occupants within a space. The recommended operative temperatures during winter and summer at 60% RH, activity level 1.2 met and low air movement are as follows:

Season	clo	Recommended operative temperature
Winter	0.8 to 1.2	68 °–74 °F (20 °–23.5 °C)
Summer	0.35 to 0.6	73 °–79 °F (22.5 °–26 °C)

The other recommendations (ASHRAE, 1989) are:

- For each 0.1 clo of increased clothing insulation, the acceptable temperature range is lowered by 1 °F (0.6 °C).
- As the temperature decreases, comfort depends more and more on maintaining a uniform distribution of clothing insulation over the entire body, especially the hands and feet.
- For sedentary occupancy of more than an hour, the operative temperature should not drop below 65 °F (18 °C).

- When radiant surfaces are too cool for comfort, the air temperature must be increased from 0.3 ° to 1 ° for every 1 ° reduction in MRT, depending on room conditions.
- Ill or nude people may require considerably different conditions for comfort than those indicated in the standard.
- People engaged in physical work need a lower effective temperature for comfort than do sedentary ones. The greater is the activity and the more clothing worn, the lower must be the effective temperature for comfort.

At low air temperature, convective heat loss increases with the motion of air associated with increased activity, thereby decreasing the heat load on the body evaporative system and resulting in a wider range of activity before discomfort is felt. At low dry-bulb temperature and RH the comfort is maintained with the maximum range of activity while compensating with an MRT sufficient to maintain comfort.

Short-term acclimatisation plays an important part in determining the best conditions for comfort. In summer, a period of time is required for the body's heat-controlling mechanisms to readjust to the lower temperature and humidity of a conditioned indoor environment. During acclimatisation to the cooler interior, the rate of perspiration decreases and the blood vessels recede from the skin surface.

A conditioned space may be comfortable for the person occupying for longer periods, but may be too cool for a person who has just entered the space from the hotter outdoors. The reverse situation will cause more discomfort – while quickly moving from the conditioned space to the hot, humid outside conditions perspiration cannot increase immediately and blood vessels do not move quickly to the skin surface to promptly balance the body heat production and loss. The spaces destined for short-time occupancy, such as stores or public lobbies, are advisable to be maintained as a relatively warm, dry climate in which the perspiration rate will change only a small amount. For occupancy times of less than one hour, the difference between indoor and outdoor temperature difference should be within 5 °F (2.8 °C) so that the body's thermal control system can easily revert to meet the outside condition, by increasing moisture loss and heat loss from the blood vessels already near the skin. For longer occupancy, as in offices, the body becomes accustomed to the conditioned environment, and the comfort chart may be used to establish appropriate conditions. In many commercial establishments the temperature and humidity are maintained between those necessary to ensure the comfort of employees and those necessary to avoid too great a contrast between indoor and outdoor temperatures for the customers. In restaurants, in spite of short occupancy, a cooler temperature is required as the digestion of food increases the metabolic rate.

The economical selection and operation of conditioning equipment require that indoor conditions should be close to the outside environment while maintaining comfort. In winter, the equipment is to raise both air temperature and humidity, while in summer, cooling and dehumidification must be accomplished and the changes must be small. Therefore, in winter, conditions near the left-hand corner of the comfort zone should be selected, while in summer those near the upper right-hand corner should be chosen.

Other distinctions between winter and summer that must be taken into consideration in selecting design conditions are:

- The MRT of the boundary and top-floor spaces are lower in winter than in summer due to contact with outdoor air.
- Conventions of clothing insulation are different in winter and summer.
- Expectations (psychological and physiological acclimatisation) of indoor thermal conditions vary according to season.

Individual preferences

No set of conditions can provide comfort for everyone. Thermal comfort standards of ASHRAE or other recommended conditions are for general application and the individual preferences are not considered. Under the same thermal conditions, some individuals may feel too hot, while others wearing identical clothing feel too cold; the condition that is slightly warm for a healthy young man may be too cool for an elderly woman; and the condition inside an air-conditioned store may be very comfortable for a pedestrian, but a bit too warm for an active clerk working for several hours.

Men generally feel warmer than women on initial exposure to a given temperature but later feel cooler, approaching women's thermal sensations after 1 to 2 hours in the environment. Comfort conditions seem independent of the time of day or night.

Comfort standards maintained in different countries may be different depending on their clothing customs and local climate extremes and on the relative economics of providing and running heating and cooling systems. Each person has a distinct perception of too hot, too cold, and comfortable. The objective in designing a common thermal environment is to satisfy a majority of occupants and to minimise the number of people who will inevitably be dissatisfied.

1.9 Response to extreme temperature

As conditions become warmer or colder than the comfort zone, individuals become increasingly thermal sensitive. Discomfort increases and the body's thermal regulation systems, namely the cardiovascular system, respiratory system, etc., are put under strain. With intense thermal exposure, individuals may feel pain and the body's thermal regulation capability may fail, which can eventually lead to death.

The physiological consequences at different effective temperature (ET*) signify that regulation of the thermal environment is more important than just for providing comfort; there are serious health purposes, too. At the high end of the scale, there is heat stress to contend with and, in extremely cold conditions, a variety of respiratory ailments, incapacity, and heart failure. The effect of exposure to extreme cold is decided by the way thermal balance is maintained. People can withstand extreme cold for only a limited time, as under the condition body heat is lost faster than that produced by metabolism. As a result of such exposure, body temperature drops and acute discomfort is caused when it drops by 4.7 °F (2.6 °C). In any given cold environment, heat loss can be reduced and thermal balance can be maintained by using clothing having appropriate insulation (clo value). The extent of clothing adjustment is dictated by clothing customs and the restriction of mobility for the intended activity.

If vascular constriction is unable to prevent body heat loss, an automatic, more efficient defence against cold is shivering. It may initially increase total heat production by 50 to 100% from resting levels. Ultimately, violent whole body shivering may result in a maximum of about six times heat production from resting levels and the individual becomes totally ineffective.

Acclimatisation is a physiological process in response to long exposure to cold which involves the following:

- Hormonal changes to metabolise free fatty acids released from fat tissue.
- Maintaining circulatory heat flow to the skin, resulting in a greater sensation of comfort.
- Improved regulation of heat to the extremities, reducing the risk of cold injury.

The selection of adequate protective clothing, controlling the body's microclimate by some other means or increase of metabolic heat production through proper activity are generally more useful than dependence on the above physiological changes.

As far as the insulating material for protective clothing is concerned, radiation-reflective materials which can seal the body from cold air currents are most effective. The greater the fibre thickness, the greater is the thickness of the insulating trapped air. The fingers and toes pose more problem than the torso, because, as thin cylinders, they lose heat much more rapidly and are difficult to insulate without hampering mobility.

1.10 Development of heat stress and its control

When a body gains heat at a rate higher than the rate at which it can lose heat, the body is under heat stress. If this condition persists without relief, there is the danger that workers, such as those in hot industries, can experience heat prostration. Personal tolerance to high temperature may be limited and he or she may be unable to:

- sense temperature
- lose heat by regulatory sweating, or
- lose heat by blood flow from the body core to the skin surface where cooling can occur.

Pain receptors in the skin are normally triggered by a skin surface temperature of 115 °F (46 °C). Although direct contact with a metal surface at this temperature is painful, much higher dry air temperatures can be tolerated since the layer of air adjacent to the skin acts as a thermal insulator.

The tolerance of increase in body temperature actually represents the body heat storage limit. The voluntary tolerance limit for an average-sized man is about 2.5 °F (1.4 °C) above body temperature. Individuals who remain in the heat and continue to work beyond this point increase their risk of heat exhaustion and collapse can occur at a 5 °F (2.8 °C) rise in internal body temperature.

The cardiovascular system also has important role to decide the heat tolerance limit. For a normal person, heart rate and cardiac output increase in a hot environment in an attempt to maintain blood pressure and supply blood to the brain. When the heart rate is as high as 180 beats per minute, there may not be enough time between contractions to fill the chambers of the heart as completely as required. As the heart rate increases further, cardiac output may fall to a level when an insufficient amount of blood is provided to the skin for heat loss and, perhaps more importantly, an inadequate blood supply to the brain as less blood is pumped. At some point, the individual will faint or black out from heat exhaustion.

An accelerated heart rate may also result from inadequate blood return to the heart caused by the accumulation of blood in the skin and lower extremities. In this case, cardiac output is limited because not enough blood returns to fill the heart between beats. This may happen frequently when an overheated person working hard in the heat stops working suddenly. The muscles suddenly stop massaging the blood past the valves in the veins back toward the heart and dehydration through sweating reduces the fluid volume in the vascular system.

If the body core temperature increases above 106 °F (41 °C), proteins in the delicate nerve tissues in the hypothalamus of the brain responsible for regulation of body temperature may be damaged. Inappropriate vascular constriction, cessation of sweating, increased heat production by shivering, or some combination of these may result. Such heat stroke damage is frequently irreversible and carries a high risk of fatality. A final problem is hyperventilation, or overbreathing, which occurs predominantly in hot-humid conditions. This exhaling of more carbon dioxide from the blood than is desirable can lead to tingling sensations and skin numbness (lack of sensation), possibly resulting in vascular constriction in the brain with occasional loss of consciousness.

The Heat Stress Index (HSI) provides an indication of the severity of the ambient environment on workers.

Table 1.6 Effect of various extents of heat stress on physiology and health on eight hour exposure

Heat stress index	Effect on male individual
−10	Mild cold strain
0	No thermal strain
+10 to +30	Mild to moderate heat strain, substantial decrease in intellectual functions, little decrease in performance of heavy physical work
+40 to +60	Severe heat strain, a threat to health, the conditions unsuitable for people with cardiovascular or respiratory impairment, or with chronic dermatitis, unsuitable for activities requiring sustained mental efforts
+70 to +90	Very severe heat strain, medical examination and trial necessary for such job, adequate water and salt intake should be ensured
+100	Maximum strain tolerable daily by fit, acclimatised young men

$$\mathrm{HSI} = \frac{\text{Evaporation loss required for thermal equilibrium}}{\text{Maximum evaporation rate}} \times 100$$

$$= \frac{E_{req}}{E_{max}} \times 100 \qquad [1.8]$$

When HSI is greater than 100, body heating occurs; when it is less than 0 (negative), body cooling occurs. The upper limit reported for E_{max} is 6 mets = 110 Btuh/ft^2 = 350 W/m^2. For an average-sized man, this corresponds to 2,388 Btuh = 700 W = 602 kcal/hr, or approximately 17 g/min of sweating. The effects of heat stress on human health are shown in Table 1.6 (Blankenbaker, 1982).

1.11 Protective clothing

Protective textiles are a part of technical textiles that are defined as comprising all those textile-based products which are used principally for their performance or functional characteristics rather than their aesthetic or decorative characteristics (Byrne, 2000). Protective clothing refers to garments and other fabric-related items designed to protect the wearer from harsh environmental effects that may result in injuries or death (Adanur, 1995). Extensive research is being done to develop protective clothing for various regular and specialised civilian and military occupations (Adanur, 1995; Bajaj *et al.*, 1992, Holmes, 2000) and providing protection for the whole population has also been taken seriously considering the anticipated disaster due to terrorism or biotechnical attacks (Holmes, 2000; Koscheyev and Leon, 1997).

1.11.1 Various types of protective clothing

Personal protective textiles can be classified as industrial, agricultural, military, civilian, medical, sports and space protective textiles, depending on the end use.

They can be further classified according to the end-use functions as fire protection, thermal protection, chemical protection, mechanical impact protection, radiation protection, biological protection, electrical protection and wearer visibility. Scott (2005) described various types of protective clothing.

Protection to heat and cold

One of the major functions of clothing is to protect the wearer against extremes of environmental temperature, i.e. excessive ambient heat as well as cold. The fundamental parameters of the development of materials for protective clothing against either heat or cold are fibre selection, blend ratios and weave structure.

Incorporation of flame-resistant fibres or finishes is the basic requirement for making heat protective clothing. Apart from that, the main approach in providing heat protection is to make use of energy-reflecting surfaces as part of the garment. Most types of heat-protective clothing are uncomfortable to wear as they are impervious to water vapour and the use of various organic coatings has been reported to overcome this problem. The optimum results can be obtained by aluminising an open-mesh structure fabric (Milenković *et al.*, 1999) and this concept has assumed great significance in military clothing applications all over the world.

Heat generated by metabolism can be life-saving or fatal depending on the surrounding atmosphere and existing circumstances. Normally, human bodies are comfortable in a very narrow temperature range of 28–30 °C (82–86 °F) (Fourt and Hollier, 1970). In summer, the excess heat generated inside the body by metabolism is to be released as soon as possible, while in winter, especially in extremely cold conditions, the heat loss from our body is to be prevented. Heat stress, defined as the situation where the body cannot dissipate its excess heat to the environment, is a serious problem especially during physical working (Bajaj *et al.*, 1992; Richardson and Capra, 2001).

The transfer of heat to or away from the body depends on the source of heat, the atmosphere, the heat-absorbing material and the extent of protection to heat (Bajaj *et al.*, 1992; Fourt and Hollier, 1970). The heat transfer is through conduction, convection and/or radiation and the heat protection is the method by which to decrease or increase the rate of heat transfer. The fabric thickness and density are the major parameters for heat transfer by conduction, since air trapped between fibres has the lowest thermal conductivity of all materials (Morton and Hearle, 1997). Heat transfer by convection, especially through flame, largely depends on the flame-retardant properties of the fabric. Metallised fabrics (such as aluminised fabrics) with high surface reflection and high electrical conductivity are used to protect against radiant heat (Adanur, 1995; Bajaj *et al.*, 1992), while the ideal clothing for protection from heat transfer is a fabric with thermo-regulating or temperature-adaptable properties (Bajaj *et al.*, 1992; Pause, 2003). Phase changing materials (PCM) can absorb heat and change to a high-energy phase in a hot environment, but can reverse the process to release heat in cold situations (Choi *et al.*, 2004b).

To survive and operate in temperatures below –30 °C, specifically designed protective clothing is necessary. Such low-temperature conditions are aggravated in the presence of wind, rain or snow leading to cold stress that may be fatal (Rissanen and Rintamaki, 2000). The most effective method of cold protection is to avoid or decrease conductive heat loss. Clothing designed to protect from cold is usually multi-layered, consisting of a non-absorbent inner layer, a middle insulating layer capable of trapping air but transferring moisture, and an outer layer that is impermeable to wind and water. Temperature-adaptable clothing that can protect from both heat and cold has been developed by fixing polyethylene glycol to cotton at different curing temperatures (Bajaj *et al.*, 1992).

Fire protection

Human skin is very sensitive to heat, pain is felt at 45 °C and the skin is completely burnt at 72 °C (Bajaj *et al.*, 1992; Panek, 1982). Fire protective clothing reduces the rate of heating of human skin so that the wearer gets enough time to react and escape. Normally only 3–10 seconds are available for a person to escape from a place of fire with a heat flux of about 130–330 kW/m^2 (Holmes, 2000). Most of the textile fibres are easy to burn and untreated cotton will either burn with flame or smoke whenever it is in the presence of oxygen and the temperature is high enough to initiate combustion (360–420 °C) (Wakelyn, 1997).

Protective clothing must be flame resistant and should form a heat barrier. The latter is very important if the wearer needs to stay near flames for a fairly long time. Flame-retardant clothing is generally used for occupation uniforms (Holmes, 2000). It may also be noted that the main cause of death in a fire accident is not direct burning but suffocation due to the smoke and toxic gases released during burning. Therefore, the use of non-toxic or low-toxic burning materials is very important for fire protection. Considering safety, the government regulations says that certain classes of garments and home textiles such as children's sleepwear, carpets, upholstery fabrics and bedding should be made flame retardant or flame resistant (Wakelyn *et al.*, 1998). Clothing and textiles are made flame retardant by using inherently flame-retardant materials such as Kevlar and Nomex or by applying a flame retardant finish.

Chemical protection

The chemical industry has been facing an ever-increasing degree of regulation to avoid workers being exposed to chemical hazards (Bajaj *et al.*, 1992). Chemical protective clothing (CPC) should be considered the last line of defence in any chemical-handling operation and every effort should be made to use less hazardous chemicals where possible, or to develop and implement engineering controls that minimise or eliminate human contact with chemical hazards (Carroll, 2001; Adanur, 1995).

The same protective clothing cannot be used for all chemical applications, since chemicals vary in most cases and a particular CPC can protect only against a limited number of specific chemicals. While designing chemical protective clothing, factors like the amount of chemical permeation, breakthrough time for penetration, liquid repellency, and physical properties of the CPC in specific chemical conditions are to be considered (Carroll, 2001; McQueen *et al.*, 2000). Based on the specific requirements and type of clothing, CPC is classified in different ways and may be of the encapsulating type or non-encapsulating type (Adanur, 1995). The encapsulating system covers the whole body and includes respiratory protection equipment and is generally used where high chemical protection is required. The non-encapsulating clothing is assembled from separate components and the respiratory system is not a part of the CPC. The Environmental Protection Agency (EPA) in the United States classifies protective clothing based on the level of protection from highest to normal protection. CPC is rated for four levels of protection, levels A, B, C and D from highest protection to normal protection (Adanur, 1995; Carroll, 2001). As per European standards, CPC are classified as types 1 to 7, depending on the type of exposure of the CPC such as gas-tight, spray-tight, liquid-tight, etc. (Carroll, 2001). Traditional disposable clothing offers resistance to a wide range of chemicals and some disposable clothing can be repaired using adhesive patches and reused before being disposed (Adanur, 1995; Carroll, 2001). Compared to solid chemicals, liquid chemicals are more often used, hence chemical protective clothing should be repellent or impermeable to liquids.

The exposure of skin to pesticides is a major health hazard to farmers. The currently available pesticide protection clothing does not provide adequate protection, especially to the hands and thighs, even if farmers use tractor-mounted boom sprayers with a closed cabin and wear protective clothing with gloves and rubber boots (Fenske *et al.*, 2002).

The other type of chemical protective clothing protects from chemicals present in the air such as toxic and noxious gases or fumes from automobiles, dust and microorganisms present in the air. Safety masks containing activated carbon particles which can absorb the dust present in the atmosphere are commonly used against air pollution.

Radiation protection

The people working in radioactive environments must wear special clothing to prevent exposure to radiation. α-, β- and γ-radiations are the major modes of nuclear radiation. Irradiation injuries by α- and some β-radiation can be prevented by keeping the radioactive dirt off the skin and out of the eyes, nose and mouth. Goggles, respiratory masks, gloves and lightweight protective clothing may be adequate for protection from some α- and β-radiation which have weak penetration

(Adanur, 1995). However, some γ-radiation have sufficient energy to penetrate through textiles and can affect the human tissue even if the radioactive substance does not contact the human skin. Protection from transmitted radiation depends on the level of contamination control, exposure time, distance from radiation source and the type of radioactive shield available (Adanur,1995). Shielding is done by placing a dense (heavy) radiation barrier such as lead between the radioactive dirt and the worker.

Woven cotton, polyester/cotton or nylon/polyester fabrics with a twill and sateen weave are the major types of fabric forms used for nuclear protective clothing (Adanur, 1995). Non-woven fabrics used as over- and transit-garments in nuclear radiation protection act as a barrier against dangerous particles, shields the main garment against contamination and are disposable when contaminated (Bajaj *et al.*, 1992).

UV radiation protection

The wavelength of solar radiation reaching the Earth's surface spans from 280 to 3000 nm (Reinert *et al.*, 1997). Ultraviolet radiation (UVR), may be divided into UV-A and UV-B, having radiation at 320–340 nm and 280–320 nm respectively. Excessive exposure of the skin to UV-A radiation can be carcinogenic resulting in chronic reactions and injury, accelerated ageing of the skin, promotion of photodermatosis, etc. (Reinert *et al.*, 1997). An overdose of UV-B can lead to acute and chronic reactions, skin reddening (erythema) or sunburn, increasing the risk factor of persons susceptible to melanoma and skin cancer (Gies *et al.*, 1998; Reinert *et al.*, 1997). In the last decade, attempts to reduce the incidence of skin cancer have been mainly focused on decreasing solar UVR exposure (Gies *et al.*, 1997).

The amount of solar UVR protection of fabrics is expressed by several terms like sun protection factor (SPF), clothing protection factor (CPF) and the most commonly used index, ultraviolet protection factor (UPF) (Gies *et al.*, 1997; Xin *et al.*, 2004). The UPF for clothing with an excellent ultraviolet protection should be higher than 40. But from a clinical viewpoint, a UPF greater than 50 is entirely unnecessary (Gies *et al.*, 1997). Sunscreens, sunglasses, hats and clothing are the main accessories used to protect from UVR. Textiles are excellent materials for UVR protection and most UV can be blocked by common clothing (Reinert *et al.*, 1997). As shown in Table 1.7, the UVR protection of a fabric depends on fibre content, weave, fabric colour, finishing processes, the presence of additives, and laundering.

Electromagnetic-radiation protection

Electrical and electronic devices are capable of emitting radio frequencies that are potential hazards to health. Examples are cell phones with frequencies from 900 to 1800 MHz, microwave ovens with 2,450 MHz, radar signal communication systems extending from 1 to 10000 MHz, and so on (Su and Chern, 2004). Many

Table 1.7 Main factors affecting UVR protection

Factors	Effectiveness
1. Fibre	Cotton – high permeability to UVR, Polyamides – fairly permeable Wool – high UV absorption Polyester – high UV-B absorption
2. Weave	Most important factors – porosity and type of weave, the tighter the weave, the less UVR transmission
3. Colour	Dark colours absorb UVR more, have high UPFs
4. Weight	Thicker and heavier fabrics transmit less UVR
5. Stretch	Greater the stretch, lower is the UPF rating
6. Moisture	Generally, fabrics provide less UVR protection when wet or moist
7. Finishing	UVR absorbing additives increase the protection of lightweight summer garments

Source: Scott, 2005.

countries are legislating new regulations so that the manufacturers of electrical and electronic equipment comply with the electromagnetic (EMC) requirement standards.

When electromagnetic waves enter an organism, they vibrate molecules producing heat that could obstruct a cell's capability for regeneration of DNA and RNA. Furthermore, electromagnetic waves can cause abnormal chemical activities that produce cancer cells leading to leukaemia and other types of cancer (Su and Chern, 2004).

Traditionally, sheet metals are used for shielding radio frequencies, but in recent years, conductive fabrics have been used for shielding electromagnetic and static charges in the defence, electrical and electronic industries. General textile fibres have sufficient insulating properties with resistivities of the order of 10^{15} Ω/cm^2, much higher than the desirable resistivity for electromagnetic shielding applications. The desired resistivities for anti-electrostatic, statically dissipated and shielding materials are 10^9 to 10^{13} Ω/cm^2, 10^2 to 10^6 Ω/cm^2 and lower than 10^2 Ω/cm^2 respectively (Cheng and Lee, 2001). The various techniques for producing conductive fabrics are:

- Applying conductive coatings onto the surface of the fabric – zinc arc sprays, ionic plating, vacuum metallised sputtering, and metal foil binding (Adanur, 1995; Bajaj *et al.*, 1992; Cheng and Lee, 2001).
- Adding conductive fillers such as conductive carbon black, carbon fibres, metal fibres (stainless steel, aluminium, copper) or metal powders and flakes (Al, Cu, Ag, Ni) to the insulating material (Cheng and Lee, 2001).
- Incorporating conductive fibres and yarns into a fabric, thereby providing flexibility in designing the conductive garments (Adanur, 1995; Bajaj *et al.*, 1992; Cheng and Lee, 2001; Su and Chern, 2004).

Electromagnetic protection

People who work close to power lines and electrical equipment have the possibility of being exposed to electric shocks and acute flammability hazards. Therefore, protection from electromagnetic sources is very much important. Generally, rubber gloves, dielectric hard hats and boots, sleeve protectors, conductive Faraday-cage garments, rubber blankets and non-conductive sticks are used for electromagnetic protection (Adanur, 1995). Conductive protective clothing with flame resistance, known as 'Live line' garments, is necessary for people who work in the vicinity of very high-voltage electrical equipment (Adanur, 1995).

Radiation from electro-magnetic fields (EMF) generated by power lines is another potential risk to people working near the lines and there have been reports about the relation between exposure to electromagnetic fields and health hazards like leukaemia and brain cancer (Adanur, 1995). A typical electromagnetic protective fabric is woven from conductive material such as spun yarns containing a mixture of fire-retardant textile fibres and stainless steel fibres (8–12 micron diameter). It has been shown that fabrics made of 25% stainless steel fibre/75% wool (or aramid) blend can protect the wearer from electromagnetic fields generated by voltages of up to 400 kV (Adanur, 1995). Protection at even higher voltages can be obtained by using a combination of these fabrics in two or more layers.

Electrostatic protection

Electrostatic charges accumulate on textile fibres, especially at low humidity and in dry conditions (Morton and Hearle, 1997) and accumulated charges are difficult to dissipate. The dissipation of an high electrostatic charge occurs through shocks and sparks which can be hazardous in a flammable atmosphere. Therefore, the presence of a static charge in textiles can be a major hazard in explosives, paper, printing, electronics, plastics, and the photographic industry (Bajaj *et al.*, 1992). There was evidence of initiated explosions in hospitals due to static electricity, before the advent of nonflammable anaesthetics and anti-static rubber components in operating theatre equipment (Scott, 1981). The charge present in a garment may be over 60 kV depending on the balance between the rate of generation and the rate of dissipation of the static charges and the body potential (Holme *et al.*, 1998).

The body-sticking of garments is a common problem caused by the presence of electrostatic charges. Electrostatic attraction may impede the opening of parachutes and even lead to catastrophic failure under certain circumstances (Holme *et al.*, 1998), so anti-electrostatic finishes are used for textiles both in civilian and non-civilian applications. The basic principle of making an antistatic garment is to decrease the electrical resistivity or the chance of electrostatic accumulation in a fabric. Examples of the former are spinning yarns containing conductive materials, producing a composite fibre in which at least one element is a conductive material or a fibre containing a conductive material such as metallic or carbon coatings. Examples of the latter are the addition of a mixture of lubricants and surfactants to the textiles, or antistatic finishing. It should, however, be noted that electrostatics can be very useful for practical industrial applications. In the textile industry, electrostatic charge is used as a means of spinning fibres and yarns (Holme *et al.*, 1998; Morton and Hearle, 1997).

Ballistic protection

Ballistic protection involves protection of body and eyes against projectiles of various shapes, sizes, and impact velocities (Adanur, 1995). Such protection is generally required for soldiers, policemen and general security personnel. Historically, ballistic protection devices were made from metals and were too heavy to wear, but textile materials now provide the same level of ballistic protection as metals but have relatively low weight and are therefore comfortable to wear. Most of the casualties during military combat or during unintended explosions are from the flying matter caused by the explosion hitting the body. It is reported that during military combat, only 19% of casualties are caused by bullets, as high as 59% of casualties are caused by fragments, and about 22% are due to other reasons. The number of casualties due to ballistic impact can be reduced 19% by wearing helmets, 40% by wearing armour and 65% by wearing armour with helmet (Scott, 2000). High-performance clothing used for ballistic protection dissipates the energy of the flying particles by stretching and breaking the yarns and transferring the energy from the impact at the crossover points of yarns (Scott, 2000).

The ballistic protection of a material depends on its ability to absorb energy locally and on the efficiency and speed of transferring the absorbed energy (Jacobs and Van Dingenen, 2001). One of the earliest materials used for ballistic protection was woven silk that was later replaced by high-modulus fibres based on aliphatic nylon 6,6 having a high degree of crystallinity and low elongation. Since the 1970s, aromatic polyamide fibres, such as Kevlar® (Du Pont) and Twaron® (Enka) and ultra-high-modulus polyethylene (UHMPE) are being used for ballistic protection (Scott, 2000).

Other mechanical impact protection

According to the US Labor Department, each year more than one million workers suffer job-related injuries and 25% of these injuries are to the hands and arms (Adanur, 1995). Gloves, helmets and chain-saw clothing are the main protective accessories used by personnel working in the chemical, construction and other industries (Adanur, 1995). Some examples of non-combat impact protection are the seat belts and air bags used in automobiles. Air bags have reduced the death rate in accidents by 28%, serious injuries by 29% and hospitalisation by 24% and seat belts can reduce fatal and serious injuries by 50% (Adanur, 1995; Fung, 2000). A typical seat belt should restrain a passenger weighing 90 kg in collision with a fixed object at 50 km/h (about 30 mph), the tensile strength of a seat belt should be at least 30 kN/50 mm (Fung, 2000).

Although relatively few deaths are caused by sports and recreational injuries (0–6% of deaths to those under age 20), these activities are associated with 17% of all hospitalised injuries and 19% of emergency admission to hospitals (Mackay and Scanlon, 2001). Modern sports clothing are made of high-performance fabrics and these are designed to operate at high speed but are still safe and comfortable to wear (O'Mahony and Braddock, 2002). The most common protective textiles used in sports are in knee braces, wrist braces, ankle braces, helmets and guards.

Biological protection

Most natural textile fibres such as wool, silk and cellulosics are subject to biological degradation by bacteria, dermatophytic fungi, etc. Fortunately, various chemicals and finishing techniques are available that can protect the textile and the wearer from biological attacks. Textiles designed for biological protection have two functions:

1. those that protect the wearer from being attacked by bacteria, yeast, dermatophytic fungi, and other related micro-organisms which cause aesthetic, hygienic, or medical problems; and
2. the textile itself is protected from bio-deterioration caused by mould, mildew, and rot-producing fungi and from being digested by insects and other pests (Bajaj *et al.*, 1992).

The antimicrobial properties of silk have been used for many years in medical applications (Choi *et al.*, 2004a). Natural fibres contain lignin, pectin and other substances that have inherent antimicrobial properties and generally, textiles made from natural fibres have better anti-microbial properties than man-made fibres. Chemical finishing is most commonly used for imparting anti-microbial properties to natural and man-made textiles by applying functional finishes onto the surface of the fabric or by making fibres inherently resistant to micro-organisms. The entire surface of inherently anti-microbial high functional fibres is made from a bioactive material and the bioactivity remains undiminished throughout the useful life of the fibres (Bajaj *et al.*, 1992). In some cases, just providing an antimicrobial finish to the fabrics may not prevent the infection.

Fabrics designed for microbial protection should act as barriers to bacteria and other microorganisms that are believed to be transported from one location to another by carriers such as dust or liquids (Belkin, 2002). Films generally have high barrier properties against microbes and chemicals, however when used with fabrics to provide antimicrobial properties they make fabrics impermeable to airflow leading to heat stress and other physiological problems that may be fatal (Wilusz *et al.*, 1997). New membrane structures called 'breathable' membranes can prevent airflow through the fabric layer but have high water-vapour permeability. Using these membranes with fabrics provides effective protection from hazardous materials or microbes without causing heat stress (HAZMAT) (Schreuder-Gibson *et al.*, 2003).

Reduced visibility protection

Reduced visibility contributes to fatal pedestrian accidents. It is reported that night-time vehicles hit and kill more than 4000 pedestrians and injure more than 30000 pedestrians annually in the United States (Adanur, 1995). High-visibility materials (HVM) are capable of assisting workers and pedestrians in avoiding death or serious injury. HVMs are used by pedestrians, highway and airport workers, cyclists, joggers, hikers, policemen, firemen and other professionals. The clothing is made highly visible by sewing high-visibility materials or by chemical finishing. There are three major types of high-visibility products (Adanur, 1995):

1. Reflective materials which shine when struck by light; e.g., reflective microprism.
2. Photoluminescent materials that can absorb daylight or artificial light, store the energy and emit a green yellow glow in darkness.
3. Fluorescent materials.

In some cases, combinations of these methods are used to provide optimum visibility during the night.

1.11.2 Human comfort in various protective clothing

Protective clothing protects the body from external influence like heat, chemicals, mechanical hazards, foul weather, etc., by shielding the human body from the environment. From a physiological point of view, the human body feels comfortable at about 29 °C in an unclothed state (Wenzel and Piekarski, 1982) and at about 26 °C with a clothing insulation of 0.6 clo (Olesen and Fanger, 1973). However, protective clothing usually has a higher insulation. The bulkiness and the weight of the clothing lead to higher metabolic heat production. The British Standard (BS 7963 2000) provides estimation for the increase in metabolic rate due to the wearing of different types of protective clothing. This increase can be as high as 155 W/m^2 when wearing highly insulating firelighters' personal protective equipment (including helmet, clothing, gloves and boots). Furthermore, many items of protective clothing have to be watertight, and the ensuing moisture barrier will reduce the transfer of sweat to the environment. Therefore the protective function of protective clothing can be achieved only by sacrificing some degree of comfort. Hence, a balance between protection and comfort has to be found for every type of protective clothing, depending on the foreseen metabolic heat production and climatic conditions.

Heat protective clothing has to perform two opposite functions – on one hand, it should prevent the external heat from flowing towards the body while on the other hand, it should allow the metabolic heat to escape to the atmosphere. Working at higher temperatures quickly leads to heat stress and stress-related heart attacks. In the United States, about 50% of the lethal accidents of firefighters on duty are due to heat stress (LeBlanc and Fahy, 2004). But even at moderate temperatures, the burden of increased weight and reduced permeability of this type of clothing can lead to heat stress. The results can, however, vary greatly depending on the set parameters: measurements at low temperatures and/or low relative humidity (Ftaiti *et al.*, 2001) resulted in different degrees of increases in the skin and core temperature for different jacket types (PVC, Neoprene or leather vs. breathable materials). As soon as the temperature or the humidity of the environment approaches the conditions near the skin, the differences become smaller. Schopper-Jochum *et al.* (1997) stated that the increase of body core temperature in an environment of 30 °C and 50% RH was independent of the jacket type, while Rossi (2003) did not find any significant differences between different types of protective clothing (breathable vs. non breathable) during exercises at higher temperatures.

The weight of the equipment represents an additional load for the firefighter. The size of the clothing and the number of textile layers (Lotens, 1983) also increase the energy consumption of the wearer and thus the required heat loss to maintain thermal comfort. With chemical protective clothing and other types of impermeable clothing, the comfort problems are mainly caused by the lack of water vapour permeability, as the protection against chemicals often imply that the materials used are totally liquid-tight and sometimes also vapour-tight. Different studies report heat stress as the primary limitation of use of chemical protective clothing (Töpfer and Stoll, 2001). Problems of body dehydration can also occur, as the sweat production cannot be easily compensated due to the wearing of a gas mask or self-contained breathing apparatus. Therefore, new developments of such clothing often aim at increasing the thermophysiological comfort primarily by improving the water vapour permeability (Wilkinson *et al.*, 1997). For totally impermeable protective clothing, technical aids like ice-, liquid-, or air-cooling can help to reduce the thermal strain of wearing such equipment. However, the additional weight causes additional metabolic heat production and lowers the benefits of such systems.

Foul weather and cold protective clothing often contains a waterproof, moisture permeable barrier. This barrier ensures the water tightness, but at the same time still hinders the free flow of part of the water vapour produced by the body to the environment, as even the most breathable membrane or coating adds a resistance to the vapour flow. If the outside temperature is low, there is a certain risk of water condensation within the clothing layers when the water vapour pressure increases beyond saturation. The presence of condensation can then change the thermal and moisture transport properties of the clothing.

Ballistic protective clothing should also have a compromise between comfort and protection. Because of weight and bulkiness problems, the users sometimes refuse to wear these garments continuously. Furthermore, ballistic protective clothing using aramid-based panels has a significantly lower ballistic resistance when water enters the structure. For this reason, the panels have to be packed into impermeable outer shells, which have unfavourable consequences on the clothing physiology of the garment.

1.12 Future trends and further information and advice

The growing application of emerging technologies (nano, bio, info and computing) has provided a much needed impetus for a paradigm shift in textiles. Apart from apparels, textile materials are now being used for many technical applications including industry (Indutech), medical (Medtech), automobiles (Mobiltech), construction (Buildtech), agriculture (Agrotech) and protection (Protech). Use of improved material, surface modification, structural design and value addition are integral parts of manufacturing these high end technical products.

The late twentieth century saw an unprecedented increase in emphasis on protection of the human form. Health and safety at work requires protective textiles for certain jobs and the threat of biological and chemical terrorist attacks is currently a topical issue. The range of hazards and the means of combating them continue to grow and become ever more complex. A consequence of this is the development and exploitation of new textile fibres, structures and clothing systems whose purpose is to provide improved protection, while maintaining comfort, efficiency and well being.

Human sensory perception of clothing involves a series of complex interactive processes, including physical responses to external stimuli, neuro-physiological processes for decoding stimuli through the biosensory and nervous systems inside the body, neural responses to psychological sensations, and psychological processes for formulating preferences and making adaptive feedback reactions.

In 1986, the concept of sensory engineering (kansei engineering) was first developed by the Mazda company in Japan as a development of human factors. Clothing biosensory engineering is a systematic and integrative way of translating consumers' biological and sensory responses, and psychological feelings and preferences about clothing, into the perceptual elements of design. It is a link between scientific experimentation and commercial application to develop economic solutions to practical technical problems. Sensory engineering has been applied with great success in the automobile industry and is being extended to other product domains including the development of new fibres (Li *et al.*, 2006).

Clothing biosensory engineering quantifies the decision-making processes through which physics, mathematics, neurophysiological and engineering techniques are applied to optimally convert resources to meet various sensory requirements – visual/thermal/mechanical. It includes theoretical and experimental observations, computer simulations, test methods, illustrations and examples of actual product development.

The technical developments in the sports clothing industry have resulted in the use of engineered textiles for highly specialised performances in different sports. With high-functional and smart materials providing such a strong focus in the textile industry generally, companies are increasingly looking for 'value added' textiles and functional design in sportswear as well as intelligent textiles which monitor performance with in-built sensors. Combining clothing functions with wear comfort is a growing market trend, and for all active sportsmen this constitutes one of the vital factors for achieving high level of performance.

The use of intelligent textiles in clothing is an exciting new field with wide-ranging applications. Development of phase change and shape memory materials and their role in clothing is a major thrust area of research. Cool Biz, a combination of the English words cool and business, is the catchword in Japan. It is the title of a government campaign to persuade office workers to dispense with their ties and jackets as an environmentally friendly way of staying cool without lowering the air conditioner thermostat. And now an innovation called kuchofuku (air-conditioned clothing) is taking the Cool Biz concept one step further. This new type of garment lets people stay cool even in long sleeves.

1.13 References

Adanur S (1995), *Wellington Sears Handbook of Industrial Textiles*, Lancaster, Pennsylvania, Technomic Publishing Company, Inc.

Aschoff J, Günther B and Kramer K (1971), *Energiehaushalt and temperaturregulation*, München, Germany, Urban & Schwarzenberg.

ASHRAE (1989), *Handbook of Fundamentals*, American Society of Heating, Refrigerating and Air-conditioning Engineers Inc.

ASTM (2003), ASTM D-123-03, Standard terminology relating to textiles, West Conshohocken, PA, ASTM.

Bajaj P, Sengupla A K and others (1992), Protective clothing, *Textile Progress* **22**(2/3/4), 1–117.

Belkin N L (2002), A historical review of barrier materials, *Aorn Journal*, **70**(4), 648–653.

Blankenbaker J. (1982), Ventilating systems for hot industries, *Heating/Piping/Air Conditioning*, **54**(2), February.

Bornais P (1997), Analysis and characteristics of comfort in clothing, *Canadian Textile Journal*, **114**(4), 12–14.

Braddock S E and O'Mahony M (2002), *Sport Tech: Revolutionary Fabrics*, Thames & Hudson, London.

BS 7963 (2000), Ergonomics of the thermal environment – Guide to the assessment of heat strain in workers wearing personal protective equipment, British Standard Institute, London.

Byrne C (2000). 'Technical textiles market – an overview', in Horrocks A R and Anand S C, *Handbook of Technical Textiles*, Cambridge, Woodhead, pp. 462–489.

Carroll T K (2001), Chemical protective clothing, *Occupational Health and Safety*, **70**(8), 36–46.

Cheng K B and Lee M L (2001), Electromagnetic shielding effectiveness of stainless steel/ polyester woven fabrics, *Textile Res J.*, **71**(1), 42–49.

Choi H, Bide M, Phaneuf M, Quist W and Logerfo F (2004a), Antibiotic treatment of silk produces novel infection-resistant biomaterials, *Textile Res J.*, **74**(4), 333–342.

Choi K, Cho G, Kim P and Cho C (2004b), Thermal storage/release and mechanical properties of phase change materials on polyester fabrics, *Textiles Res J.*, **74**(4), 292–296.

Das S (2005), Comfort characteristics of apparels, www.expresstextile.com

Denton M J (1970), Fit, stretch and comfort, in *3rd Shirley International Seminar: Textiles for Comfort*, Manchester, England: The Cotton Silk and Manmade Fibres Research Association.

Denton M J (1971), Fit, stretch and comfort in *Proceeding of the 3rd Shirley International Seminar: Textiles for Comfort*, Manchester, England.

EN ISO 7730 (1984), Moderate thermal environments – determination of the PMV and PPD indices and specification of the conditions for thermal comfort. ISO 9920, Ergonomics of the thermal environment–estimation of the thermal insulation and evaporative resistance of a clothing ensemble.

Fanger P O (1970), *Thermal Comfort-Analysis and Applications in Environmental Engineering*, Danish Technical Press, Copenhagen, Denmark.

Fenske R A, Birnbaum S G, Methner M M, Lu C and Nigg H N (2002), Fluorescent tracer evaluation of chemical protective clothing during pesticide application in central Florida citrus groves, *Journal of Agricultural Safely and Health*, **8**(3), 319–331.

Fourt L and Hollies N R S (1970), *Clothing: Comfort and Functions*, Marcel Decker Inc., New York.

Ftaiti F, Duflot J C, Nicol C and Grelot L (2001), Tympanic temperature and heart rate changes in firefighters during trademill runs performed with different fireproof jackets, *Ergonomics*, **44**, 502–512.

Fung W (2000), Textiles in transportation, in Horrocks A R and Anand S C, *Handbook of Technical Textiles*, Cambridge, Woodhead, pp. 490–528.

Gagge A P, Fobelets A P and Berglund L G (1986), A Standard Predictive Index of Human Response to the Thermal Environment, *ASHRAE Trans.*, Vol. 92, Pt 2.

Garnsworthy R K, Gully R L, Kandiah R P, Kennis P, Mayfield R J and Westerman R A (1985), Identification of the physical stimulus and the neural basis of fabric-evoked prickle, *Journal of Neurophysiology*, **59**(4), 1083–1097.

Ghai O P (2004), *Essential Pediatrics*, 6th edn, Published by Dr. O.P. Ghai, Delhi-92, p. 152.

Gies P H, Roy C R, and others (1997), UV protection by clothing: An intercomparison of measurements and methods, *Health Physics Society*, **73**(3), 456–464.

Gies P H, Roy C R, Toomey S and Melennan A (1998), Protection against solar ultraviolet radiation, *Mutation Research*, **422**, 15–22.

Hatch, K L (1993). *Textile Science*, West Publishing Co., Minneapolis, MN, USA, p. 26.

Holme L, McIntyre J E and Shen Z J (1998), Electrostatic charging of textiles, *Textile Progress*, **28**(1), 29.

Holmes D A (2000), Textiles for survival, in Horrocks A R and Anand S C, *Handbook of Technical Textiles*, Cambridge, Woodhead, pp. 462–489.

Hui S C M, Thermal Comfort, MEBS6006 Environmental Services I, http://www.hku.hk/ mech/msc-courses/MEBS6006/index.html

Ishiaque S M (2001), Engineering comfort, *Asian Textile Journal*, **10**(11), 36–39.

Jacobs M J N and Van Dingenen J L J (2001), Ballistic protection mechanism in personal armour, *Journal of Materials Science*, **36**(13), 3137–3142.

Kamata Y T, Kato A I and Yahata N (1988), Condiuctive heat transfer from human body. II. Effect of fabric on heat transfer, *Sen-i-Gakkaishi*, **44**(2), 78–87.

Kawabata S and Niwa M (1989), Fabric performance in clothing manufacture, *Journal Textile Institute*, **80**(1), 19–50, doi: 10.1080/00405008908659184.

Kirk W J and Ibrahim S M (1966), Fundamental relationship of fabric extensibility to anthropometric requirements and garment performances, *Textile Research Journal*, **57**, 37–47.

Koscheyev V S and Leon G R (1997), Rescue worker and population protection in large-scale contamination disasters, *Clinical & Health Affairs*, **80**(1), 23–27.

LeBlanc P R and Fahy R F (2004), *Firefighter Fatalities in the United States – 2003*, Quincy, MA, National Fire Protection Association.

Li Y (1988), The objective assessment of comfort of knitted sportwear in relation to psycho-physiological sensory studies, in *The Effect of Thermal Insulation of Clothing on Human Thermal Comfort*, Department of Textile Industries, The University of Leeds, Leeds, p 213.

Li Y A (1998), Wool sensory properties and product development, *Textile Asia*, **29**(5), 35–40.

Li Y (2001), The science of clothing, in J M Layton, ed., *Textile Progress*, 31 (112), The Textile Institute, Manchester.

Li Y, Wang Z and Zhang X (2006), Mechanical and thermal sensory engineering design in: *Clothing Biosensory Engineering*, Woodhead and CRC, Cambridge, England.

Lotens W A (1983), Clothing, physical load and military performance, *Aspects médicaux et biophysiques des vêtements de protection*, Centre de Recherches du Service de Santé des Armées, Lyon, France, 268–279.

Mackay M and Scanlan A (2001), *Sports and Recreation Injury Prevention Strategies: Systematic Review and Best Practices Executive Summary*, British Columbia, Injury Research and Prevention Unit.

McQueen R H, Laing R M, Niven B E and Webster J (2000), Revising the definition of satisfactory performance for chemical protection for agricultural workers, *Performance of Protective Clothing: Issues and Priorities for the 21st Century*, **7**, 102–116.

Milenković L, Škundrić P, Sokolović R and Nikolić T (1999), Comfort properties of defense protective clothings, UDC 614.8.086, University of Niš, *The Scientific Journal*, Facta Universitatis, series: Working and Living Environmental Protection Vol. 1, No 4, pp. 101–106.

Morton W E and Hearle J W S (1997), *Physical Properties of Textile Fibres*, The Textile Institute, UK.

Ogulata R Tugrul (2007). Fibres and textiles in Eastern Europe, Apr/Jun, Vol.15, No.2 (61), p. 67.

Olesen B W and Fanger P O (1973), The skin temperature distribution for resting man in comfort, *Arch. Sci. Physiol.*, Vol.27, 385–393.

Panek C (1982), *Protective Clothing*, Shirley Institute, Manchester.

Pause B (2003), Nonwoven protective garments with thermo-regulating properties, *Journal of Industrial Textiles*, **33**(2), 93–99.

Peirce F T (1930), The handle of cloth as a measurable quantity, *Journal of the Textile Institute*, **21**, T377–416.

Pontrelli G J (1977), Partial analysis of Comfort's Gestalt, in *Clothing Comfort* (eds N.R.S. Hollier and R.F. Goldman), Ann Arbor Science Publishers Inc., Michigan, USA, pp. 71–80.

Reinert G, Fuso F, Hilfiker R and Schmidt E (1997), UV-protecting properties of textile fabrics and their improvement, *Textile Chemist and Colorist*, **29**(12), 36–43.

Richardson J E and Capra M F (2001), Physiological responses of firefighters wearing level 3 chemical protective suits while working in controlled hot environments, *Journal of Occupational and Environmental Medicine*, **43**(12), 1064–1072.

Rissanen S and Rintamaki H (2000), Prediction of duration limited exposure for participants wearing chemical protective clothing in the cold, *International Journal of Occupational Safety and Ergonomics*, **6**(4), 451–461.

Rossi R (2003), Firefighting and its influence on the body, *Ergonomics*, **46**, 1017–1033.

Schopper-Jochum S, Schubert W and Hocke M (1997), Vergleichende Bewertung des Trageverhaltens von Feuerwehreinsatzjacken (phase 1), Arbeitsmed. *Sozialmed. Umweltmed.*, **32**, 138–144.

Schreuder-Gibson H L, Truong O, Walker J E, Owens J R, Wander J D and Jones W E Jr. (2003), Chemical and biological protection and detection in fabrics for protective clothing, *MRS Bulletin*, **28**(8), 574–578.

Scott R A (1981) Static electricity in clothing and textiles, *Thirteenth Commonwealth Defence Conference on Operational Clothing and Combat Equipment (Malaysia)*, Colchester, UK: Stores and Clothing Research and Development Establishment.

Scott R A (2000), Textiles in defence, in Horricks A R and Anand S C, *Handbook of Technical Textiles*, Cambridge, Woodhead.

Scott R A (2005), *Textiles for Protection*, Cambridge, Woodhead.

Slater, K. (1985). *Human Comfort*, Charles C. Thomas Publisher, Springfield, IL, USA, p. 4.

Su C and Chern J (2004), Effect of stainless steel-containing fabrics on electromagnetic shielding effectiveness, *Textiles Res J.*, **74**(1), 51–54.

Thirty M C (2003), Feelin' fine: Textiles bring a sense of comfort, *AATCC Review*, **3**(3), 9–14.

Threlkeld J L (1970), *Thermal Environmental Engineering*, 2nd edn, Prentice Hall, Inc., New Jersey.

Töpfer H J and Stoll T P (2001), Physiological assessment of permeable NBC protective clothing for hot climate conditions, RTA/HFM Symposium 'Blowing hot and cold: Protecting against climatic extremes', Dresden, Germany.

Wakelyn P J (1997), Overview of cotton and flammability, in Conference on recent advances in flame retardancy of polymeric materials: materials, applications, industry developments and markets, Morwalk, CT. USA: Business Communication, Co.

Wakelyn F J, Rearick W and Turner J (1998), Cotton and flammability – overview of new developments, *American Dyestuff Reporter*, **20**(2), 13–21.

Wenzel H G and Piekarski C (1982). *Klima und Arbeit*, München, Germany, Bayerisches Staatministerium für Arbeit and Sozialordnung.

Wilkinson M C, Scott R A, Williams J T and Lovell K V (1997), Consideration design and construction of novel integrated NBC ensembles, *Proceedings of the fifth Scandinavian symposium on protective clothing*, Elsinsinore, Denmark, 121–126.

Wilusz E, Truong Q T, Rivin D and Kendrick C E (1997), Development of selectively permeable membranes for chemical protective clothing, *Polymeric Materials Science and Engineering*, **77**, 365.

Wong A S W and Li Y (2006), Prediction of clothing sensory comfort, in: *Clothing Biosensory Engineering*, Woodhead and CRC, Cambridge, England.

Xin J H, Daoud W A and Kong Y Y (2004), A new approach to UV-blocking treatment for cotton fabrics, *Textiles Res J.*, **74**(2), 97–100.

2

Properties of fibers and fabrics that contribute to human comfort

S. A. HOSSEINI RAVANDI, Isfahan University of Technology, Iran and M. VALIZADEH, University of Guilan, Iran

Abstract: Comfort characteristics of cloths are made by a collection of interactive properties of fibers, yarns and the fabric which have contributed in construction of the clothing. This chapter focuses on the main affective properties of fibers, yarns and fabrics. It starts with fiber specification, comparing two natural and synthetic sources of fibers used in apparel textile products, and then investigates physical treatments to modify fiber properties. The chapter continues with yarns and fabrics as the intermediate products to cloths, investigating the producing parameters which create and affect garment comfort.

Key words: comfort, clothing, fibrous structures, physical properties.

2.1 Introduction

As a key parameter of clothing, comfort of the garment is the complex effect of textile properties which are basically dependent on the chemical structure and morphology of the constituent fibers. Comfort properties of textile products such as yarns, fabrics, mats and any other product that is used for wearing purposes embrace different mechanical properties as well as heat and moisture transfer (or isolation). Due to interactive effects of numerous parameters which are involved in comfort formation and the subjective nature of most of these parameters, there is an agreement between textile comfort researchers: *comfort cannot be quantitatively defined!* The same cloth that in a specific situation sounds very comfortable to one person can be absolutely uncomfortable to another person in the same time and situation. Moreover, when the situation changes even the same person may not have the same comfort feeling.

Comfort can be considered as the consistency of the clothing with the human surroundings or environment determined by human pleasantness or relief. This consistency includes physical aspects such as heaviness, thickness, thermal transmission or heat transfer, air permeability, moisture absorbency and moisture diffusion, handle, ease of movement (flexibility), drape as well as aesthetic aspects such as color, luster, fashion, style, and fit and in addition to these aspects we have to consider the subjectivity of the user, culture, etc.

As comfort is definitely an individualistic sense it is very difficult to define, design or determine it. Discomfort sense, which is the opposite point of comfort, may be easier to define using terms of prickle, hot (cold), tight, moist, etc., that happen when we are consciously aware of the unpleasantness of the worn clothes. Two major types of discomfort sensation are psychological discomfort and physiological discomfort. Psychological discomfort appears when the cloth is inappropriate for the person who is using it, personally or for an occasion. Physiological discomfort appears when the body feels uncomfortable, as for example when one feels too cold, feels itchy, or the garment is too tight (Smith, 1986). Slater mentions that 'discomfort will typically be experienced if the body is exposed to a surface which subjects it to frictional contact outside a tolerable range. If friction is high, or if the external surface contacted is too stiff, then the skin may rub, subsequently causing severe pain with possibly an open sore developing. Conversely, if the contacted surface is too soft, a tickling sensation may occur, bringing with it discomfort of a different type' (Slater, 1991).

Comfort itself has been described in different terms by researchers. Hatch (1993) defined comfort as: 'Freedom from pain and from discomfort as a neutral state.' Comfort can be considered as 'an experience that is caused by integration of impulses passed up the nerves from a variety of peripheral receptors smell, smoothness, consistency and color etc in the brain' (Das and Ishtiaque, 2004).

Saville has mentioned thermo-physiological wear comfort and skin sensational wear comfort as two aspects of wear comfort of clothing. The former concerns the heat and moisture transport properties of clothing and the way that clothing helps to maintain the heat balance of the body during various levels of activity, and the latter concerns the mechanical contact of the fabric with the skin, its softness and pliability in movement and its lack of pricking, irritation and cling when damp (Saville, 1999). However, the major definition of clothing comfort describes it as a decision made by the brain according to psychological, physical and thermo-physiological condition of the body. Psychological comfort is mainly related to the subjectivity of the wearer and aesthetic parameters such as the latest fashion trend, color harmony, fit, and acceptability in society (which returns to the culture) (Slater, 1991; Li and Wong, 2006; Das and Ishtiaque, 2004; Li and Dai, 2006).

Physiological comfort is considered as the interaction of the body with the surroundings and the physiological feeling of a person with respect to the environment. Physiological comfort is mainly related to maintenance of thermal and moisture balance of the body; in other words, it is the proper relationship between body heat and moisture production and loss.

Although the comfort concept has not yet been clearly identified, there are several remarkable parameters revealing direct or indirect effects on comfort specifications. These parameters can be categorized in chemical and morphological identities of material constituents, physical properties of material constituents, constructional properties, and finishing treatment. These parameters can be counted for each of the constituents, i.e. fibers, yarns, fabric.

This chapter aims to introduce the effects of the properties and responses of the constructional elements of the wear (fiber, yarn and fabric) on comfort.

2.2 Comfort properties of fibers

For thousands of years, the presence of textile products in the life of men was limited by the inherent qualities available naturally: cotton, wool, silk, linen, hemp, ramie, jute and many other natural resources. Since 1910, with the production of rayon as 'artificial silk', man-made fibers started to be developed and to be used. Nowadays man-made fibers are found in different features of life with countless applications, from apparel, sporting clothes and furnishings to industrial, medical, aeronautics and energy. The main advantage of the commercial man-made fibers is their low cost with respect to natural fibers and man-made fibers have enormous advantages with respect to natural fibers. Considering their remarkable mechanical and chemical properties, modern life cannot survive without man-made fibers. However in the field of comfort, man-made fibers can not overcome the natural fibers.

The behavior of fabric is affected by chemical and physical properties of its constituent fibers, fiber content, physical and mechanical characteristics of its constituent yarns, and the finishing treatments which are applied on it. Cost, quality, care and comfort are the essential properties that the customer considers before buying the cloth, all of which are different aspects of comfort.

The most important parameter that determines the comfort of a cloth is the material. Tables 2.1 and 2.2 contain the main properties and the advantages and disadvantages of fabrics of the most used fibers for apparel applications. Considering all these properties, blending the fibers in a fabric seems to be a powerful solution to modify the properties and cost of garments. For example, cotton has a high moisture retention. It absorbs the sweat and lets it evaporate, therefore cotton is highly water-vapor permeable or breathable. Polyester has low moisture retention and it is not breathable, but is crease resistant; cotton is not. A garment with a blend of cotton and polyester is wrinkle resilient and does not need to be ironed or will require less ironing, while retaining much of the comfort

Table 2.1 Properties of fabrics according to their constituent fibers

Property	Cotton	Wool	Polyester	Acetate	Rayon	Acrylic	Nylon
Durability	Very good	Moderate	Excellent	Poor	Moderate	Moderate	Excellent
Liquid absorbency	Excellent	Slow/ good	Poor	Good	Good	Poor	Poor
Wrinkle resistance	Poor	Excellent	Excellent	Poor	Moderate	Good	Excellent
Care	Machine wash	Dry clean	Machine wash	Dry clean	Machine wash gently	Machine wash	Machine wash

Table 2.2 Advantages and disadvantages of fabrics made of different fibers

Material	Advantages	Disadvantages
Cotton	Strong, soft, durable, comfortable, absorbent, washable	Shrinks, wrinkles, expensive
Wool	Warm, strong, wrinkle resistant	Shrinks, can be damaged by moths, care difficulties, expensive
Polyester	Strong, blends well with other fibers, washable, wrinkle resistant, dries quickly, cheap	Holds oily stains, does not absorb sweat, weak handle
Acetate	Soft, drapes well, looks like silk, cheap	Wrinkles, fades, heat-sensitive, loses strength when it is wet, poor abrasion resistance, has to be kept away from perfume and nail polish remover (dissolves in acetone), care difficulties
Rayon	Soft, comfortable, highly absorbent	Loses strength when wet, wrinkles easily, care difficulties
Acrylic	Soft, lightweight, warm, often resembles wool, wrinkle resistance, blends well with other fabrics, non-allergenic, cheap	May pill with abrasion, sensitive to heat
Nylon	Strong, holds shape well, washable, dries quickly, resilient, elastic	Sensitive to heat, does not absorb moisture, can pick up dyes when washed with colored items, static electricity

provided by cotton. Polyester is strong and rayon has shininess. A fabric produced of cotton/polyester/rayon offers durability, ultra-softness, and excellent resilience so that if wrinkled, the fabric bounces back. Spandex is stretchy and durable. Its blend with cotton is a very good choice for sports cloth. Silk has luster and good drapability and does not crease easily. Linen creases easily, but is strong and breathable. A fabric made of a blend of silk and linen would not crease as readily and will be shiny and drape better.

2.2.1 Type of fiber

The type of fiber is the most crucial specification which determines important properties such as strength, durability, handle, elasticity, dyeability, luster, friction properties, moisture absorbance, heat isolation and abrasion resistance; all the physical and chemical properties of fibers and their end-products. Fiber type is the most effective parameter in defining the comfort of the end-product. Here another term appears: fiber content.

Fiber content is the ratio of presentation of different types of fibers in textile production. This factor determines the moisture absorbency which in turn affects the thermal balance of the product, the moisture vapor and liquid permeability, the durability and the electrostatic properties in particular and all the other aforementioned properties of textile material in general.

Concerning comfort, the fibers are mainly divided into two groups of natural and synthetic fibers. Natural fibers are generally believed to provide better comfort sense. This is an old idea that comes from the higher ability of natural fibers in providing better moisture absorbency, heat isolation, handle, luster, etc. Synthetic fibers are mostly deficient in providing warmth, adequate bulkiness, soft feeling, thermal isolation and moisture absorbency due to their physical properties and chemical structure. For example in terms of thermal comfort, an ideal cloth should have high thermal resistance for cold protection, low water vapor resistance to be efficient in heat transfer under a mild thermal stress condition, and has to have rapid liquid transport to eliminate unpleasant tactile sensations due to water under a high thermal stress condition. Compared to cotton products, polyester fabric has lower water vapor resistance, but is inferior in both thermal resistance and liquid water transport (Yoon and Buckley, 1984). However, synthetic fibers have high strength, durability, dimension stability, abrasion resistance coupled with their thermoplastic properties and good resistance against heat. All these properties motivate textile producers to employ them in apparel products, especially considering their low cost. Nowadays, the growth of knowledge of fiber science and technology has come in the service of fiber designing. Using the technology of fiber production it is possible to produce the synthetic fibers with a cross-section of natural fibers. It is also possible to produce modified synthetic fibers which present closer behavior to natural fibers and it is possible to employ special finishing treatments to modify the properties of produced synthetic fibers according to the preferable behavior of natural fibers. For example if water absorbency is needed, absorbent fibers by various physical and chemical modifications could be applied, examples of which are presented in Fig. 2.1 (Schmidtbauer, 2009).

However it has to be noticed that two kinds of fibers, even natural or synthetic, may never behave absolutely alike. The aforementioned points do not mean that natural fibers behave better in all the comfort aspects with respect to synthetic fibers. The following paragraphs mention the properties of the most comfortable natural fibers known, and the solutions for synthetic fibers to be produced to represent comfort feeling as well as natural fibers.

Cotton

Cotton fibers are natural hollow fibers; they are soft, cool, known as breathable fibers and absorbent. Cotton fibers can hold water 24–27 times their own weight.

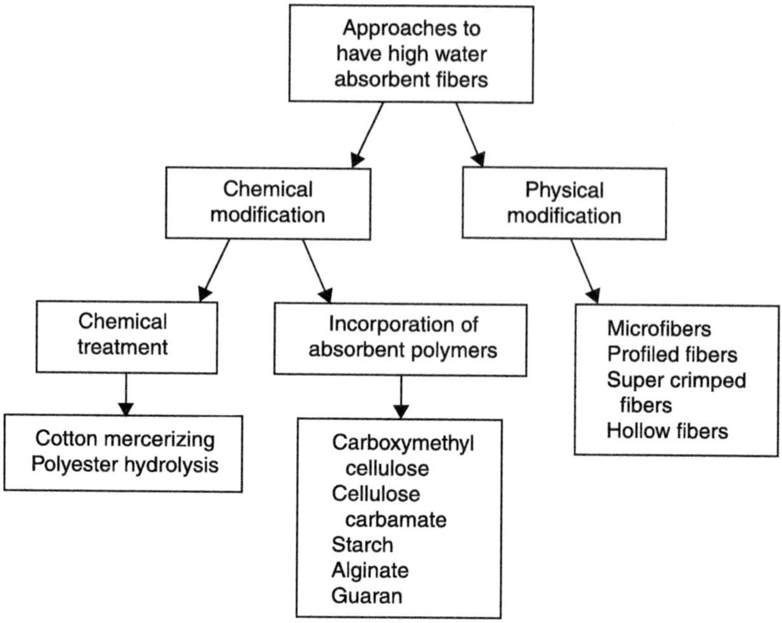

2.1 Approaches to have high water absorbent fibers.

They are strong, dye absorbent and can stand up against abrasion wear and high temperature. In one word, cotton is comfortable. Since cotton wrinkles, mixing it with polyester or applying some permanent finish gives the proper properties to cotton garments. Cotton fibers are often blended with other fibers such as nylon, linen, wool, and polyester, to achieve the best properties of each fiber.

Wool

The word 'wool' brings warmth to mind. The unique property of wool is its superior thermal comfort that comes from its high hygroscopicity. Wool fibers have crimps or curls which create pockets and give the wool a spongy feel that makes for heat isolation. Wool fibers have a high resiliency and crease recovery and can absorb moisture up to 30 times more than their weight. This property makes them a remarkably comfortable clothing material. Wool fibers are also flame resistant, dirt resistant and with high dimensional stability which make them one of the favorite choices for tailors. Wool is highly breathable and is the proper choice for both cold and hot weather. Wool is considered a prickly fiber; however studies have shown that the prickle sense is the direct effect of wool fiber diameter, and finer wool fibers are not prickly (Wang *et al.*, 2003).

Silk

Silk is the best known fiber for its luster, strength, light weight and elegance. Silk cloths are prized for their versatility, wearability, washability and comfort, and it does not shrink like other fibers. Silk fabric is well-known as an excellent cloth that has material characteristics to fit human's health. From the physiological and psychological viewpoint of clothing comfort, underwear of spun silk yarn is the best for clothes in summer time (Kamijo *et al.*, 2009).

Considering, the important roles of fiber properties in both heat and moisture transfer in textiles and with regard to the dynamic physical, physiological and psychological interactions in the perception of comfort concept, fiber engineering design finds a particular role in developing functional garments with special effects and characteristics that can be dictated for clothing comfort. Fiber designing gives a powerful tool to the producers to present their garments according to environmental (warm or cold, humid or dry), physiological, sensorial and any special requirement of the applicant playing with crucial fiber properties. Comfort is a term that is influenced by three main properties of fibers: type, fineness and length.

2.3 Physical modification of fibers

The cross-section of synthetic fibers depends on the shape of the spinneret and the behavior of fiber dough when it comes out of the spinneret and solidifies. Although fiber solidification is a quite complicated phenomenon depending on many parameters such as solidification bath or air flow direction and velocity, the temperature of solidification and the physical chemistry of fiber dough, however it is possible to achieve special cross-sections playing with the shape of the holes of the spinneret and the mentioned parameters. This technique results in a very important category of fibers called 'profiled fibers'. The original and simplest shape for the spinneret holes is the round or circular cross-section that is the most manufactured shape of synthetic fiber producers. For polymeric fiber dough, a circular spinneret hole gives a round cross-sectional fiber as well. However there are fibers such as viscose or acrylic which quench in the solidification bath and present non-circular crenulated or dumbbell-shape cross-sections. This means that the designed synthetic fiber is able to mimic the luster, friction, aberration resistance, drape and handle of natural fibers. Moreover it presents higher liquid absorbency according to the designed cross-section.

2.3.1 Profiled fibers

Properties of noncircular cross-sectional shape fibers, including softness, luster, handle, comfort, the bending stiffness, coefficient of friction, pilling, bulkiness, and performance are different from those of fibers with a circular cross-sectional shape. The development of melt spun fibers noncircular cross-sections started in the 1960s, mimicking the trilobal cross-section of silk fibers to achieve the same gloss properties of these more expensive fibers. Since that time, different shapes of cross-sections have been developed for synthetic and semi-synthetic fibers to gain closer properties to natural fibers. If the cross-sectional shape of the nozzle is varied, the spinning behavior can change significantly because the heat-transfer and air-friction effects for the noncircular cross-sectional fibers are not the same as those for circular fibers. These effects may lead to the alteration of the structure and properties of as-spun fibers (Karaca and Ozcelik, 2006; Warner, 1995; Bueno *et al.*, 2004). Viscose fibers made of sustainable, renewable wood resources combine their natural properties with the advantages of man-made fibers. They are fully biodegradable and provide natural absorbency and softness to nonwoven products. Due to their inherent water-holding capacity and being fully biodegradable, Viscostar® trilobal cross-sectional cellulosic fibers are a very good choice for nonwoven applications where absorbency is the primary key parameter, such as washing products, medical care, body protection products, and so on. Unlike synthetic fibers, the diameter of cellulosic fibers increases significantly in the wet state. Producing trilobal viscose fiber is a physical modification to improve both bulkiness and water absorbency of viscose fibers by 17% and 40% respectively (Schmidtbauer, 2009).

Karaca and Ozcelik presented a study that compared the physical and structural properties of four types of cross-section for polyester fibers, including round, hollow round, trilobal and hollow trilobal fibers. Their study showed that the enthalpy and entropy of hollow round fibers are lower than the round fibers and hence they present higher take-up stress and amorphous orientation. The full fibers were tough and ductile, whereas the hollow fibers were stiffer and more resistant to plastic deformation. The hollow fibers have lower maximum strain values and higher values for the modulus, yield stress, take-up stress, and shrinkage in boiling water with respect to the round fibers. The hollow-round fiber had the highest value for unevenness and the lowest value for crystallinity and the change in the cross-sectional shapes had only a small effect on the crystallinity and maximum stress of the fibers produced in the FDY melt-spinning process (Karaca and Ozcelik, 2006).

2.3.2 Microfibers

Microfiber is defined as a staple fiber or filaments of linear density approximately 1 dtex or less, and above 0.3 dtex. Although acrylic, viscose and polypropylene are available for the production of microfibers, polyester and polyamide are the main source. The fabrics made from them can be 100% microfiber or in blends with wool, cotton or viscose. Microfibers are half the diameter of a fine silk fiber, one-third the diameter of cotton, one-quarter the diameter of fine wool, and one hundred times finer than human hair. In order to be classified as a microfiber, the fiber must be less than 1 dtex in width. Fabrics made of microfibers are generally lightweight, resist wrinkling, have a luxurious drape on the body, retain shape, and resist pilling. They are also relatively strong and durable in relation to other fabrics of similar weight, and they are more breathable and more comfortable to wear. Fabrics made of microfibers show lower heat conductance and therefore higher thermal insulation properties. Microfiber fibers exhibit a warmer feeling than conventional fabrics depending on pressure, which may be due to the difference in the fiber and fabric surface in contact with the human skin (Schacher *et al.*, 2000; Purane and Panigrahi, 2007).

The first 'micro-denier' products were introduced by Japanese fiber manufacturing companies during the 1970s. The production of microfibers followed in Europe during the 1980s and since the 1990s by American manufacturers. Toray was the first company in the world to introduce microfibers, followed by Teijin, Hoechst, ICI, DuPont, and others. Recently Toray has introduced an ultra-fine polyester microfiber with a linear density of filament of about 0.05 dtex. This may be called the finest synthetic fiber so far produced commercially. At present, polyester and nylon are generally used for manufacturing microfibers. However, 'micro-denier' versions of rayon and acrylic products are on the horizon.

Three conventional spinning methods, i.e. melt spinning, dry spinning, and wet spinning can be used to manufacture microfibers. However for producing microfibers by these methods, the polymerization process, polymer spinning and drawing conditions have to be selected and executed very carefully. The technology involved in the extrusion of microfibers is more sophisticated and costly than that of conventional deniers as microfibers are delicate products that require great attention in handling during textile mill processing.

Microfiber spinning is now possible by many major fiber producers on their existing equipment, however economical production of high quality microfibers will require significant changes in future machine design and operation. Generally speaking, there are two techniques to produce microfibers:

- direct spinning (conventional POY spinning), and
- bi-component process (segment and island-in-sea type).

Different procedures have been presented and employed to produce microfibers:

1. Dissolved type
2. Split type
3. Direct spun type
4. Super-drawing technique
5. Sheath-core spinning method
6. Flash-spinning method
7. Solution flash-spinning
8. Emulsion-spinning method
9. Jet-spinning method
10. Centrifugal-spinning method
11. Turbulent forming method
12. Conjugate-spinning method.

In the early development of microfibers, researchers were interested in finding suitable fields of application as the microfibers had properties which had not existed in previous clothing and technical textile concepts.

Microfibers are specified by their remarkable properties such as luster, pleasant softness and handle, good drapability, bulk and outstanding surface properties. Woven fabric was produced from a 0.1 dtex UFF hollow microfiber combined with a single hollow staple fiber. This product offers a sense of warmth, dry handle, softness, bulk, good recovery, and is lightweight. Microfibers have high strength properties, are very soft, have luxurious hand with a silken or suede touch, extreme drapability, ultra-fine linear density (less than 0.1 dtex/f) and are finer than the most delicate silk. They are shrink resistant, washable, dry-cleanable, non-electrostatic and are hypoallergenic, therefore they do not create problems for those suffering from allergies. Anti-microbial agents help to protect the wearers from the dangers of the bacteria that cause odor and mildew. Microfibers are super-absorbent, absorbing over seven times their weight in water and they dry in one-third of the time of ordinary fibers. They insulate well against wind, rain and cold and furthermore are environmentally friendly.

The higher absorption surface of microfibers results in a dyeing rate four times higher than that of normal fibers. Therefore, to reach the same depth of shade they require more dyestuff than standard fibers, which can cause unevenness in the dyeing. Their larger external surface means an increase in the number of threads exposed to light which, on destruction of dye, is expressed as the lower light fastness rating. Because of the fineness, the total surface area of microfiber yarn or fabric is much bigger than ordinary fibers. Therefore the quantity of required size needed to be applied on microfilament warp yarns is higher. Since microfibers have very small interstitches, with consequent difficulties of size accessibility and diffusibility, desizing becomes quite difficult and costly. The most useful machine for microfiber fabric processing is a Jet dyeing machine, as it allows the fabric to develop a desirable bulk. The difficulty in processing microfibers can be overcome by proper selection of dyestuffs, using appropriate dyeing machinery (air jet type) and choosing suitable processing parameters. Proper dye selection eliminates problems regarding build-up and fastness properties. Staple microfibers offer difficulty in carding, but the emerising effect, which imparts a slightly napped, peach-like surface and a pleasant soft handle, has grown in importance for microfiber fabrics. The emerising treatment must always be carried out before pre-setting to prevent an uneven surface (Fibronet, 2004).

2.3.3 Hollow fibers

Polyester hollow fiber was introduced to the market in 1980. The cross-section of this fiber has a tubular form that contains one or more holes and gives the following advantages over solid fibers:

- more resilience/recovery
- more bulky and fluffy
- higher heat insulation
- better cover
- lighter in weight
- higher absorption of water and perspiration.

Nowadays hollow fibers are being produced with various geometries such as rounded, trilobal, triangular and squared. Hollow fibers trap air, providing loft insulation characteristics better than solid fibers, and when used in carpet show less soil and dirt.

Hollow polypropylene microfibers are used because of their high breathability, light weight and softness. These fibers that are used for underwear are highly elastic and have perfect temperature control and thermal isolation. Their seamless construction ensures superb comfort next to the skin with excellent moisture delivery (Up and Under, 2010; Biemme, 2009; Northwave, 2009). Hollow fibers are also used in different domestic, industrial and medical fields for reverse osmosis, hemodialysis, microfiltration, water cleaning systems, pervaporation, gas separation, bedding fabrics such as sheets and cushions, carpets and garments. Application of hollow fibers results in lower extensibility and lower tensile and compressional resilience, but higher bending and shear stiffness and hysteresis. They have a pleasant handle, good water vapor transmission, high thermal isolation and resistance, lower fabric thickness and considerably lower pilling extent (Khoddami *et al.*, 2009). The difference between the spinning processes of hollow fibers and round fibers is due to an additional dimensional variable (inner radius) which is mainly caused by the shape of the spinneret (Pal, 1993).

2.4 Comfort properties of yarns

Yarn properties are first of all created by physical and chemical properties of their constituent fibers: the chemical nature of the fibers, surface tension, fiber diameter and cross-section. However, the spinning technique, yarn linear density, pore size in the yarn, the distribution of pore size and blend ratio are the other parameters influencing the properties of yarns such as strength, bending rigidity, evenness, frictional properties, wickability, thermal insulation, liquid vapor permeability, air permeability and the properties of fabrics and clothes which are made of them.

Many researchers have focused on the effects of yarn on the comfort properties of clothes, but mainly on the effects of yarn spinning techniques, fiber content ratio, and yarn structure.

2.4.1 Effect of yarn structure characteristics

Ozguney and his colleagues have investigated the effect of yarn linear density on comfort properties of the fabrics made of them. They reported that the bending stiffness and compression of the fabrics produced of yarns with higher linear density are more than the fabrics with lower linear density yarns (Ozguney *et al.*, 2009). Increase in yarn coarseness results in enhancement of all low stress mechanical parameters of the fabric. The investigations of Raj and Sreenivasan on different cotton fabrics show that increase in yarn fineness, yarn twist and providing a less dense structure improves the air permeability of the fabric (Raj and Sreenivasan, 2009).

Ozdil and her coworkers investigated the effect of twist number and yarn linear density of rib cotton fabrics. Their study shows that yarn diameter and fabric thickness decrease as linear density decreases. This specification results in lower thermal resistance and thermal conductivity, but with a higher water vapor permeability value. Therefore fabrics produced of finer yarns have warmer filling and lower thermal absorptivity. By increasing the twist number of the yarns, water vapor permeability and thermal absorptivity increase, producing lower thermal resistance and cooler feeling. They also showed that thermal conductivity, thermal absorptivity and water vapor permeability of combed cotton yarn knitted fabrics are higher than carded cotton yarn knitted fabrics. Therefore combed yarn fabrics provide warmer feeling with respect to carded yarn fabrics (Ozdil *et al.*, 2007).

Hairiness properties of yarns have been studied by Aliouche and Viallier. They have proved the important effect of hairiness on tactile feeling such as surface roughness, fabric compression and handle (Aliouche and Viallier, 2000).

As a very important parameter, there are numerous studies on wicking properties of yarns and fabrics (Doakhan *et al.*, 2007; Hong, 2007; Simoncic, 2007). Wicking is the necessary item for cloth breathablity. Asayesh and Maroufi reported the effect of yarn twist on wickability of cotton interlocked weft knitted fabric. A higher yarn twist number results in lower wickability in such fabrics (Asayesh and Maroufi, 2007). Wickwire and his colleagues stated that a loose garment reduces the effectiveness of wicking the sweat and moisture away from the skin, but it could possibly increase the comfort of the wearer independent of its ability to wick moisture away (Wickwire *et al.*, 2007).

2.4.2 Effect of spinning technique

Kumar Tyagi and his coworkers made a comparison between air jet spun and ring spun polyester/cotton and polyester/viscose yarns, with round and trilobal cross-section polyester fibers. Their study proves that thermal insulation of fabric is highly dependent on yarn structure, the shape of fiber cross-section and fiber content in yarn blend. Woven fabrics of air-jet spun yarns have slightly higher thermal insulation in respect to ring-spun yarns. Trilobal polyester content yarns have higher thermal insulation compared to round polyester, and polyester/cotton fabrics show higher thermal insulation with respect to polyester/viscose blends. They have also reported the wickability properties of these fabrics. Air jet yarn fabrics display lower wickability and the wickability of polyester/cotton fabrics is lower than polyester/viscose fabrics. Fabrics with air jet yarns exhibit higher air permeability, higher water vapor permeability and higher absorbency than those of ring yarn fabrics. A lower polyester content increases water and air permeability, however generally polyester/viscose blends are superior to polyester/cotton fabrics in permeability properties (Kumar Tyagi *et al.*, 2006). In similar studies Ozdil, Das and their colleagues investigated the comfort properties of fabrics knitted of pure polyester and pure cotton yarns in two different ring and compact spinning systems. They reported that knitted fabrics made of compact yarns have particular visual properties with respect to classic ring spun yarns. Fabrics made of compact yarns are more brilliant and glossy. Furthermore they have finer and smoother handle and they reveal better pilling behavior. The fabrics knitted of compact spun yarns have also higher bursting resistance comparing to ring yarn knitted fabrics. The difference between the results of these two spinning systems is because of the incorporation of most of the fibers in constructing the compact yarn structure that leads to optimal exploitation of the raw material with respect to ring spun yarns. In the compact spinning system, fibers are mostly aligned and parallel under relatively equal tension. The twist diffusion in this system is more than in the ring spinning system, also involving the short fibers. Therefore the produced yarns have higher evenness and lower hairiness with respect to the ring spinning method (Ozdil *et al.*, 2005; Das and Ishtiaque, 2004).

Radhakrishnaiah and Tejatanalert compared the effect of spinning techniques on comfort properties of cotton/polyester yarns. They used 60/40 random cotton/polyester blend yarns and 60/40 cotton covered polyester yarns in plain pattern fabrics with fixed weave characteristics for both types of fabrics. Their study shows that handle quality, dry and wet energy dissipation, and the warm–cool contact sensation of cotton covered polyester yarn fabrics are better than random cotton/polyester yarn fabrics (Radhakrishnaiah and Tejatanalert, 1993).

2.4.3 Texturizing

Since synthetic fibers do not have the proper handle and physical properties required for comfortable clothing, they need to be proceeded by treatments to modify their comfort properties and provide a closer feeling to natural fibers. Texturizing is one of the most important treatments to provide crimped and bulky structures for synthetic filament yarns. False-twist, air-jet and steam-jet are the most used methods to produce textured yarns.

Texturizing is a procedure to increase the bulkiness and the elasticity of the filament yarn, through making permanent crimped or coiled filament yarns. The essential properties caused by texturizing are softness, thermal insulation, natural fiber touch, warmth, fullness, a high degree of elasticity and moisture-transporting properties. The bulky structure of textured yarns causes a more porous construction and good thermal insulation. The less organized surface of textured yarns gives dispersed light reflection and a matte appearance. The coiled and crimped structure of yarns gives a lower effective elastic modulus and the sponge-like structure of textured yarn is felt to be softer than the flat or more ordered yarns (Ozeelik *et al.*, 2007).

The studies on comfort properties of textured filament yarns have proved that the thickness and thermal resistance of fabrics made of textured yarns are higher than fabrics with non-textured filaments. By texturizing the value of thermal absorptivity decreases due to less contact area between the cloth and the skin and the fabric feels warmer. Comparing the false-twist and air-jet texturizing methods has shown that the yarn textured by the false-twist method is more crimped and bulky and provides a warmer feeling with respect to air-jet textured yarn. Furthermore, the fabrics produced from false-twist textured yarns have higher cover factor and lower air permeability (Ozeelik *et al.*, 2007; Rengasamy *et al.*, 2009).

Physical and comfort properties of nylon fabrics, knitted from intermingled nylon elastomeric yarns, have been studied by Ucar and his colleagues. Their study includes the effect of different draw ratio and different number of knots per unit length of intermingled yarns, revealing that the increase of draw ratio and number of knots lead to an increase in fabric weight and stitch density, but decrease its spirality, wickability, porosity and abrasion. Fabric thickness decreases by increasing the number of knots, but increases by increasing the draw ratio of yarns (Ucar *et al.*, 2007).

2.5 Comfort properties of fabric structures

Cloth is made from fabrics which are even knitted (interlocked loops), or woven (interlacing threads), or nonwoven (matted fibers). Fabric physical properties create the comfort characteristics of the cloth. The major properties are fabric density, porosity, bulkiness, thickness, structure and pattern.

2.5.1 Fabric constructional parameters

One of the most important characteristics of the fabric which affects the comfort properties of cloth is the constructional specification. Parameters such as thickness, weight per square meter, pattern of weave and yarn count can be counted as the most effective parameters. The fabric pattern creates its thickness and weight properties, but on the other hand it determines the air and liquid permeability and thermal insulation. Furthermore, it should not be forgotten that fabric pattern also has aesthetic value for the customers. For example, twill weave fabrics are tightly woven diagonal line patterns that are strong, hard-wearing and drape well. They do not get dirty quickly, but if they get dirty, it is more difficult to clean them. As another weave, satin does not show the diagonal lines as in twill weave. It is flat, smooth and has a lustrous surface, but is a weaker fabric. Warp yarns are more visible on the right side of the fabric and the floats are more susceptible to snagging; the longer the float, the more likely it is to snag. Satin weave shows signs of wear quickly, but it drapes well and is warmer with respect to twill weave as it traps air between the threads. Weave pattern has a significant effect on the dimensional behavior of woven fabrics. Dense weave patterns, such as plain weave with a high number of interlacings, reveal lower shrinkage values. At the same time, lower yarn crimp values restrict the fabric shrinkage and result in better dimensional stability (Topalbekiroglu, 2008). With respect to plain weave, twill weave is more flexible and has lower shear and bending resistance. Twill pattern is smoother than plain pattern and has higher fullness, softness and compressibility which reduces the hysteresis effect.

Fabric yarn count is an important parameter affecting the mechanical properties of fabrics. Woven fabrics have generally smoother surface compared to knitted and crepes. A smooth fabric surface provides a bigger contact area with the skin and heat flow, while a more rough fabric surface has less contact area. Therefore, a smoother fabric surface provides better heat conduction and a cooler feeling (Raj and Sreenivasan, 2009; Schnieder *et al.*, 1996). Whereas air permeability depends on yarn and fabric structures and the shape and volume of airflow channels in the fabric, to explain the thermal insulation, warmness and heaviness, fabric thickness and weight are the most effective parameters (Raj and Sreenivasan, 2009; Mehrtens, 1962; Holcombe and Hoschke, 1983; Yoon and Buckley, 1984).

2.5.2 Finishing

Finishing is a mechanical or chemical treatment applied on fabric (yarn or fibers) to improve the quality of the product for a chosen use. There are several chemical finishing treatments such as bleaching, softening, water-resistance, antibacterial, stain resilience, antistatic, wrinkle resistance, flame retardant, etc., but all finishing treatments affect the fabric handle and comfort (usually the stiffening effect is observed). For example, mercerizing is the treatment to improve the mechanical properties, wettability, dye absorption and luster of cotton fibers, but on the other hand it stiffens the fabric handle and increases its bending rigidity. However, among different common treatments for fabrics, printing with pigments is the most stiffening treatment (Ozguney *et al.*, 2009).

Coating treatments are used for different purposes. For example waterproof finishing is a coating process which closes the capillary channels of fabrics which are used for water vapor transmission; therefore coating usually decreases the breathability of the cloth. Silicon softener-treated polyester blended fabrics are warmer to the touch, but less comfortable regarding their hydrophobic tendency and reduced water-vapor permeability. The finishing stage of the fabrics has considerable influence on their thermal touch sensation and water-vapor permeability (Tzanov *et al.*, 1999). Yan and his colleagues have investigated the effect of bleaching on mechanical and comfort properties of fabrics. They proved that bleached fabrics have coarser handle and lower fullness with respect to unbleached fabrics (Yan *et al.*, 2000).

The effect of durable flame retardant finishing on mechanical properties of knitted cotton fabrics has been investigated by Mavruz and Ogulata. They found that this treatment increases the compression resistance of the fabric and the fabric becomes rather hard in compressional mode (Ozguney *et al.*, 2009).

2.6 Conclusions

In this chapter, the comfort characteristic of clothing has been investigated focusing on the physical and constructional specifications of garments. The chapter points to the importance of fiber as the basic constituent of the fabric, and the two main categories of fibers, i.e. natural fibers and man-made fibers, have been mentioned and explained briefly. The material of fiber, fiber blend and fiber content and physical techniques to modify the comfort properties of synthetic fibers such as profiled fibers, microfibers and hollow fibers have been mentioned and investigated. Following the fibers influences, the characteristics of yarns have been considered and the effects of parameters such as yarn structural properties, spinning technique and texturizing have been presented.

The chapter ends with the influence of fabric structural and finishing treatments as the last procedures to produce the clothing raw material, counting the parameters induced by these ultimate processes affecting the comfort sense of the garment.

2.7 References

Aliouche, D., & Viallier, P. (2000) Mechanical and tactile compressions of fabrics: Influence on handle. *Textile Research Journal*, **79**, 734–738.

Asayesh, A., & Maroufi, M. (2007) Effect of yarn twist on wicking of cotton interlock weft knitted fabric. *Indian Journal of Fibers & Textile Research*, **32**, 373–376.

Biemme (2009) Biemme 09 winter collection. Appeared in www.biemmesport.com

Bueno, M. A., Aneja, A. P., & Renner, M. (2004) Influence of the shape of fiber cross section on fabric surface characteristics. *Journal of Materials Science*, **39**, 557–564.

Das, A., & Ishtiaque, S. M. (2004) Comfort characteristics of fabrics containing twist-less and hollow fibrous assemblies in weft. *Journal of Textile and Apparel, Technology and Management*, **3**, 1–7.

Doakhan, S., Hosseini Ravandi, S. A., Gharehaghaji, A. & Mortazavi, S. M. (2007) Capillary rise in core-spun yarn. *Iranian Polymer Journal*, **16**, 397–408.

Fibronet (2004) Microfibre. Appeared in www.fibernet.org

Hatch, K. L. (1993) *Textile Science*, Minneapolis, MN: West Educational Publishing.

Holcombe, B. V., & Hoschke, B. N. (1983) Dry heat transfer characteristics of underwear fabrics. *Textile Research Journal*, **53**, 368–373.

Hong, C. J. K. (2007) A study of comfort performance in cotton and polyester blended fabrics. I. Vertical wicking behavior. *Fibers and Polymers*, **8**, 218–224.

Kamijo, M., Uemae, T., Kwon, E., Horiba, Y., Yoshida, H. & Shimizu, Y. (2009) Comfort evaluation of T-shirt type underwear made of spun silk yarn. *Biometrics and Kansei Engineering International Conference, ICBAKE 2009*. Cleszyn.

Karaca, E., & Ozcelik, F. (2006) Influence of the cross-sectional shape on the structure and properties of polyester fibers. *Journal of Applied Polymer Science*, **103**, 2615–2621.

Khoddami, A., Carr, C. M., & Gong, R. H. (2009) Effect of hollow polyester fibres on mechanical properties of knitted wool/polyester fabrics. *Fibers and Polymers*, **10**, 452–460.

Kumar Tyagi, G., Goyal, A., Mahish, S., & Madhusoodhanan, P. (2006) Effect of fiber cross-section on comfort characteristics of ring and MJS yarn fabrics. *Melliand International*, **1**, 29–32.

Li, Y., & Dai, X-Q (2006) *Biomechanical Engineering of Textiles and Clothing*, Cambridge, England, The Textile Institute, Woodhead Publishing.

Li, Y., & Wong, A. S. W. (2006) *Clothing Biosensory Engineering*, Cambridge England, The Textile Institute, Woodhead Publishing Limited.

Mehrtens, D. G. M. (1962) Fiber properties responsible for garment comfort. *Textile Research Journal*, **32**, 658–665.

Northwave (2009) Underwear. Appeared in www.northwave.com

Ozdil, N., Marmarah, A., & Kretzschmar, S. D. (2007) Effect of yarn properties on thermal comfort of knitted fabrics. *International Journal of Thermal Sciences*, **46**, 1318–1322.

Ozdil, N., Ozdogan, E., Demirel, A., & Oktem, T. (2005) A comparative study of the characteristics of compact yarn-based knitted fabrics. *Fibers & Textiles in Eastern Europe*, **13**, 39–43.

Ozeelik, G., Cay, A., & Kirtay, E. (2007) A study of the thermal properties of textured knitted fabrics. *Fibers & Textiles in Eastern Europe*, **15**, 55–58.

Ozguney, A. T., Taskin, C., Ozcelik, G., Gurkan Unal, P., & Ozerdem, A. (2009) Handle properties of the woven fabrics made of compact yarns. *Tekstil ve Konfeksiyon*, **2**, 108–113.

Pal, S. K. (1993) Microfibre – production, properties & application. *Textile Asia*, **24**, 53–58.

Purane, S. V., & Panigrahi, N. R. (2007) Microfibres, microfilaments & their applications. *AUTEX Research Journal*, **7**(3), 148–158.

Radhakrishnaiah, P., & Tejatanalert, S. (1993) Handle and comfort properties of woven fabrics made from random blend and cotton-covered cotton/polyester yarns. *Textile Research Journal*, **63**, 573–579.

Raj, S., & Sreenivasan, S. (2009) Total wear comfort index as an objective parameter for characterization of overall wearability of cotton fabrics. *Journal of Engineered Fibers and Fabrics*, **4**, 29–41.

Rengasamy, R. S., Das, B. R., & Patil, Y. B. (2009) Thermo-physical comfort characteristics of polyester air-jet-textured and cotton-yarn fabrics. *Journal of The Textile Institute*, **100**, 507–511.

Saville, B. P. (1999) *Physical Testing of Textiles*, Cambridge, The Textile Institute, Woodhead Publishing.

Schacher, L., Adolphe, D. C., & Drean, Y. (2000) Comparison between thermal insulation and thermal properties of classical and microfibers polyester fabrics. *International Journal of Clothing Science and Technology*, **12**, 84–95.

Schmidtbauer, J. (2009) Viscostar – a star-shaped viscose fiber for improved absorbency, Appeared in www.bioway.com.tw/Files/DownloadFile/VISCOSTARPlus.pdf

Schnieder, A. M., Holocombe, B. V. & Stephens, L. G. (1996) Enhancement of coolness to the touch by hygroscopic fibers. *Textile Research Journal*, **66**, 515–520.

Simoncic, B. R. (2007) Wettability of cotton fabric by aqueous solutions of surfactants with different structures. *Colloids and Surfaces A: Physicochem. Eng. Aspects*, **292**, 236–245.

Slater, K. (1991) *Textile Objective Measurement and Automation in Garment Manufacturing*, London, Horwood.

Smith, J. E. (1986) The comfort of clothing. *Textiles*, **15**, 23–27.

Topalbekiroglu, M. K. (2008) The effect of weave type on dimensional stability of woven fabrics. *International Journal of Clothing Science and Technology*, **20**, 281–285.

Tzanov, T., Betcheva, R. & Hardalov, I. (1999) Thermophysiological comfort of silicone softeners-treated woven textile materials. *International Journal of Clothing Science and Technology*, **11**, 189–197.

Ucar, N., Karakas, H., & Sen, S. (2007) Physical and comfort properties of the hoisery knit product containing intermingled nylon elastomeric yarn. *Fibers and Polymers*, **8**, 558–563.

Up and Under (2010) Clothing (and Clothing Accessories). Up and Under (Outdoor Gear) Ltd. Appeared in www.upandunder.co.uk

Wang, G., Zhang, W., Postle, R. & Phillips, D. (2003) Evaluating wool shirt comfort with wear trials and forearm test. *Textile Research Journal*, **73**, 113–119.

Warner, S. B. (1995) *Fiber Science*, Upper Saddle River, NJ, Prentice Hall.

Wickwire, J., Bishop, H. A., Green, J. M., Richardson, M. T., Lomax, R. G., Casaru, C., Curther-Smith, M. & Doss, B. (2007) Physiological and comfort effects of commercial 'wicking' clothing under a bulletproof vest. *International Journal of Industrial Ergonomics*, **37**, 643–651.

Yan, K., Hocker, H. & Schafer, K. (2000) Handle of bleached knitted fabric made from fine yak hair. *Textile Research Journal*, **70**, 734–738.

Yoon, H. N., & Buckley, A. (1984) Improved comfort polyester. Part I: Transport properties and thermal comfort of polyester/cotton blend fabrics. *Textile Research Journal*, **54**, 289–298.

3 Wool and garment comfort

J. STANTON, Department of Agriculture and Food (Western Australia), Australia

Abstract: This chapter discusses the contribution of wool fibres to the comfort sensation evoked by a garment and the provision of a link between wool fibre quality and garment quality. It reviews wool fibre characteristics that might contribute to a definition of wool quality and a test protocol to measure the difference in comfort scores for garments made of different wool fibre specifications is discussed. The chapter then develops a prediction for comfort scores using environment and fibre information, and discusses the level of differentiation between garments achieved by the untrained participants.

Key words: garment testing, wool fibre quality, garment preference, controlled environmental testing, comfort prediction.

3.1 Introduction

Comfort has always been an important attribute of wool garments. Prior to the introduction of man-made and synthetic fibres the wide choice of natural fibre available allowed fabrics to be prepared with contrasting properties and applications. The contributions of the fibre to the properties of the garment were acknowledged in the choice of fibre relative to the garment application. Natural fibres were categorised in relation to their contribution to the garment properties and thus perceptions such as 'warm wool' and 'cool silk' garments developed.

The introduction of man-made and synthetic fibres has blurred the boundaries of fibre choice in garment applications, thus increasing the options for fibre selection in garments. For wool fibre producers to remain competitive in this new commercial environment, the contribution of wool to the comfort of the wearer needs to be better understood, in a way that allows improvement in comfort to be achieved through manipulation of the wool fibre specifications. Interest is growing in defining the comfort response of the wearer to a garment made of wool fibres and the application of this knowledge in areas such as wool fibre selection, processing and marketing.

A critical element in the discussion of comfort is the environment in which the garment is worn. Previously, extensive research into garment performance in extreme and hazardous environments (Norman *et al.*, 1985; Havenith, 1999; Meinander *et al.*, 2004) has been undertaken resulting in standards for clothing designed to improve the probability of survival in such environments. However, environmental research into garment comfort relevant to the retail consumer shopping in the high street has not been as extensive. These consumers are more likely to be exposed to short periods of activity throughout the day in which they will have elevated skin temperatures and feel hot, before returning (perhaps many times) to a 'domestic' environment, most probably air-conditioned to about 22 degrees Celsius. The consumer wants to be comfortable during each of these very different environmental periods.

3.1.1 Definition of comfort

In this chapter, comfort is defined as a sensation *experienced by the wearer, evoked by the garment*. In this definition, comfort is not a property of the garment *per se* and is not considered as a mechanical property of the garment or fabric. However the comfort sensation, or some component of comfort, may be predicted from a variety of mechanical and objective measurement inputs relating to the environment, garment construction, fabric and yarn types, and fibre properties (Paek and Davis, 1975; Bartels and Umbach, 1999; Lau *et al.*, 2002; Wong *et al.*, 2003, 2004; Yoo and Barker, 2005a, b).

3.2 Wool quality

A useful definition of wool quality that could integrate with the comfort definition above would include improvement of garment comfort by changing wool fibre specifications used in the garment. In the absence of evidence from consumer testing, the wool industry however has looked to fibre measurements to define wool quality (Rogan, 1995), and price differentials in raw wool fibre auctions as evidence of wool quality. In addition to wool being a highly specified natural fibre, it has a raw fibre trading sector that uses these measurements to estimate the price for fibre lots that they buy at auction (Stott and Hanson, 1993).

Price differentials in raw wool can be expected to appear in wool fibre characteristics that improve the quality of the final product, and those that improve processing efficiency in the making of that product. Characteristics in the latter group include washing yield, contamination levels, fibre strength and fibre length, and generally have small price differentials (see for example Gleeson *et al.*, 1993).

Fibre diameter is strongly linked to the expectations of wool quality (Whiteley and Rottenbury, 1990), and probably because of this, has the strongest influence on the auction price. Fibre crimp, seen as the wave form in the wool staple, was traditionally used as an indicator of wool quality before the introduction of fibre diameter measurements (Whiteley, 1967). Fibre curvature (closely related to fibre crimp) is now measured in the raw wool, and has been shown to be a significant determinant of the auction price in the fine wool types (Curtis and Stanton, 2002a, b). The form of the price differential suggests that fibre crimp interacts with fibre diameter in setting the quality of the retail wool product. Therefore if higher auction prices are an indication of quality, higher quality wool will have a lower average fibre diameter and a higher crimp (or curvature measurement).

3.2.1 Intrinsic wool properties

Wool fibres and fabrics perform well in comparative objective measurements for moisture and thermal properties. Substitution of other fibres in a fabric with wool will change fabric moisture and thermal properties. It is not known, however, whether wool fibres can be specified with extreme values in these properties so that wool fibre batches can be assembled to produce significant and predictable differences in these properties in fabric or garment. Thus although a comparative advantage in these properties exists for wool it is currently not possible to change the wool fibre selections to improve the thermal and moisture performance of pure wool fabrics.

3.2.2 Wool processing effects

Countries which produce apparel wool fibre tend to be geographically removed from the affluent retail markets that consume the final wool product. The number and complexity of fibre transformations in production of yarn, fabric and garments also separates the quality contribution of the input material (wool fibre) from the quality assessment of the retail garment. Some processors claim to be able to use low grade fibre and yarn to make high grade fabric. However significant price differentials exist for the higher quality wool fibres that go into the high end product. Evidently processors, as a group, value some wool properties highly for their contribution to product quality.

The use of wool fibre diameter measurements as an estimate of fibre quality is helped by the ability to extract and measure wool fibres from intermediate textile products using standard test procedures developed by the International Wool Textile Organization (IWTO, 2007a, b). Similarly, using the same measurement techniques, wool fibres can be extracted from retail garments made of wool and measured for average fibre diameter and fibre curvature.

3.3 Benchmarking: wool quality in retail garments

Before embarking on wholesale changes to retail garment collections, wool quality being used in commercial retail garments was investigated (Stanton and Ladyman, 2003). One hundred and sixty-eight garments were purchased at retail in London, Milan, Tokyo, Beijing, New York and Sydney. These garments were reverse engineered to estimate the fabric and yarn structures, and to allow measurement of fibre diameter and fibre curvature. The garment selection criteria for this project limited retail purchases to wool knitwear worn 'next to skin'. Garment choice was therefore limited to base layer or single layer jumpers, t-shirts and underwear.

The retail garments were made with a wide range of fabric weights and finishes, different knitting technologies and yarn specifications (Fig. 3.1). However, each

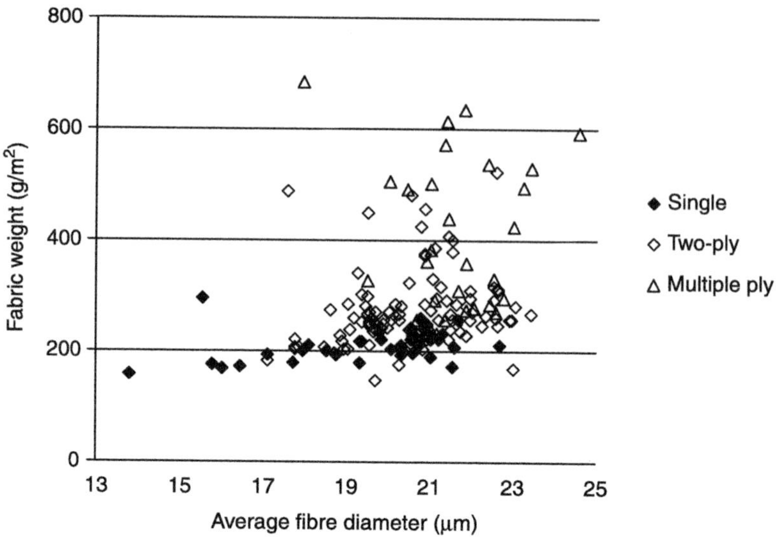

3.1 Fabric weight, yarn structure and fibre diameter in wool garments bought at retail. The 168 garments were a knitted next-to-skin design, bought in major retail centres of London, Milan, Beijing, Tokyo, New York and Sydney. Fabric samples were removed from the garments, weighed, and reverse engineered to measure fabric, yarn and fibre properties for each garment. Source: Stanton and Ladyman 2003.

garment represented a commercial solution to the issue of quality and technology at each stage, from fibre production through processing to garment finishing. In garments with fabric weights under 220 g/m^2, the single ply yarns show a wide range of wool being used, up to a fibre diameter of 23 μm. Two-ply yarns tended to appear in higher weight fabrics and with a higher fibre diameter. In each case, there was a wide range of wool fibre diameters used in each fabric/yarn combination.

The wide range of fibre diameter used in these different garments raises several possibilities for linking wool fibre quality and garment quality. The first possibility is that the wool fibre diameter characteristic contributes to the quality of yarn or fabric, but outside these yarn or fabric effects fibre quality does not contribute to garment comfort. The second is that the wool fibre characteristic contributes to the garment comfort in addition to its contribution to yarn and fabric effects, but this contribution is small or overlooked in one or more of these retail products.

Lack of supply of suitable wool fibre may be another explanation for the wide range of fibre diameter values seen in these garments. The objective measurement of the Australian wool clip (which produces about 100,000 tonnes of fibre annually and supplies about 70% of wool traded internationally and possibly 50% of

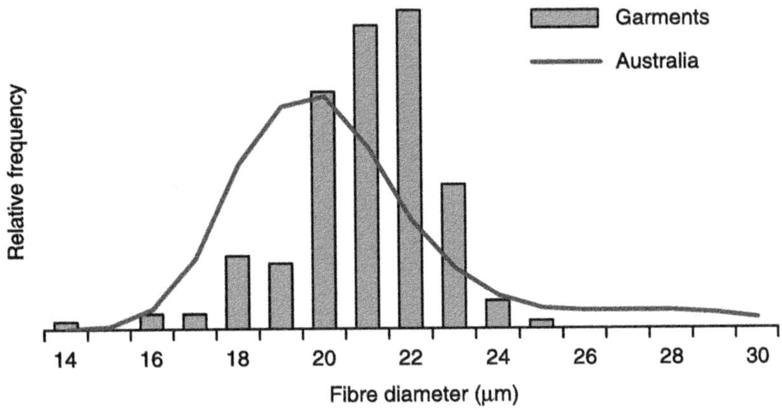

3.2 Average fibre diameter used in wool garments is compared to the supply profile of Australian wool clip. The garment collection of 168 garments bought in major retail centres was reverse engineered, and the average fibre diameter measured for each garment. The garment results are presented as a frequency distribution based on the fibre diameter. The Australian wool clip is measured before being auctioned, and the relative weight of wool in each fibre diameter class is used to develop a supply profile. Australia is the largest single producer of apparel wool. Source: Stanton and Ladyman 2003.

apparel wool production) allows the supply of fibre for each fibre diameter to be quantified. Thus the incidence of wool fibre diameter in the benchmark garments can be mapped over the fibre diameter distribution of the Australian wool clip (Fig. 3.2). The offset of the Australian wool fibre supply to the left (i.e. finer fibres) shows that there are significant volumes of 'higher quality' wool available to these retailers and processors, should they need (or choose) to use it.

3.3.1 Fabric appraisal of benchmark garments

Attempts were made to appraise the handle, skin comfort, appearance and overall appeal of fabrics cut from these benchmark garments. The objective was to identify the more appealing fabrics, and extract the fibre and fabric properties for them (Stanton and Ladyman, 2003). While the study gave some insight into the haptic and visual components of the overall fabric rating, and some evidence of the fibre diameter being used in an international collection, it was difficult to separate the fibre effects from the yarn and fabric processing technology when explaining the fabric rating.

The study also showed that generating a solution for fabric appraisal would not automatically progress into a solution for the linkage between wool quality and garment comfort. It would also be risky to develop a prediction of garment comfort without ever recording 'an evoked sense of comfort' from a wearer.

3.4 Comfort in wool garments: a new assessment protocol

One contentious issue of testing for fibre effects in garments centres around measuring the garment comfort score well enough to reveal the presence of the fibre effects. In particular, the fibre changes are not based on fibre substitution (see Mehrtens and McAlister, 1962; Hollies *et al.*, 1979, 1984; Demartino *et al.*, 1984) but rather a change in fibre specification in the same fibre type–wool fibre diameter.

Another issue is raised by Wong *et al.* (2003) and others who contend that comfort cannot be defined as it differs from person to person. One solution could be to design a trial to measure a change in comfort across a population of testers and rely on an aggregate response.

It was assumed that finding wool specification effects in garments will require:

1. a sensitive protocol;
2. a large number of people to be tested;
3. a suitable garment design; and
4. a wool characteristic that can initiate significant changes in wearer response.

If wool fibre diameter can be manipulated in the manufacture of a garment and the quality of the garment is changed as a result of the fibre diameter change, then a link between fibre quality and garment quality can be considered. If garments made with the same construction, colour, fabric and yarn specifications are used in consumer testing then some of the complexity in defining garment quality can be avoided. The opportunity thus exists to define a change in garment quality solely in terms of a change in garment comfort. If there is a link between fibre quality and garment quality, then changes in wool fibre diameter will result in a change in the sensation of comfort of the wearer evoked by the garment.

In a significant review of clothing comfort, Hollies and Goldman (1977) listed comfort components and methods for measuring the assessment of comfort. In work on the comfort of knit t-shirts, Fuzek (1981) listed three components of garment comfort: tactile, thermal and moisture. These components have been carried through into an extensive review of clothing comfort by Li (2001). In mid layer and outer layer applications, the comfort sensation might be dominated by thermal components and possibly to a lesser extent by moisture components in those applications where the source of the moisture is body sweat. Tactile component contributions to mid layer garment comfort are unclear. Once the garment has been selected (during which there is some tactile input based on both fabric and garment properties) and is being worn, it is unclear if the mid layer garment continues to provide tactile input to the sensors of the wearer.

If comfort sensations are to be evoked by the garment, the comfort proposition should be tested where all the tactile, thermal and moisture components are challenged. This can be achieved using base layer garments worn as a single layer.

The tactile component is continuously present in the base layer garment. Thermal challenges can be provided by raising the skin temperature and/or raising the environmental temperature during the test protocol. The moisture component can be challenged by raising the level of activity causing the body to sweat as part of the body's response to maintain a physiological equilibrium, and to remain 'comfortable'.

A protocol to test all the components of the comfort proposition was prepared at the Department of Agriculture and Food (Western Australia) (DAFWA) Design for Comfort Laboratory (Stanton, 2008). The protocol incorporated environmental protocols used by Hollies (1971), Hollies *et al.* (1979) and Barker (2002) in which thermal and moisture challenges were introduced. Although the marketing proposition is generally geared towards the perception of product as 'more comfortable', Mehrtens and McAlister (1962) suggest that recording the level of *discomfort* (rather than the level of comfort) is more reliable. The recording of discomfort scores is also consistent with the concept of testing wearers in challenging environments.

3.5 Wool garment comfort assessment

Garment wearer tests were undertaken with a garment set of long-sleeved black t-shirts with a round neck weighing about 150 g, using a single jersey fabric of 170 g/m^2 made on a 24 gauge circular knit machine with 1/40 Nm wool yarns. Garments were made from fibre batches of 16.5 μm, 18.5 μm and 20.5 μm raw wool. The tests were undertaken on a population of 43 untrained 25 to 35-year-old female participants selected from the urban population.

The test protocol was based around 4 stages: a change room, a hot room, a hot active session, and a return to the cool room after the hot room. After 15 minutes in the change room environment (24 °C and 55% RH) the participant is taken into the hot environment (38 °C and 24% RH). The level of discomfort increased and the discrimination between the garments in terms of the recorded discomfort sensation increased. At the end of 30 minutes in the hot room (including the last 15 minutes on a treadmill), the participants returned to the cool environment and the test finished 15 minutes later.

A sample of the physiological changes that can occur during the test protocol is shown in Fig. 3.3 to illustrate the changes that occurred while the comfort records were being taken. The temperature profile shows the initial resting temperatures in the change room with the garment on, then a sudden rise in skin temperature as the participant transfers into the hot room. The humidity under the garment rises as the body begins to sweat. The skin sensor goes to saturation, and the sensor facing the inner face of the test garment begins to record rising water vapour pressure. As the participant finishes the treadmill activity, the skin temperature is at a maximum, but begins to fall as the participant returns to the change room.

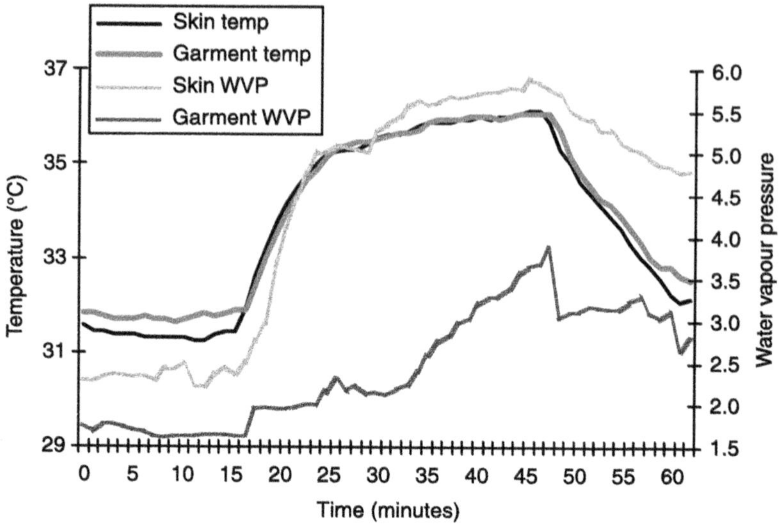

3.3 Temperature and water vapour pressure changes at the C4 position on the neck for an individual during the test period. Two sensors were used to record temperature and relative humidity at (a) the skin level, and (b) the inner surface of the garment at the neck position. The results up to 15 minutes show the response in the change room. At 15 minutes they enter the hot room, and the skin temperature rises. The water vapour pressure in the skin rises rapidly as sweat appears on the skin. At 30 minutes, they start on a treadmill, and the temperatures rise to the maximum, and the vapour pressure on the inner face of the garment starts to rise. At 45 minutes, they return to the change room and the temperature and vapour pressures begin to return to the starting values.

3.5.1 Garment evoked sensation of comfort

At the start of the test procedure the participants are acclimatised in a 'change room' environment of 24 °C and 55% RH. The '2 in 5 test' (Meilgaard *et al.*, 2007) was used and participants asked to put on five test garments and identify which two garments were different. Statistically, these participants could not detect any difference between the 16.5 µm and the 18.5 µm wool garments.

Average comfort scores, calculated from the 15 scores taken over the 90 minutes of testing, show significant differences between garments (Fig. 3.4). These differences between garments show that fibre specification effects in garment comfort are:

(a) observable; and
(b) relate to changes in wool fibre diameter.

The change in the level of comfort across the test period raises some application issues. For a retailer, the results from the '2 in 5' test, and the low discrimination

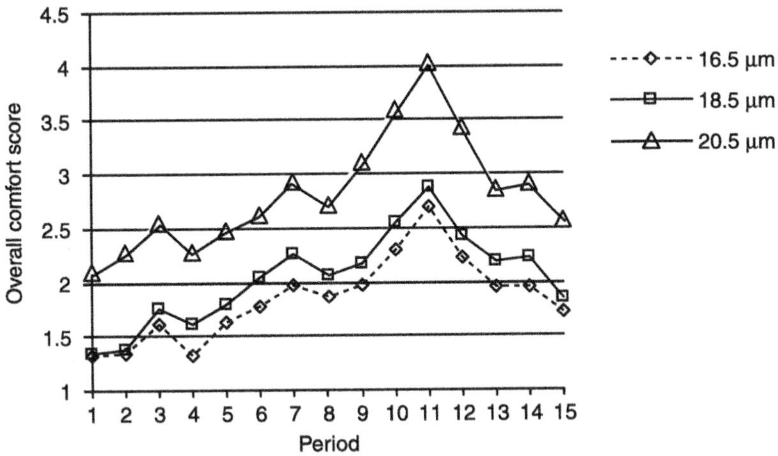

3.4 Average overall comfort score for each garment in each period (43 test participants). Periods 1 to 4 are in the change room. Periods 5 to 8 are in the hot room, but inactive. Periods 9 to 11 are in the hot room on a treadmill. Periods 12 to 15 are on their return to the change room.

between garments in the initial change room environment, suggest the comfort levels from these different garments are the same. However, when the consumer wears the garment in a stressful application, significant differences between these garments become evident. Care therefore needs to be taken when defining the conditions under which the comfort of the garment is categorised. Care also needs to be taken to design garments for performance in specific applications, rather than solely for successful selection in the change room.

3.5.2 Population response in average garment comfort score

As the differences between the comfort scores were significant, it was possible to prepare a population response for the comfort score of a garment from the average comfort score during the test (Fig. 3.5). Thus it is possible to report to a manufacturer or retailer that 80% of the population will give a score between 1 and 4 to the 20.5 μm garment.

Differences in the distribution of the average score for each garment can be seen in Fig. 3.5. A simple differential equation was fitted to each of these curves for each garment.

$$\frac{dP}{dS} = k(P - P_a) \qquad [3.1]$$

where S is the comfort score, P is the proportion of the population predicted to report that comfort score or less, P_a is the maximum proportion (or 100%) of the

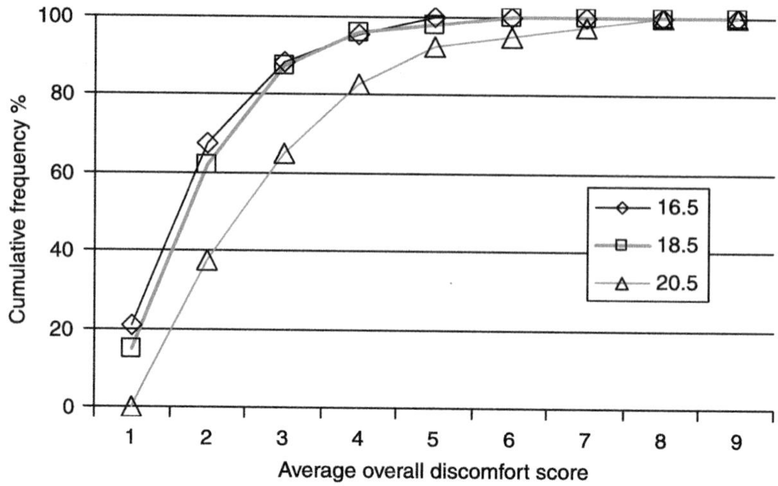

3.5 Estimated population response using the cumulative relative frequency to individual garments for average (dis)comfort scores for each garment (16.5 μm, 18.5 μm and 20.5 μm fibre diameter).

population, k is the rate constant and P_o is the proportion of the population that gave the garment the lowest (most comfortable) score.

In expanded form, this equation becomes

$$P = 1 + (P_o - 1)e^{-kS} \quad [3.2]$$

Figure 3.6 shows that P_o is heavily dependent on the average fibre diameter, and is depressed by the changes in the testing environment. The rate constant k is related to fibre diameter in the change room, but less so in the hot environment or after return to the change room.

These garment results show that differences in the comfort sensation can be generated in garments made from wool fibres of different fibre diameters. The results therefore support the contention presented above that wool quality can be defined in terms of garment quality, and on that basis, average fibre diameter is a component of wool quality. The development of a prediction of comfort for these garments, and the dependence of the prediction terms on the average fibre diameter, confirms a relationship exists between wool quality and garment comfort.

3.6 Comfort response of individuals

At the end of the test activities, the participants were asked a series of questions about their acceptance and preferences for the individual garments they had just tested. While it is instructive to average their individual responses to a garment (as seen above), there is also valuable information in the response of the individual

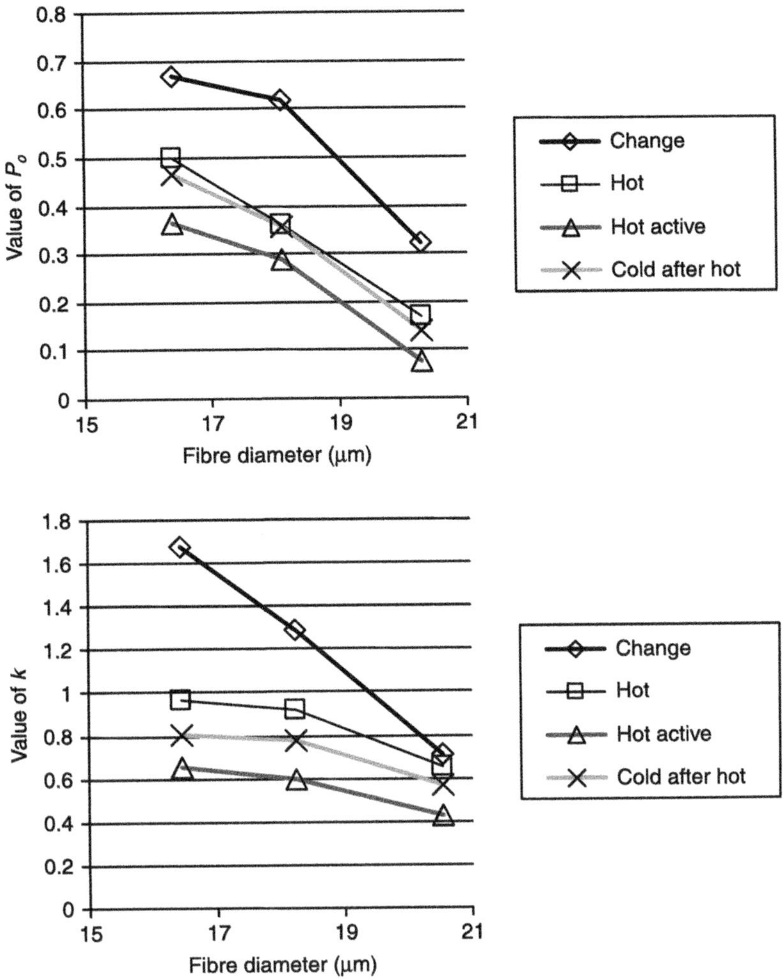

3.6 Values for P_o and k for each garment in each test environment. The horizontal axis is the average fibre diameter measured in each of the wool garments.

to each of the garments. If the error level of the average comfort response is too high, then the analysis of the individuals' responses would be of little value. However the average comfort responses were significantly different between garments, so a close study of the individual responses was undertaken.

3.6.1 Comfort score and garment acceptance

At the end of the individual test, the participant was asked if they liked the garment they were wearing. The information from this response allowed the average

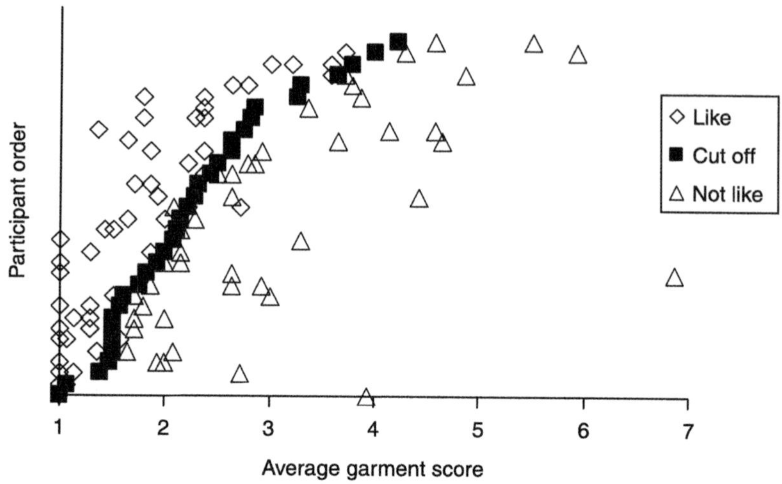

3.7 Estimation of cut-off scores for individual participants. Each point represents the average garment score for a test. Participants were asked at the end of the test if they liked the garment. Average garment scores were classified into 'like' or 'not like' categories. The cut-off score lies midway between these categories for each participant.

comfort score from each garment to be allocated into the participant's 'like' or 'not like' category. The cut-off comfort score was estimated for each participant. The cut-off score was set midway between the highest average comfort score in the 'liked' garment set, and the lowest average comfort score in the 'not liked' set. Average comfort scores above that point relate to the rejection of the garment, and scores below that point, to garment acceptance.

The individual participant's scores for each garment and their cut-off score are shown in Fig. 3.7. The participant order was achieved by sorting the participants' results in order of increasing cut-off value. Note the very low incidence of results which are placed on the incorrect side of the personal cut-off score. When these cut-off scores are collected for each participant, the population distribution of cut-off scores can be generated.

The consistency of both participants' comfort scoring and the very low error rate in their use of the cut-off point suggests that the comfort sensation for the wearer is quite accurately perceived, and is linked to accepting or rejecting the garment. Trials were separated by 1 to 32 days, with an average of 5 days. Therefore it can also be suggested that the comfort scoring is based on a long term recollection of sensations rather than a transient response.

3.6.2 Fabric handle, comfort and garment preference

To illustrate the strength of the discrimination between the garments in the fabric handle, garment comfort and preference results were used. Fabric handle results were reanalysed to generate a preference table based on the rank of the fabric. The handle test was undertaken by each participant as part of the garment testing. Similarly, the average garment comfort scores were used to generate a preference table based on comfort.

The participants were also asked if they preferred the current test garment to the garment they wore in the previous trial. The participants did not have both garments available to them when answering this question, only the current trial garment. Their responses are recorded in Table 3.1.

Fabric handle scores show the expected order for fabrics made from different wool qualities. The actual scores are closer for the 16.5 μm to 18.5 μm pair than traditionally expected. The fabric handle test uses a pure haptic response – and so can be expected to be accurate – but relies on a limited set of sensory inputs compared to the garment tests.

The extension of testing from fabric to garment introduces a large number of variables that may impact on the comfort response. These variables relate to the garment itself, the response of the person to the garment on the body, and the modification of the environment in the space between the body and the garment. On this basis it is unlikely that the differences shown here between the fabrics will be driven by the same variables that are behind the differences seen in the garment

Table 3.1 Differentiation between garments based on various comparisons

Paired comparison based on	Wool fibre diameter	Preferred garment		
		16.5	18.5	20.5
Fabric handle scores[1]	16.5		1	1
	18.5	2.0		1
	20.5	8.0	2.4	
Average garment comfort scores[2]	16.5		1	1
	18.5	1.4		1
	20.5	5.7	3.4	
Recall of garment preference[3]	16.5		1	1
	18.5	4.0		1
	20.5	11.0	9.0	

Fabric handle and garment comfort comparisons were based on value of scores given to each fabric or garment by each participant. Garment preference was taken directly from a request for preference at the end of the test session.

Notes:
1: participants gave 16.5 a better fabric handle score than 18.5 at a frequency of 2:1
2: participants gave 16.5 a better garment comfort score over 18.5 at a frequency of 1.4:1
3: participants preferred the 16.5 garment over the 18.5 garment at the rate of 4:1

comparison. The paired comparison of garment comfort scores shows a common theme with both fabric and garment responses suggesting a wool fibre effect. The comfort response illustrates an effect of wool fibre diameter, with preference to the finer fibres.

The garment preference shows increased discrimination between the garments over both the garment comfort scores and the fabric handle scores, and yet the ranking effect of fibre diameter is retained. These responses link comfort responses from trials run from 1 to 32 days apart. The accuracy of these responses in a paired comparison is highly significant, higher than when using the fabric handle scores or the garment comfort scores alone. The garment preference is strongly wool fibre-related.

Several possibilities may explain this increased discrimination:

1. The use of factors other than comfort to generate a preference for the garment. Assuming that the tactile, moisture and thermal components are sub-components of comfort then factors such as garment fit may be involved. However care had been taken with sizing the garment on each participant, and as a result each participant had the same size garment in each test. It is possible that garment fit and the tactile component of comfort could combine in an elastic fabric to modify the evoked comfort sensation. However, as the comfort sensation was recorded, such an interaction should have been captured in the comfort record.
2. The use of average comfort scores being a less than optimal metric for expressing the differences in comfort between garments. Alternatively the average comfort score has a level of error that becomes evident when discrimination on garment comfort is compared to discrimination on garment preference.
3. The differences in fibre specification in the garments are contributing to garment discrimination in ways not identified in these trials. These could include contributions from other fibre characteristics such as fibre crimp which is correlated to fibre diameter, but could have a large role in the bulk properties of fabrics. It is also possible that there are interactive effects between the fibre and yarn or fabric properties, even though these were held constant in these garments. The identification of these factors deserves further work.

3.7 Wool quality and garment comfort

Garment testing produced a comfort rating for garments made from a range of wool fibre diameters. Auction price differentials which support traditional definitions of wool quality and the comfort rating are shown together in Fig. 3.8 for the first time.

Superimposing these curves shows that for both curves the anticipated quality increases as the fibre diameter decreases. However there is an apparent declining

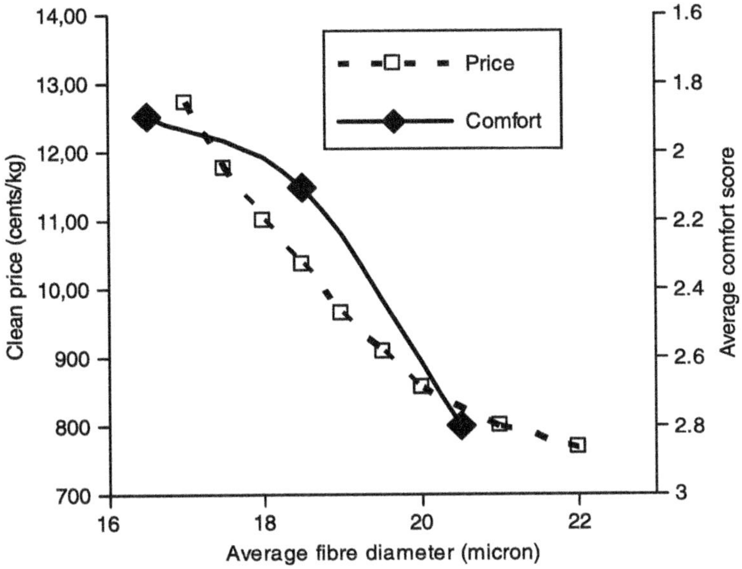

3.8 Relationships between average fibre diameter, the Australian wool auction price, and average garment comfort for each garment tested. The Australian wool auction information was taken from the DAFWA wool portal web site. The implied marginal change in comfort score with fibre diameter is not consistent with the corresponding marginal change in auction price.

marginal return for comfort as the diameter is reduced. The price curves show an increasing marginal return as the fibre diameter decreases. At this point the comfort curve relates to one garment type tested in one market demographic, and might be seen as a special case compared to the price curve which would accommodate supply for all garment types.

3.8 Conclusions

A link between wool quality and garment comfort has been proposed and discussed in this chapter. Garment comfort has been defined as a sensation *experienced by the wearer, evoked by the garment,* and wool quality as the ability to improve comfort by changing wool fibre specifications used in the garment.

Reverse engineering of 168 knitted garments bought at retail showed a large range of wool fibre diameters were used in the garments, loosely linked to changes in fabric weight and yarn count. The analysis allowed the market segment to be well defined, but it did not allow for the separation of fibre effects from processing effects in an appraisal of fabrics taken from these garments.

An alternative approach was to set up garment appraisal testing on a set of garments that were identical except for changes in the wool fibre diameter. By combining a sensitive protocol, a large number of selected participants, a suitable garment and suitable changes in wool specification, significant differences in comfort scores were achieved between garments, and by deduction, between wool fibre diameter specifications.

Population responses to the comfort scores of the garments were shown to be predictable, and the terms used in the prediction were themselves related to the test environment and fibre diameter. A link between wool quality and garment comfort has also been shown to exist, at least in this garment type, and participants from this demographic. Further work will attempt to expand the range of garments and wool types tested for comfort response.

3.9 Sources of further information and advice

- International Wool Textile Organization http://www.iwto.org/
- Australian Wool Innovation http://www.wool.com.au/
- CRC for Sheep Industry Innovation http://www.sheepcrc.org.au/
- Department of Agriculture and Food Western Australia Wool portal http://www.agric.wa.gov.au/servlet/page?_pageid=213&_dad=portal30&_schema=PORTAL30

3.10 Acknowledgments

The author acknowledges the support of the Department of Agriculture and Food (Western Australia) for use of the Design for Comfort Laboratory, and the Cooperative Research Centre (CRC) for Sheep Industry Innovation for project support through the Sheep CRC Wool Program.

3.11 References

Barker, R. L. (2002) From fabric hand to thermal comfort: The evolving role of objective measurements in explaining human comfort response to textiles. *International Journal of Clothing Science and Technology*, **14**, 181–200.

Bartels, V. T. & Umbach, K.-H. (1999) Assessment of the physiological wear comfort of garments via a thermal manikin. In Nilsson, H. O. & Holmer, I. (eds.) *Third International Meeting on Thermal Mannikin Testing 3IMM at the National Institute for Working Life*, Stockholm.

Curtis, K. & Stanton, J. H. (2002a) *Analysis of the Curvature of Wool Offered for Auction in Australia*. Nice, International Wool Textile Organization (IWTO): Technology & Standards Committee.

Curtis, K. & Stanton, J. H. (2002b) *The Relationship between Curvature and Clean Price in Australian Fine Wool*. Barcelona, International Wool Textile Organization (IWTO): Technology & Standards Committee.

Demartino, R. N., Yoon, H. N., Buckley, A., Evins, C. V., Averell, R. B., Jackson, W. W., Schultz, D. C., Becker, C. L., Booker, H. E. & Hollies, N. R. (1984) Improved comfort polyester. Part III: Wearer trials. *Textile Research Journal*, **54**, 447–458. doi: 10.1177/004051758405400707

Department of Agriculture and Food (Western Australia) (DAFWA). Wool Portal. Available from: http://www.agric.wa.gov.au/servlet/page?_pageid=213&_dad=portal30&_schema=PORTAL30 [31 January 2009].

Fuzek, J. F. (1981) Some factors affecting the comfort assessment of knit t-shirts. *Industrial and Engineering Chemistry, Product Research and Development*, **20**, 254–259. doi: 10.1021/i300002a008

Gleeson, T., Lubulwa, M. & Beare, S. (1993) *Price Premiums for Staple Measurement of Wool*. Canberra, Australian Bureau of Agriculture and Resource Economics (ABARE).

Havenith, G. (1999) Heat balance when wearing protective clothing. *Annals of Occupational Hygiene*, **43**, 289–296.

Hollies, N. R. (1971) The comfort characteristics of next-to-skin garments, including shirts. *Third Shirley International Seminar: Textiles for Comfort*, Manchester, England.

Hollies, N. R., Custer, A. G., Morin, C. J. & Howard, M. E. (1979) A human perception analysis approach to clothing comfort. *Textile Research Journal*, **49**, 557–564. doi: 10.1177/004051757904901001

Hollies, N. R., Demartino, R. N., Yoon, H. N., Buckley, A., Becker, C. L. & Jackson, W. (1984) Improved comfort polyester. Part IV: Analysis of the four wearer trials. *Textile Research Journal*, **54**, 544–548. doi: 10.1177/004051758405400807.

Hollies, N. R. & Goldman, R. F. (eds.) (1977) *Clothing Comfort: Interaction of Thermal, Ventilation, Construction and Assessment Factors*. Ann Arbor, Michigan, Ann Arbor Science Publishers.

International Wool Textile Organization (IWTO) (2007a) Measurement of the mean and distribution of fibre diameter using an optical fibre diameter analyzer (OFDA). *IWTO Specifications*, Brussels.

International Wool Textile Organization (IWTO) (2007b) Measurement of the mean and distribution of fibre diameter using the Sirolan-Laserscan fibre diameter analyzer. *IWTO Specifications*, Brussels.

Lau, L., Fan, J., Siu, T. & Siu, L. Y. C. (2002) Comfort sensations of polo shirts with and without wrinkle-free treatment. *Textile Research Journal*, **72**, 949–953. doi: 10.1177/004051750207201103

Li, Y. (2001) *The Science of Clothing Comfort*, Manchester, UK, The Textile Institute.

Mehrtens, D. G. & McAlister, K. C. (1962) Fibre properties responsible for garment comfort. *Textile Research Journal*, **32**, 658–665. doi: 10.1177/004051756203200807

Meilgaard, M., Civille, G. V. & Carr, B. T. (2007) *Sensory Evaluation Techniques*. Florida, CRC Press.

Meinander, H., Anttonen, H., Bartels, V., Holmer, I., Reinertsen, R. E., Soltynski, K. & Varieras, S. (2004) Manikin measurements versus wear trials of cold protective clothing (Subzero project). *European Journal of Applied Physiology*, **92**, 619–621. doi: 10.1007/s00421-004-1139-9

Norman, C. J., Street, P. J. & Thompson, T. (1985) Flame protective clothing for the workplace. *Annals of Occupational Hygiene*, **29**, 131–148.

Paek, S. L. & Davis, S. G. (1975) The wear-comfort prediction of specified knit garments. *Textile Research Journal*, **45**, 763–766. doi: 10.1177/004051757504501101

Rogan, I. M. (1995) A quality profile of the Australian wool clip. *Wool Technology and Sheep Breeding*, **43**, 295–306.

Stanton, J. H. (2008) *Sheep CRC Project 2.1.1., Task R4.1.1.3 Report*. Perth, Design for Comfort Laboratory, Department of Agriculture and Food (Western Australia) (DAFWA).

Stanton, J. H. & Ladyman, M. (2003) *Garment Benchmark Project Final Report*. Perth, Department of Agriculture and Food (Western Australia) (DAFWA).

Stott, K. & Hanson, P. (1993) *Wool Prices: Analysis of Wool Sold at Auction Fourth Quarter 1992/93*. Melbourne, Australian Wool Corporation.

Whiteley, K. J. (1967) Quality control in wool commerce. The threat of man-made fibres. *Wool Technology and Sheep Breeding*, **14**, 65–66.

Whiteley, K. J. & Rottenbury, R. A. (1990) Research into specification and marketing of Australian greasy wool. *Wool Technology and Sheep Breeding*, **38**, 83–88.

Wong, A. S. W., Li, Y. & Yeung, P. K. W. (2004) Predicting clothing sensory comfort with artificial intelligence hybrid models. *Textile Research Journal*, **74**, 13–19. doi: 10.1177/004051750407400103

Wong, A. S. W., Li, Y., Yeung, P. K. W. & Lee, P. W. H. (2003) Neural network predictions of human psychological perceptions of clothing sensory comfort. *Textile Research Journal*, **73**, 31–37. doi: 10.1177/004051750307300106

Yoo, S. & Barker, R. L. (2005a) Comfort properties of heat-resistant protective workwear in varying conditions of physical activity and environment. Part I: Thermophysical and sensorial properties of fabrics. *Textile Research Journal*, **75**, 523–530. doi: 10.1177/0040517505053949

Yoo, S. & Barker, R. L. (2005b) Comfort properties of heat-resistant protective workwear in varying conditions of physical activity and environment. Part II: Perceived comfort response to garments and its relationship to fabric properties. *Textile Research Journal*, **75**, 531–539. doi: 10.1177/0040517505054190

4

How consumers perceive comfort in apparel

F. S. KILINC-BALCI, Auburn University, USA

Abstract: Human with clothing is a system that is always in a dynamic interaction with the environment. In this system, subjective perception of comfort involves very complex processes in which numerous stimuli (visual, thermal, tactile, etc.) from clothing and the environment communicate to the brain, and sensory perceptions are formulated with the evaluations against past experiences. It is very important to understand the working mechanism of the brain and the sensory system in order to understand the comfort perception. This chapter is a brief look into the brain, sensory system, skin and their functions. It also discusses the perception of different senses. In addition, the chapter outlines the overall comfort perception and highlights the importance of the psychological factors.

Key words: brain, nervous system, sense, skin, stimuli, sensory receptor, touch, prickle, itch, comfort perception.

4.1 Introduction

Comfort is a multidimensional subject which is very difficult to define. In general, clothing comfort refers to how the human feels. It is difficult to describe clothing comfort positively while discomfort can be easily defined by wearers with terms including: hot, cold, wet, prickly, itchy, heavy, not breathing, non-absorbent, chill, stiff, sticky, clammy, clingy, and rough.

Among all aspects associated with human feelings and desires, comfort represents a central concern. Indeed, just about every activity a human performs in life involves a process of seeking comfort or relief from environmental and/or mental constraints. Since a human is always exposed to some environmental media, it is natural that he/she will attempt to interact with this environment. The options that humans will typically have are: to forcefully stay in, to get out, or to adopt. These options are driven by numerous factors, which can be explained in three main categories: environmental factors (air temperature, radiant temperature, humidity, etc.), physical factors (health and physical condition, activity level, etc.), and psychological factors (human psychological condition, past experiences, future desires, etc.). These factors typically interact with each other in a nonlinear manner. Furthermore, human beings hardly experience a still environment or body condition, in other words, there is a continuous change over time that leads to transitional effects.

There are several aspects of clothing comfort. One of these aspects, thermo-physiological comfort, is associated with how cold or how hot the wearer feels. People reach this type of comfort when they don't need to add or remove clothing in order to be satisfied with the temperature. This type of comfort is influenced by the changes in physiological variables of the body, such as skin and core temperatures, activity level of the wearer as well as fabric thermal and moisture transfer properties (e.g., thermal resistance and moisture vapor transmission). Environmental variables, such as temperature and humidity, also significantly affect the thermal comfort. Another aspect of comfort, namely neurophysiological, is associated with tactile sensations that result from the fabric/skin contact. Fabric tactile properties (stiffness, friction, softness, etc.), skin properties, environmental conditions, activity level, and garment fit are some of the parameters influencing this type of comfort. The third aspect of comfort, namely psychological, is associated with many factors, such as garment design, fashion, cultural and social factors, price, brand, past experiences, beliefs, and psychological status of the wearer. Psychological factors are very critical for comfort since these factors may outweigh the actual physiological and other factors and become the primary determinants of consumer behavior.[1,15] Because of its subjective nature, psychological comfort differs from one person to another and it is very difficult to analyze. Under the same environmental conditions and using the same type of clothing, while one person feels hot, the other may feel cold. Likewise, even though the core and skin temperatures of the two people are equal, they may not perceive the same comfort level with the garments they wear. Further, even though all of the conditions and physical results may seem equal, two people may not feel equally comfortable or equally uncomfortable. This is mainly because of the psychological factors and physiological differences. Psychological factors significantly impact not only the comfort level but also the purchase decisions of consumers and it can become even more critical for the protective clothing. Cardello points out the impact of soldier attitudes and beliefs regarding the efficacy of the protective aspects of the clothing on the psychological comfort and explains that if the soldier does not have confidence in the protective clothing in terms of its protection, then he/she may experience a psychological discomfort.[1] Moreover, when this discomfort reaches a certain level, he/she may even avoid using the protective clothing in a hazardous environment.

Humans with clothing is a system that is always in a dynamic interaction with the environment. In this system, subjective perception of comfort involves very complex processes in which numerous stimuli (visual, thermal, pressure, tactile, etc.) from clothing and the environment communicate to the brain and then relevant sensory perceptions are formulated, weighed, and evaluated against the past experiences. Overall comfort status can only be assessed after completion of all these processes. In addition, the brain can also influence the physiological status of the body by several functions, such as sweating, blood flow, shivering, etc. It is very important to understand the working mechanism of the brain and the sensory system to be able to define and understand the comfort perception. In this chapter, the brain, sensory system, skin and their functions are explained. Perception of different senses is described and research results in these areas are briefly presented. Finally, the chapter focuses on the overall comfort perception and the importance of the psychological factors.

4.2 How humans sense comfort

The overall comfort perception is a result of a complex combination of inputs from various sensory organs. Sensory organs such as the skin, eyes, ears, nose, and mouth typically react to physical stimuli including visual stimuli (color, light, etc), thermal stimuli (heat and moisture), and tactile stimuli (touch, pressure, etc) by attempting to adopt or adjust. These organs send neurophysiological impulses to the brain, which processes them, initiates human perception to the stimuli, and performs comparative evaluation with past experience references or inherent media. The way humans sense various stimuli can be oversimplified using the step-wise mechanism shown in Fig. 4.1.

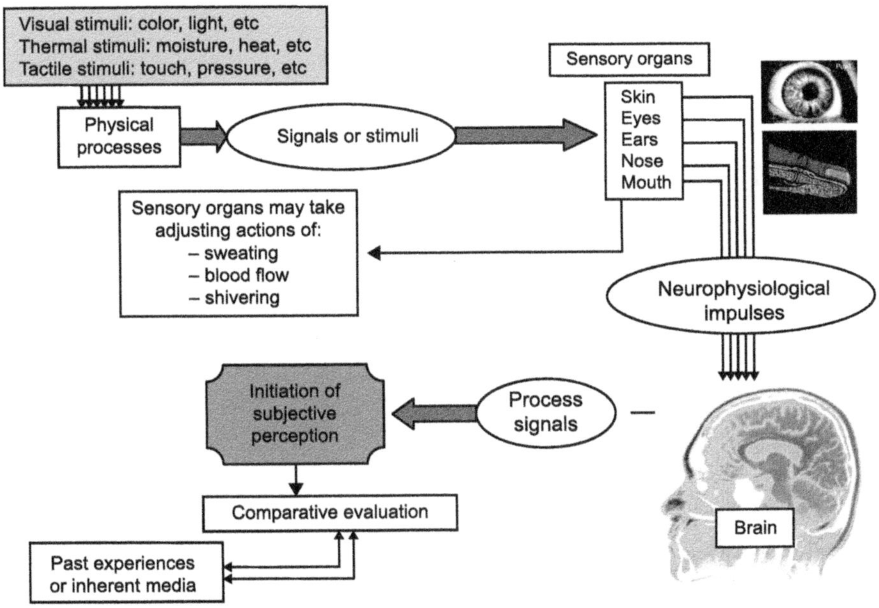

4.1 Simple illustration of human sensory mechanism.[2,15,16]

4.3 The Nervous System

The Human Nervous System (NS) is simply the system that reflects human emotion and reaction to all kinds of external and internal stimuli. The basic functions of the Nervous System are: sensation, motion, homeostasis, reproduction, and adoption. These functions were explained extensively in the literature.[2–14] The Human Nervous System is divided into two subsystems: the Central Nervous System (CNS) and the Peripheral Nervous System (PNS). The Central Nervous System is also divided into two major parts: the brain and the spinal cord (Fig. 4.2).

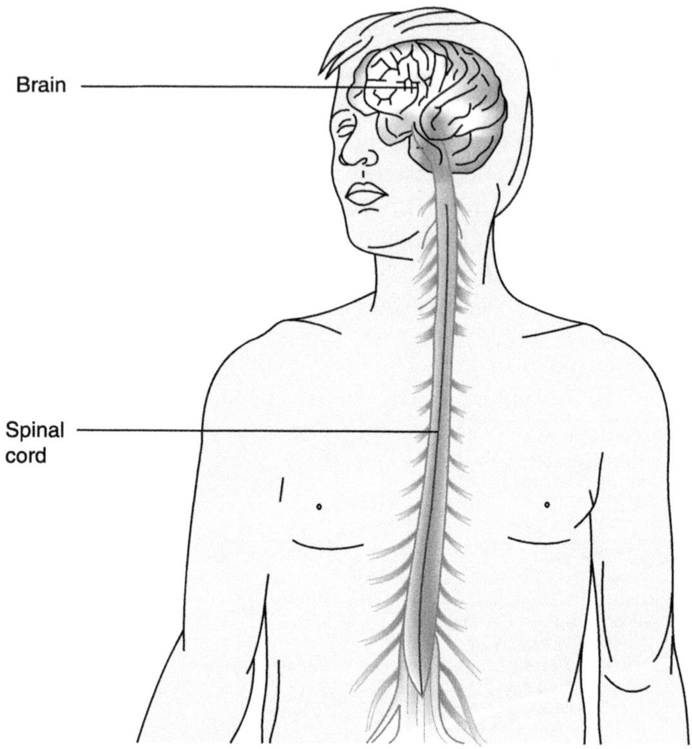

4.2 Spinal cord and brain.

The brain contains about 100 billion 'nerve cells' called neurons, and trillions of 'support cells' called glia. The spinal cord contains the nerves that carry messages between the brain and the body. Nerves transmit impulses from brain to body and from body to brain. The brain and the body are also interconnected chemically, by substances such as hormones which course in the bloodstream.[5]

The Central Nervous System is connected to every point of the body by nerves, which are bundles of axons originating in the cell body of neurons. The collection of all nerves connecting the CNS with the periphery and vice versa constitutes the Peripheral Nervous System. The PNS connects the CNC to sensory organs (e.g., eyes, ears), other organs of the body, muscles, blood vessels, and glands.

The spinal cord represents a key component in controlling the entire nervous system. It represents the main pathway for information connecting the brain and Peripheral Nervous System. The spinal cord receives sensory information from the skin, joints, and muscles of the trunk and limbs, and contains the motor neurons responsible for both voluntary and reflex movements. It also receives sensory information from the internal organs and controls many visceral functions. Receptors in the skin send information to the spinal cord through the spinal nerves.

4.4 Human brain

The brain is the most fascinating and perhaps the least understood organ in the human body. Scientists and philosophers have pondered the relationship between behavior, emotion, thought, consciousness, and the brain for centuries. As early as in the 17th century, Thomas Willis proposed that various areas of the cortex of the brain had specific functions, in particular the circle of vessels at the base of the brain, which now bear his name. In the 19th century, Gall explained that the brain was the seat of all intellectual and moral faculties and particular activities could be localized to some specific region of the cerebral cortex.[5]

Penfield claimed that he found exactly where each part of the body that was touched or moved was represented in the brain.[10–11,17] He then showed it in his famous 'homunculus' cartoons of the somatosensory and motor areas. These maps do not accurately reflect the size of body parts, only their sensitivities. For example, arms and legs take up very little space despite their size, while the face and hands are given more space since they have greater sensitivity and complexity, especially the tips of the fingers. Figure 4.3 illustrates Penfield's classic picture of the brain in which different relationships between where a stimulus is applied to the skin and where neural activity occurs in the somatosensory cortex are shown.

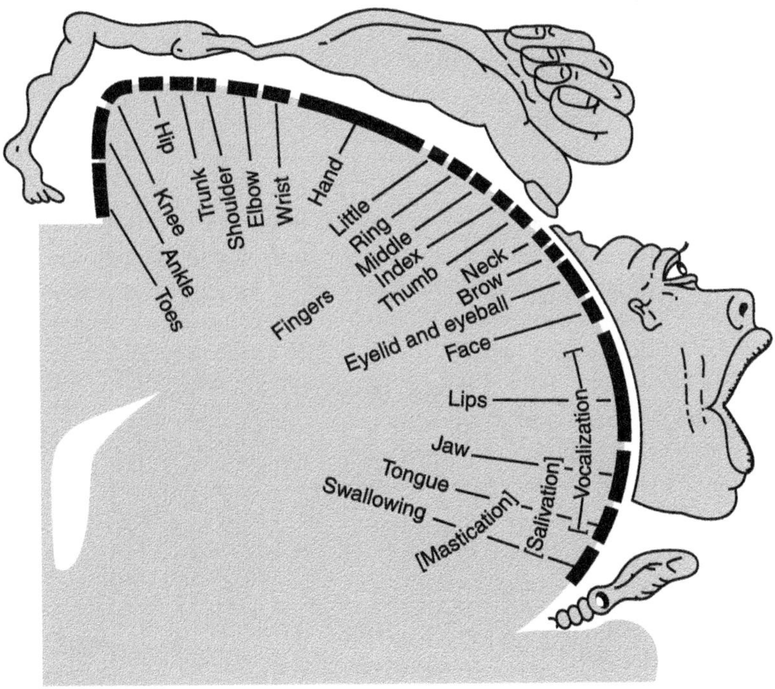

4.3 Illustration of Penfield Map.[10,11]

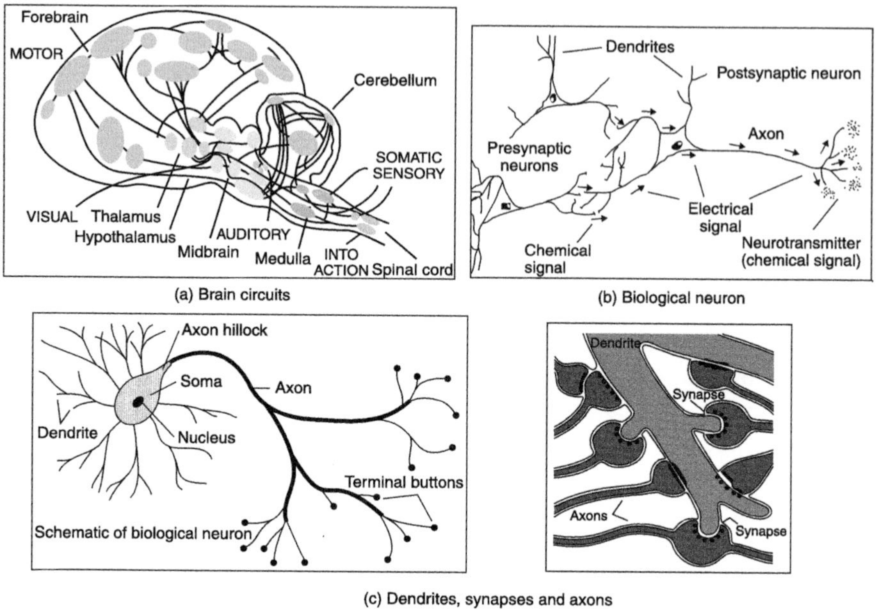

4.4 Brain circuits and neuron structure.[5,15]

As indicated earlier, the human brain is the central element of the nervous system and contains billions of neurons and trillons of support cells (Fig. 4.4). Communication of information between neurons is accomplished by movement of chemicals across a small gap, called a 'synapse'. When neurons become active, an electric current is propagated away from the cell body and down the axon (Fig. 4.4b). When this current arrives at a synapse, it triggers the release of chemicals known as neurotransmitters. In an excitatory neuron, the cooperative interaction of many other neurons whose synapses are adjacent determines whether or not the next neuron will become active, that is, whether it will produce its own action potential, which will lead to its own neurotransmitter release, and so forth.[5]

The discussions in the above sections clearly illustrate that the human nervous system is an extremely complicated system that even top neurologists are hardly beginning to understand. When the issue is how humans utilize this system to feel, desire, or react, further complexity is added. In the following sections, the mechanisms associated with different human sensory components will be reviewed in a simplified way.

4.5 Skin and its functions

Human skin is the largest and one of the most fascinating organs in the human body. Indeed, if the skin of a person is taken off and laid flat, it would approximately cover an area of about 21 square feet and weigh about 7 pounds (3 kilograms). Skin protects the human body from external forces and holds humans in place.

The most widely accepted function of the skin is to define the organism; that is to send boundaries in space, inside of which exists the organism and outside of which exists the environment.[17] There are five major functions of the skin: protection, temperature regulation, sensory perception, excretion, and vitamin production. The skin helps the body to regulate its own temperature, so when the body surface is cold, blood vessels in the skin contract and force the blood deeper into the body. This prevents the body from losing too much heat by radiation. Also, hair muscles pull hairs on end and erected hairs trap air. Muscles can also receive signals from the hypothalamus when humans feel too cold, and they respond by shivering. The rapid contraction of muscles during shivering results in heat being produced during respiration and this heat then warms up surrounding tissues. When the body is too warm, the same blood vessels expand and bring more blood to the surface of the skin. This allows the body to lose heat by radiation. Also, the sweat glands start perspiration. The perspiration evaporates and since evaporation is a cooling process, the skin is further cooled. Millions of microscopic nerve endings are distributed throughout the skin. These nerve endings serve as receptors for pain, touch, temperature, pressure, etc. They keep the body informed of changes in the environment.

The skin, which is less than a millimeter thick in places, is composed of three layers. The outermost layer, called epidermis, is the bloodless layer. The middle layer, called dermis, includes collagen, elastin, and nerve endings. The innermost layer, subcutaneous fat, contains tissue that acts as an energy source, cushion, and insulator for the body. These characteristics of skin result in one of the profound sensations, the touch. As Swerdlow indicates in his famous article in *National Geographic*,[18] the human can live without seeing or hearing, or other senses. However, babies born without effective nerve connections between skin and brain can fail to thrive and may even die. He also added that in laboratory experiments decades ago, when baby monkeys were kept from being touched by their mothers, it made no difference that the babies could see, hear, and smell their mothers; without touching, the babies became apathetic (indifferent or bored) and failed to progress. They did not explore as young primates normally do according to this article. Studies of a variety of cultures also showed a correspondence between high rates of physical affection in childhood and low rates of adult physical violence.[18]

Ironically, as easy as it may seem to understand the impact of touching, the underlying mechanics of touch are hardly understood. This may be due to the complex response of the millions of nerve cells that the human skin has that are of various shapes and at different depths. According to Bolanowski,[18] when nerve cells are stimulated, physical energy is transformed into energy used by the nervous system and passed from the skin to the spinal cord and brain. It is called transduction, and no one knows exactly how it takes place.

In general, human skin is believed to sense three different external effects: pressure, temperature, and pain. Yet, the way these effects are translated to the brain is often very complex and very different by common human descriptors. Bolanowski demonstrates this point with a simple example in which a subject with closed eyes may describe how he/she feels, by saying 'my skin feels wet' or 'something cold and wet is on my forehead' or 'I wait for water to start dripping down my cheeks'. This wet sensation can indeed be a result of a chilled, but dry, metal cylinder contacting the skin. Here, humans sense a combination of pressure and cold, and the skin perceives this as skin wetness. The same sensation is felt when a person wearing a surgical glove puts their finger in a glass of cold water. Here, the finger will feel wet, even though it is not touching water. That is how often the feeling of touch can be quite deceiving. Indeed, when one showers or washes hands, the skin primarily feels pressure and temperature; it is the brain that says it feels wet.

Swerdlow[18] also explains that perceptions of pressure, temperature, and pain manifest themselves in many different ways. Gentle stimulation of pressure receptors may result in ticklishness, gentle stimulation of pain receptors in itching. Both sensations result from a neurological transmission, not from something that physically exists. Scratching puts a quick end to a variety of itches by creating a counter-irritation on the skin which diverts itch perception of the brain. Even though no one has identified exactly what part of the brain receives itch signals, itches trigger activity in areas of the brain that prompt arm movement, presumably initiating a scratch response. In 2001, Leok and his colleagues reported some patterns related with itch. For example, temperature can inhibit an itch. Also, if a finger on one hand itches and the same finger on the other hand is put in cold water, the itch on the first finger goes away. As a result, imposing pain in one place can inhibit an itch in another place.[18]

4.6 Structure of the skin

In humans, the skin has a very complex structure. Figure 4.5 is a diagram of the most important structures in hairy skin, which covers most of the human body. A different kind of skin, which is called glabrous skin and found on the palms of the hands, soles of the feet, parts of fingers and toes, and other places, has no hairs protruding from it. Although there is a thick outer layer of dead cells in glabrous skin, there are also many free nerve endings embedded in this layer. This makes such skin effective protection but also extremely sensitive to stimulation.[17]

As indicated earlier, there are two basic layers to the skin: the outer, thinner layer called the epidermis, and the inner, thicker layer called the dermis (see Fig. 4.5).

The epidermis is made up of numerous cells that are placed side-by-side and arranged one above the other in several layers. This topmost layer is composed of dead cells which are continually being worn off. There are some nerve cells in the epidermis, but there are no blood vessels. The epidermis contains the pigment (melanin) that is responsible for the color of our skin, suntan, and freckles.

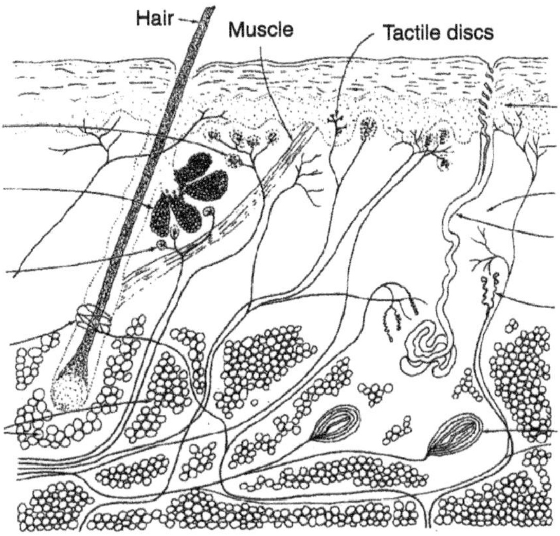

4.5 A piece of hairy skin in cross-section.[17]

The dermis, which is also known as the 'true skin', lies under the epidermis. Into the upper layer of the dermis come thousands of capillaries, the small blood vessels that bring food and oxygen to the cells and remove waste. This inner layer is strong and elastic, and it contains nerve fibers; receptor organs for sensations of touch, pain, temperature, muscular elements, hair follicles, and oil and sweat glands (see Fig. 4.5). Deeper in the dermis are the roots of skin glands and hair, as well as more blood and lymph vessels, more nerves and larger and tougher fibers.

4.7 Senses and sensory receptors

The skin responds to a variety of physical stimuli. When an object is pressed against the skin, it deforms the surface and the sensation of touch, or pressure is experienced. When an object makes contact with a hair, causing it to bend, touch can be experienced as well. The skin is very sensitive to the temperature of the object in contact with it as well as to electrical stimulation.[17]

As indicated earlier, sensation involves detection of an environmental event, the stimulus. The sensory organs are collections of receptor cells and sensory receptor neurons convert stimuli (light, pressure, heat) into action potentials. These receptor cells include mechanoreceptors (sense of touch), photoreceptors (intensity and color of light), thermoreceptors (heat and cold), chemoreceptors (taste and smell), and pain receptors (nociceptors) which react to tissue damage. A stimulus, such as a cold breeze, a hot plate, or a needle, causes receptor cells to produce electrical activity which is taken by means of specialized nerve cells, called conductors, to the brain. The brain then sends signals to nerve cells, called effectors, which carry the messages to targets such as muscles and glands. Muscles contract and glands secrete hormones or other substances in response to the signals. These cause a feeling or a sensation. Even though the brain interprets the impulses, the actual feeling or sensation happens in the sense organ.

4.8 Skin and senses

The sense organ for touching is the skin and the nerve endings under the skin detect touch and register pain. The sense of feeling is made up of messages that the brain receives from the skin. There are numerous nerve endings in the skin and some progress has been made in associating various receptor types with nerve fibers that have different response characteristics (e.g., receptors that respond selectively to mechanical, thermal, or painful stimuli). There are two types of nerve endings: corpuscular and noncorpuscular (or free nerve) endings. Corpuscular nerve endings respond to touch stimuli, while free nerve endings in subcutaneous fat are associated with pain fibers, and those projecting into the epidermis can be associated with cold fibers or pain fibers.[17] As an example of how a skin receptor responds to stimulation, when the Pacinian corpuscle is investigated, it would be seen that a mechanical action causes numerous tiny holes in the corpuscle's membrane to open, allowing electrically charged particles (ions) to flow from one side of the membrane to the other, which causes an electrical current to flow between the point of stimulation and another point on the axon. Then, this electrical current generates a spike potential that jumps along the axon and carries the message of stimulation to the brain.[17]

Recent studies have show that Meissner corpuscles (which are found in large numbers on the skin of the fingers and the face) detect touch, while Ruffini corpuscles detect pressure, Pacinian corpuscles detect vibration and texture and nociceptors evoke pain. Each receptor type is specialized for responding to different types of stimuli. The encoding of specific sensory information is started in the skin by these receptors. The central nervous system further analyzes this sense through neural pathways by transferring the information to the brain.

4.9 Sensations and fabrics

While most people can tolerate the placement of a fabric on their skin without experiencing unpleasant sensations, contact with certain types of fabrics can cause discomfort for some people. Sensitivity to fabrics differs widely, both within a population and between populations and also from one age group to another. Some unpleasant sensations, such as prickliness and itchiness, are experienced when fabric irritates sensory receptors and nerve fibers in the skin.[20]

Some of the most common factors affecting the comfort of the wearer, such as prickliness, pressure, itchiness, roughness, clinginess, warmness/coolness are described here.

4.9.1 Perception of touch and pressure, and mechanical stimuli

As indicated earlier, touch is a basic and powerful necessity. The sense of touch has the ability to discriminate and recognize complex stimulus patterns. Each touch sensation is located at a particular place on the skin and is directly related to the amount of neural representation at each area in the touch cortex.[17] During fabric–skin contact and mechanical interaction during wear, the garment applies a pressure and dynamic mechanical stimulation to the skin and this triggers various mechanoreceptors and generates different sensations of touch and pressure. There have been a number of research projects conducted in touch and pressure perceptions and clothing. Makabe *et al.* (1991)[21] found that the brassiere wearer preferences were related to pressure distribution, Amano *et al.* (1996)[22] reported that fluctuation in the clothing pressure is related to the comfort of the clothing, while Johansson *et al.* (1999)[23,24] determined the discomfort and pain thresholds at the finger, the palm and the thenar area. It has also been found that pressures of less than 60 grams per square meter exerted by the fabric on the body are usually judged to be comfortable, pressures of 60–100 grams per square meter to be uncomfortable.[23,24]

4.9.2 Perception of fabric prickle and itch

Fabric-evoked prickle has been identified as one of the most irritating discomfort sensations for garments worn next-to-skin. Itch is usually a component of the prickle sensation which stimulates the pain group of sensory receptors. Prickliness is experienced when the fabric is patted or pressed onto hairy skin, but it is not felt on the hairless skin such as on palms and fingers since pain nerve endings are very close to the surface in hairy skin but not in glabrous skin. It has been found that moisture on the skin can significantly increase the prickliness sensation.[20]

Itch sensation is the result of the activation of some superficial skin pain receptors. According to recent research results, the sensation of itching is mediated by the same nerves as the sensation of pain. The difference between these two sensations is a function of the degree of stimulation: itching results from a mild stimulus and pain from a more severe stimulus. It was found that itches trigger activity in areas of the brain that prompt arm movement, and that temperature can inhibit an itch.[18]

Since itch and prickle sensations are one of the most reported reasons of discomfort, numerous research has been carried out to study the mechanism of these sensations. Some of the findings can be summarized as follows:[24]

- Garnsworthy *et al.*[24,25] identified the special type of pain nerve responsible for prickle sensation, triggered by a threshold of force of about 0.75 mN. They also found that the neurophysiological basis for fabric-evoked prickle is not caused by skin allergic reaction, or by chemicals released from wool. They identified the cause as the mechanical stimulation of fabric to the skin that induces low-grade activity in a group of pain nerves. When fabric begins a contact with skin, at first the protruding fibers of the fabric take all the force. As the fabric moves closer to the skin, the forces increase and the protruding fibers bend. When the forces from the individual fibers reach certain levels, large shear forces in the skin are generated and pain nerve endings in the skin are activated. The activated nociceptors, i.e. the pain receptors, cause a low rate of discharge from nociceptors over a skin. As a result, individual protruding fiber ends from a fabric surface are responsible for triggering the pain nerve endings when contacting the skin. Garnsworthy *et al.* (1988)[26] also reported that the intensity of fabric prickle perception is a function of the density of high load bearing fiber ends at the fabric surface and the area of contact between fabric and skin. This indicates that fiber mechanical properties and fabric surface characteristics are important for fabric-evoked sensations.[23]

- Naylor *et al.* (1997)[27] explored that the prickle can be predicted from the density of coarse fiber ends per unit area.
- Kennis (1992)[28] reported that the fiber diameter, cover factor and finishing are the most influential factors for fabric prickle.
- Itchiness was found highly correlated with prickliness and Li (1988)[23] reported that itchiness and prickliness are correlated with fiber diameter, fabric thickness at low and high pressures, and fabric surface roughness.
- In several studies, it was found that the sensitivity to fabric prickle is influenced by a number of factors such as gender, age, the area in the body (hairiness), and moisture content of the skin.[23]

4.9.3 Perception of fabric smoothness, roughness and scratchiness

Roughness occurs when fabric moving across the skin stimulates the touch group of sensory receptors. Displacement of skin takes place, and as more skin is displaced under the fabric, the perception of fabric roughness becomes greater. Roughness causes friction between fabric and skin. Moisture also increases the friction causing larger amounts of skin to be displaced under the moving fabric and therefore triggers more touch receptors.[20]

Extensive research has been carried out to study the physical and neural bases of roughness perception. It has been found that perception of fabric roughness is correlated with fabric surface roughness, compression properties, fiber diameter and fiber tensile properties. Scratchiness, which is another term to define discomfort, was found highly related to the sensation of roughness in both consumer surveys and the sensory responses of subjects in wear trials.[23] Mehrtens and McAlister[29] reported that scratchiness perception decreases as the filament flexural rigidity and friction decreases. Winakor *et al.*[30] also reported that all four main properties (fiber content, fabric stiffness, roughness, and thickness) significantly affect sensory responses of judges.

4.9.4 Thermal and moisture stimuli

Thermal senses tell us about our internal and external thermal state. Thermic comfort is the effective perception of temperature and depends on the difference between the thermoregulatory central set point and body temperature. Any measures which help to reduce this difference are felt as pleasant, and vice-versa.

Moisture in clothing has been accepted as one of the most important factors contributing to discomfort. Studies found that the sensation of humidity is correlated with skin wetness.[24] As indicated earlier, after a lot of research, there is a consensus of opinion which indicates that there are no specific moisture detectors in the human body and humidity might be perceived through some indirect methods. Furthermore, the dampness sensation might be a synthetic sensation that consists of a number of components such as fabric temperature, pressure, and distribution of pressure during the contact between skin and fabric.[18,24]

4.9.5 Fabric warmness and coolness

The warm or cool feelings of textiles are another important aspect. When fabric is placed on the skin, there is a momentary sensation of warmness or coolness. The faster the heat transfer occurs between the fabric and the body, the greater is the cold feel of the fabric. The thermal character of the fabric determines the apparent difference between the temperature of the fiber and the temperature of the skin. The differences in cold feel between fabrics are mainly determined by their surface structure rather than by the fiber type. The area of contact between the skin and fabric may be responsible for the rate of heat flow. As the surface area of the contact increases, heat flow from the skin also increases, so the fabric feels cooler. In general, fabrics with fuzzy surfaces feel warmer than smooth-surfaced fabrics of the same fiber composition (e.g., cotton percale bedsheets vs. cotton flannel bedsheets).[20]

4.9.6 Fabric charging and cling

Charged fabrics cling to the body and result in another unpleasant feeling and charged fabrics may cause shocks when the wearer touches metal. Fabric cling results from the formation of an electrostatic charge on the fabric and the induction of this charge on the body. During wear, adjacent layers of garment fabric and surrounding fabrics are frequently pressed and rubbed together when the wearer is sitting or walking. Positive charges are produced on one surface and negative charges on the other surface during this contact. When the wearer moves, since fabrics separate during this move, fabrics may become charged, one positively and the other negatively. When an electric field is generated from the charges, the fabric induces an opposite charge on the skin and fabric cling may occur. The degree of the cling depends on the fabric types worn next to each other. Moisture will increase the electrical conductivity of most textile fabrics. The actual area of fabric/skin contact, which is influenced by fabric structure, is one of the key parameters impacting the fabric cling.[20]

4.10 Psychological factors and overall comfort perception

Wear is a part of our daily life and comfort of the clothing can significantly affect human performance. During the wear, as is described in depth in the previous parts of the chapter, the contact between skin and clothing produces a number of mechanical, thermal, electrical and chemical stimuli. The sensations obtained from these stimuli influence human comfort status. The type of sensation heavily depends on the fabric/skin interaction and the sensory receptors triggered.

The comfort level of clothing does not solely depend on the fabric properties and design features. The perception of clothing comfort is a function of garment, environment, body, and psychological factors. The perception of comfort can be influenced by a variety of cognitive factors, such as beliefs, social and cultural factors, past experiences, and present desires. As is explained in the beginning of the chapter, once the attitudes and beliefs toward the fabrics and clothing are formed, they may outweigh the actual physiological factors and become the primary determinant of the consumer behavior.

It is very important for manufacturers and retailers to understand how consumers perceive the clothing and formulate the preferences. There has been extensive research carried out in order to understand consumer behavior. The overall sensory perception and preferences of the consumer are the result of a complex combination of sensory factors that come from various receptors. These sensory receptors are influenced by the psychological and physiological state of the wearers. All sensory factors have two psychological dimensions: quality and magnitude (intensity). Sensations combined with past experiences, beliefs, attitudes, and present desires form the overall perception.

A number of behavioral approaches have been used in order to assess consumer attitudes and it has been found that consumers have well-defined attitudes toward fibers and fabrics.[1]

In order to understand the psychological processes, measurement studies have been carried out in which subjective perceptions were obtained by psychological scaling.[1,23,24] There are a lot of problems involved in psychological scaling, such as wide variations in opinions, statistical analysis problems, inconsistencies due to physiological, psychological, social, cultural, and environmental, etc. factors affecting the data.[19]

Many people, including some experts, would like to think of the comfort phenomenon as entirely psychological or human-preference based. Although it seems hard to argue against this thought on the basis that human judgment is often influenced by the psychological status at the time of judgment, a true evaluation of the comfort phenomenon should neutralize variable personal psychological effects; or personal preferences to factors that are not associated directly with the source of comfort/discomfort being examined. This point brings up a serious issue associated with most psychological scales that have been developed in many studies to quantify human judgment. When an individual is asked to evaluate a certain fabric in relation to handle performance while the fabric under evaluation is visible, the human preference may be divided into two categories:

(a) Object-related personal preference such as color, fashion, design, brand, and style. Earlier experiences with the fabric (or with similar fabric that person used) and price, if it is known or declared, may also affect the human preference. Previous experiences with the fabric are very important since in human sensory mechanism, the brain performs comparative evaluation with past experience references.
(b) Internal personal preference such as internal feelings at the time of judgment (happiness, sadness, excitement, depression, neutral, etc.), skin sensitivity to touch (typically a factor of age, job, and normal inherent environment), ability of self expression or ability to relate to a given descriptor of the object, ability to clearly distinguish between different descriptors or different objects, and the extent of sensitivity to the environment surrounding the judgment location.

Statistically speaking, one cannot evaluate the effect of a certain factor using a number of experiments or sample size that is less than the number of factor levels. Indeed, the number of levels and factor interaction levels should represent the minimum number of points being analyzed or the minimum sample size. Taking the above two categories of factors and the multiplicity of levels into consideration will require examination of their interactions. We may have an infinite number of experimental or judgment combinations, which would require an infinite number of people to pass a truly reliable judgment.

Traditionally, researchers attempted to overcome the problems discussed above through elimination of the effects of some of these factors by selecting more or less unique panels of judges with many human characteristics in common. Other researchers used randomization or random stratification as a way to resolve the complexity associated with comfort evaluation. These approaches are still largely deficient in revealing true evaluation of this critical phenomenon. As a result, some researchers accepted the fact that these factors will inevitably coexist and continued their research using different judgment panels and different evaluation techniques (different questionnaire forms, different sample representation, and different types of psychological scaling). Other researchers relied on the so-called expert panels to yield reliable evaluation of comfort. Obviously, the problem with an expert panel lies in the need to verify the extent of its experience, whether yesterday's experience can provide much for today's practice, and the extent of how spontaneous is its judgment. As a result, it is often the case that panel expert judgment only reflects the panel evaluation and not the majority of people that rely on spontaneous feelings.[15]

4.11 Conclusions

Comfort is a fundamental and universal need for consumers. Hence, research on clothing comfort has a fundamental meaning for the survival of human beings and the improvement of the quality of life. It is very important to understand the working mechanism of the brain and sensory system to be able to define and understand the comfort perception. In this chapter, the brain, sensory system, skin and their functions were explained. Perception of difference senses was described and research results in these areas were briefly presented. Finally, the chapter focused on the overall comfort perception and the importance of the psychological factors.

4.12 References

1 Cardello, A.V. (2008) 'The sensory properties and comfort of military fabrics and clothing', in *Military Textiles*, ed. E. Wilusz, Cambridge, Woodhead Publishing, pp. 71–103.
2 Nieuwenhuys, R., Ten Donkelaar, H.J., and Nicholson, C. (1998) *The Central Nervous System of Vertebrates*, Vol. 3, Berlin, Springer.
3 Berta, A., *et al.* (1999) *Marine Mammals. Evolutionary Biology*, San Diego, Academic Press.
4 Best, J. Boyd (1963) 'Protopsychology', *Scientific American*, February, pp. 55–62.
5 Damasio, A. (1999) *The Feeling of What Happens*, Harcourt Inc., Florida, USA.
6 Demski, L.S., and Northcutt, R.G. (1996) 'The brain and cranial nerves of the White Shark: an evolutionary perspective', in *Great White Sharks, The Biology of Carcharodon Carcharias*, San Diego, Academic Press.
7 El Mogahzy, Y.E., Kilinc, F.S., Hassan, M. (2004) Design-Oriented Fabric Comfort Model, presentation at the National Textile Center Research Forum, Hilton Head, SC, February 2004.
8 Kety, Seymour S. (1960) 'A biologist examines the mind and behavior', *Science*, **132**: 1861–9.
9 Mink, J.W., Blumenschine, R.J. and Adams, D.B. (1981) 'Ratio of central nervous system to body metabolism in vertebrates: its constancy and functional basis', *Am. J. Physiology*, **241**: R203–12.
10 Penfield, W., and Perot, P. (1963) 'The brain's record of auditory and visual experience: a final summary and discussion', *Brain*, **86**: 595–696.
11 Penfield, W. (1968) 'Engrams in the human brain', *Proceedings of the Royal Society of Medicine*, **61**: 831–40.
12 Postle, R. (1990) 'Fabric objective measurement technology, present status and future potential', *Int. J. Clothing Sci. Technol.*, **2**: 7–17.
13 Rehkamper, G., Frahm, H.D. and Zilles, K. (1991) 'Quantitative development of brain and brain structures in birds (galliformes and passeriforms) compared to that in mammals (insectivores and primates)', *Brain Beh. Evol.*, **37**: 125–43.

14 Ridgway, S.H. and Harrison, S. (1985) *Handbook of Marine Mammals*, Vol. 3, London, Academic Press.
15 Kilinc-Balci, F.S. (2004) A Study of the Nature of Fabric Comfort: Developing a Design-Oriented Comfort Model, Ph.D. thesis, Auburn University, Auburn, AL.
16 Kilinc-Balci, F.S., and Elmogahzy Y. (2008) 'Testing and analyzing comfort properties of textile materials for the military', in *Military Textiles*, ed. E. Wilusz, Cambridge Woodhead Publishing, pp. 107–36.
17 Coren, S., and Ward, L.M. (1989) *Sensation and Perception*, third edition, New York, Harcourt Brace Jovanovich Publishers.
18 Swerdlow, J.L. (2002) 'Unmasking skin', *National Geographic*, **2**(5): 36–63.
19 Slater, K. (1986) 'Assessment of comfort', *J. Textile Inst.*, **77**: 157–71.
20 Hatch, K. L. (1993) *Textile Science*, West Publishing Company, MN, USA.
21 Makabe, H., Mamota, H., Mitsuno, T., and Ueda, K. (1991) 'Study of clothing pressure developed by the brassiere', *J. Japan Res. Assoc. for Textile End-Uses*, **32**: 416–23.
22 Amano, T., Minakuchi, C., and Takada, K. (1996) 'Spectrum analysis of clothing pressure fluctuation', *Sen-i Gakkaishi*, **52**: 41–4.
23 Li, Y., Wong, A.S.W. (eds) (2006) *Clothing Biosensory Engineering*, Cambridge, Woodhead Publishing,
24 Li, Y. (2001) 'The science of clothing comfort', Textile Progress, *J. Textile Inst.*, **31**(1/2): 1–135.
25 Garnsworthy, R.K., Gully, R., Kennis, P., Mayfield R.J., and Westerman, R.A. (1988) 'Identification of the physical stimulus and the neural basis of fabric-evoked prickle', *J. Neurophysiology*, **59**: 1083–97.
26 Garnsworthy, R.K., Gully, R., Kandiah, R.P., Kennis, P., Mayfield, R.J., and Westerman, R.A. (1988) 'Understanding the causes of prickle and itch from the skin contact of fabrics', *Australian Text.*, **8**: 26–9.
27 Naylor, G.R.S., Philips, D.G., and Veitch, C.J. (1997) 'Fabric-evoked prickle in worsted spun single jersey fabrics Part I: The role of fiber and diameter characteristic', *Text. Res. J.*, **67**: 288–95.
28 Kennis, P., (1992) 'The cause of prickle and the effect of some fabric construction parameters on prickle sensations', *Wool Technol. Sheep Breeding*, **40**: 19–24.
29 Mehrtens, D.G. and McAlister, K.C. (1962) 'Fiber properties responsible for garment comfort', *Text. Res. J.*, **32**: 658–65.
30 Winakor, G., Kim, C.J., and Wolins, L. (1980) 'Fabric hand: tactile sensory assessment', *Text. Res. J.*, **50**: 601–10.

5 Laboratory measurement of thermophysiological comfort

L. HES, Technical University of Liberec, Czech Republic and
J. WILLIAMS, De Montfort University, UK

Abstract: Clothing interposes a barrier between the body and the environment and therefore affects the person's wellbeing in terms of protection, skin sensations and heat and moisture management. The heat and moisture balance between the body and the environment determines the thermophysiological comfort of an individual. Although human wear trials and manikin evaluations can be undertaken on whole garments, this chapter reviews some of the methods for evaluating the heat and moisture management properties of small textile samples in a laboratory.

Key words: comfort, water vapour, thermophysiological, thermal insulation, hotplate, permeability, resistance.

5.1 Introduction

The choice of materials along with the design and fit of a garment or clothing system greatly affects the comfort of the wearer. Often the feeling of comfort is subjective, i.e. it is too tight, too hot, too long, etc., and can vary greatly depending on gender, size, race, age, fitness, of the wearer. The human body strives to maintain a constant core temperature of 37 °C by thermoregulation in a range of climatic conditions and levels of activity. Thermal comfort is normally achieved when the body is in heat balance and the body does not store or lose heat to the surroundings (Fanger 1970).

This heat transfer takes place in three different forms: first conduction, convection and radiation in dry conditions; second, diffusion of insensible perspiration; and third the diffusion of liquid perspiration (Greenwood *et al.* 1970). Clothing interposes a barrier to both evaporative cooling and the heat transfer to or from the environment, i.e. from a high temperature region to a low temperature region. The rate of transfer depends on the temperature difference and the resistance to transfer of the clothing.

It follows that a measure of both the thermal resistance of clothing (section 5.3) and the ability of a fabric to transmit the water vapour emitted from the body (section 5.4) are important factors in assessing the thermophysiological comfort of the wearer (Greenwood *et al.* 1970; Weiner 1971). In addition, the air penetration through clothing generally increases body cooling which can be very detrimental at low environmental temperatures (section 5.5) and the movement of liquid sweat away from the skin also plays an important role in comfort (section 5.6).

5.2 Thermo-physiological comfort

Controlled wearer trials are the ultimate assessment of clothing performance including the affects of layering and ventilation, work-rest routines and climate, but can be expensive to control all the variables (Holmer 2009). A step back from this approach is the use of life size thermal or sweating manikins to evaluate a complete clothing system. Due to the wide variety of material combinations possible for a given clothing system, it would not be feasible to construct and test them all in garment form. Although organisations producing or evaluating clothing can use either method they may also need to assess individual fabrics and even combinations during both development and production and therefore must rely on methods utilising small samples in a laboratory.

In terms of thermophysiological comfort the two most important parameters are concerned with the movement of heat and sweat away from the body and can be measured in terms of thermal insulation and water vapour resistance. For full garment assemblies one laboratory approach is to use a life size manikin in a controlled climatic chamber (Runbai and Hang 1991). The manikin can be heated to assess both the thermal insulation of the total clothing and the total air layers trapped between the fabrics. The insulation of specific garment areas can also be measured in terms of the required heat input to maintain a constant temperature at various sites on the manikin surface. The use of a jointed manikin allows the effects of ventilation to be measured as it simulates an actual level of work (ASTM F1291-80). The thermal resistance of the clothing system can be compared to the resistances of the fabric layers and the resultant thickness of the trapped air determined. Some workers also calculate the overall water vapour resistance of the clothing system from separate fabric measurements and the calculated value of the air layer on the manikin (Umbach and Mecheels 1976).

Coppelius, a sweating thermal manikin, has been used to assess the properties of cold weather clothing (Meinander 1995). It again has a segmented body with individual heating elements but also has an inbuilt sweating mechanism with the sweat produced at a constant rate from 187 sites on its surface with evaporative losses being measured by weighing the clothed manikin. Water uptake by the individual layers is determined by weighing each item of clothing before and after the trial period. There are a number of variants around the world, for example the sweating fabric manikin Walter used in Hong Kong (Fan and Qian 2004) and SAM, a sweating agile thermal manikin (Richards and Mattle 2001). Both manikin techniques are initially expensive to set up and are therefore relatively exclusive such that there is an ongoing need for the evaluation of small fabric samples for both quality control and material development.

There are a number of standard tests on heat and vapour resistance available for comparison of fabric samples, e.g. in ISO, EN or ASTM standards. The most popular techniques used in Europe in the late 1990s were the guarded hot plate and the Togmeter (ISO 5085-1 1989 and BS 4745 2005) for thermal resistance and the Canadian Control Dish for water vapour resistance (CGSB 1977) (BS 7209 1990 is a simplified form), and the sweating guarded hot plate (Skin Model) (ISO 11092 1993) which measures both parameters. Other alternatives include the Gore Cup (1986), ASTM E96 (2005) and the Permetest Skin Model (Hes 1993).

These laboratory tests normally measure the properties of individual fabrics in a controlled atmosphere and do not evaluate the air gaps and ventilation associated with clothing when it is worn. A number of these have been compared with each other and also with manikin and human wear trials to aid predictions of comfort from small sample evaluations (Williams 1994, 1997; McCullough *et al.* 2003). The dynamic moisture permeation cell (DMPC) (Gibson *et al* 1997) allows measurements at varying humidity gradients and assesses the effect of air pressure on water vapour transmission rates (ASTM 2298).

The following sections review some of the test methods available for the measurement of thermal insulation and water vapour transport on small fabric samples in the laboratory.

5.3 Thermal resistance

The thermal insulation of small fabric specimens can be readily measured in the laboratory by testing dry heat transfer. Hot-plate methods based on two-plate, single plate and guarded plate are the most popular and give results of reasonable accuracy (Fig. 5.1) (ISO 5085-1 1989 and BS 4745 2005).

1. The two-plate method of test is employed when the fabric to be tested would normally be shielded from the ambient air by an outer layer, for example a shirt when worn beneath a suit. This method utilises the heat flow principle, where measurement of the temperature gradient through the fabric is made using thermocouples.
2. The single-plate method is used when the outer surface of the fabric will be exposed to the ambient air, for example, outerwear clothing. The testing procedure and apparatus are similar to those of the two-plate method, but the cold-plate is now placed adjacent to the apparatus to measure ambient air temperature and the outer surface of the fabric is therefore left uncovered. However, reproducibility is not as good as the two-plate method.

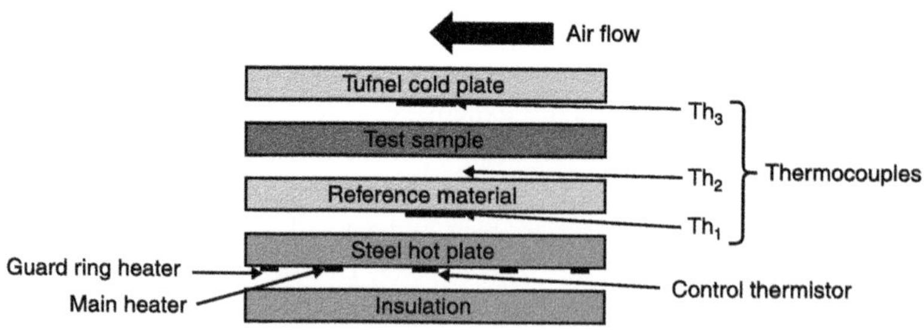

5.1 Togmeter – guarded hot plate.

3. The guarded hot-plate method is similar to the single-plate method, but is particularly suited to thicker fabric samples, requiring more sophisticated temperature control but achieving greater accuracy (e.g. ISO 11092). A metal guard plate surrounds all but the upper surface of the test-plate and is maintained at the same temperature by a separate power supply to eliminate the effect of heat loss from the edges and bottom of the test-plate. There is also a thin layer of insulating material between the guard-plate and the test-plate to keep the temperature control of the two independent of one another. The temperature, humidity and air speed can be controlled when the apparatus is placed in an environmental chamber.

Using a guarded hot plate for heat resistance measurement to measure the amount of heat lost through a sample with a temperature gradient between the plate and the environment, the insulation of the fabric can be calculated as:

$$R_{ct} = \frac{\overline{t}_{plate} - \overline{t}_a}{H_{DRY}} - R_0 \quad [5.1]$$

where,

R_{ct} = heat resistance of fabric sample (m^2 K W^{-1})
t_{plate} = mean hot plate surface temperature (C)
t_a = ambient temperature (C)
H_{DRY} = dry heat loss per m^2 of plate (wm^{-2})
R_0 = heat resistance measured without sample (m^2 K W^{-1})

5.4 Water vapour transport

The more efficient the fabric is at allowing this water vapour to reach the ambient air the more breathable it is said to be. Breathability as described here is therefore regarded as a measure of the water vapour migration through a textile and not a measure of its air permeability or windproof characteristics. The water vapour transport properties of a textile are commonly expressed in one of three ways:

1. The water vapour flow in unit time through unit area of fabric under specific conditions of temperature and relative humidity, for example g m^{-2} day^{-1}. This is commonly defined as either water vapour permeability (WVP) or moisture vapour transmission rate (MVTR).
2. The water vapour resistance, a measure of the resistance to water vapour migration expressed as the equivalent thickness of still air which has the same resistance to water vapour diffusion as the textile (mm still air). Resistance data are generally more useful than the permeability data because they are additive for clothing layers in the same way as electrical resistances in series and the thermal resistances of clothing. The overall resistance of a clothing assembly can therefore be estimated by adding together the

resistances of the individual textile layers and the resistances of the air gaps between the layers. The resistance of an ensemble bears an inverse relationship to its water vapour permeability.

3. The resistance to evaporative heat flow (R_{et}) (Umbach 1988) is the quantity which determines the latent or evaporative heat flux of a textile layer under steady state conditions effected by a partial water vapour pressure gradient perpendicular to the fabric. The resistance of a textile to water vapour flow has values for individual materials commonly expressed in units of $m^2mbar\ W^{-1}$ or $m^2Pa\ W^{-1}$.

5.4.1 Water vapour permeability

There are many different methods described in the literature for measuring the water vapour permeability of breathable textiles: some are national standards (CGSB49 BS, 7209, etc.) and others have been independently developed by clothing manufacturers, often to show their own products more favourably. The most common and simplest approach is that of the cup or dish method, of which there are two basic types, both based on weight change (Dohlan 1987) as shown in Fig. 5.2.

1. *The Desiccant Method*, where a sample is sealed to the open mouth of a dish containing a desiccant, such as calcium chloride or calcium acetate, and the assembly placed in a controlled atmosphere. The weight gain of the assembly with time as moisture is attracted to the desiccant is used to determine the permeability (ASTM E96-80A/C/E, 1980).
2. *The Water Method*, where a sample is sealed to the open mouth of a vessel containing water. This is then placed in a controlled atmosphere, either upright or inverted depending on realism and sample type. A further modification is to interpose a microporous membrane or an air gap between the water and the sample. The weight loss with time is used to determine the rate of water vapour migration.

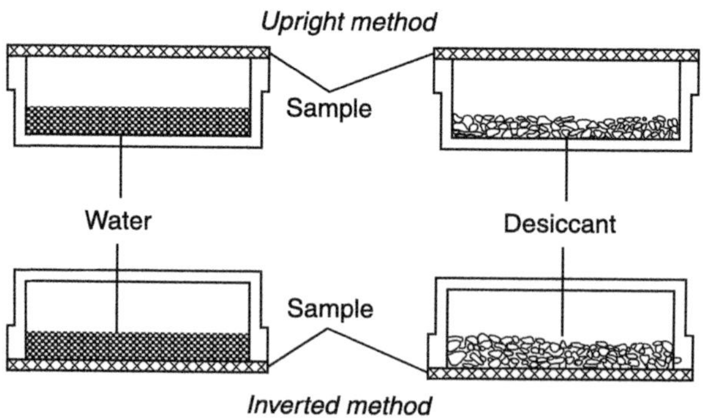

5.2 Water vapour permeability – cup method.

The main test variables which influence water vapour permeability are the temperature and relative humidity at each surface of the sample. These are also influenced by the still air layers on both sides of the assembly. In general, the water vapour permeability is given by:

$$WVP(gm^{-2}day^{-1}) = \frac{M_0 - M_1 * 24}{A * t} \quad [5.2]$$

where,

M_0 = weight of the test assembly before test (g)
M_1 = weight of the test assembly after test period t (g)
t = time between successive weighings of the test assembly (hours)
A = area of the exposed test fabric (m^2).

The water vapour content of air is usually stated as per cent relative humidity (RH). However as the term suggests, relative humidity is not a measure of the absolute concentration of water molecules, but a percentage of the maximum amount of water vapour the air can hold at that temperature (Tanner 1977). The

Table 5.1 The relationship of temperature and relative humidity to the concentration of water vapour in the air

Temperature (°C)	Relative humidity (%)	Actual volume of water vapour as % of total air volume	Water vapour pressure (mbar)
0	100	0.6	6.09
0	90	0.5	5.48
0	65	0.4	3.96
0	50	0.3	3.04
0	35	0.2	2.13
0	10	0.1	0.61
20	100	2.3	23.37
20	90	2.1	21.04
20	65	1.5	15.19
20	50	1.15	11.69
20	35	0.8	8.18
20	10	0.2	2.34
40	100	7.4	73.77
40	90	6.6	66.39
40	65	4.8	47.95
40	50	3.7	36.89
40	35	2.6	25.82
40	10	0.7	7.38

Source: Tanner 1977.

same RH at different temperatures can have markedly different water vapour pressures (as shown in Table 5.1) from:

$$Vapour\ pressure = \frac{\%RH}{100} * 10e^{16.6536} - \frac{4030.183}{235+T} \qquad [5.3]$$

Table 5.1 shows the actual amount of water vapour in air for a range of temperatures and humidities expressed as a percentage of the total air volume as well as the vapour pressure at those conditions. The direction of water vapour diffusion depends on the driving force, the absolute difference in water vapour concentration.

The water vapour permeability of textiles can also be assessed in relative units. 'Relative wvp' is calculated by comparing the masses of water vapour that are evaporated from either an open dish or a dish covered with a standard fabric with a dish with specimen as in BS 7209 (1990).

$$WVP_R = \left(\frac{m_{op}}{m_{sp}}\right).100 \qquad [5.4]$$

where,

WVP_R = relative wvp (%)
m_{op} = mass of water vapour evaporated from the open dish or dish with standard fabric (g)
m_{sp} = mass of water vapour evaporated from the dish with specimen (g).

5.4.2 Water vapour resistance

A relationship between water vapour permeability and water vapour resistance is derived from Fick's first law of diffusion which states that the flux in the x-direction, Fx, is proportional to the concentration gradient Dc/Dx (Glasstone 1960; Fourt and Harris 1947) such that:

$$R = \frac{1}{Q} D(\Delta C)At \qquad [5.5]$$

where,

R = resistance of the system (cm)
Q = weight change of the test assembly during test period t (g)
t = test period (s)
A = area of the exposed test fabric (cm^2)
D = the diffusion coefficient (cm^2 s^{-1})
ΔC = difference in water vapour concentration across the test assembly (g cm^{-3}).
(Q/At, the water vapour permeability, can be measured directly.)

The diffusion coefficient, D, varies with the average absolute temperature, T, and the barometric pressure, P by:

$$D = 0.220\left(\frac{T}{273}\right)1.75\,\frac{760}{P} \qquad [5.6]$$

where P = the barometric pressure (mm Hg). Between 0 and 50 °C this can be simplified to:

$$D = 0.220 + 0.00147 * T_1 \qquad [5.7]$$

where T is the temperature (°C).

The concentration difference, ΔC, in g cm^{-3} can be obtained from the relative humidities, absolute temperatures and corresponding saturated vapour pressures each side of the resistance as:

$$\Delta C = \frac{18}{22400} * \frac{273}{760}\left(\frac{P_1 RH_1}{T_1} - \frac{P_2 RH_2}{T_2}\right) \qquad [5.8]$$

where,

RH = the relative humidity
T = the absolute temperature (K)
P = the saturated vapour pressure on each side of the assembly (mm Hg)
18 = molecular weight of water
22400 = molar volume of gas at STP ($cm^{-3}mol^{-1}$).

At a fixed temperature this becomes:

$$\Delta C = 2.89 * 10^{-4}\,\frac{\Delta P}{T} \qquad [5.9]$$

where, ΔP is the difference in actual vapour pressures.

To find the permeability under any conditions, Equation 5.5 can be solved for Q/At and evaluated from the appropriate value of D, ΔC and R.

$$\frac{Q}{At} = \frac{\Delta C}{R} * D \qquad [5.10]$$

The water vapour resistance can be calculated from measurements of water vapour permeability under a known water vapour concentration gradient. This gradient is usually estimated from measurements of temperature and humidity away from the sample. Therefore, a resistance value obtained from the permeability value is a composite of the sample resistance and the resistances of the two air layers either side of the sample. The resistance may also be affected by local atmospheric conditions above the dish and the air velocity over the sample.

For multiple layers of clothing, such as likely in a complete clothing ensemble, the total resistance of the system is the arithmetic sum of the resistance of each component layer together with an assessed value for each air gap between layers.

In this way, estimates of various clothing combinations can easily be made and the likely benefit of substituting different components deduced quickly without the need, in the first instance, for actual composites to be made.

5.4.3. Water vapour resistance measurements

The Canadian Government Specification Board (CGSB) Method 49, Control Dish method for textiles (also known as the Turl Dish) (Whelan *et al.* 1955), gives results expressed as the equivalent resistance to water vapour migration of the fabric sample in millimetres of still air (Fig. 5.3).

The test itself compares the rate of permeation of water vapour from dishes containing water and covered by the test fabric and a standard cover fabric with that from a control dish, also containing water, but covered only by the cover fabric. The rate at which water vapour diffuses from the dishes is governed by the vapour pressure difference existing between the water surface inside the dish and the external conditions, the ambient temperature and pressure and the total thickness of the still air layer between the water surface and the cover fabric. This thickness comprises the air layers above and below the sample and the equivalent resistance of the sample and cover fabric. The latter is a high porosity fabric, for example British Standard fabric BSF118/854, which maintains a layer of still air above the sample material, but is of negligible resistance to water vapour migration.

The interior walls of the dishes are treated with a PTFE or silicone spray prior to the test to ensure a uniform vertical gradient of water vapour pressure over the entire area of the dish. In placing assymetric samples, for example coated fabrics and laminates, in the test dishes, care is taken to ensure that the samples are placed with the correct side towards the water surface so as to imitate the behaviour of the material on the human body as closely as possible. The whole apparatus is placed in a relative humidity and temperature controlled chamber, at a standard condition of 20 °C ± 0.2 °C and 65% RH ± 1%.

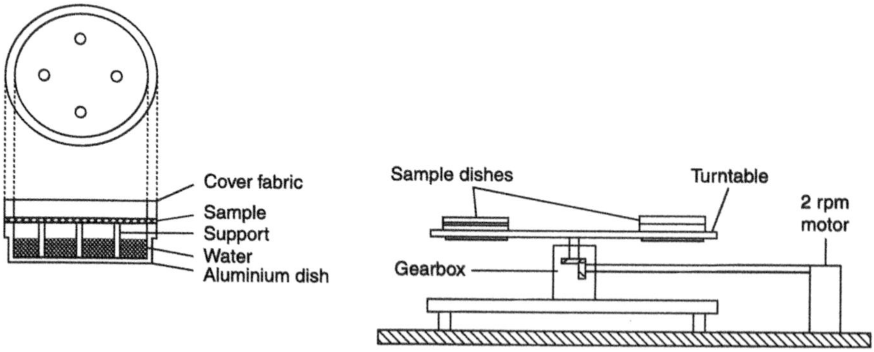

5.3 Turl Dish apparatus.

The control dishes, without samples, are made up with 35, 50 and 65 ml of water in each, establishing different thicknesses of air between the water surface and cover fabric. The relationship between these distances and their respective rates of water loss under controlled conditions forms the basis of the apparatus calibration. When a sample is interposed between the water surface and cover fabric, as in the sample dishes, the rate of water loss is reduced. This reduction is expressed as an additional thickness of still air existing between the water surface and cover fabric and represents a resistance to water vapour permeation.

Local inhomogenities in RH are alleviated by rotating the dishes on a turntable at 2 rpm for up to 20 hours. From the rate of water loss over the test period a value of the total effective air layer thickness inside each of the dishes is calculated. Since the thickness of the air layers in the dish are known, the equivalent thickness of the sample can be found and expressed as an equivalent thickness of still air. A similar method has been used in Russia by Dimitrieva (1958) but with a turntable rotating in excess of 5 rpm to yield results in approximately 5 hours.

Furthermore, it is possible to convert this resistance to water vapour migration into a water vapour permeability value using the method of Fourt and Harris (1947). The water vapour permeability, WVP, in g $m^{-2}day^{-1}$ can be expressed in terms of the resistance R in mm by the following:

$$WVP = \frac{RH * D * C * 10^{-3}}{R} \qquad [5.11]$$

where RH = relative humidity difference across the sample (%)
D = diffusion coefficient of water vapour in still air ($cm^2\ sec^{-1}$)
C = saturated water vapour concentration ($g\ m^{-3}$).

If, under the experimental conditions, RH, D, C and the test temperature are constant, then Equation 5.11 can be simplified;

$$WVP = k\frac{1}{R} \qquad [5.12]$$

However, if the relative humidity and temperature are not constant, the value of k will vary slightly each time the test is run giving rise to inaccuracy in converting from units of resistance to permeability.

A modified version of the Control Dish method has been adopted as BS7209 (1990) as a quality control procedure. This has been achieved by retaining the dishes and turntable and the controlled environment (20 °C, 65% RH), but with the cover fabric removed and simply measuring the weight loss in unit time through a sample. The weight loss through the fabric sample is then compared to the weight loss through an apertured control fabric. Results can then be expressed as a percentage of the control fabric and are given three grades (i) better than 80%, (ii) between 50 and 80% and (iii) below 50%.

Farnworth and Dolhan's DND method sandwiches a sample between two PTFE membranes separating it from a dish of water on one side and a stream of

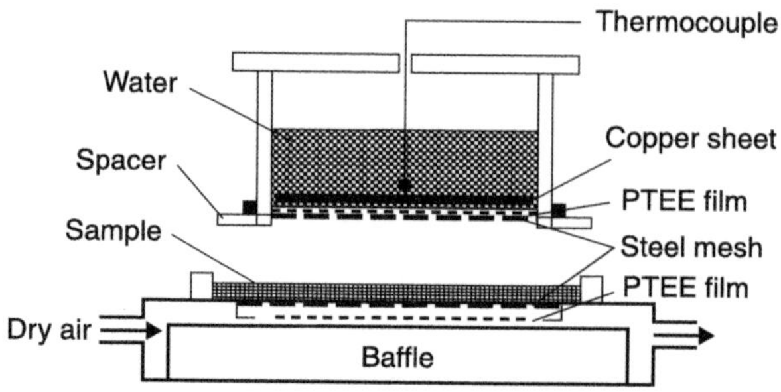

5.4 DND apparatus.

dry air on the other as shown in Fig. 5.4. The resistance of the sandwich is calculated from the water mass transfer from which the resistance of the PTFE is subtracted (Farnworth and Dohlan 1984).

The apparatus described by Van Beest and Wittgen (1986) is similar in principle to the DND apparatus but is based upon a volume change of water permeating through a sample. The test specimen is placed either on a perforated sintered disc connected to a water reservoir and a graduated capillary tube or on spacers above the sintered disc. Dry air is passed over the sample and the volume of water permeating the specimen noted as shown in Fig. 5.5. By comparison to a controlled air gap a resistance to water vapour migration is determined.

The controlled atmosphere for these tests is normally air at prescribed values of temperature and humidity with no temperature gradient across the sample.

ASTM E96 has many modes of operation including both upright and inverted water methods and desiccant methods. Each allows different test conditions and air gaps between the sample and water surface. There are many variants of the methods to measure WVP, ranging from the upright MOD test (SCRDE 1986) and the Gore Cup (Bekleidung Physiologisches Institut EV Hohenstein (BPI 1.4 1986). The former places the test dishes on a hot plate at 33 °C to simulate skin temperature and produce a temperature gradient across the sample. The Gore Cup

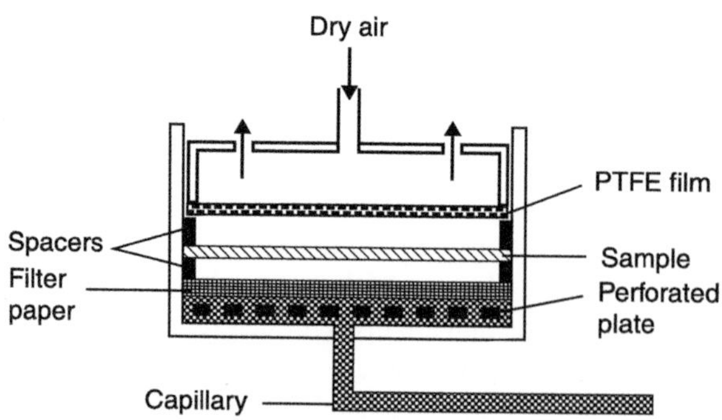

5.5 Van Beest and Wittgen apparatus.

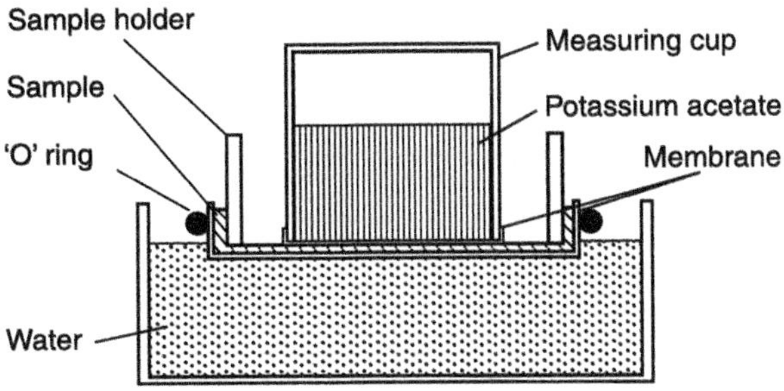

5.6 Gore Cup method.

is an inverted desiccant method but with saturated potassium acetate, at ~19% RH, in the dish. This is then placed in a water bath, as shown in Fig. 5.6, at ~26 °C for 20 to 30 minutes. Slight modification to the test procedure in terms of temperature (23 °C) and humidity (23%) with two layers of PTFE membrane sandwiching the sample fabric is described in JIS L1099 method B2 (JIS 1999).

However, all the variants use different water vapour pressure gradients across the fabric sample and therefore give different values for the same fabric. Differing methods and conditions will give rise to considerable variations in the measured WVP for a sample and may even change the ranking order of a set of fabrics. For example a 2-ply PTFE laminate measured by its manufacturer with its in-house method may give a value of 20000 g m^{-2} day^{-1} whilst on another test only 5000 g m^{-2} day^{-1} (Williams 1994). Both values are correct, the difference being due solely to a change in the test conditions. Comparisons made using literature values are very prone to error and it is therefore useful to specify the exact method and the actual test conditions used for any reported values along with the air gap if appropriate.

A more recent development in the evaluation of water vapour resistance is the dynamic moisture permeation cell (DMPC) (Gibson *et al.* 1997) which has been adopted by ASTM F2298 (2003) (Fig. 5.7). The apparatus allows the humidity either side of the inverted fabric under test to be controlled by means of dry or saturated nitrogen gas, typically 95% RH above the sample and 5% RH below the

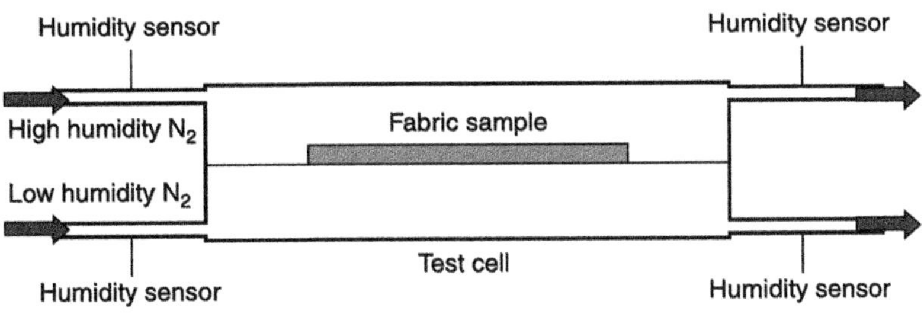

5.7 DMPC apparatus (ASTM F2298, 2003).

sample. Knowing the relative humidities of the entering and exiting gas flows along with the temperatures and flow rates, the water vapour diffusion resistance is determined by

$$R_{dtot} = \frac{A(\Delta\varphi)}{Q(\delta\varphi)} = \frac{\Delta C}{m} \qquad [5.13]$$

where,

R_{dtot} = total water vapour diffusion resistance (sm^{-1})
A = area of sample (m^2)
$\Delta\varphi$ = relative humidity difference between top and bottom incoming gas in decimal format
Q = volumetric flow rate (m^3s^{-1}) at temperature T
$\delta\varphi$ = relative humidity difference between incoming and outgoing gas at bottom of cell in decimal format
m = mass flux of water vapour across sample ($kg\ m^{-2}s^{-1}$)
ΔC = log mean concentration difference between top and bottom gas streams ($kg\ m^{-3}$).

The volumetric flow rate is

$$Q = Q_s(T_a/T_s)(P_s/P_a) \approx Q_s(T_a/T_s) \qquad [5.14]$$

where,

Q = actual volumetric flow rate at bottom outlet (m^3s^{-1})
Q_s = indicated volumetric flow rate (m^3s^{-1})
T_a = ambient temperature (K)
T_s reference temperature used by mass flow meter (K)
P_a = ambient pressure of gas flow (Pa)
P_s = reference pressure used by flow meter (Pa).

The water vapour flux can then be calculated

$$M = \frac{\delta\varphi P_{sat}.M_w.Q_s}{ART_s} \qquad [5.15]$$

where,

m = mass flux of water vapour across sample ($kg\ m^{-2}s^{-1}$)
A = area of sample (m^2)
$\delta\varphi$ = relative humidity difference between incoming and outgoing gas at bottom of cell in decimal format
P_{sat} = water vapour saturation vapour pressure at test temperature (Nm^{-2})
M_w = molecular weight of water vapour (18.015 kg $mole^{-1}$)
Q_s = indicated volumetric flow rate (m^3s^{-1})
R = universal gas constant (8314.5 N m kg^{-1} K^{-1})
T_s = reference temperature used by mass flow meter (K).

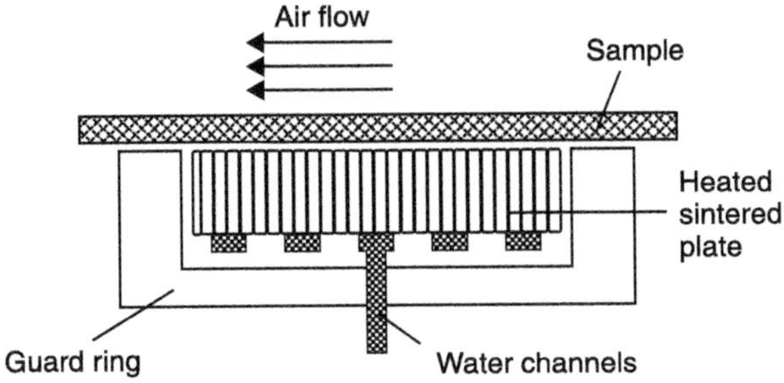

5.8 Sweating guarded hot plate.

A test developed at the German Hohenstein Institute which simulates the thermoregulation in human skin – the Skin Model or Sweating Hot Plate, determines the water vapour resistance of fabrics (Umbach and Mecheels 1976, CEN/pr EN340 1993). The Hohenstein Skin Model is represented by a 20 cm × 20 cm porous plate of sintered stainless steel and by means of electrical heating is brought to the temperature of human skin (35 °C) and is completely covered by the textile sample to be tested. A guard ring, heated to the same temperature as the porous metal plate, prevents heat loss at the sides and underneath, so that the amount of heat given up by the 'measuring head' can only flow through the test sample, shown in Fig. 5.8. Underneath the metal plate, the head contains engraved channels along which distilled water is supplied, which vaporises through the numerous pores of the plate. In this way, sweating human skin is said to be simulated. The Skin Model is contained in a climatic chamber in which air of defined temperature, moisture content and velocity flows over the textile sample.

A sweating guarded hot plate or one of the other water vapour tests can be used to determine the vapour resistance of a fabric by:

$$R_{et} = \frac{p_{\text{plate}} - p_a}{H - H_{DRY}} - R_{et0} \qquad [5.16]$$

where,

R_{et} = vapour resistance of fabric sample (m^2PaW^{-1})
p_{plate} = mean hot plate surface vapour pressure (Pa)
p_a = ambient vapour pressure (Pa)
H = total heat loss per square metre of wet plate area (Wm^{-2})
R_{et0} = vapour resistance measured without sample (m^2PaW^{-1})
H_{DRY} = dry plate heat loss at t_a (Wm^{-2}), 0 if $t_a = t_{\text{plate}}$.

In general, the lower the vapour resistance the better, but as heat resistance increases the thickness usually also increases and so will the vapour resistance. To be more meaningful a comparison of the ration between heat and vapour resistance is often expressed by the water vapour permeability index:

$$I_{mt} = S\,(R_{ct}/R_{et}) \qquad [5.17]$$

where, $S = 60$ Pa K^{-1}.

This number varies between zero (totally impermeable) and one (air).

With some water vapour permeable membranes, i.e. solid film hydrophilics, moisture transfer can be affected by fabric orientation (Williams 1997), the humidity at the membrane and by the local temperature.

To determine the heat and moisture transport properties of textiles a differentiation has to be made between stationary and non-stationary wear situations such as occur in practical use. In the stationary normal wear situation there is a flow of heat and moisture from the body to the surroundings which is constant with time. When the climate or level of activity are moderate then the moisture flow is entirely due to insensible perspiration and therefore occurs in the form of moisture vapour.

Non-stationary wear situations are characterised by intermittent pulses of moderate or heavy sweating by the wearer of the garment. This can be caused by moderate to strenuous body activity. In order to give good wear comfort under these conditions, the textiles have to have not only good moisture transport properties but also a good moisture buffer action.

In the stationary measurements, which cover normal wear situations, the water vapour permeation resistance R_{et} of a textile is determined with an air temperature in the chamber set at 35 °C and with a relative humidity of 40%. The Skin Model is supplied with distilled water via the channels in the measuring head with the plate itself covered with a Cellophane® film (thickness ~20 εm). This film is permeable to water vapour but holds back liquid water and thus prevents the sample from becoming wet. The vaporisation heat flow from the measuring head through the test specimen is a measure of its water vapour permeation resistance. A blank value is also subtracted in this case so that any resistance effects on the measuring apparatus are eliminated.

$$R_{eT} = R_{et^1} + R_{et^2} + R_{et^i} \qquad [5.18]$$

For a multilayer system, the resistances of the individual layers can be summed where, R_{eT} = water vapour resistance of the clothing ensemble, and $R_{et}^{\,i}$ = water vapour resistance of the ith textile layer.

A simpler version of the Hohenstein Skin Model, the Permetest (Fig. 5.9), has been developed by Hes and co-workers (Hes 1993, 2002, 2006; Hes and Bajzik 1993; Hes and Dolezal 2003). Measurements are expressed in units R_{et} and R_{ct} as defined in ISO 11092 (1993). A slightly curved porous surface is moistened and exposed in a wind channel to parallel air flow of adjustable velocity. The sample to be tested is located on the wetted area with a diameter of about 80 mm and the amount of evaporation heat taken away from the active porous surface is measured. The instrument body can be heated above room temperature or kept at room temperature to maintain isothermal working conditions.

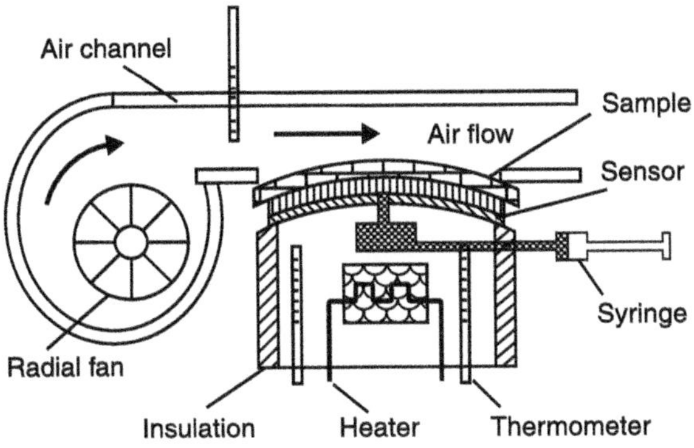

5.9 Permetest apparatus.

The measuring head is first covered by a semi-permeable film, to keep the measured garment dry. Then, heat flow value q_0 without a sample is registered. In the next step, the full-size garment with thickness up to 5 mm (without being cut to special shape) is inserted between the head and the orifice in the bottom of the channel. When the signal is steady, the level of q_s, which quantifies heat losses of wet measuring head covered by a sample, is recorded. Both values then serve for automated calculation of mean value and variation of water vapour resistance R_{et} and thermal resistance R_{ct}, along with the average values of these parameters and their variation coefficients.

The term 'absolute water vapour permeability' means the evaporation resistance defined in ISO 11092. The relative water vapour permeability P% is given by the 100 multiple of the ratio of heat flow level q_s measured with the inserted sample divided by the heat flow level q_0 passing from the free measuring surface. The 100% relative water vapour permeability P then corresponds to human skin. The short measurement time allows the instrument to be used for the determination of the heat of moisture absorption generated in fabrics and for the measurement of thermophysiological properties of fabrics in wet state (Hes and Dolezal 2007).

One of the main drawbacks with the determination of water vapour transport properties is that different methods tend to use differing test conditions and express results in different units making direct comparison between fabrics difficult as seen in Table 5.2. Furthermore ranking of materials by the different methods also show differences especially when fabrics containing semi permeable membranes are evaluated (McCullough *et al.* 2003; Williams 1994, 1997).

5.5 Air permeability

A measure of the windproof properties of a textile is its resistance to penetration by air, that is its air permeability, and determines the extent of the air exchange, its related moisture content and thermal insulation, between the microclimate of a

Table 5.2 Water vapour transmission comparison

Fabric	Control dish (mm still air)	BS7209 (%)	Gore cup (g m^{-2} day^{-1})	Skin model (m^2 mbar W^{-1})	Permetest (%)
Knitted cotton	0.9	96	19500	0.0415	22.48
Woven cotton	0.9	96	20000	0.0343	17.17
Woven polyester cotton	1	95	18350	0.0386	20.14
Nonwoven insulation	1.4	94	14100	0.0527	16.83
3-ply PTFE hybrid	3.3	85	7300	0.0789	10.46
3-ply Microporous PU	6.4	74	4400	0.1383	8.65
2-ply Impermeable	267	1	250	2.818	5.52

garment and the environment (Larose 1974). When wind speed increases, a fabric's heat and vapour resistance may decrease and in general, the lower the air permeability, the more windproof the materials is said to be.

In calculating the air permeability of materials at given wind speeds, the method given by Douglas (1975) is often used. The method is analogous to the impact of a fluid on a flat plate using:

$$P = r\,a\,v \qquad [5.19]$$

where, P = force exerted on plate (Pa)
r = mass density of air
v = wind velocity in ms^{-1}.

The pressure drop across the plate increases with increasing wind velocity.

5.5.1 Air permeability measurement

The two principal methods for determining the air permeability of fabrics are both related to the air flow through a known area of the sample and the pressure drop across the sample. In one case, the pressure drop is constant and the flow rate measured, and in the other the flow rate is constant and the pressure drop measured. The most commonly used method to determine a fabric's air permeability is by EN ISO 9237 (1995).

Standard tests have traditionally used flat samples but Keighley *et al.* (1989) and Watanabe *et al.* (1991) looked at the air flow characteristics around cylindrical cells to simulate the shape of limbs or torso. Much of this latter work has considered the reduction of thermal insulation by air penetration into air permeable clothing in windy conditions. In such conditions air pressure is exerted on the outer surface of a clothing system and due to the geometry, pressure varies over the surface such that in air permeable clothing air penetrates where the outside air pressure is higher than that inside the clothing. The air migrates around the body to areas of lower pressure and then diffuses back into the environment where the air pressure outside is lower than in the clothing (Fig. 5.10). Since the temperature

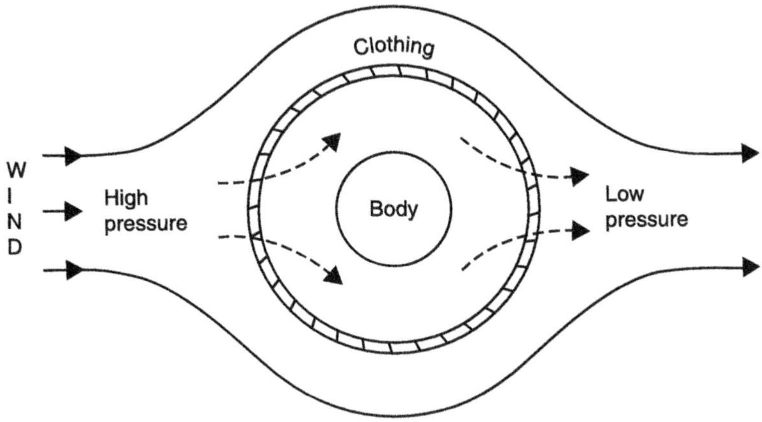

5.10 Wind penetration of air permeable clothing.

of the air passing from the clothing system to the environment is usually higher than that entering the system, the air movements remove heat and hence the body will be cooled. The most common method of reducing such effects is by the use of low air permeable outer fabrics, but with the advent of breathable windproof materials many of the problems of windchill can be alleviated. As the system becomes less air permeable, then air exchange is dependent on closures and ventilation of the garment.

However, opinion is divided on whether or not good air permeability is desirable in protective clothing. In temperate conditions, it is likely that high air permeability results in better management of physiological stresses in conditions of strenuous exercise. The microclimate between the various layers of clothing will be more homogenous if air movement is relatively unrestricted and locally high amounts of perspiration can be more easily dispersed. The comfort offered by a garment will then be, to some extent, related to its air permeability. The beneficial effects of venting in warm environments have been shown by several workers (Gooderson 1991; Danielsson 1992; Umbach 1981; Vokac *et al.* 1973; Lamb and Yoneda 1990).

For materials containing a low air permeability polymer film, the measurement of air permeability readily detects defects in the film as observed by a marked increase in permeability. It is therefore important to screen candidate materials by such a test to assess imperfections in any film.

5.6 Wicking, buffering and absorbency

Liquid sweat on the surface of the skin often gives rise to a feeling of discomfort. Umbach and Mecheels (1976) have suggested that the average sweat wetted area of skin for the average male should be less than 30% to remain comfortable. It follows that clothing next to the skin should aid removal of sweat from the skin by either evaporation or wicking.

5.6.1 Wicking

Wicking can take place either around the body to allow a greater area for evaporation or through the fabric away from the skin. If the clothing becomes saturated from exertion then the wicking ability is limited. The wicking/buffering effect has been measured by Mecheels and Umbach (1976) on the sweating guarded hot plate by allowing liquid water to be in contact with the fabric sample and looking at overall moisture loss or alternatively measuring the microclimate response to a short sweating burst.

The simplest approach is to measure the height of liquid wicking into vertically hanging strips of fabric after fixed time periods (BS 3424 1996; DIN 53924), or by visually looking at dispersion of a drop of liquid on a fabric. One end of a strip (25 mm wide × 170 mm long) is clamped vertically with the dangling end immersed to about 3 mm in distilled water at 21 °C and can be coloured for non white fabrics to aid observation. The height to which the water is transported up the strip is measured at typically 1, 5 and 10 minute intervals and reported in centimetres. Higher wicking values show greater liquid water transport ability.

More objective methods rely on electrical conductivity changes in wet fabrics to define the water absorption speed (Van Langenhove and Kiekens 2001) and Williams and Davies (2009) looked at liquid movement in three dimensional fabrics.

5.6.2 Absorbency

The two most commonly used parameters to characterize the properties of absorbent materials are the rate of absorbency and the total absorbent capacity. The former determines the rapidity with which fluid is imbibed while the latter determines the total capacity of the material to absorb and retain fluid. The Gravimetric Absorbency Testing System (GATS) is shown schematically in Fig. 5.11. Fluid is absorbed radially outward along the plane of the sample (62 cm^2) from a single point in the bottom of the test plate and a plot of the amount of fluid absorbed as a function of time is produced, Fig. 5.12.

A test specimen is positioned on the sample plate which has a 2 mm diameter hole at the centre connected to a fluid reservoir. The level of the cell is adjusted to give zero hydrostatic head guaranteeing that absorbency takes place on demand. A solenoid valve supplies fluid from the fluid reservoir equal to the amount the specimen can absorb. A fluid sensor automatically weighs the amount of water supplied, maximum absorbent capacity, V. This resultant absorbency rate is measured in the range between 20% and 80% of the maximum absorbent capacity. From the output, specific maximum absorbent capacity, C, and flow rate, Q_o, are determined in the unit of grams fluid/minute.

$$\text{Specific absorbent capacity } C = V/W \text{ (grams of fluid/grams dry fabric weight)} \quad [5.20]$$

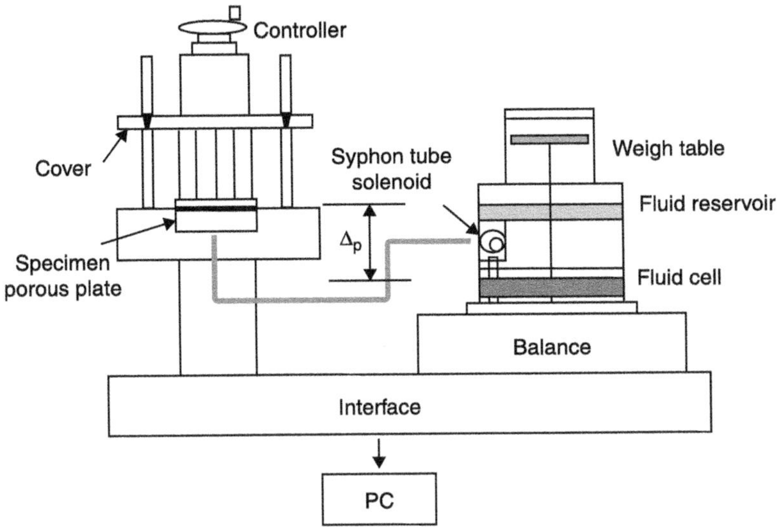

5.11 GATS apparatus.

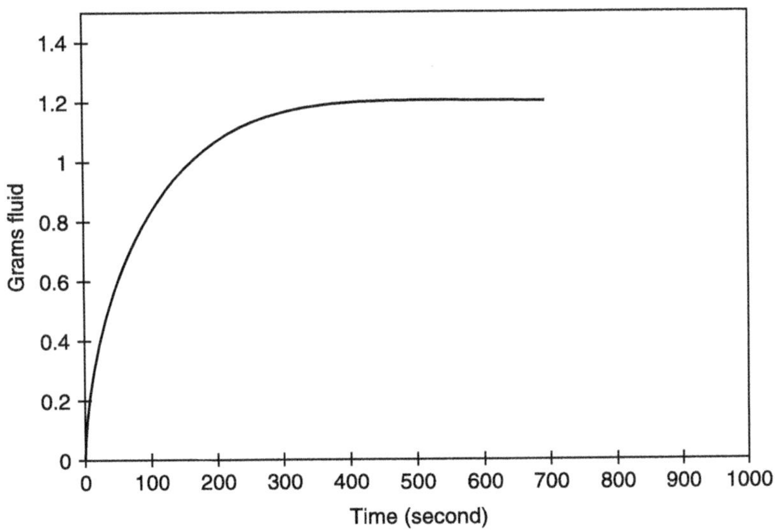

5.12 GATS print out.

where,

V = maximum absorbent capacity, total amount of fluid absorbed in grams for a given sample area of 62 cm^2

W = weight of the dry fabric (grams)

Flow rate, Q_o = V/T (grams fluid/min) (where T = Time (seconds)) [5.21]

Specific Flow Rate $Q = Q_o$/W (grams fluid/grams dry fabric·min). [5.22]

AATCC 195(2009) is a test method for the measurement, evaluation and classification of liquid moisture management properties of textile fabrics in a moisture management tester (MMT). It assesses spreading outwards on the inner surface of a fabric normally worn next to the skin, transfer through the fabric from the inside to the outer surface and spreading outward on the outer surface and evaporating. The method measures wetting time, absorption rate, maximum wetted radius and spreading speed on both surfaces to provide an accumulative one way transport index and an overall moisture management capability (OMMC). The results obtained with this test method are based on water resistance, water repellency and water absorption characteristics of the fabric structure, including the fabric's geometric and internal structure and the wicking characteristics of its fibres and yarns. Fabrics can then be classified as suitable for a variety of end uses including waterproofs, quick drying or moisture management fabrics.

5.7 New developments and future trends

Davies and Williams (2009) have developed an electronic method to map the spreading of liquids in three dimensional structures, for example in spacer fabrics.

The method is based on conductive sensors similar to the ATS600 absorbency testing system positioning sensor pins at set distances away from a liquid source but setting them at specific heights using spacer plates also allows fabrics to be assessed through their thickness. Taking distance measurements every 5 seconds (5–50s) through the fabric in horizontal and vertical directions give profiles similar to Fig. 5.13 where the direction and rate of flow can be seen in three dimensions.

As electronic measurement systems get smaller and more refined along with increased computational power, devices for laboratory assessment will become more sophisticated both for differentiated sweating manikins but also fabric testers to simulate conditions closer to real life situations. With new developments in smart materials and clothing, methods will also need to be established to assess their functionality in variable climatic conditions and with intermittent work loads. Whereas small differences in fabric properties can be measured for quality control

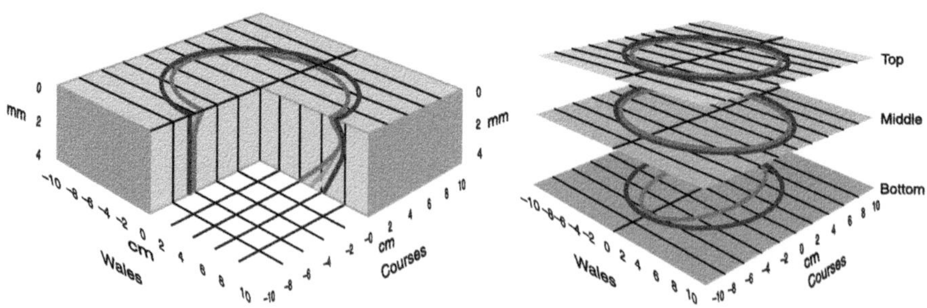

5.13 3D liquid spreading.

purposes, these may not be seen on the wearer as size, degree of fit, fitness, acclimatisation, gender, etc., can all mask such differences. As with performance swimmers and elite athletes any improvement in speed is an advantage and the same is true for thermo-physiological comfort, being able to perform at optimum temperature or survive minutes longer gives the wearer an advantage. It follows that these small differences need to be measured by whatever means available.

5.8 References

AATCC Test Method 195 (2009), Liquid Moisture Management Properties of Textile Fabrics.

ASTM E96-05 (2005), Standard test methods for water vapour transmission of materials, American Society for Testing and Materials, West Conshohocken, PA.

ASTM-F1291-80, Standard Method of Measuring the Thermal Insulation of Clothing using a Heated Thermal Manikin, American Society for Testing and Materials, West Conshohocken, PA.

ASTM F2298 (2003), standard test methods for water vapour diffusion resistance and air flow resistance of clothing materials using the dynamic moisture permeation cell, American Society for Testing and Materials, West Conshohocken, PA.

BPI 1.4 (1986), testing of textiles – determination of stationary water vapour resistance by means of the cup method.

BS 3424-18 (1986), Testing coated fabrics. Methods 21A and 21B. Methods for determination of resistance to wicking and lateral leakage, British Standards Institute, London.

BS 4745 (2005), Determination of Thermal Resistance of Textiles, British Standards Institute, London.

BS 7209 (1990), Water Vapour Permeable Apparel Fabrics, British Standards Institute, London.

CEN/pr EN340, (1993), Measurement of Stationary Thermal and Water Vapour Resistances by Means of a Thermoregulatory Model of Human Skin.

CGSB method 49 (Can 2-4.2-M77) (1977), Resistance of materials to water vapour diffusion (control dish method), Canadian General Standards Board.

Danielsson U (1992), 'Convection in clothing air layer', *Proceedings of the 5th International Conference on Environmental Ergonomics*, Maastricht, November, 70–71.

Davies AM and Williams JT (2009), 'Use of spacer fabrics for absorbent medical applications', *Journal of Fiber Bioengineering and Informatics*, **1**(4): 321–30.

Dimitrieva IA (1958), *Research Works of Central Research Institute of Silk*. Moscow, Rostehizdat,

DIN 53924 (1997) -03 Prüfung von Textilien – Bestimmung der Sauggeschwindigkeit von textilen Flächengebilden gegenüber Wasser (Steighöhenverfahren), German Standardisation Institute, Berlin.

Dolhan PA (1987), 'Comparison of apparatus used to measure water vapour resistance', *J. Coated Fabrics*, **17**: 96–109.

Douglas JF (1975), *Solution of Problems in Fluid Mechanics, Part 1*, Pittman Publishing Co Ltd, London, p. 127.

EN ISO 9237 (1995), Textiles. Determination of the Permeability of Fabrics to Air, International Organisation for Standardisation, Geneva.

Fan J and Qian X (2004), 'New functions and applications of Walter, the sweating fabric manikin', *European Journal of Applied Physiology*, **92**: 641–4.

Fanger PO (1970), *Thermal Comfort*, Danish Technical Press, Copenhagen.

Farnworth B and Dohlan P (1984), 'Apparatus to measure the water vapour resistance of textiles', *Journal of Textile Institute*, **2**: 142–5.

Fourt L and Harris M (1947), 'Diffusion of water vapour through textiles', *Textile Research Journal*, **17**, 256–63.

Gibson PW, Kendrick CE, Rivin D and Charmchi M (1997), 'An automated dynamic water vapour permeation test method', in *Performance of Protective Clothing* (ASTM STP1273 vol 6), eds J O'Stull and AD Schwope, pp. 93–107, ASTM, West Conshohocken, PA, USA.

Glasstone S (1960), Textbook of Physical Chemistry, Macmillan, London.

Gooderson C Y (1991), Physiological Evaluation of NBC Clothing Ensembles, 4th International Symposium on Military Protective Clothing Systems, Munich, June.

Greenwood K, Rees WH and Lord J (1970), Problems of Protection and Comfort in Modern Apparel Fabrics. Studies in Modern Fabrics: Papers of the Diamond Jubilee Conference of the Textile Institute, Manchester, pp. 197–218.

Hes L (1993), Water Vapour Permeability of Wool Blended Fabrics, International Conference on Textile Science, Liberec.

Hes L (2002), The Effect of Planar Conduction of Moisture on Measured Water Vapour Permeability of Thin Woven Fabrics, Fall Fiber Society Conference, Lake Tahoe.

Hes L (2006), Alternative Methods of Determination of Water Vapour Resistance of Fabrics by Means of a Skin Model. European Conference on Protective Clothing, Gdynia.

Hes L and Bajzik V (1993), International Conference on Textile Science, Liberec.

Hes L and Dolezal I (2003), A New Portable Computer-Controlled Skin Model for Fast Determination of Water Vapour and Thermal Resistance of Fabrics. Asian Textile Conference (ATC 7), New Delhi.

Hes L and Dolezal I (2007), 'Precise measurement of water vapour permeability of wet fabrics', *Proc. of the AUTEX International Textile Conference, Tampere, Finland.*

Holmer I (2009), 'Human wear trials for cold weather protective clothing systems', in *Textiles for Cold Weather Protective Apparel*, ed. JT Williams, Cambridge, Woodhead Publishing.

ISO 5085-1 (1989), Textiles – Determination of Thermal Resistance – Part 1: Low Thermal Resistance

ISO 11092 (1993), Textiles – Physiological effects – Measurement of Thermal and Water-vapour Resistance Under Steady-state Conditions (Sweating Guarded-hotplate Test).

JIS (1999), Method B2, Testing Methods for the Water Permeability of Clothes.

Keighley JH and Fan J (1989), 'The design of effective clothing for use in windy conditions', *International Journal of Clothing Science and Technology*, **1**(2): 28–32.

Lamb GER and Yoneda M (1990) 'Heat loss from a ventilated clothed body', *Textile Research Journal*, July, 378–83.

Larose P (1947), 'The effect of wind on the thermal resistance of clothing with special reference to the protection given by coverall fabrics of various permeabilities', *Canadian Journal of Research*, **25**: 169–90.

McCullough EA, Huang J and Kim CS (2004), 'An explanation and comparison of sweating hot plate standards', *Journal of ASTM International*, **1**(7), 13 pp.

McCullough EA, Kwon M and Shim H (2003), 'A comparison of standard methods for measuring water vapour permeability of fabrics', *Meas. Sci. Technol.* **14**(8): 1402–8.

Meinander H (1995), 'Heat stress in sportswear, assessment of heat and moisture transmission', *Proceedings of International Conference of Textiles in Sports and Sportswear, Huddersfield*, pp. 46–52.

Richards M and Mattle N (2001), Development of a Sweating Agile Thermal Manikin (SAM), 4th International Conference on Thermal Manikins, Switzerland.

Runbai W and Hang Y (1991), 'The measures of clothing insulation using a heated manikin', *Proceedings of the 11th Congress of the International Ergonomics Association, Paris*, pp. 1146–8.

SCRDE (1986), Development of a New Moisture Vapour Transmission Test, 10DL-1-13, Colchester, UK.

Tanner JC (1979), 'Breathability comfort and GORE-TEX laminates', *Journal of Coated Fabrics*, **8**: 312–22.

Umbach KH (1981), Improvement in the Wearer Comfort of Men's Outer Clothing by Optimised Ventilation in Clothing, Hohenstein Research Report AIF-Nr3937, January.

Umbach KH (1988), 'Physiological tests and evaluation models for the optimisation of performance of protective clothing', in: I.B. Mekjavik (ed), *Environmental Ergonomics*, Taylor and Francis, pp. 139–61.

Umbach KH and Mecheels J (1976), 'Thermophysiologische Eigenschaften von Kleidungssystemen (Thermophysiological properties of clothing systems)' *Melliand Textilberichte*, **57**(12): 1029–32 and **58**: 73–81.

van Beest CA and Wittgen PPMM (1986), 'A simple apparatus to measure water vapour resistance of textiles', *Textile Research Journal*, **57**(9): 566–8.

Van Langenhove L and Kiekens P (2001), 'Textiles and the transport of moisture', *Textile Asia* 32–43.

Vokac Z, Kopke V and Keul P (1973), 'Assessment and analysis of the bellows ventilation of clothing', *Textile Research Journal*, August, pp. 474–82.

Watanabe T, Kato T and Kamata Y (1991), 'The velocity distribution in the inner flow field around a clothed cylinder', *Sen I Gakkaishi* **47**(6): 271–5.

Weiner LI (1971), 3rd Shirley International Seminar, Manchester, England.

Whelan ME, Machattre LE, Goodings AC and Turl LH (1955), 'The diffusion of water vapour through laminate with particular reference to textile fabrics', *Textile Research Journal*, **25**: 197–223.

Williams JT (1994), Physiological considerations of NBC protective clothing, PhD Thesis, Cranfield University, UK.

Williams JT (1997), *A Comparison of Techniques Used to Assess the Thermal Burden of Protective Clothing, Performance of Protective Clothing*, 6th Volume, ASTM STP1273, pp. 303–13.

Williams JT and Davies AM (2009), 'Use of spacer fabrics for absorbent medical applications', *JFBI*, **1**(4): 321–30.

6
Testing, analyzing and predicting the comfort properties of textiles

F. S. KILINC-BALCI, Auburn University, USA

Abstract: The multiplicity and complexity of factors influencing the comfort phenomenon has resulted in numerous studies dealing with different comfort-related aspects. This chapter provides an overview of the different approaches used for characterizing thermophysiological and neurophysiological comfort attributes. The chapter also highlights a recently developed design-oriented comfort model that integrates comfort related factors into a single index.

Key words: thermal comfort, tactile comfort, handle, stiffness, roughness, softness, testing, design-oriented comfort.

6.1 Introduction

Understanding and meeting consumers' needs and expectations towards products have become essential for the long-term survival of any enterprise in the competitive textile and apparel market. Nowadays, consumers demand clothing which not only looks good but also feels comfortable. Consumers choose comfort as the most important attribute that they seek in apparels, followed by easy care and durability. On the other hand, rapidly increasing development of information technology and the internet has made a huge impact on consumer shopping habits. In consequence of inevitable increase in e-shopping, consumers increasingly demand more objective characterization of fabric and clothing comfort. Thus, there is a growing need for characterization tools which can substitute for the physical feel and touch of traditional intimate shopping. Although comfort has been one of the key attributes of consumers' desirability of apparel products in all markets, there is neither an objective way to characterize comfort reflecting the overall human perception nor standard procedures to produce comfort by design rather than by guesswork.

The word comfort refers to how the individual feels. Under the same environmental conditions and with the same type of clothing, one person may feel comfortable and another may not. Besides, those who feel comfortable may not be in the equal comfort level. This is because of the complex interaction between all of the parameters affecting the comfort of humans, e.g., clothing parameters, environmental parameters, physiological factors, psychological factors, etc. These factors typically interact in a very complex manner. Furthermore, there is a continuous change over time that leads to transitional effects.

The multiplicity and complexity of factors influencing the comfort phenomenon has resulted in numerous studies dealing with different comfort-related aspects.

Many of these factors and impacts on the comfort status of humans have been addressed in various independent research efforts. However, what is missing in the literature is an integrated effort in the form of comprehensive modeling of the interactive nature of comfort factors supported by psychological effects.

In this chapter, different approaches used for characterizing thermophysiological and neurophysiological comfort attributes are discussed. Furthermore, a design-oriented the comfort model that integrates comfort related factors into a single index is described.

6.2 Characterization of comfort

Since Peirce's well-known study in 1930,[1] there have been numerous studies conducted with many experimental and analytical approaches, each providing an insight into the nature of the comfort phenomenon. However, a complete evaluation of the comfort phenomenon requires a substantial multidisciplinary approach.

Kilinc et al.[3,28] groups comfort analysis into three main categories:

1. Objective analysis, in which quantitative measures characterizing comfort can be determined (tactile and thermal parameters).
2. Subjective analysis, in which psychological evaluation is made by surveys, ratings and scales.
3. Correspondence analysis, in which the subjective and objective analyses are combined to develop quantitative measures.

Although comfort or discomfort is a well-realized mental status by humans there is no objective output parameter that can fully describe this realization. There are hundreds of parameters, each emphasizing one comfort-related aspect; however none of these parameters truly reflects the whole comfort or discomfort realization. Since humans are different in their perceptions of comfort, relying totally on subjective evaluation will not be adequate. Besides, as indicated earlier, comfort-related factors typically interact in a nonlinear fashion and that makes traditional linear, discrete, or by-polar physiological scaling automatically deficient. Moreover, most of the subjective analyses rely on descriptors developed by few experts. This makes comfort analysis a form of multiple choice questionnaire rather than a collective analysis leading to an integrated index of comfort.[3] Furthermore, while the initial feel of comfort or hand mostly depends on the initial perception, a reliable overall comfort evaluation may require a significant amount of time and experience with the clothing. If these factors are not fully considered, any comfort study will be limited by constraints such as people's attitude, familiarity with the clothing system being tested, psychological factors, prejudices, quality assumptions, and stereotypes. These constraints can mask the particular factors determining the comfort status of a particular clothing system and lead to misleading results.

6.3 Testing, analyzing and predicting neurophysiological comfort

Neurophysiological (tactile or sensory) comfort refers to the feel of fabric against the skin or interaction of the clothing with the skin. This feel is triggered by sensory receptors in the skin, which are connected to the brain by a network of nerve fibers. Overall comfort perception is a result of very complicated multi-step process, clothing, environment, body, and past experiences are some of the key factors affecting this complex process.

In regards to fabric and body, the skin/fabric interaction is one of the most important factors which impact the overall comfort status. The skin/fabric interaction is stimulated by many mechanically-related (bending rigidity, surface roughness, etc.) and thermally-related (warm/cool sensation, moisture transfer, etc.) factors. The fabric/skin contact area changes over time during wear and as indicated earlier, in Chapter 4, different parts of the body have different sensitivities. These changes result in the generation of new mechanical stimuli. Moreover, due to the frequent changes of the human physiological state, skin temperature and humidity at the skin surface change very frequently and these changes also generate new thermal stimuli. Besides, new mechanical stimuli are generated as the body moves. All these mechanical and thermal stimuli trigger responses from sensory receptors lead to the formulation of various tactile and thermal related perceptions. This process defines the comfort status of an individual. Many research studies have been carried out to analyze various neurophysiological aspects of comfort, particularly hand-related factors. Perhaps the most notable outcome of the neurophysiological comfort studies are the numerous systems developed for testing fabric tactile properties (e.g., bending, drape, handle, friction, etc.). Peirce pointed out three basic determinants of fabric handle: bending stiffness, bulk compressibility, and surface friction.[1] Fabric drape was added as another component of fabric hand by later studies. Other parameters such as shear, crease recovery, and fabric thickness have also been considered as determinants of fabric handle. The Shirley Bending Tester, Cusick Drape Tester, Handle-o-Meter, and laser hairiness devices are a few examples of those systems which were developed to measure fabric tactile properties. Many standards were also established for evaluation of fabric tactile properties such as the American Society for Testing and Materials (ASTM) D1388 'Standard Test Method for Stiffness of Fabrics', ASTM D6828 'Standard Test Method for Stiffness of Fabric by Blade/Slot Procedure', and ASTM D1894 'Standard Test Method for Static and Kinetic Coefficients of Friction of Plastic Film and Sheeting'.

In the following subsections, studies conducted in the area of fabric stiffness, fabric softness, fabric roughness, and fabric handle are summarized. In addition, some of the predictive models using the sensory data are briefly reviewed.

6.3.1 Fabric stiffness

Fabric stiffness is one of the key parameters determining fabric tactile properties. A comprehensive review on the objectively measurable physical properties associated with stiffness was presented by Bishop in 1996.[4] Studies in the area of theoretical modeling of the bending stiffness can be categorized in three groups: descriptive modeling, numerical modeling, and predictive modeling. Research conducted in this area was also extensively reviewed by Ghosh et al. in 1990.[74–76]

Elder et al.[5] studied stiffness using different objective measurement techniques. In their first study, they compared the objective measurement results obtained by the Shirley Cantilever, the Cusick Drapemeter and the Shirley Cyclic Bending Tester with the subjective assessment results using woven and non-woven fabrics. As a result of this study, they found a logarithmic–linear relationship between the subjective stiffness estimation and the flexural rigidity. Later, they also verified this conclusion by using different woven fabrics and knitted fabrics.[6] They reported that three objective measurements, bending length, flexural rigidity and drape coefficient, were related to each other and also highly correlated with the subjective stiffness perception. In another study, Elder et al.[7] found a significant relationship between the subjective stiffness estimation and the drape coefficient. In this study, the authors selected the drape coefficient obtained by the Cusick Drapemeter as the objective measure of stiffness instead of the flexural rigidity measured by the cantilever method since cantilever measurements were directional and have greater variability. They concluded that the drape coefficient obtained by the Cusick Drapemeter gives an integrated measure of stiffness as good as human handle judgment.

Several studies have shown significant correlation between fabric stiffness and various fabric properties. Hu et al.[9] found a correlation between fabric stiffness and a number of objectively measured parameters obtained using Kawabata systems, including coefficient of friction, linearity of compression thickness curve, bending rigidity, and energy in compression fabric under 5 kPa. Li[8] also showed a high correlation between subjective ratings of garment stiffness and fiber diameter, tensile breaking load, fabric compression properties such as thickness at low and high pressure and fabric frictional properties.

6.3.2 Fabric softness

Fabric softness is another important characteristic determining the tactile property of a fabric. Peirce[1] defined softness as the opposite of stiffness which is measured by bending length. Several researchers have defined softness in different ways, such as the opposite of firmness or hardness measured by thickness tests or ease of yielding to pressure. Bishop[4] summarized the physical properties associated with softness as bending, compression and tensile properties, shear stiffness and hysteresis, areal density and friction in his extensive review. The relationship between subjective fabric softness assessment and fabric compression properties using the Instron Tensile Tester was investigated by Elder et al.[10] and a strong correlation was found. The same study also shows that fabric thickness in woven and non-woven fabrics, fabric density and specific volume only in woven fabrics are critical for softness. Fabric compression properties, fabric tensile properties, fiber diameter and breaking load are also important factors related to fabric softness.[8]

6.3.3 Fabric roughness

Many researchers have also conducted studies on the fabric roughness and relationship between roughness and yarn/fabric properties as well as the subjective perceptions. It was found that fabric roughness and several fiber-to-fabric properties including fiber diameter, fiber tensile properties, yarn diameter, yarn twist, and fabric thickness are essential factors in determining roughness. It has been shown that fabric roughness is associated with a number of objectively measured physical properties, such as prickle, shear stiffness, friction, bending stiffness, thickness, areal density and the level of smoothness.[4] Perception of roughness was found to be dependent on the textile construction parameters and roughness of woven and knit fabrics and can be determined by the roughness spacing. The roughness perception decreases logarithmically with yarn diameter and, with the same yarn diameter, the knitted fabrics were perceived as 'rougher' than woven fabrics.[11]

Behmann[11] found that the thickness of woven fabrics and the diameter of yarns are correlated with subjective roughness evaluations. Li[8,65] also found a good correlation between perception of roughness and fabric surface roughness, compression properties, fiber diameter, and fiber/fabric tensile properties. Hu et al.[9] reported that fabric smoothness is related to many objective Kawabata measurements, including fabric thickness at low pressure, geometric roughness, bending rigidity, linearity of compression thickness curve, and fabric mass per unit. Ramgulam et al.[12] compared the method of measuring fabric surface roughness using a laser sensor with the conventional Kawabata Evaluation System contact method and found a relatively good correlation between the two methods. Okur[13] also found that yarn type (combed or carded), yarn linear density, twist factor and fabric setting parameters impact fabric frictional properties in knitted cotton fabrics. Okur also added that protruding fibers on the fabric surface was the key factor impacting fabric surface smoothness and frictional properties.

6.3.4 Fabric hand/handle

As early as 1930, Peirce[1] suggested correlations between fabric handle and fabric mechanical properties and initiated the measurement of the bending stiffness of fabrics. Following Peirce's work, many other researchers have tried to correlate fabric handle with fabric mechanical properties. These investigations provided a great deal of insight into the fundamental nature of fabric deformational behavior, particularly under low stresses.

Fabric hand has been defined as 'subjective assessment of a textile obtained from the sense of touch. It is concerned with the subjective judgment of roughness, smoothness, harshness, pliability, thickness, etc.'[16] Peirce[1] described hand as being the judgment of the buyer which depends on time, place, fashion and personal predilections. Thus, replacing human assessors with physical testing would be worthless. However, what human fingers sense depends on the physical properties of fabrics, therefore objective measurements can provide a basis upon which to exercise judgment. Judgments of fabric handle are used as the basis for evaluating quality and comfort and thus for determining fabric value within the textile clothing and related industries. The term 'comfort' is a more general term that implies many aspects of human-related clothing performance including fabric handle. While comfort implies experience over time with the clothing prior to passing judgment on its performance, handle implies an initial evaluation of the apparel prior to passing judgment on its appeal. Handle basically reflects a mechanical interaction between human skin and fabric in which both the fabric surface and the material bulk are being spontaneously tested by exerting external body movement. Therefore, handle is a reaction to positive mechanical actions. Comfort, on the other hand, is a more complex phenomenon because it involves physical interactions between the human body, the fabric, and the external environment.

Fabric hand (handle) evaluation systems can be divided into two main categories: indirect systems, and direct methods.[17] The difference between these two categories is the type of parameters produced and their interpretations. Indirect systems produce instrumental parameters that are believed to represent basic determinants of fabric handle such as fabric stiffness, fabric roughness, and compressibility. Only through parallel subjective assessment and cross-correlations are some parameters that simulate fabric handle estimated. Two of the most common methods of this category are the Kawabata Evaluation System for Fabrics (KES-F®) and the Fabric Assurance by Simple Testing System (SiroFAST™). In these systems objective and subjective evaluations are combined to develop quantitative measures of fabric hand, and reviews of them have been discussed in numerous studies.[4,18–20] Although the initial purpose of these systems was to replace subjective hand assessment with objective means, they relied heavily on subjective scaling to produce objective hand characteristics.

On the objective side, Kawabata developed a set of instruments to measure appropriate handle-related fabric properties, including tensile, shearing, bending, compression, surface, weight and thickness. Fabric handle is assessed by measuring 16 objective mechanical and surface parameters, all at low levels of force, and correlating these parameters with the subjective assessment of handle using linear regression equations.

The SiroFAST™ system was developed by the Commonwealth Scientific and Industrial Research Organization (CSIRO) in Australia in the late 1980s for use by goods manufacturers to detect and diagnose problems associated with the process of conversion from fabric to garments. The system consists of three instruments and a test method: a compression meter for measuring fabric thickness, a bending meter for measuring fabric bending length, an extension meter for measuring fabric extensibility, and a test procedure for measuring dimensional properties of fabrics.

The arguments against the indirect systems are associated with the time consumed for testing, analyzing, and interpreting the results, and the cost, in terms of labor and maintenance.[21–24] However, these systems can provide useful quantitative guidelines particularly in the area of fabric and garment design.

Direct methods of fabric hand evaluation represent creative techniques that are intended to simulate two or more aspects of hand evaluation and produce quantitative measures that are labeled as hand force or hand modulus. It should be pointed out that the term 'direct' does not necessarily mean more representative or more accurate in comparison with the indirect systems. These methods include the ring test, the slot test (Handle-o-Meter), and the Elmogahzy-Kilinc handle measurement system (see Fig. 6.1 and Fig. 6.2). These methods were developed to measure the handle properties of fabric through simulations of the various mechanisms reflecting fabric hand.[17,25–28,30] The first two methods were based on pulling a fabric sample through a ring or pushing a fabric sample through a slot and measuring the resistances to the pull-through or push-through mechanisms. This is believed to simulate how a person tends to handle a piece of fabric when

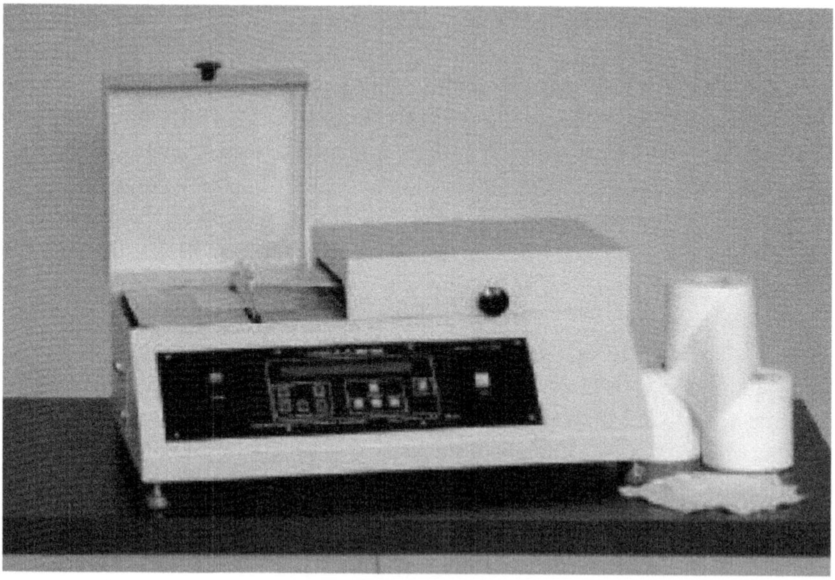

6.1 Handle-o-Meter, by Thwing-Albert Instrument Company.[77]

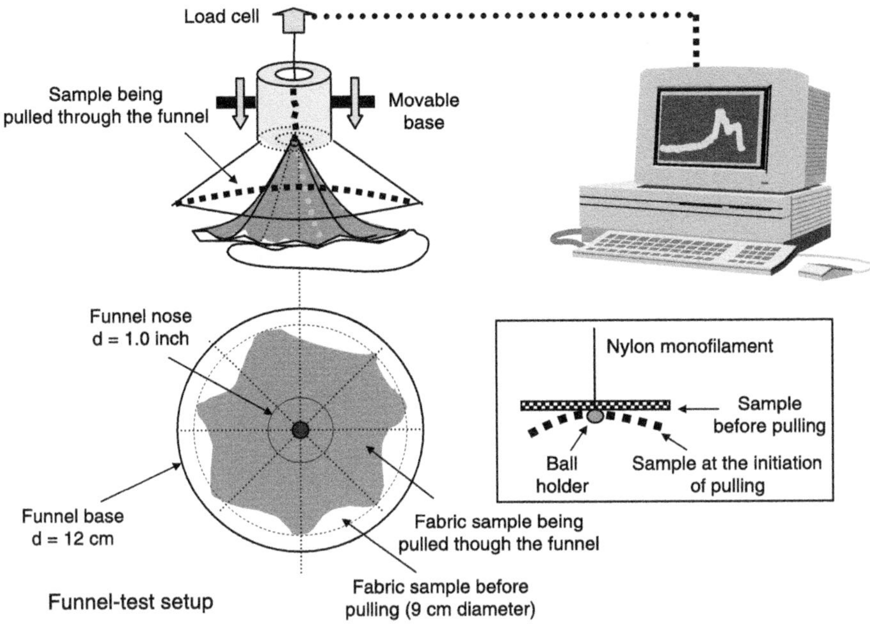

6.2 Elmogahzy-Kilinc Fabric Hand Measurement System.[17]

evaluating it. A more recently patented method is the Elmogahzy-Kilinc hand method.[17,28–30] The underlying concept of this method was inspired by the theoretical and experimental efforts made by many previous investigators. In addition, the method aimed at overcoming some of the problems associated with the statistical reproducibility and characterization parameters found in previous methods. In this method, a flexible light funnel is used to represent the media through which the fabric sample is pulled (see Fig. 6.2). The idea of using a funnel media instead of a ring or a slot arrangement is to provide multiple configurations of fabric hand that closely simulate the various aspects of the hand phenomenon. Besides, the funnel media is believed to allow both constrained and unconstrained fabric folding or unfolding, a key aspect of fabric handle. Also the total area under the hand profile produced by this system is proposed as an integrated parameter of fabric hand. In a detailed study, this parameter was found to be highly correlated to subjective hand assessments of many woven and knit fabrics.[30]

6.3.5 Other test methods for measuring fabric tactile properties

As indicated earlier, many testing instruments have been developed to evaluate the fabric tactile properties, such as the Shirley Bending Tester and Cusick Drape Tester (see Fig. 6.3). According to Peirce,[1] bending length of the fabric cantilever

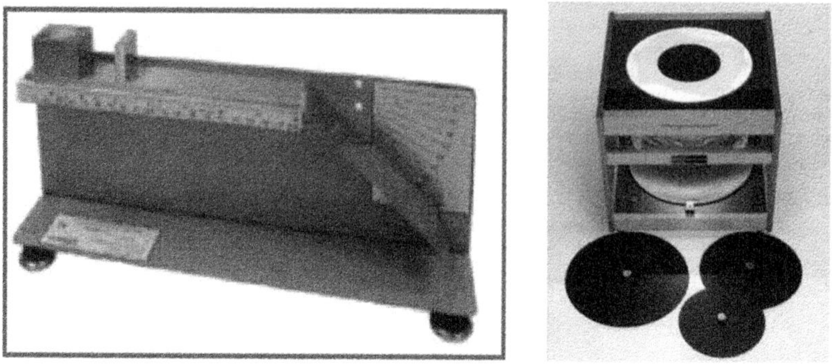

(a) (b)

6.3 Fabric bending and drape testers, (a) Cantilever bending tester by Testing Machines Inc., (b) Cusick drape tester by Hanatek Services Ltd.[69,70]

can be evaluated either by measuring the extended fabric length under a fixed angle or by measuring an angle from the extension of a fixed length. A bending tester from the Shirley Institute was designed in 1970 for measuring extended length with a fixed angle. Bending length is calculated from tested extended length and the SiroFAST™ bending meter uses this same concept. In the cantilever test, using ASTM D1388 'Standard Test Method for Stiffness of Fabrics', fabric is slid in a direction parallel to its long dimension on a cantilever, until its leading edge projects from the edge of a horizontal surface (see Fig. 6.3a). The length of the overhang fabric is measured when the tip of the specimen is depressed under its own mass to the platform. From this measured length, the bending length and flexural rigidity are calculated. Several other methods have also been developed to measure fabric bending properties[71–73] and, in addition, the relationship between fabric drape and mechanical properties has been studied by many researchers.[73]

Drape is the term used to describe the way a fabric hangs under its own weight. It is one of the most important fabric characteristics since it shows how good a garment looks in use. A couple of drape measurement instruments have been developed, the principle being the same (see Fig. 6.3b).[73] In the drape test, the specimen deforms with multi-directional curvature and consequently the results are dependent to a certain amount upon the shear properties of the fabric, but are mainly dependent on the bending stiffness.

Many researchers have developed models to predict the tactile comfort or one of the constituents of fabric tactile behavior. Raychaudhuri and Das[14] developed a mathematical model for the physical testing of fabric stiffness by the cantilever method in order to express the stiffness in terms of the physical parameters. Pandita and Verpoest[8] compared different stiffness prediction models, including the Krenchel, Voigt, Reuss, and Inclusion models and Kregers' weight average model (WAM), and preferred the Inclusion model. In 2000, Peykamian and Rust[15] used linear and non-linear models and yarn parameters such as CV%, hairiness and surface softness to classify the softness of knitted fabrics for comparison with subjective evaluations. They were able to classify 91% of the samples accurately based on the human data using surface response parameters and measured yarn properties to sort T-shirt softness into three classes, with tree modeling.

6.4 Testing, analyzing and predicting thermophysiological comfort

Thermophysiological comfort relates to the way the clothing buffers and dissipates metabolic heat and moisture. The American Society of Heating Refrigeration and Air Conditioning Engineers (ASHRAE) standard 55–66 defines thermal comfort as 'that condition of mind, which expresses satisfaction with the thermal environment'. This type of comfort is reached when there is no need to add or remove clothing in order to be satisfied with the temperature. This aspect of comfort becomes particularly more important for some of the clothing, such as sportswear, where the human activity level is expected to be high, and for protective clothing (e.g., military, firefighting) where high insulation properties are largely desired due to the protection needs. A considerable amount of research has been carried out to measure and analyze fabric thermal and moisture related properties.[31–36] Many researchers including Fourt and Hollies,[37,39] Slater,[38] and Li et al.[8] have carried out extensive reviews of the thermal comfort properties of textiles and the measurement of thermal resistance of textiles, water vapor transmission, moisture transmission and air permeability.

6.4.1 Human thermoregulatory system

The human body has an excellent thermoregulatory system which regulates its internal temperature with a certain level of precision during changes in external and internal conditions. A healthy human body system typically maintains a constant core temperature of about 37 °C (98.6 °F) varying about 0.5 °C either side of this over a day for a non-active person. A comfortable mean skin temperature for a resting person is in the range of 33–34.5 °C. The body core temperature value varies slightly from person to person. Both the central and peripheral nervous systems continuously monitor temperature fluctuations in the body to keep them in balance by means of several biological responses. The human thermoregulatory system has three primary mechanisms for maintaining core body temperature: the skin blood flow, sweating, and shivering (Fig. 6.4). The skin blood flow is adjusted to change skin temperature for reducing or increasing heat loss, which is controlled by temperature signals from the skin and the central body core.

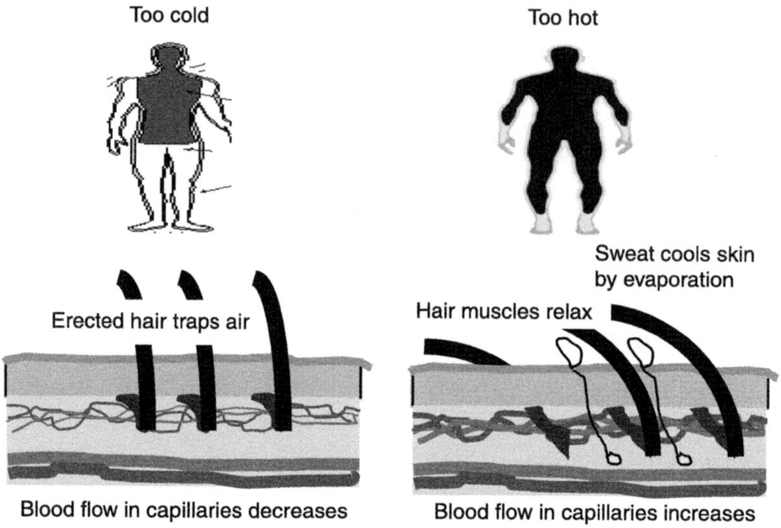

6.4 Body temperature control system.

Thermophysiological comfort depends on the capability of clothing to transmit heat and evaporated sweat from the skin into the environment. Heat transmission from the body to the surrounding environment depends on the temperature difference between the skin and the environment and equilibrium must be found between how much heat is produced and lost within the body. When the clothing–body system maintains a balance between heat produced by the body and the heat lost from the skin, a human being feels comfortable.

The heat generated by the body is called metabolic heat and is produced during aerobic metabolism in which nutrients are processed for conversion to energy. It is generally agreed that 75% of the total metabolic energy is released as heat. The metabolic heat generated by the body together with the heat received from external sources must be matched by the loss from the body of an equivalent amount of heat. Heat is transferred from a hotter object to a cooler object in one of three ways: by conduction, convection, or radiation. Metabolic heat generated by the body is also transferred to the environment by respiration and perspiration. The body loses heat during respiration because the cold air is drawn into the lungs, heated, then exhaled with high moisture content. The volume of air inhaled is humidified by the respiratory tract to saturation in order to be used efficiently. In an environment at a lower temperature than skin temperature, approximately 80% of metabolic heat of an unclothed body is emitted from the body by conduction, convection, and radiation; 10% is lost by evaporation; and 10% by respiration.

In cold conditions, the minute muscles under the surface of the skin, called erector pili muscles, contract and hair follicles are lifted upright, which traps heat within the layer of air created by the hair follicles. Blood vessels carrying blood to superficial capillaries under the surface of the skin constrict, thereby rerouting blood away from the skin and towards the warmer core of the body thus reducing heat loss from the skin surface to the surroundings and preventing the core temperature from dropping further. Muscles can also receive messages from the hypothalamus, the thermoregulatory center of the brain, to cause shivering. Shivering thermogenesis increases heat production by increasing cellular respiration and is an exothermic reaction in muscle cells. During shivering thermogenesis, metabolism can increase an average of three times above resting metabolism (4–5 times for short durations) during which an increase in body temperature occurs. The increase in body temperature is accompanied by a parallel increase in oxygen consumption required for the increase in aerobic cellular metabolism. Lastly, the body starts to convert fat directly into heat energy.

In hot conditions, sweat glands under the skin secrete sweat which causes heat loss via evaporative cooling. Sweating is an effective mechanism to release extra heat from the body and is determined by sweat glands. The glands are controlled by the mean body temperature signal and the skin temperature signal. Also, the hairs on the skin are laid flat by erector pili muscles, which prevents heat from being trapped by the layer of still air between the hairs. These flat hairs increase the flow of air next to the skin which increases heat loss by convection. Due to the relaxation of smooth muscle in blood vessel walls, blood flow is increased through the arteries. This redirects blood into the superficial capillaries in the skin increasing heat loss by convection and conduction.

On a cold day, the body loses heat mainly due to radiation, while on a warm day the body cools down primarily through evaporation of perspiration (see Fig. 6.5). Not only the clothing but also several environmental factors impact thermal comfort. Fanger[31] identifies the six variables that influence thermal comfort of humans as: air temperature, mean radiant temperature, relative air velocity, water vapor pressure in the ambient air, activity level, and thermal resistance of clothing (clo).

6.4.2 Testing of thermophysiological comfort

As indicated earlier, thermal comfort is one of the most important aspects of the wearer comfort. If the body is not in a thermal comfort status, it may also induce the tactile discomfort or increase this type of discomfort. Numerous research conducted in this area resulted in many testing systems developed to evaluate one or more aspects of thermal comfort, including thermal insulation (resistance to dry heat loss from the body), thermal conductivity (the thermal transfer behavior of the heat flow through a fabric due to a combination of conduction and radiation), sweat evaporation, and water vapor permeability. Guarded sweating hot plate, togmeter, heat flowmeter, Gore cup, and sweating thermal manikin are a few examples of these testing systems.

Transmission of heat through a fabric occurs both by conduction through the fibers and entrapped air and by radiation. Methods developed for evaluation of

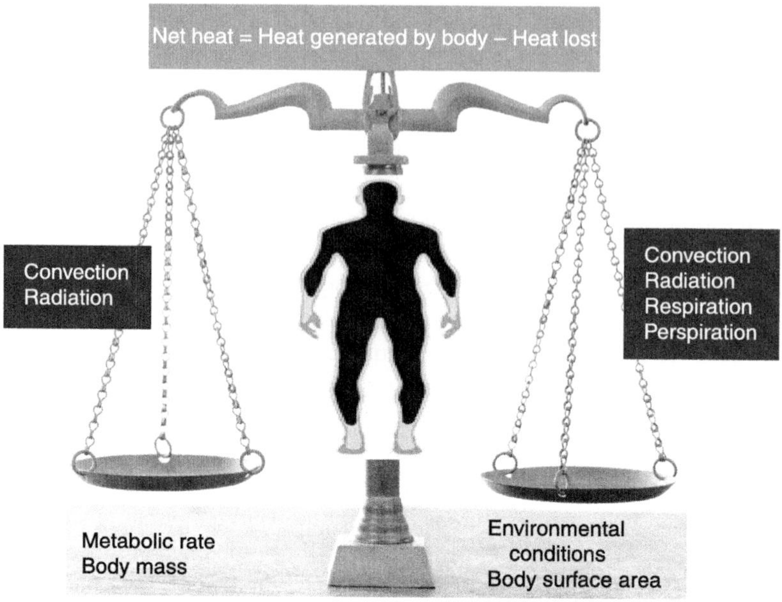

6.5 Thermoregulatory system of the human body.

thermal conductivity measure the total heat transmitted by both mechanisms. The insulation of a fabric is measured by its thermal resistance, which is the reciprocal of thermal conductivity, and this resistance is defined as the ratio of the temperature difference between two faces of the fabric to the rate of heat flow per unit area normal to the faces. The thermal resistance to dry heat transfer (i.e., fabric insulation value) of a fabric is usually measured with a guarded hot plate. In this test, a fabric sample is placed on a square plate, which is heated to a constant temperature that approximates the human skin temperature (e.g., 35 °C). A guard heater and bottom heater prevent lateral and downward heat loss and these extra heaters force all of the heat generated in the main heater to flow through the fabric sample. The plate is housed in a controlled environmental chamber and the system measures the amount of the electrical power required to maintain the plate at a constant temperature under steady-state conditions in a cooler environment. The power is proportional to the heat loss through the fabric. Several testing standards including ASTM D1518 'Thermal Transmittance of Textile Materials', ASTM F1868 'Thermal and Evaporative Resistance of Clothing Materials Using a Sweating Hot Plate Test', and ISO 11092 'Textiles–Determination of Physiological Properties–Measurement of Thermal and Water-Vapour Resistance' require use of this method for determination of the thermal resistance value.

A sweating hot plate is the most widely used instrument to measure the heat and moisture vapor transfer properties of materials. This instrument measures the dry thermal resistance (insulation) and the evaporative resistance of fabrics. Sweating hot plate test methodologies have been available for many years and are standardized by the American Society for Testing and Materials (ASTM) and the International Standards Organization (ISO). ASTM F1868 and ISO 11092 use this method for determination of evaporative resistance of fabrics. In this test, to measure the evaporative resistance of fabrics, water is fed to the surface of the test apparatus that is covered by a liquid barrier (e.g., untreated cellophane or microporous polytetrafluorethylene film) in order to shield the fabric from liquid water. The conditions in the environmental chamber create thermal heat loss through the test fabric, or heat loss that is influenced by both the dry thermal insulation and evaporative resistance of the test material. As it is described earlier, dry tests are conducted to determine conductive thermal resistance. Wet tests are conducted to determine apparent evaporative thermal resistance. The total heat loss of the test material is calculated using an equation that combines both conductive and evaporative heat transfers. Numerous international standards have been developed for determining these factors using different instruments (see Table 6.1).

Sweating hot plate tests, conducted on fabric samples, cannot provide information critical to garment design since these tests cannot validate the effects of garment-related factors such as garment fit, design, and seams and activity-related effects such as body movement. Sweating manikin tests provide opportunity to address these shortcomings in instrumented approaches (see Table 6.1). Use of thermal/sweating manikins in research and measurement standards has grown rapidly over the years and they are now widely used in textiles and clothing research laboratories all over the world for analyzing the thermal interface of the human body and its environment. The sweating manikin system (Coppelius type) consists of a computer controlled heating system with 18 individually controlled body sections, a computer controlled sweating system with 187 individually controlled sweating glands, and prosthetic joints to permit movements and different postures. The manikin is housed in a climatic chamber and water is supplied from a reservoir. A micro-valve system distributes the water to sweat glands and condensed water on the dressed manikin is recorded by measuring the change in the weight of the clothed manikin during the test. The garments are also weighed before and after the test to estimate the amount of moisture condensation in the individual clothing layers. Studies have shown the potential value of using sweating manikins as tools for assessing the heat stress and comfort of clothing.

The ability of a clothing material to transport moisture from the skin is crucial for thermophysiological and neurophysiological comfort. Many laboratory testing technologies have been developed to characterize the ability of fabrics to wick liquid moisture from sweating skin and one such test is the Gravimetric Absorbency Testing System (GATS). This test indicates the lateral wicking ability of the fabric, or the ability of the material to take up liquid in a direction perpendicular to the fabric surface. The GATS apparatus incorporates a special test cell and

Table 6.1 Thermal comfort relevant standards[8]

Standard	Description
ASTM C518	Standard Test Method for Steady-state Thermal Transmission Properties by Means of the Heat Flow-meter Apparatus
ASTM D1518	Standard Test Method for Thermal Transmittance of Textile Materials
ASTM E1530	Standard Test Method for Evaluating the Resistance to Thermal Transmission of Materials by the Guarded Heat Flow-meter Technique
ASTM F1868	Standard Test Method for Thermal and Evaporative Resistance of Clothing Materials Using a Sweating Hot Plate Test
ASTM F1291	Standard Test Method for Measuring the Thermal Insulation of Clothing Using a Heated Manikin
ASTM F2370	Standard Test Method for Measuring the Evaporative Resistance of Clothing Using a Sweating Manikin
ASTM F1720	Standard Test Method for Measuring Thermal Insulation of Sleeping Bags Using a Heated Manikin
ASTM F2732	Standard Practice for Determining the Temperature Ratings of Cold Weather Clothing
ASHRAE 3739	Evaluating Thermal Environments by Using a Thermal Manikin with Controlled Skin Surface Temperature
ASHRAE HI-02-17-4	Measurement of Thermal Comfort and Local Discomfort by a Thermal Manikin
ISO 7345	Thermal Insulation – Physical Quantities and Definitions
ISO 9288	Thermal Insulation – Heat Transfer by Radiation – Physical Quantities and Definitions
ISO 9346	Thermal Insulation – Mass Transfer – Physical Quantities and Definitions
AATCC 79	Absorbency of Textiles
AATCC 193	Aqueous Liquid Repellency: Water/Alcohol Solution Resistance Test
ASTM D737	Standard Test Method for Air Permeability of Textile Fabrics
AATCC 127	Water Resistance: Hydrostatic Pressure Test
ISO 8302	Thermal Insulation – Determination of Steady-state Thermal Resistance and Related Properties – Guarded Hot Plate Apparatus
ISO 11092	Textiles – Determination of Physiological Properties – Measurement of Thermal and Water-Vapour Resistance
ISO 2582	Cork and Cork Products – Determination of Thermal Conductivity – Hot Plate Method
ISO 9920	Ergonomics of the Thermal Environment – Estimation of the Thermal Insulation and Evaporative Resistance of a Clothing Ensemble
ISO 11079	Evaluation of Cold Environments – Determination of Required Clothing Insulation (IREQ)
ISO 7730	Moderate Thermal Environments – Determination of the PMV and PPD Indices and Specification of the Conditions for Thermal Comfort

Table 6.1 Continued

Standard	Description
ISO 7933	Ergonomics of the Thermal Environment – Analytical Determination and Interpretation of Heat Stress Using Calculation of the Predicted Heat Strain
ISO 7920	Estimation of the Thermal Characteristics of Clothing
EN-ISO 15831	Thermal Manikin for Measuring the Resultant Basic Thermal Insulation
ISO 15496	Textiles – Measurement of Water Vapor Permeability of Textiles for the Purpose of Quality Control
ENV 342	Protective Clothing Against Cold
EN 511	Protective Gloves Against Cold (Thermal Hand Model)
ISO 14505	Evaluation of the Thermal Climate in Vehicles, Parts 1 and 2
EN 13537	Requirements for Sleeping
BPI 1.4	Determination of Stationary Water Vapor Resistance by Means of the Cup Method

cover to assess absorption behavior in the presence of evaporation. Vertical wicking, Gore cup (Bekleidungs Physiologisches Institut E. V. Hohenstein Standard Test Specification BPI 1.4 'Determination of Stationary Water Vapor Resistance by Means of the Cup Method'), drop absorption (AATCC 79 'Absorbency of Textiles'), contact angle, and drying rate tests are a few examples of methods available for determining the water transport properties of fabrics. Very recently a new method and instrument developed by Li et al.,[40] called the Moisture Management Tester, was introduced to quantitatively measure multi-directional liquid moisture transfer in a fabric. Air permeability is also a very important characteristic of fabrics which affect the thermal comfort in addition to thickness, and ASTM D737 'Standard Test Method for Air Permeability of Textile Fabrics' has been widely used for determining air permeability of fabrics.

6.4.3 Thermophsyiological comfort studies

Over the years, much research has been carried out in order to predict clothing thermophysiological comfort on the basis of fabric properties. There has been considerable research effort in this area and reviews of some of these studies have been presented by several researchers. Also, a number of thermal models have been developed to predict thermal comfort, and the capabilities and limitations of these models have been discussed in many studies.[31,41–50] Predictions of heat loss by the models were compared with experimental data from thermal manikin tests or human subject tests, and reasonable agreements were found.

In 1971, Gagge et al.[51] developed a two-node model for describing the thermoregulatory system of the human body. Fanger[31] developed a mathematical model in 1970 to define the neutral thermal comfort zone of man in different combinations of clothing and at different activity levels. Mean skin temperature and sweat secretion rates were used as physical measures of comfort. ASHRAE developed comfort charts and indices of thermal sensations based on Fanger's study, and by using these indices it is possible to predict the comfort acceptance under different combinations of clothing insulation, metabolic level, air temperature, and wet-bulb temperature (or radiant temperature). In 1985, Fanger et al.[52] presented an international standard related to thermal comfort, which is named ISO 7730 (see Table 6.1). The standard was intended to specify conditions that are predicted to be acceptable in thermal comfort for a given percentage of the population.

Mecheels and Umbach[60] reviewed the psychrometric range of clothing systems. They pointed out that the thermal properties of a clothing system are determined by its resistance to heat transfer and its resistance to moisture transfer. They also explained that through these two values of resistance, the minimum and the maximum ambient temperature could be determined. The difference between the maximum and minimum ambient temperatures is called the psychrometric range of the clothing system. The resistance to heat and moisture transfer and the psychrometric range can be measured using a thermal manikin and skin model, which were developed by the Hohenstein Institute. These parameters are dependent on clothing design and the wear type, textile materials, and wind velocity.

Barker et al.[2] developed a theoretical model to integrate the various measured comfort related physical properties of the fabrics into a prediction of human comfort limits for given climatic and metabolic work load conditions. The model is based on rates of heat loss and storage and their effect on body core temperature. Sensible and evaporative heat loss, as well as percent of skin area wetted by sweat, is considered in the model, which predicts the range of body activity within which an individual wearing a clothing system is thermo-physiologically comfortable. Above these limits, heat stress is likely and below them hypothermia may occur.

The coupled heat and moisture transfer in textile fabrics has been widely recognized as being very important for understanding the dynamic thermal comfort of clothing during wear and a number of models were developed to simulate the dynamic heat and moisture transfer in fabrics.[54–58] It is shown that theoretical prediction and experimental measurements are in good agreement.

Li and Holcombe[48] also developed a mathematical model to describe the dynamic heat and moisture transfer behavior of the body–clothing–environment system under transient conditions. With specification of the physical activity and ambient conditions, their model was able to predict the thermoregulatory responses of the body, together with the temperature and moisture profiles in the clothing. After a series of experimental measurements, they concluded that the model is able to predict the transient heat and moisture transport behavior of garments made from high and weak hygroscopic fibers in dynamic wear situations.

Later, Hamdi et al.[53] applied fuzzy logic to estimate thermal comfort levels depending on the state of air temperature, mean radiant temperature, relative humidity, air velocity, activity level of occupants and their clothing insulation. The fuzzy comfort model is deduced on the basis of learning Fanger's equations.

Wang et al.[59] applied mathematical models, which describe the physical mechanisms of heat and moisture transfer in fabrics, to simulate the perception of thermal and moisture sensations. These two sensations have been shown to be related to skin and fabric temperature changes, which are associated with changes in moisture content and temperature in the fabrics. Results show that this mathematical simulation, based on heat and moisture transfer in fabrics, neurophysiological mechanisms of the thermoreceptors and psychophysical relationships, is able to predict the perception of coolness and dampness sensations to the touch with reasonable accuracy.

Recently, Jones[47] discussed the capabilities and limitations of the thermal models developed earlier, including the Fanger, Gagge, Wissler, Smith, Jones, Lotens and Fu models, and concluded that even though they have some limitations, thermal models are powerful tools for predicting the thermal comfort.

Another important aspect of the thermal clothing comfort is the transient behavior of the whole system. Since the human body is subject to changes in environmental variables, transient thermal response of the human clothing system becomes more relevant during these changes. Woodrock[61–62] investigated the after exercise chill. More recently, Dent[63] studied the analogous buffering effect due to changes in temperature and humidity and questioned the impact of fiber moisture regain in all transient cases. Toftum et al.[64] studied the effect of fabric type on comfort or on perceived humidity of skin or fabric using woven and knitted polyester and cotton clothes at controlled levels of skin-relative humidity and found no impact. They reported predictive model and found that acceptance of skin humidity decreases with the increase in the relative humidity of skin.

6.5 Design-oriented comfort model

An integrated approach that coordinates all these efforts to produce very reliable, yet very simplified ways of characterizing comfort is required. Several parameters can be used to characterize clothing comfort and, furthermore, there are hundreds of different design guidelines from fiber-to-garment that can assist in producing clothing with different comfort levels. What is missing in the comfort area is an integrated approach that ties together all these aspects associated with fabric comfort. Recently, Kilinc[30] developed a 'Design-Oriented Comfort Model' to establish highly objective design criteria for production of fabrics with different comfort levels for certain consumer applications.

The contributors to clothing comfort are the environment, human health status, physical activity and applications. For the development of a design-oriented fabric comfort model, these parameters have to be analyzed in detail. The model developed in Kilinc's study is based on three aspects: the first is the psychological aspect which means to provide complete unawareness of clothing and good protection; the second is the thermo-physiological aspect which means to provide optimum accommodation to the surrounding media; and the third is the neurophysiological aspect which means to provide optimum interaction with the skin. These aspects were established by Kilinc et al.[28,30,65] in consequence of an extensive survey as the basis for a design-oriented comfort model. In this model, the common factor that can truly tie all aspects associated with fabric comfort was proposed as fabric/skin interaction.

Translation of this factor into a characterization index of fabric comfort yielded itself to the concept of area ratio. The area ratio is defined by the ratio between the true area of fabric/skin contact and the corresponding apparent area:

$$\text{Area ratio } (AR) = \frac{\text{True area of contact}}{\text{Apparent area of contact}} = \frac{A_t}{A_a} \qquad [6.1]$$

The apparent area of contact between a textile fabric and another surface is typically much greater than the actual area of contact since the fabric never exhibit a complete flatness when it comes into contact with other solid surfaces because of structural and deformation effects inherent to fibrous structures.[66,67] There is a substantial gap between the skin and the fabric in contact. In the context of comfort, this can be considered as a unique added-value phenomenon by virtue of the advantage that it provides, with respect to the level of the wearer, awareness of the fabric/skin contact.

On a macroscopic level, fabric natural irregularities created by the fabric pattern (valleys and troughs) prevent the fabric from a pure contoured contact. On a microscopic scale, surface disturbances such as projecting hairs or pills further decrease the true contact. When one considers lateral effects such as applied pressure, multiple fabric layers (weight and thickness), and human dynamics, one will see that the area ratio represents a key variable that can determine fabric performance in different applications.

Although the area ratio is defined by pure geometrical parameters, it is actually a function of a complex interaction between a host of parameters, some of which are geometrical and others mechanical (see Fig. 6.6). This point is demonstrated by the general expression of the area ratio:

$$\frac{A_t}{A_a} = (C_M)^{\frac{\alpha+1}{\alpha}} K^{-\frac{1}{\alpha}} M_a P^{\frac{1}{\alpha}} \qquad [6.2]$$

where

A_t/A_a = the ratio between the true area of fabric/skin contact and the corresponding actual area

K = surface resilience, or hardness, constant (SI units = $N/m^{2(\alpha+1)}$)

α = a constant which lies between 0 for purely plastic behavior and 1.0 for purely elastic behavior

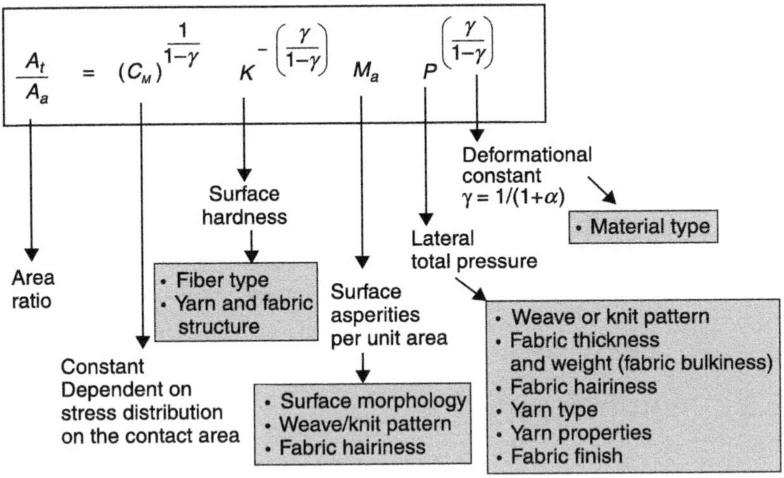

6.6 Structural model of area ratio.[30]

M_a = the number of contacting asperities at the fabric/skin interface (an index of surface roughness)

C_M = a parameter determined from the type of load distribution applied on the interface

P = the lateral pressure applied on the area of fabric/skin contact.

Based on the relationship between the lateral pressure and fabric thickness, a more detailed expression of area ratio was also derived:[30]

$$\frac{A_t}{A_a} = (C_M)^{\frac{\alpha+1}{\alpha}} K^{-\frac{1}{\alpha}} M_a \left[K_t.E.V^3.\left(\frac{1}{t^3} - \frac{1}{t_o^3}\right)\right]^{\frac{1}{\alpha}} \qquad [6.3]$$

where K_t is a constant, E is the Young's modulus of fibers, V is the volume of fibers, and t and t_o are thicknesses of the fiber mass at two different levels of pressure.

The above expressions clearly indicate that the area ratio is related to many deformational and surface parameters that collectively reflect the tactile mechanical and surface behavior of fabrics. Empirical analysis performed in Kilinc's study[28,30] also revealed that the area ratio is related to non-tactile parameters such as thermal insulation, thermal absorptivity, air permeability, wicking parameters, and pore size distribution.

In Kilinc's study, the area ratio was measured using a reference method in which fabric samples were coated with high-viscosity ink and the true imprint area against a flat metallic surface was determined using image analysis.[30] Measures were taken at different levels of lateral pressure. In addition, two empirical indices of area ratio, namely 'a' and 'm', were estimated for each fabric from the general power function:

$$\frac{A_t}{A_a} = a P^m \qquad [6.4]$$

Based on the experiments conducted in Kilinc's study, it was found that different fabric structures exhibit different values of area ratio.

Extensive empirical analysis of the area ratio parameter proved that it relates to all tactile (mechanical and surface) comfort parameters and many thermal comfort parameters.[30] An exploratory partitioning model of area ratio, developed over different fabric patterns and a wide range of tactile and thermal properties, clearly proved that the area ratio is significantly related to most of these factors.

6.6 Future trends

Clothing comfort has been identified as one of the key attributes in consumers' perception of the desirability of apparel products in all markets. Despite the extensive research in the area of comfort perception and clothing, there are still uncertainties with regard to fabric or garment design. Requirements of consumers are changing along with products and wear situations. In a highly competitive textile and apparel market, in order to succeed in the market place, the market players have to meet or even exceed consumers' needs and expectations.

In this chapter, tactile and thermal related factors affecting human comfort status were discussed and predicting models were briefly reviewed. Also a new approach to characterizing comfort was described. The comfort phenomenon is typically described by hundreds of parameters. However, this approach is based on using a single index of comfort expressed by the ratio between the actual area of fabric/skin contact and the corresponding apparent area. Extensive analysis showed that this index correlates extremely well with all main aspects of comfort, namely psychological, thermophysiological, and neurophysiological.

Clothing comfort is expected to continue to interest researchers of different sectors of the textile/apparel market and the subjective nature of this phenomenon will probably remain an essential aspect of research and implementation. This is primarily due to the critical importance of human judgment, which is highly variable and often psychologically driven. Unfortunately, subjective evaluation does not yield precise design guidelines except for extreme conditions. It is our opinion that an objective comfort evaluation coupled with subjective assessment seems to be the appropriate approach.

6.7 References

1 Peirce, F.T. (1930) 'The handle of cloth as a measurable quantity', *J. Text Inst.*, **21**: T377.

2 Barker, R.L., Woo, S.S., Radhakrishnaiah, P., Hatch, K.L., Markee, N.L., and Maibach, H.I. (1990) 'In vivo cutaneous and perceived comfort response to fabric, Part II:

mechanical and surface related comfort property determinations for three experimental knit fabrics', *Textile Res. J.*, **60**(8): 490–94.

3 El Mogahzy Y., Kilinc F.S., Hassan M., Farag R., and Kamel A. (2004) 'Protective Clothing – The Unresolved Ultimate Trade-off between Protection and Comfort – The Concept of Area Ratio, 4th International Conference on Safety & Protective Fabrics, IFAI, Pittsburgh, PA, USA.

4 Bishop, D.P. (1996) 'Fabrics: Sensory and mechanical properties', *Textile Progress*, **26**(3): 1–62.

5 Elder, H.M., Fisher, S., Armstrong, K., and Hutchison, G. (1984) 'Fabric stiffness, handle and flexion', *J. Text. Inst.*, **75**(2): 99–106.

6 Elder, H.M. (1984) 'Fabric stiffness', *J. Text. Inst.*, **75**(4): 307–11.

7 Elder, H.M., Fisher, S., Hutchison, G., and Beattie, S. (1985) 'A psychological scale for fabric stiffness', *J. Text. Inst.*, **76**(6): 442–9.

8 Li, Y. and Wong, A.S.W. (editors) (2006) *Clothing Biosensory Engineering*, Cambridge, Woodhead Publishing.

9 Hu, J., Chen, W. and Newton, A. (1993) 'A psychophysical model for objective fabric hand evaluation: An application of Steven's Law', *J. Text. Inst.*, **84**(3): 354–63.

10 Elder, H.M., Fisher, S., Armstrong, K., and Hutchison, G. (1984) 'Fabric softness, handle, and compression', *J. Text. Inst.*, **75**(1): 37–46.

11 Behmann, F.W. (1990) 'Tests on the roughness of textile surfaces', *Melliand Textilberichte*, **71**(6): 438–40.

12 Ramgulam, R.B., Amirbayat, J. and Porat, I. (1993) 'Measurement of fabric roughness by a non-contact method', *J. Text. Inst.*, **84**(1): 99–106.

13 Okur, A. (2002) 'Frictional properties of plain-knitted cotton fabrics', *Textile Asia*, **33**(8): 32–4.

14 Raychaudhuri, B.D.S. and Das, S.P. (2000) 'Physical testing and mathematical modelling for stiffness characteristics of fabrics using cotton–polyester blends', *Man Made Textiles in India*, **43**(9): 427–30.

15 Peykamian, S. and Rust, J.P. (2000) 'Fabric softness classification using linear and nonlinear models', *Textile Res. J.*, **70**(3): 201–4.

16 The Textile Institute (1988) *Textile Terms and Definitions*, 8th edition, Manchester, UK.

17 El Mogahzy, Y., Kilinc, F., and Hassan, M. (2005) 'Developments in measurement and evaluation of fabric hand', in, *Effect of Mechanical and Physical Properties on Fabric Hand*, H. Behery (ed.), Cambridge, Woodhead Publishing, pp. 51–57.

18 Kawabata, S., Niwa, M., and Wang, F. (1994) 'Objective hand measurement of non woven fabrics', *Textile Res. J.*, **64**(10): 597–610.

19 Kawabata, S. and Niwa, M. (1989) 'Fabric performance in clothing and clothing manufacture', *J. Text Inst.*, **80**, 19–50.

20 Minazio, P.G. (1995) 'FAST–Fabric Assurance by Simple Testing', *Int. J. Cloth. Sci. Tech.*, **7**(2/3): 43–8.

21 Behery H M. (1986) 'Comparison of fabric hand assessment in the United States and Japan', *Text Res. J.*, **56**: 227–40.

22 Pan, N., Yen, K.C., Zhao, S.J., and Yang, S.R. (1988) 'A new approach to the objective evaluation of fabric handle from mechanical properties Part II: Objective measure for total handle', *Textile Res. J.*, **58**(9): 531–7.

23 Pan, N., Zeronian, S.H., and Ryu, H.S. (1993) 'An alternative approach to the objective measurement of fabrics', *Textile Res. J.*, **63**(1): 33–43.

24 Grover, G., Sultan, M.A., and Spivak, S.M. (1993) 'A screen technique for fabric handle', *J. Textile Inst.*, **84**(3): 1–9.
25 Kim J.O. and Slaten, B.L. (1996) 'Objective assessment of fabric handle in fabrics treated with flame retardants', *J. Test. Eval.*, **24**(4): 223–8.
26 Kim, J.O. and Slaten, B.L. (1999) 'Objective evaluation of fabric hand: Part I: Relationships of fabric hand by extraction method and related physical and surface properties', *Textile Res. J.*, **69**(1): 59–67.
27 Alley, V.L., Jr., and McHatton, A.D. (1977) Quantitative Measurement of the Feel of Fabric, NASA Tech. Brief LAR-12147.
28 Kilinc-Balci, F.S. and Elmogahzy, Y. (2008) 'Testing and analyzing comfort properties of textile materials for the military', in *Military Textiles*, E. Wilusz (ed.), Cambridge, Woodhead Publishing, pp.107–36.
29 El Mogahzy, Y.E., Kilinc, F.S., Hassan, M., and Farag, R. (2005) 'An innovative method for measuring objective total fabric hand', *Proceedings of Cotton Beltwide Conference*, U.S. Cotton Council.
30 Kilinc-Balci, F.S. (2004) A Study of the Nature of Fabric Comfort: Developing a Design-Oriented Comfort Model, Ph.D. thesis, Auburn University, Auburn, AL.
31 Fanger, P.O. (1970) *Thermal Comfort*, Danish Technical Press, Copenhagen, Denmark.
32 Binns, H. (1926) 'The discrimination of wool fabrics by the sense touch', *Br. J. Psychol.*, **16**: 237–47.
33 Houghton, F.C. and Yaglou, C.P. (1923) 'Determining lines of equal comfort', *J. ASHVE*, **29**: 165–76.
34 Winslow, C-E.A., Herrington, L.P., and Gagge, A.P. (1937) 'Physiological reactions of the human body to varying environmental temperature', *Am. J. Physiol.* **120**, 1–20.
35 Winslow, C-E.A., Herrington, L.P., and Gagge, A.P. (1937) 'Relations between atmospheric conditions, physiological reactions and sensations of pleasantness', *Am. J. Hyg.* **26**: 103–15.
36 Rohles, F.H. (1971) 'Psychological aspects of thermal comfort', *ASHRAE J.*, **13**: 86–90.
37 Fourt, L. and Hollies, N.R.S. (1970) *Clothing: Comfort and Function*, Marcel Dekker, NY.
38 Slater, K. (1977) 'Comfort properties of textiles', *Textile Progress*, **9**(4): 1–91.
39 Hollies, N.R.S. and Goldman, R.F. (1977) *Clothing Comfort: Interaction of Thermal, Ventilation, Construction and Assessment Factors*, Ann Arbor, MI, Ann Arbor Science Publishers Inc.
40 Hu, J.Y., Li, Y., Yeung, K.W., Wong, A.S.W., and Xu, W.L. (2005) 'Moisture management tester: A method to characterize fabric liquid moisture management properties', *Textile Res. J.*, **75**(1): 57–62.
41 de Dear, R.J., Knudsen, H.N., and Fanger, P.O. (1989) 'Impact of air humidity on thermal comfort during step-changes', *ASHRAE Trans.*, **95**: 336–50.
42 Fu, G. (1995) A transient, 3D Mathematical Thermal Model for the Clothed Human, PhD thesis, Kansas State University.
43 Gagge, A.P., Fobeletes, A., and Berglund, L.G. (1986) 'A standard predictive index of human response to the thermal environment', *ASHRAE Trans.*, **92**(2B): 709–31.
44 Hayward, J.S., Eckerson, J.D., and Collis, M.L. (1977) 'Thermoregulatory heat production in man: Prediction equation based skin and core temperatures', *J. Appl. Physiol.*, **42**: 377–84.
45 Hensel, H. (1973) 'Neural processes in thermoregulation', *Physiol. Rev.*, **53**: 948–1017.

46 Jones, B. and Ogawa, Y. (1992) 'Transient interaction between the human body and the thermal environment', *ASHRAE Trans.*, **98**(1): 189–95.
47 Jones, B.W. (2002) 'Capabilities and limitations of thermal models for use in thermal comfort standards', *Energy and Buildings*, **34**(6): 653–9.
48 Li, Y. and Holcombe, B.V. (1998) 'Mathematical simulation of heat and moisture transfer in a human–clothing–environment system', *Text Res. J.*, **68**(6): 389–97.
49 Smith, C. (1991) A Transient Three-Dimensional Model of the Human Thermal System, PhD thesis, Kansas State University.
50 Pan, N. and Gibson P. (eds) (2006) *Thermal and Moisture Transport in Fibrous Materials*, Cambridge, Woodhead Publishing.
51 Gagge, A.P., Stolwijk, J.A.J., and Nishi, Y. (1971) 'An effective temperature scale based on a simple model of human physiological regulatory response', *ASHRAE Trans.*, **77**: 247–62.
52 Fanger, P.O. (1985) 'Thermal environment–human requirements', *Sulzer Tech. Rev.*, **67**(3): 3–6.
53 Hamdi, M., Lachiver, G., and Michaud, F. (1999) 'A new predictive thermal sensation index of human response', *Energy and Building*, **29**(2): 167–78.
54 Farnworth, B. (1986) 'Numerical model of the combined diffusion of heat and water vapor through clothing', *Text. Res. J.*, **56**(11): 653–65.
55 Li, Y. (1986) 'Fabric wetting factors', *Textile Asia*, **30**(6): 39–41.
56 Li, Y. and Luo, Z.X. (2000) 'Physical mechanisms of moisture diffusion into hygroscopic fabrics during humidity transients', *J. Text. Inst.*, **91**(2): 302–16.
57 Li, Y., Zhu, Q., and Yeung, K.W. (2002) 'Influence of thickness and porosity on coupled heat and liquid moisture transfer in porous textiles', *Text. Res. J.*, **72**(5): 435–46.
58 Wehner, J.A. (1987) Moisture Transport through Fiber Networks. PhD thesis, Princeton University.
59 Wang, Z., Li, Y., and Kwok, Y.L. (2002) 'Mathematical simulation of the perception of fabric thermal and moisture sensations', *Text. Res. J.*, **72**(4): 327–34.
60 Mecheels, J.H. and Umbach, K.H. (1977) 'The psychrometric range of clothing systems', in *Clothing Comfort: Interaction of Thermal, Ventilation, Construction and Assessment Factors*, N.R.S. Hollies and R.F. Goldman (eds), Ann Arbor, MI, Ann Arbor Science, pp. 133–151.
61 Woodcock A.H. (1962) 'Moisture transfer in textile systems, Part I', *Text. Res. J.*, **32**, 628–33.
62 Woodcock A.H. (1962) 'Moisture transfer in textile systems, Part II', *Text. Res. J.*, **32**, 719–23.
63 Dent, R.W. (2001) 'Transient comfort phenomena due to sweating', *Textile Res. J.*, **71**(9): 796–806.
64 Toftum, J., Jorgensen, A.S., and Fanger, P.O. (1998) 'Upper limits for indoor air humidity to avoid uncomfortably humid skin', *Energy and Buildings*, **28**(1): 1–13.
65 Kilinc, F.S., El Mogahzy, Y.E., Hassan, M., Farag, R., Elbiely, R., ElDieb, A.S., and Tolba, A. (2004) 'The tactile behavior of textile materials: New perspectives – Part I: A study on the Nature of Fabric Handle', *Proceedings of Cotton Beltwide Conference*, U.S. Cotton Council.
66 El Mogahzy, Y. and Gupta, B.S. (1993) 'Friction in fibrous materials Part II: Experimental study of the effects of structural and morphological factors', *Text. Res. J.*, **63**(4): 219–30.
67 Gupta, B.S. and El Mogahzy, Y. (1991) 'Friction in fibrous materials Part I: Structural Model', *Text. Res. J.* **61**(9): 547–55.
68 Li, Y. (2001) 'The science of clothing comfort', *Textile Progress*, **31**(1/2): 1–135.

69 http://www.testingmachines.com
70 http://www.hanatek.co.uk
71 Clapp, T. G., Peng, H., Ghosh, T. K., and Eischen, J. W. (1990) 'Indirect measurement of the moment-curvature relationship for fabrics', *Text. Res. J.*, **60**(9): 525–33.
72 Potluri P., Atkinson, J., and Porat, I. (1996) 'Large deformation modelling of flexible materials', *J. Text. Inst.*, **87**, Part I (1): 129–51.
73 Hu, J. (2004) *Structure and Mechanics of Woven Fabrics*, Cambridge, Woodhead Publishing.
74 Ghosh, T.K., Barker, R.L., and Batra, S.K. (1990) 'On the bending behavior of plain woven fabrics, Part I: A critical review', *J. Textile Inst.*, **81**: 255–71.
75 Ghosh, T.K., Barker, R.L., and Batra, S.K. (1990) 'On the bending behavior of plain woven fabrics, Part II: The case of linear yarn bending behavior', *J. Textile Inst.*, **81**: 272–87.
76 Ghosh, T.K., Barker, R.L., and Batra, S.K. (1990) 'On the bending behavior of plain woven fabrics, Part III: The case of bi-linear yarn bending behavior and the effect of fabric set', *J. Textile Inst.*, **81**: 288–99.
77 http://thwingalbert.thomasnet.com/Asset/HandleOMeter.pdf
78 Kilinc-Balci, F.S. (2003) Fabric Handle as a Critical Aspect of Comfort, INFORMS Annual Meeting, Atlanta, GA, USA.

Part II

Improving comfort in apparel

7
Improving thermal comfort in apparel

C. P. HO, J. FAN, E. NEWTON and R. AU,
The Hong Kong Polytechnic University, P.R. China

Abstract: Comfort is a key factor to be considered in clothing design. Among all the comfort factors, thermal comfort is the primary one, as an important function of clothing is to provide aids in maintaining the thermal balance of the human body and ensure that the heat loss, skin temperature, air movement and humidity at the body surface produce a sensation of comfort. There are three main approaches in the development of clothing for improved comfort, i.e. the appropriate use of textile materials, garment design and attachment of special wearable devices into the garment system. In this chapter, different design approaches to enhance the thermal comfort, particularly in terms of body cooling, are discussed.

Key words: thermal comfort, heat and moisture transfer, ventilation, clothing design.

7.1 Introduction

Comfort is a key factor to be considered in clothing design. Among all the comfort factors, thermal comfort is the primary one,[1] as an important function of clothing is to provide aids in maintaining the thermal balance of the human body and ensure that the heat loss, skin temperature, air movement and humidity at the body surface produce a sensation of comfort.[2] In general, there are three main elements that affect the thermal comfort of human body: the condition of the environment; the person themselves; and the clothing. The condition of the environment includes air temperature, mean radiant temperature, relative air velocity, and water vapour pressure in ambient air. The body exchanges heat with the surrounding environment through radiation, conduction, convection and evaporation.

Humans normally release body heat and moisture periodically during states of sleeping, standing still or moving. Large amounts of sweat and higher body temperature are the outcome of intensive activity, such as sports and exercises. If heat and moisture are not released effectively from the body, heat stress may occur and the wearer's performance will be negatively affected.[3] If the moisture vapour cannot transfer out from the garment layers, the accumulation of the vapour will lead to the increase of humidity until the vapour condenses and develops to liquid form. The undergarment may soak up this liquid and the wearer may feel discomfort at this stage. Therefore, contemporary functional garments are designed to improve the heat and moisture transfer from the wearer, no matter whether they are T-shirt, jacket or protective coat. Of course, thermal comfort may be improved by the use of functional materials such as moisture management fabrics. Garment design, in terms of style, fit and design details, can also play a crucial role in achieving the ideal performance of thermal comfort, as clothing is a 3D object that can also influence thermal comfort.

7.2 Different approaches for improving the thermal comfort of clothing

In order to maximize body heat release and sweat evaporation, some people might keep the body nude when they are exercising. However, this approach should not be encouraged especially in hot, humid weather because direct and reflected radiation from the sun or artificial lighting can increase heat gain. In this case, a layer of garment can act as a barrier to heat absorption from solar radiation.[4,5] Thus a layer of functional sport clothing is still necessary, especially for outdoor exercise. For professional athletes, a nude body is not allowed when they are participating in a match formally.

When designing sportswear, fashion designers should not only consider the style, colour, fabrication and design details, but also the functionality. Some forms of sportswear are designed primarily for wind-proof, while others are for water-repellency purposes. However, the first priority for all designs should be the retention of heat in a cold climate and the easy release of excess heat in a hot climate or under strenuous body activity. There are different approaches in the development of functional clothing, including the appropriate use of textile materials, creativity in design and attachment of special wearable devices into the garment system. The last approach is not common for daily wear, but primarily used for special garments such as fire fighters' uniforms and space suits for astronauts. In this chapter, different design approaches to enhance the thermal comfort, particularly in terms of body cooling, are discussed.

7.2.1 Appropriate use of textile materials

Clothing thermal comfort relates to many different factors of textile materials, including fibre types, yarn type, yarn smoothness, fabric structure, fabric thickness, and special material like GORE-TEX®, which contains a layer of membrane that provides wind and water protection but still allows water vapour passing through. Textile technologists can make use of available technologies to optimize the heat and moisture transfer properties of textile materials to improve clothing thermal comfort. Clothing designers should choose appropriate textile materials and place them suitably within the clothing system.[6]

Appropriate use of different fibres

Past studies investigating the body-cooling effect of hydrophilic and hydrophobic fibre have shown that the hydrophilic properties of fabric or fibre reduce the heat strain during exercise and rest, especially when influenced by wind. Thermal comfort, particularly the dryness of skin, is also related to the ability of the fibre or fabric to absorb sweat or moisture, viz. hygroscopic properties of the fibres. The greater the ability of the fibre in absorbing sweat or moisture from the skin of the wearer, the better it is at keeping the skin dry. The moisture absorbency of the fibre is directly related to the moisture regain of the fibre or fabric. In the normal situation, fibre with higher moisture regain could absorb more moisture and heat from the body. Natural fibres like wool and cotton have been shown to have higher ability to absorb large amounts of moisture due to their hygroscopic properties.[7]

Many researchers have indicated that thermal resistance is determined largely by constructional factors, primarily thickness, rather than fibre type.[8] However, some fibres are more suitable for producing certain types of fabric structure. Cotton and silk can be produced in fine, lightweight weaves and have been traditionally favoured for clothing in hot climates. Wool has long been associated with protection in cold weather because it can be used to produce thick and lofty structures with good wind resistance and can therefore be utilized for overcoat garments.[9]

Appropriate fabric structure and thickness

Fabric structure and its thickness influence the heat and moisture transfer and hence thermal comfort. This is because fabric structure and thickness affect the air and moisture permeability, which play a significant role in heat and moisture transfer.

Air has low thermal transmittance. Air entrapped in small spaces experiences low convection and is therefore 'still'. Generally, more still air in the textile structure can improve the insulation value of the textile and keep the body warm,[10] and thus entrapped air has been considered a vital factor affecting heat and moisture transfer. Air flow through a fabric occurs when the air pressure is different on each side of the fabric. Air permeability is the rate of air flow through the fabric when there is a different air pressure on either surface of the fabric, and it is affected by the fabric porosity. Fabric porosity is the total volume of void space within a specified area of the fabric.[11] Apart from the fabric structure, hairy fibre and fabrics also provide a greater surface volume of still air than smooth ones, which have an insulating barrier to heat and moisture.[12]

During walking, the air spaces between the fabric layers of a porous clothing system and the skin will change according to the walking speed and rhythm. This movement of fabric will cause air penetration in and out of the clothing system and thus reduce the heat and moisture transfer resistance of the clothing system,[13] depending on the fabric's pore size. Designers can utilize mesh fabric in sportswear for better ventilation, as the pores facilitate heat and moisture transfer. The porous structure of mesh fabric can improve the air flow between the ambient environment and the body surface and this condition is more remarkable once the wearer is moving or during windy conditions.

Moisture permeability is one of the key properties affecting the comfort. Water evaporated at the skin surface that passes as vapour through a fabric will not clog the fabric's pores and thus, there will be continuous movement of air along with heat transfer through the fabric. On the other hand, the wearer's comfort is reduced if skin moisture cannot be efficiently transported to the garment surface.

Liquid water transport properties of fabrics are also important to comfort. The wearer will experience discomfort if sweat accumulates at the skin surface. Fabrics with poor water absorption and wicking properties, mean that sweat cannot be transported to the outer surface for evaporation, which is necessary for the release of body heat.[14] Inspired by the branching xylem conduits of the plants, woven and knitted plant structured fabrics have been developed in Fan's group[15,16,17] to improve the water transport properties for improved comfort.

Use of single or multiple fabric layer(s) in the garment construction

Apart from fabric structure and its thickness, the number of fabric layers also contributes to the thermal insulation and moisture vapour resistance of the clothing system. Previous sections have noted the importance of still air as a barrier to the exchange of warm and cool air between the skin surface and the environment. The same theory can be applied to clothing systems with more than one single layer of fabric. In general, more layers create additional layers of still air resulting in higher thermal insulation and moisture vapour resistance compared to a single layer of fabrics.[12]

As mentioned in the previous section, the warmth of the fabric is governed by entrapped air, with thicker fabrics having more room to keep still air within the fibres or the fabric structure itself.[9] Hence, in designing the clothing that facilitates the easy release of body heat and sweat or moisture, a thin and single layer of fabric should be considered as the first priority.

Special undergarment construction to improve ventilation

For some functional outerwear, use of windproof or waterproof shell fabrics is required, which tend to be less permeable to perspiration. To solve this problem, one may use breathable fabrics; however, the breathability of these fabrics is limited. To further improve the release of moisture through the windproof or waterproof clothing, some designers have developed special constructions between fabric layers.

Gioello[18] developed a ribbed ventilating undergarment (see Fig. 7.1) which was worn beneath a non-porous outer garment. In this design, he put a series of parallel raised ribs or cords to form the air channels. The channels would contact the base of the outer garment so that a wider air gap between the channels was formed. By creating a wider gap between the undergarment and the outer garment, the body

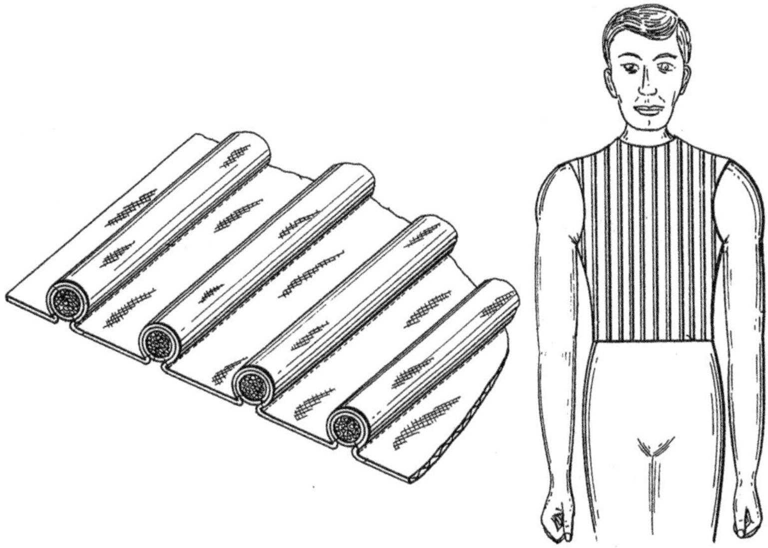

7.1 Ribbed ventilating undergarment for protective garments. Source: Gioello, 1984.[18]

heat and moisture transferred from the undergarment could be released from this air gap. The inventor further improved this idea. He applied two layers of ribs or cords together, so as to ensure that the outer garment was separated from the under garment. By using porous material in the under garment touching the skin, moisture vapour could be more easily transferred to the air channels.[19]

Frim et al.[20] invented a cooling vest for aircrew. This garment contained an inlet and outlet port for a gaseous fluid for the cooling purpose. This cooling vest should be worn under the flight overall. In this design, the inventor created a bigger air gap inside the vest to improve the circulation of gaseous fluid. The first inner layer was made of a spacer material, the second of an impermeable material with holes, and the third with more spacer materials. The inventor constructed it into a wave shape, so that it could increase the air gap by extending the distance between the second and the fourth layer which was made of impermeable materials. In this construction, the second and fourth layers should be fastened together along the edges. As the spacer materials in the middle created a bigger air gap, with the aid of the impermeable materials at both sides, more gaseous fluid could be incorporated into this air gap to achieve the cooling effect. Moreover, two impermeable materials could reduce the chance that the gaseous fluid would flow out through the fabric layer, thus the cooling effect could be circulated for a longer period than in the normal cooling vest.

Moretti[21] invented a breathable coat which contained several fabric layers between the shell and the lining (see Fig. 7.2). The concept of the design was to release moisture vapour from the body, and through an enlarged air gap by special

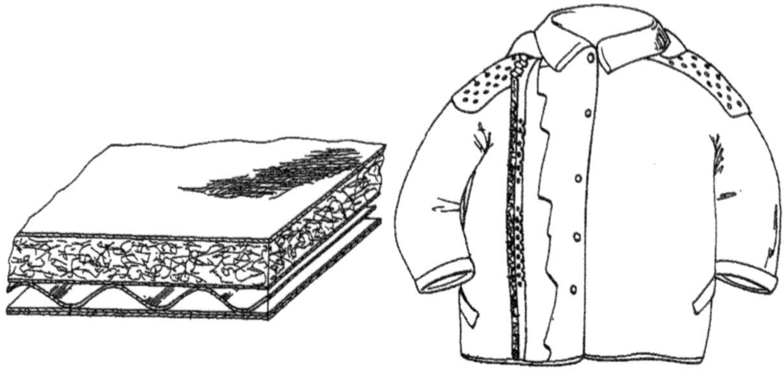

7.2 Breathable garment to be worn to improve the comfort of the human body. Source: Moretti, 2001.[21]

construction of the material, such moisture vapour could then move upward and accumulate in the shoulder areas. Holes would be created at the shoulder parts of the garment to allow the accumulated moisture vapour to be released to the ambient environment. For the construction of the material, the inventor proposed several different means to create the air space within the material. To simplify, the first layer of the material was the lining fabric, in which holes were formed on the fabric structure. Through these holes, moisture vapour could pass to the second layer of the material which was a propping up inter-space created by a sheet of rigid undulated fabric, a pile cloth-like material, or a plurality of small tubes arranged side by side. The moisture vapour would be kept within this air gap and by the aid of the opening at the shoulders, the warm air could be released to the ambient environment. The third layer was a layer of padding made of hydrophilic fibres such as cotton wool or felt. The function of this padding was to enable the still air to be retained between the fibre for maintaining insulation from outside. Mainly, this layer was used to keep the wearer warm. The final layer was the outer shell fabric combined with a membrane for enhancing its protective function to keep it impermeable to water or storms. This invention claimed that the special structure could contribute to improved air circulation inside the layers, but at the same time could protect the wearer from the cold and rain.

Development of functional textiles

Clothing comfort may also be improved by making use of recently developed functional textiles.

Moisture management fabric

Moisture management fabric is widely applied in sport clothing, high value casual wear and uniforms. The concept of this technology is to have a quicker drying rate and provision of efficient movement of moisture away from the skin. The mechanism of moisture management fabric could be divided into five states.[22]

1. Uptake of moisture from the skin surface.
2. Removal of moisture away from the skin and transport through the fabric surface.
3. Spreading of moisture within the fabric structure.
4. Absorption of moisture within suitable fibres: 'dynamic' fabrics usually contain an 'outer layer' of hydrophilic fibres to absorb and store sweat away from the skin surface.
5. Evaporation of the moisture from the fabric surface.

Coolmax® by INVISTA, Dri-FIT by Nike, and ClimaCool® are examples of this type of moisture management material.

Phase change material

Phase change material (PCM) such as paraffin can be incorporated into textile materials through spinning dope, and mixing it into the insulating foams or coating, so providing a thermal regulating function. When the environmental temperature reaches the PCM melting point, the physical state of PCM in the fabric will change from solid to liquid, resulting in the absorption of heat; heat is released when the temperature reaches the freezing point of the PCM.

Textile fabrics or garments with phase change material have been used as a pre-cool vest invented by the Australian Institute of Sport at the 2004 Athens Olympics. The vests were aimed at reducing the core temperature of the athletes so as to avoid heat stress. Some designers also tried to use the phase change material on protective garments such as fire-resistant cooling suits and cold protective garments, in which PCM is used to assist the wearers in temperature regulation under changing environmental conditions.

7.2.2 Garment design

Garment fit

Garment fit is considered as one of the elements that influence thermal insulation and evaporative resistance,[23] and many researchers have studied this factor. McCullough et al.[24] investigated how garment design influences the thermal insulation value of clothing. By comparing tight-fit and relatively loose-fit long trousers, they found that loose-fit trousers provided higher insulation than tight-fit trousers. However, the researchers also stated that movement could circulate air inside the trousers and thus increase the convective heat transfer of the loose-fit trousers more than tight-fitting trousers.

Havenith et al.[25] conducted research on the relationship of garment fit and clothing insulation. By testing three clothing ensembles on four male subjects (two wearing loose-fit and two tight-fit clothes) while sitting and walking, and under three wind speed conditions (air speed <0.1m/s; air speed = 0.7m/s; and air speed = 4.1m/s). The results showed that tight-fit clothing had 6–31% lower insulation than loose-fit ones. Additionally, the difference was highest during sitting and decreased in windy conditions.

Chen et al.[26] conducted an investigation on the effects of garment fit on thermal insulation and evaporative resistance by using a perspiring fabric manikin (Walter). Three types of jackets (woven poplin, woven denim and knitted fabric) were made in different sizes (five for each style) and tested under no-wind and windy conditions (air velocity = 2.0 ± 0.5 m/s). The results showed that, under no-wind conditions, the thermal insulation increased with the thickness of the air gap when the air gap was small. The rate of increase slowed down once the air gap was thicker. And when the air gap exceeded 1 cm (equal to the girth difference of 7.5 cm between the garment girth and the body girth), the thermal insulation decreased with the further increase in the air gap. This was because the increased air gap provided more space for natural convection. A similar trend was found in windy conditions, except that the thermal insulation values were smaller and reached maximum when the air gap was about 0.6 cm (a girth difference of 5 cm between the garment and the body). In windy conditions, tight-fitting garments could provide better performance to keep the wearer warm. The authors concluded that clothing construction and garment fit played a major role in thermal insulation and evaporative resistance compared to fabric properties.

The above studies were carried out under relatively low levels of body motion and wind, and at high levels of body motion and wind. Loose-fit clothing allows bigger spaces between the body and the material, thus more airflow is allowed and the replacement of warm and cool air should be more active and effective.[27]

Open apertures or vents in garment design

Open apertures or vents may be designed in garments so as to create ventilative cooling during body motion. Body motion can contribute to the release of body heat and moisture, and When people move, the air within the microclimate of the garment can be forced into the ambient environment through the openings of garments or pores in the fabric.[25–31] The air space between the skin and the inner fabric layer changes over time, depending on the level of activity and movement. During body movements, air must go in and out as the fabric moves outward and inward towards the skin surface, thus leading to ventilation.[32] This ventilation of the microclimate of the clothing can be called the 'pumping effect'.[33,34]

Designers tend to use materials with special porous knitted or woven structures, so that body heat and moisture vapour can be released through the pores. For examples, Zelano[35] used lace fabrics and put them into the garment for ventilation purposes (see Fig. 7.3). In his design, he put lace as stripe-like openings on various positions of the garment such as front, back panel, sleeves and trousers. For the

7.3 Ventilated clothing. Source: Zelano, 1944.[35]

suit, the lace was not only put on the shell fabric but also on the lining layer. Lace, in this case, could conduct the body heat and moisture vapour away because of the porous structure of this fabrication.

Rudman[36,37] developed a ventilated garment system which involved the concept of air circulation (see Fig. 7.4). In this design, he tried to put the vents at the top of the shoulders, in two side seams and at the back. His design concept was to allow ambient air to pass into the body skin surface from the mesh panels at the side seams, then convectively move upward and release out through the mesh panels at the shoulders. This ventilation could help to carry the body heat and moisture vapour out of the microclimate next to the body surface. The vents could be fastened using concealed zippers.

Apart from putting flat mesh or other vents on the garment, some designers also invented some three dimensional vents attached on the garment for improving the body ventilation. For example Foster[38] invented an aperture which looked like a normal pocket with the flap covered (see Fig. 7.5). The flap was designed with mesh fabric

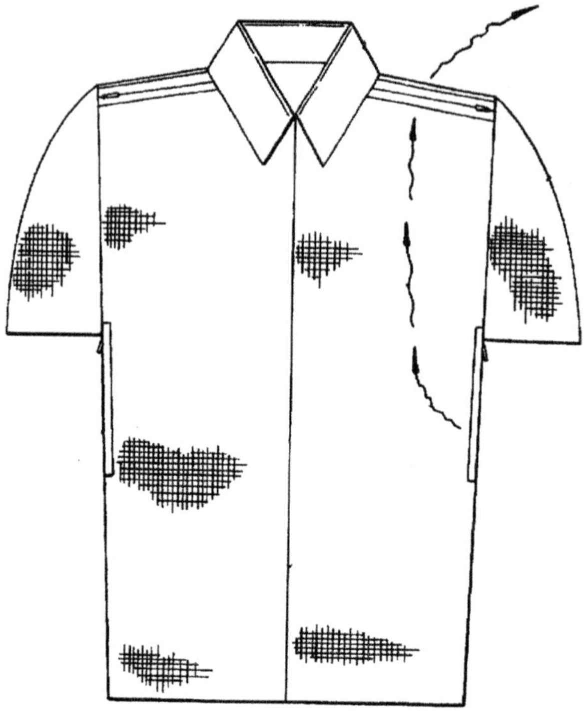

7.4 Convectively ventilated garments having protective shield layers. Source: Rudman, 1998.[36]

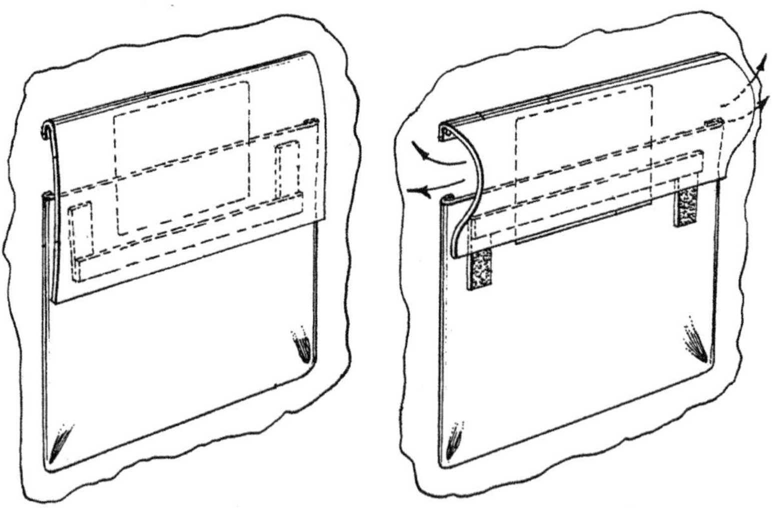

7.5 Garment ventilation apertures with cover flap. Source: Foster, 1988.[38]

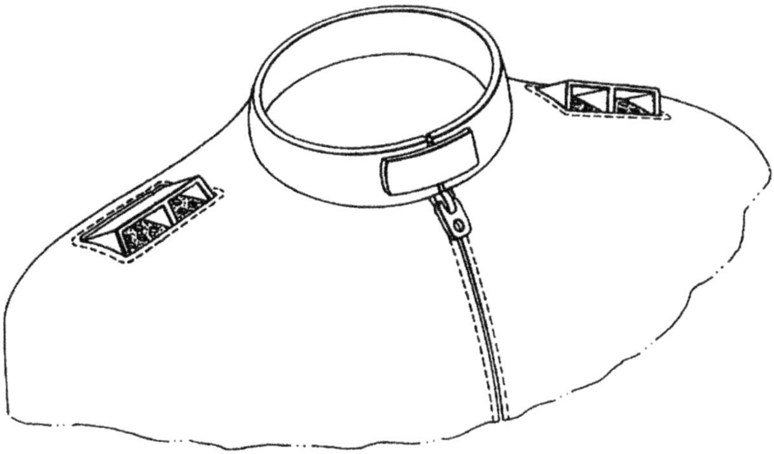

7.6 Garment with structural vent. Source: Sleesen, 1998.[39]

inserted underneath on the body. In the normal case, this flap was secured to the outer garment just like a regular pocket-closing position. If the wearer needs to have more air ventilation within the garment, he could put the flap in an arched position, fixed by the velcro fastener. The arched shape provided a tunnel over the mesh and the air could come in through the mesh covered by this channel. Sleesen[39] put a scoop-like airflow design on the jacket which was worn for motorcycling (see Fig. 7.6). The inventor put a three-dimensional scoop on the shoulders, so when the wearer bent down to control the motorcycle, the air could enter from the scoop located at the shoulders, thus taking the body heat and moisture vapour out of the back hem of the jacket.

Ruckman et al.[40] conducted an investigation on the effectiveness of zip openings (underarm sleeve opening; side seam opening; and both underarm sleeve and side opening; and without opening) of two jackets with a waterproof breathable fabric and a waterproof non-breathable fabric (polyurethane coated). The test was conducted on six subjects walking in a chamber. The authors found that the design of the pit zip openings could contribute to heat loss during an exercise period, irrespective of the use of breathable materials.

The location of openings or vents in the garment may be optimized. Ho et al.[41] investigated the effect of the location of openings on a T-shirt on the heat and moisture transfer from the wearer. The test was conducted by using a moveable thermal manikin (Walter)[42,43] in the still air condition, while the manikin was in the standing position and walking motion. The results showed that openings or mesh panels located at the two vertical side panels along the side seams provided the best ventilation effect to the wearer. This is because the air gap between the fabric layer and the body is the largest at the sides of the body trunk. The action of moving arms stretched the side seams of the T-shirt, thereby the fabric layer near to the side seams moved accordingly. Such action strengthened the pumping effect so that the body heat and moisture could be more effectively released through the openings or mesh panels located at the two vertical side panels along the side seams.

Garment design to increase the air circulation between the skin surface and the garment

Garments normally hang on the shoulders and because of the natural drape of the fabric and gravity, the garment tends to touch the skin at the chest and the upper part of the back. When doing exercise, the fabrics at these areas may stick to the skin once the wearer sweats. Some advanced materials can absorb or wick away the sweat very quickly, however it does not change the fact that the close contact between the garment and the body at the chest and the upper part of the back may block the air ventilation of the body. To improve air circulation at the body surface, garments are designed to keep a distance from the skin surface. Bengtsson et al.[44] invented a garment for vigorous physical activities in warm environments (see Fig. 7.7). The garment was comprised of cords, a

7.7 Garment for use in vigorous physical activities. Source: Bengtsson, 1980.[44]

comparatively open warp knit fabric at the inner surface of the garment. The inventor placed the cords vertically and bound them with the base knit fabric to create a plies of vertical air channels. Due to the stiffness and the thickness of the cords, it could increase air movement in those air channels closely adjacent to the skin.

Moretti[45] invented a garment in which spacer objects were put under the shoulder areas in order to keep the fabric layer of the garment away from the shoulders. On the construction of the garment, a pair of vents formed by mesh fabrics were created at the top of the shoulder area. This design could create a bigger air space between the garment and the shoulders, and the opening of the vented details could help to release accumulated warm air from the shoulders. He applied this design concept to outer garments such as coats and jackets.

7.2.3 Attachment of wearable devices

Air cooling system

During hot weather, people used electric fans to keep the body cool before the invention of air conditioning systems. The fan system has been adopted in the garment system for specific occasions. Normally these fans are lightweight and small enough to be installed on the garment.[46] Apart from using fans, some of the air cooling garments use compressors to pump pre-chilled air into a bladder, and then force the air to circulate around the body through tubes, or other engineered openings in the system.[47] Figure 7.8 is an example of a garment using the air cooling method to provide ventilation to the body.

Liquid cooling system

Liquid cooling systems require a fluid reservoir, a circulating pump and connecting hoses which attach to the garments. The cooled liquid is circulated through the garment and then returned to the reservoir for re-chilling. This circulated mechanism provides lower temperatures to the wearer.[46,47] One of the examples of this application is the Liquid Cooling and Ventilation Garment (or LCVG) which is worn by the astronauts of NASA (see Figure 7.9). The space environment is a vacuum and thus heat cannot be lost through heat conduction. In order to maintain comfortable core body temperature during extra-vehicular activity, a network of flexible tubes circulates cool water in direct contact with the astronaut's skin.

Ice or cold pack system

This system uses ice or a gelled coolant to keep the body cool. However, the duration of this cooling system is shorter than the air or liquid water-cooled

7.8 Jackets with two fans to ventilate the body. Source: *Ming Pao Daily News*, Hong Kong (2004).

garments, and the ice packs or gelled coolants require replenishment.[47] Wearing cooling garments during exercise is, however, not practical in a formal competition as the mobility of the wearer is restricted by the equipment.[47] These kinds of garments can therefore only be worn prior to the competition or during resting periods. This wearable device approach is not common for daily wear, as the electronics and extra equipment would hinder normal activities.

7.3 Conclusions

In applying the different approaches for improving the thermal comfort of clothing, the designer should take a holistic view and be creative in maximizing the functionality of the garment without compromising other important factors such as aesthetics, cost, etc.

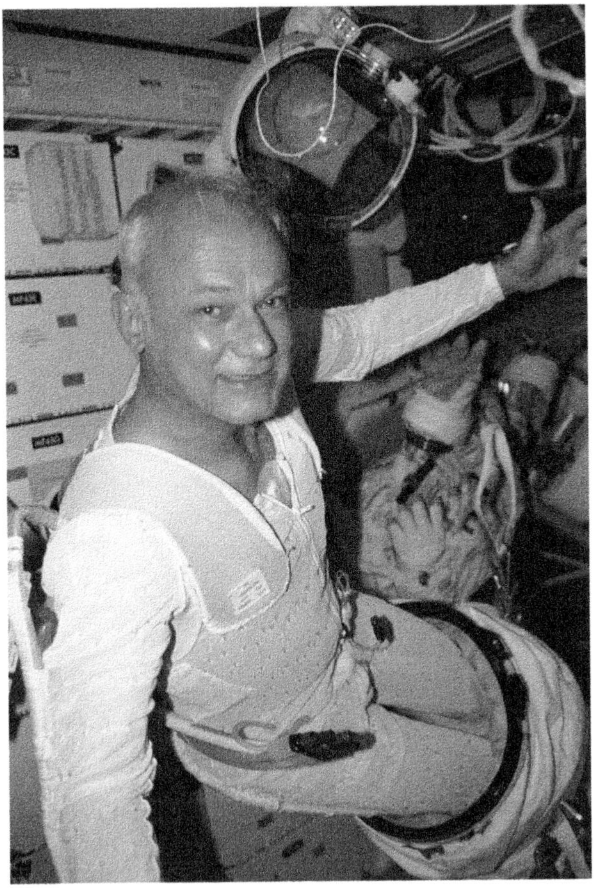

7.9 Liquid cooling and ventilation garment.

7.4 References

1. Hollies N R S and Goldman R F (eds), 1977, *Clothing Comfort, Interaction of Thermal, Ventilation, Construction and Assessment Factors*, Michigan, Ann Arbor Science.
2. Yee S L, 1988, *The Design and Manufacture of a Female Clothing System for the Hospital Environment*, PhD thesis, UK, the University of Leeds.
3. Sawka M N and Young A J, 2000, Physical exercise in hot and cold climates. In William E, Garrett W Jr and Donald T K, ed. *Exercise and Sport Science*, Philadelphia, Lippincott Williams & Wilkins, pp. 385–400.
4. Foss M L and Keteyian S J, 1998, *Temperature Regulation: Exercise in the Heat and Cold, Fox's Physiological Basis for Exercise and Sport*, Boston, McGraw-Hill.
5. Gavin T P, 2003, 'Clothing and thermoregulation during exercise', *Sports Medicine*, **33**(13): 941–7.
6. Wilensky J, 2001, 'Clever Clothing', *Human Ecology*, **29**(3): 9–10.
7. Kwon A, Kato M, Kawamura H, Yanai Y and Tokura H, 1998, 'Physiological significance of hydrophilic and hydrophobic textile materials during intermittent

exercise in humans under the influence of warm ambient temperature with and without wind', *European Jornal of Applied Physiology*, **78**: 487–93.
8. Holcombe B V, Brooks J H, Schneider A M and Watt I C (1988), *Conference Proceedings: Annual World Conference of the Textile Institute*, Manchester, The Textile Institute, p. 436.
9. Ukponmwan J O, 1993, 'The thermal-insulation properties of fabrics', *Textile Progress* **24**(4): 13–23.
10. Louis I W, 1970, 'Moisture vapour transmission in textiles fabric', Textiles of Comfort, 3rd Shirley International Seminar, Manchester, Shirley Institute, pp. 2–23.
11. Collier B J and Epps H H, 1999, *Textile Properties that Influence Performance, Textile Testing and Analysis*, New Jersey, Prentice Hall.
12. Grise K S, Goswami B C and Hassenbeohler C B, 1983, 'Heat flow characteristics of single and multiple layer fabrics', *ASHRAE Transactions*, **89**(2782): 353–60.
13. Ghali K, Ghaddar N and Jones B, 2002, 'Modeling of heat and moisture transport by periodic ventilation of thin cotton fibrous media', *International Journal of Heat and Mass Transfer*, **45**: 3703–14.
14. Tokura H, Yamashita Y and Tomioka, S, 1982, 'The effects of moisture and water absorbency of fibres on the sweating rates of sedentary man in hot ambient environments', in S. Kawabata, R. Postle and M. Niwa eds, *Objective Specification of Fabric Quality, Mechanical Properties and Performance*, Osaka, Textile Machinery Society of Japan pp. 407–17.
15. Fan J, Salar M, Szeto Y and Tao X, 2007, 'Plant structured fabrics', *Material Letter*, **61**: 561–5.
16. Sarkar M, Fan J, Szeto Y and Tao X, 2009, 'Biomimetics of plant structure in textile fabrics for the improvement of water transport', *Textile Research Journal*, **79**(7): 657–68.
17. Chen Q, Fan J and Sarkar M, 2010, 'Biomimetics of plant structure in knitted fabrics to improve the liquid water transport properties', *Textile Research Journal*, **80**(6): 568–76.
18. Gioello DA, 1984, *Ribbed ventilating undergarment for protective garments*, US patent application 4451934.
19. Gioello D A, 1996, *Multilayered ribbed ventilating garment*, US patent application 5515543.
20. Frim J and Michas R D E, 1993, *Micro-climate conditioning clothing*, US Patent No. 5243706.
21. Moretti M P, 2001, *Breathable garment to be worn to improve the comfort of the human body*, US patent application 6263511B.
22. Nyoni A B, 2003, *Liquid transport in nylon 6.6 yarns and woven fabrics used for outdoor performance fabrics*, PhD thesis, UK, The University of Leeds.
23. Fan J, 1989, *A study of heat transfer through clothing assemblies*, PhD dissertation, Department of Textile Industries, UK, The University of Leeds.
24. McCullough E A, Jones B W and Zbikowski P J, 1983, 'The effect of garment design on the thermal insulation values of clothing', *ASHRAE Transactions*, **89**: 327–51.
25. Havenith G, Heus R and Lotens W A, 1990, 'Resultant clothing insulation: a function of body movement, posture, wind, clothing fit and ensemble thickness', *Ergonomics*, **33**: 67–84.
26. Chen Y S, Fan J, Qian X and Zhang W, 2004, 'Effect of garment fit on thermal insulation and evaporative resistance', *Textile Research Journal* **74**(8): 742–8.
27. McArdle W D, Katch F I and Katch V L, 2001, *Exercise Physiology, Energy, Nutrition, and Human Performance*, 4th ed, Philadelphia, Lippincott Williams and Wilkins.

28. Vokac Z, Kopke V and Keul P, 1973, 'Assessment and analysis of the bellows ventilation of clothing', *Textile Research Journal*, **43**(10): 475–82.
29. Nielsen R, Olesen B W and Fanger P O, 1985, 'Effect of physical activity and air velocity on the thermal insulation of clothing', *Ergonomics*, **28**(12): 1617–31.
30. Bouskill L M, Havenith G, Kuklane K, Parsons K C and Withey W R, 2002, 'Relationship between clothing ventilation and thermal insulation', *AIHA Journal*, **63**: 262–8.
31. Haghi A K, 2004, 'Moisture permeation of clothing: a factor governing thermal equilibrium and comfort', *Journal of Thermal Analysis and Calorimetry*, **76**: 1035–55.
32. Ghaddar N, Ghali K and Jones B, 2003, 'Integrated human-clothing system model for estimating the effect of walking on clothing insulation', *International Journal of Thermal Sciences*, **42**: 605–19.
33. Olesen B, Sliwinska E, Madsen T and Fanger P O, 1982, 'Effect of body posture and activity on the insulation of clothing: measurement by a movable thermal manikin', *ASHARE Transactions*, **88**: 791–805.
34. Vogt J J, Meyer J P, Candas V, Libert J and Sagot J C, 1983, 'Pumping effects on thermal insulation of clothing worn by human subjects', *Ergonomics*, **26**(10): 963–74.
35. Zelano J, 1944, *Ventilated clothing*, US patent application 2391535.
36. Rudman F, 1998, *Sunlight protecting garments having convective ventilation*, US patent application 5727256.
37. Rudman F, 2005, *Convectively ventilated garment having protective shield layers*, US patent application US 2005/0273903A1.
38. Foster R W, 1988, *Garment ventilation apertures with cover flap*, US patent application 4731883.
39. Sleesen M F, 1988, *Garment with structural vent*, US patent application 5704064.
40. Ruckman J E, Murray R and Choi H S, 1998, 'Engineering of clothing systems for improved thermophysiological comfort: the effect of openings', *International Journal of Clothing Science and Technology*, **11**(1): 37–52.
41. Ho C, Fan J, Newton E and Au R, 2008, 'Effects of athletic T-shirt designs on thermal comfort', *Fibers and Polymers*, **9**(4): 503–8.
42. Fan J and Chen Y S, 2002, *Thermal manikin*, US patent application 09/811.833.
43. Fan J and Qian X M, 2004, 'New functions and applications of Walter, the sweating fabric manikin', *European Journal of Applied Physiology*, **92**(6): 641–4.
44. Bengtsson A G and Mullsjo K E, 1980, *Garment for use in vigorous physical activities*, US patent application 4195364.
45. Moretti M P, 2002, *Ventilated item of clothing*, US patent application 6442760B2.
46. Watkins S M, 1984, *Clothing: The Portable Environment*, Iowa State University, State University Publishers.
47. Laing R M and Sleivert G G, 2002, 'Clothing, textiles and human performance: a critical review of the effect of human performance of clothing and textiles', in Bookset A ed., *Textile Progress*, Manchester, Textile Institute.

8

Improving moisture management in apparel

R. S. RENGASAMY, Indian Institute of Technology, India

Abstract: It is very important for the human body to maintain a balanced temperature with its surroundings, so the heat generated by the body has to be dissipated or preserved by clothing in accordance with the environmental conditions. The secretion of sweat and its subsequent evaporation on the skin or into the atmosphere depends on the water vapour permeability of clothing, and with increased activity, sweat may occur in a liquid form which must be transported quickly by the capillaries of the clothing if it is to evaporate into the atmosphere. The transport of moisture is influenced by several factors such as the heat of sorption and the condensation of moisture into liquid. Ventilation and convection play a critical role in moisture management, which in clothing involves coupled heat and liquid moisture transport. This chapter discusses the various mechanisms and factors involved in the transport of moisture from the skin to the environment by the clothing, clothing requirements for different environmental conditions, and a review of developments in improving moisture management in clothing.

Key words: heat balance, sweat, diffusion, vapour pressure, condensation, heat of sorption, permeability, capillaries, environment, developments.

8.1 Introduction

Thermo-physiological comfort is concerned with the heat balance of the body during various levels of activity. In most environments, body temperature will be above that of the environment and in order to maintain the heat balance between the body and its surroundings, the body generates heat through its metabolism. The human body will try to maintain its core temperature at about 37 °C by thermal balance. The sum of the internal heat generated through metabolism and the external heat received by the body have to be matched by the heat loss from the body if the core temperature is to be maintained. Around 15–30% of the energy generated from food is spent in useful work and the rest is wasted as heat. Various physical activities generate different amounts of heat within the body, e.g. 90 watts of heat are developed by a body at rest; 140–350 watts by walking, depending upon the walking speed; 1400–1500 watts by sprinting (Keighley, 1985). Therefore to maintain the body core temperature, these levels of heat must be dissipated by dry and wet/latent heat transfer.

Dry heat transfer at the surface of the skin takes place by means of conduction, convection and radiation (longwave and shortwave). Wet heat transfer takes place through the evaporation of moisture secreted on the skin. Conduction of heat from the body to the environment relies on the temperature difference between the skin and environment and on the thermal resistance of the clothing. The mechanism of cooling by conduction is not always effective as it is reversible and heat can flow from outside into the body in hot summer weather, so reducing heat loss from the body. In extreme cold weather, the body loses heat rapidly, leading to a loss of thermal balance. Lower levels of physical activity will lead to a fall in body temperature due to a large reduction in heat generation so the body heat must be conserved by increasing insulation with suitable clothing, especially during cold weather.

Most fibres have a similar, high thermal resistance but this has no effect when compared to the extremely high resistance of air trapped within the clothing. The free movement of air determining the convective loss through the clothing is restricted by the fibres which form a three-dimensional network within an enclosed space. Clothing is a good insulator for thermal conduction and convection. The heat gained by the bare body through radiation during hot conditions is around 100W; and the body dissipates around 130W through radiation in cool conditions (Pan, 2008). The heat exchange between body and environment through radiation is quite high compared to that of conduction and convection.

8.2 Transport of perspiration

During normal activities and weather conditions, clothing easily dissipates heat through conduction and convection, as the level of heat generated is not high. When activity is increased, dry heat transfer becomes inadequate to dissipate heat from the skin and the body must cool through wet heat transfer, i.e. by perspiration. Whether we are aware of it or not, perspiration takes place all the time, not only during periods of intense activity. Moisture from within the body is exuded through the pores of the skin and evaporates on reaching the outer surface. As it does so, it draws the latent heat of evaporation from the body which is needed to convert liquid water to vapour. One gram of water takes 2260 joules of heat energy from the body in evaporating. We are largely unaware of the process of perspiration when water is removed at the rate of about 500 g per day (13W). This is called 'insensible perspiration'. The water exuded through the skin appears initially as a liquid which evaporates immediately and forms moisture vapour (Slater, 1999). This vapour is then removed from the vicinity of the body, either by convection or diffusion through the pores of the clothing to the outermost surface and is carried away by the air. The form of perspiration of which we are aware is known as 'sensible perspiration' and manifests as a high sweating rate during heightened physical activity. Sensible perspiration occurs in liquid form and wets the clothing which is in contact with the skin when the ability of the clothing to transport water vapour from the skin to the outer environment is lower than the rate of perspiration. The water then condenses on the skin and must be transported to the outer surface of the clothing. Consequently, the comfort of a garment is greatly affected by transport of moisture in vapour or liquid form through the clothing to the outside.

If one component of the clothing is impermeable or has a low permeability, water will condense into liquid during the transport of water vapour at the plane of that component and the transfer of additional moisture coming from the skin will enter the liquid layer so formed. The process of moisture condensation will then move through the inner layers of clothing to the skin. At the onset of sensible perspiration, the moisture exuded through the skin gives a sense of discomfort to the wearer. In this case, the clothing must have the ability to transport liquid moisture as quickly as possible to the outer surface of the clothing to be spread over a large area for evaporation into the atmosphere. The liquid water in the pores will block the passage for the diffusion of vapour and cause increased discomfort with the onset of the condensation process.

The transport of perspiration depends on several major factors, such as water vapour pressure and relative humidity of the micro climate, clothing and outside environment, absorption-desorption of the clothing medium, heat of sorption and the saturation/condensation of vapour into liquid or water in the clothing. These are briefly reviewed in the following sections.

8.2.1 Water vapour pressure and relative humidity

Moisture vapour molecules constantly vibrate and collide with each other and the consequent pressure exerted on their surroundings is known as vapour pressure. The vibration of vapour molecules is a phenomena induced by thermal energy and the energy of vibration depends on the temperature of the vapour. At higher temperatures, the kinetic energy of water vapour molecules increases, so increasing the pressure.

Relative humidity (*RH*) is the ratio of actual water vapour to saturation water vapour pressure at a given temperature (i.e. p_v/p_{sv}). The potential energy of the water or water activity a_w, is the force which drives water movement from regions of higher to lower activity. When water vapour reaches saturation (i.e. a_w or relative humidity is 1), the number of water molecules moving from liquid to vapour, and those moving from vapour to liquid, will be constant. Any further addition of water molecules into the vapour phase leads to condensation of vapour into liquid. This maximum value of one for the water activity is valid only for a flat pure water surface. For water containing salts, the maximum relative humidity is less than one, because the salts act as nuclei for the water to condense.

The saturated water vapour pressure in mmHg and absolute temperature, T in K are related as (Gibson et al., 1995),

$$P_{sv} = 4.607 \exp\left[17.06\left(\frac{T - 273.15}{T - 40.25}\right)\right] \qquad [8.1]$$

At very high humidities, liquid water may be held by the forces of surface tension in the capillary spaces between fibres or in crevices in the surface of the fibre. The equilibrium water vapour pressure p_{sv} over a concave curved surface of radius r_p is lower than that over a plane surface, due to the surface tension of the water (Morton and Hearle, 1962). There are several factors controlling water activity within a system which have been summarised mathematically in the well known Kelvin equation (Pan and Sun, 2006) as:

$$RH = a_w = \frac{p_v}{p_{sv}} = e^{-\left(\frac{2\gamma M \cos\theta}{r_p \rho RT}\right)} \quad [8.2]$$

where, γ is the surface tension of water, M is the molecular weight of water (i.e, 18), θ is the contact angle of water with the substrate, r_p is the pore radius, ρ is the density of water and R is the gas constant. The above equation is valid except $r_p \to 0$, that implies $p_v \to 0$.

8.2.2 Capillary condensation and hygroscopic behaviour

Water activity rises with increasing values of pore radius and water temperature. The surface tension of water slightly decreases with its increasing temperature, causing a further rise in water activity. The water activity may exist down to a value of r_p = 0.1 μm (100 nm), which is the size of the normal free path of water vapour molecules. According to the Kelvin equation, the saturation vapour pressure (p_{sv}) reduces inside narrower capillaries. Therefore at the same vapour pressure, the saturation point becomes lower in smaller pores, so in hygroscopic materials, as water condenses inside the pores, the tightest pores will be the first to be filled with condensed liquid water. This premature condensation in pores is termed capillary condensation.

In hydrophobic fibre assemblies, moisture simply diffuses into the molecular networks without being chemically attracted by the molecules of the fibre, and emerges at the other side at a low water vapour pressure. In hydrophilic fibre assemblies, the presence of hydroxyl, amine and carboxyl groups on the fibre molecule chains chemically attracts the water vapour diffusing through the fibre molecular networks. A mono-layer of water molecules form on the pore walls by chemically bonding with the water-attractive groups on the fibre molecule chains. These water molecules cannot easily escape to form the vapour phase. Thus they exert a much lower vapour pressure than does free water at the given temperature. Over the bound water, the newly arrived molecules are held together by hydrogen bonding which facilitates the formation of multi-molecular layers of water. When RH increases further, the pores between the fibres become completely filled with liquid water, leading to condensation in the capillaries. This hygroscopic behaviour causes condensation in the pores, even when the environment surrounding the fibres is not saturated.

At a given relative humidity, there will be no tendency for water to evaporate from capillaries in which the radius of the water meniscus (r) is less than that given by the equation when $\cos\theta$ is one, i.e. for a completely wettable fibre:

$$r = \frac{0.47}{2 - \log_{10} a_w} \qquad [8.3]$$

The number of water molecules that can be accommodated on a meniscus of radius given by the above equation is plotted in Fig. 8.1. It is clear that the capillary water can play a part only at very high humidities, i.e. RH > 99% which corresponds to a meniscus radius of 110 nm or a pore radius of 55 nm, if the pore is circular. Capillary condensation occurs only when large amounts of water vapour (high perspiration rate) are transported across clothing with a very low water vapour transport rate. Capillaries with $r_p > 55$ nm fill up with liquid only when they are in direct contact with liquid water on the skin, and capillary condensation does not occur.

8.2.3 Heat of sorption/absorption

Hygroscopic textile fibres absorb water vapour according to their chemical composition and structure. As they absorb moisture, heat is produced, which is known as heat of absorption. This is basically the heat of solution, which occurs in the case of exothermic reaction. It results from the interactive forces between fibre molecules and water molecules. If water vapour is absorbed, there is additional heat similar to the latent heat of condensation. Heat effect and the moisture absorption in hygroscopic materials, such as textile fibres, are inseparably inter-related and the evolved heat may be expressed in various ways. The

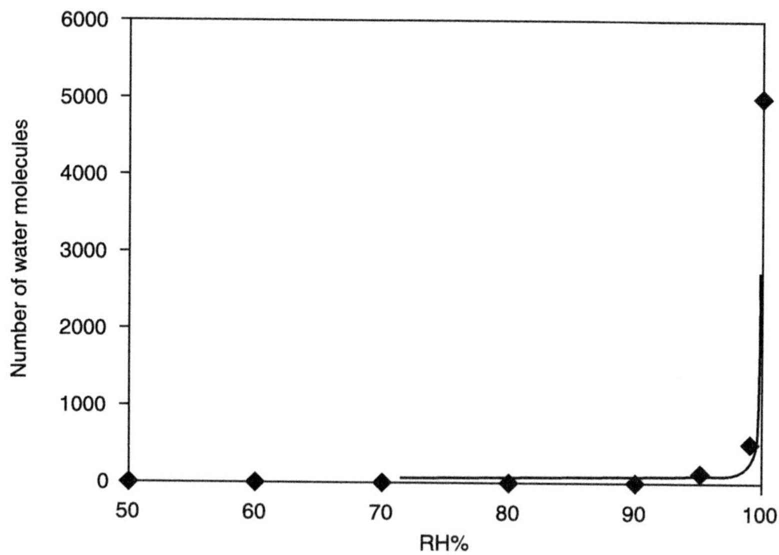

8.1 Capillary condensation: number of water molecules constricted in capillary as a function of RH.

differential heat of sorption, Q (sometimes called heat of absorption), is the heat produced when one gram of water is absorbed by the infinite mass of the material at a given moisture regain. It is expressed in joules per gram of water absorbed. The water may be absorbed from water vapour to give the value Q_v, or from liquid water to give the value Q_l. The heat produced through absorption of moisture by the fibre assemblies has a huge effect on the thermodynamics of the system as it influences the diffusion process of the moisture transmission. The heat of sorption increases the temperature of the fibrous assemblies. As the temperature of the system increases, diffusion into the fibres, coupled with the heat transfer process, will be determined by the ability of the fibres to absorb moisture. According to Cassie et al. (1939), if a textile material is surrounded by a humid atmosphere, the time required for the fibres to come to equilibrium with this atmosphere is negligible when compared to the time required for the dissipation of heat generated or absorbed when the regain changes. Crank (1975) pointed out that the water uptake rate of wool is reduced by a rise in temperature due to the heat of sorption, while Li et al. (1993) described the heat of absorption for the non-Fickian diffusion, which happens in the later stage of diffusion in case of hygroscopic material.

8.2.4 Absorption–desorption

Absorption and desorption are important processes in maintaining the micro-climate during transient conditions. A hygroscopic fabric absorbs water vapour from the humid air close to the sweating skin and releases it in dry air. This enhances the flow of water vapour from the skin to the environment when compared with a fabric which does not absorb and reduce the moisture built up in the micro-climate (Hong et al., 1988). In the absorption-desorption process, an absorbing fabric works as a moisture source to the atmosphere (Wehner et al., 1988). It also acts as a buffer by maintaining a constant vapour concentration in the air immediately surrounding it, i.e. a constant humidity is maintained in the adjoining air, though the temperature changes due to the heat of sorption. Barnes and Holcombe (1996) examined the magnitude of the differences in moisture transport caused by fabric sorption, and the perception of these differences. Adsorption of water molecules takes place below a critical temperature, due to the van der Waals forces between the vapour molecules and the solid surface of the structure. The higher the vapour pressure and the lower the temperature, the higher will be the amount absorbed. The mass of absorbed water is directly proportional to relative humidity. In a thermodynamic equilibrium, the chemical potential of the vapour is equal to that of the absorbed film. An increase in vapour pressure will cause an imbalance in chemical potential, and more vapour will transfer to the absorbed layer to restore the equilibrium (Chatterjee, 1985). The amount of water vapour which can be absorbed by the materials is dependent on the fibre regain and the humidity of the atmosphere (%). In the case of absorbent fibres, e.g. cotton or rayon, the moisture sorption is not only dependent on regain and humidity, but also on the phenomena associated with sorption hysteresis, the effect of heat and dimensional changes and elastic recovery effects caused by reduced swelling of the fibres. During swelling, the fibre macro-molecules or micro-fibrils are pushed apart by the absorbed water molecules, reducing the pore size between the fibres as well as between the yarns, thus reducing the rate of water vapour transmission through the fabric.

8.2.5 Saturation and condensation of water vapour

Water vapour will only condense onto the surface of clothing which is cooler than the temperature of the water vapour, indicating that a temperature gradient should exist across the clothing. This happens on the outer surface of the clothing when the outside temperature is very low. Localised condensation also occurs in the clothing when the water vapour equilibrium in air has been exceeded i.e., $p_a \leq p_{sv}$. When water vapour condenses onto a surface, a net warming occurs on that surface. The water molecule brings a parcel of heat with it and in turn, the temperature of the surroundings drops slightly. When a dry hygroscopic fibrous layer with a low water vapour permeability is exposed to a moist hot environment or sprayed with liquid water while the other side of the layer is subject to low temperature, condensation is likely to occur. Condensation can lead to an increase in the thermal conductivity of clothing, as the thermal conductivity of water is approximately 24 times that of air. The heat released by the rapid moisture uptake of the fibres raises the local temperature, which increases the local water vapour pressure, thus reducing the water vapour pressure gradient and water vapour transport rate. As the water uptake causes the fibres to swell, the water vapour transport rate is also reduced.

8.3 Fundamentals of moisture transfer between the human body and the environment

Moisture can be transferred in three ways: diffusion, convection, and capillary transfer; and also by a combination of these.

8.3.1 Moisture diffusion through clothing

Liquid water/sweat is secreted through the pores of the skin. On evaporation, it takes latent heat from the skin and in the process the skin is cooled. When the environmental temperature reaches skin temperature, heat loss through convection and radiation gradually come to an end. So at environmental temperatures above skin temperature, the evaporation of sweat is the only means for the body to lose heat. Sweating itself is not effective as it is the conversion of the liquid to vapour which removes the heat. This mechanism works well in a hot dry environment, but evaporation of sweat becomes a problem in hot humid climates.

When heat is supplied to the liquid water droplets, a few of the molecules present on the surface of the liquid water overcome intermolecular forces, break away from the liquid and form a vapour surrounding the liquid. The water vapour molecules kept at 0 °C and 760 mm water pressure will vibrate with a collision frequency (ratio of mean speed to mean free path) of 111×10^8 per second which is about twice the value for air (Poynting, 1997/1925). As the heat of the body surface and of sweat on the skin increases, more water molecules escape from the sweat and collide with the surrounding water vapour molecules with increasing kinetic energy, pushing the latter away from the body. The water vapour pressure in the vicinity of the skin depends on the number of water vapour molecules and their kinetic energy, both of which are defined by the temperature of the sweat/skin. Thus, through a series of collisions of water vapour molecules, water vapour may be transported from the skin to the atmosphere, provided there is a negative water vapour pressure gradient between them. In the case of the clothed body, water vapour pressure must decrease from the micro-climate region (the interface of the skin and inner layer of the clothing) to the outer layer of the clothing. The process of water vapour transport from the regions of higher water vapour pressure to the regions of lower water vapour pressure is known as 'diffusion'.

The diffusion process of vapour is analogous to heat conduction as both rely on the gradients across the clothing, the gradients being vapour pressure and temperature respectively for the diffusion and conduction of heat. In the diffusion process, the vapour pressure gradient acts as a driving force in the transmission of moisture from one side of a textile layer to the other. Water vapour can diffuse through a textile structure in two ways: simple diffusion through the air spaces between the fibres and yarns, and along the fibre itself (Fohr et al., 2002; Lomax, 1985). In the case of diffusion along the fibre, water vapour diffuses from the inner surface of the fabric to the surface of the fibres and then travels along the interior of the fibres (i.e. intermolecular pores) and its surface, to reach the outer fabric surface. The flux occurs predominantly through the air spaces of the fabric; the role of the fibres being to act as a moisture sink or source. The diffusion of water vapour follows Fickian law as

$$m = f[-D_F(p_{vsk} - p_{vo})/t] \qquad [8.4]$$

where m = mass flux of water vapour in kg/m^2.s, D_F is the diffusion coefficient of the clothing in kg/m^2.s.Pa; p_{vsk} and p_{vo} are water vapour pressures in Pa at the skin and outer surface of the clothing respectively; and t is the fabric thickness. In the clothed human body, the water vapour has to be transported from the skin to the outer layer through the normal plane of the clothing and the resistance to water vapour transport will be proportional to the thickness of the clothing. Fibres in the clothing provide barriers through which the water vapour diffuses. The measure of porosity of the fabric is less than one. The pores in the clothing vary in size and complexity (increased channel length) and are often discontinuous. Hence the water vapour resistance of fabrics can be high. The larger porosity of open fabrics, such as knitted materials, naturally offer much higher water vapour permeability and lower water vapour resistance than do woven fabrics. The diffusion coefficient of clothing is related to the porosity and complexity of the pores (Bejan, 2004) as

$$D_F = D_A\left(\frac{\varepsilon}{\tau}\right) \qquad [8.5]$$

The water vapour transport rate of clothing (kg/m^2.s) may be expressed as

$$m = f\left[\left(\frac{D_A\varepsilon}{\tau t}\right)\left(\frac{C_{vo}}{p_{vo}}\right)(p_{vsk} - p_{va})\right] \qquad [8.6]$$

where D_A is the free space diffusion coefficient of water vapour in air in m^2/s, ε and τ are the porosity and tortuosity of the pores of the clothing respectively, and C_{vo} is the concentration of the water vapour at the outer surface of the clothing in Kg/m^3. The porosity is the sum of the volume fractions of the spaces between the yarns, fibres and inter-molecular pores within the fibres/film which are available for vapour diffusion, i.e. excluding those pores which are filled with liquid water due to condensation.

The concentration of water vapour, C_{vo} is related as (Fukazawa et al., 2000):

$$C_{vo} = f\left(\frac{p_{vo}}{T_a}\right) \qquad [8.7]$$

where T_a is the absolute temperature of the air in K. The diffusion coefficient of water vapour in air at a standard atmosphere (i.e. 273 K and 101,325 Pa) is 2.2×10^{-5} m^2/s, and it varies with the absolute temperature of the air (K) and the ambient pressure, p_a.

$$D_A = 2.99\times10^{-5}\left(\frac{T_a^2}{p_a}\right) \qquad [8.8]$$

The diffusion coefficient of water vapour in air will be large at a higher temperature and lower air pressure. In high porosity clothing, the water vapour pressure gradient across the clothing and the ambient temperature will enhance the water vapour transport rate of the clothing. The resistance to water vapour transport increases with the tortuosity of the pores, with the thickness of the clothing/membrane/film and with the atmospheric pressure.

Temperature, pressure gradients and density of water vapour

The values of saturated vapour pressure (mmHg) and the density of saturated vapour (g/m^3) follow a similar pattern, both increasing with the temperature (°C) as given in Fig. 8.2. Hence, for simple diffusion of water vapour through pores following Fick's law, the gradients of pressure and the concentration of water vapour are interchangeable and are applicable for hydrophobic materials.

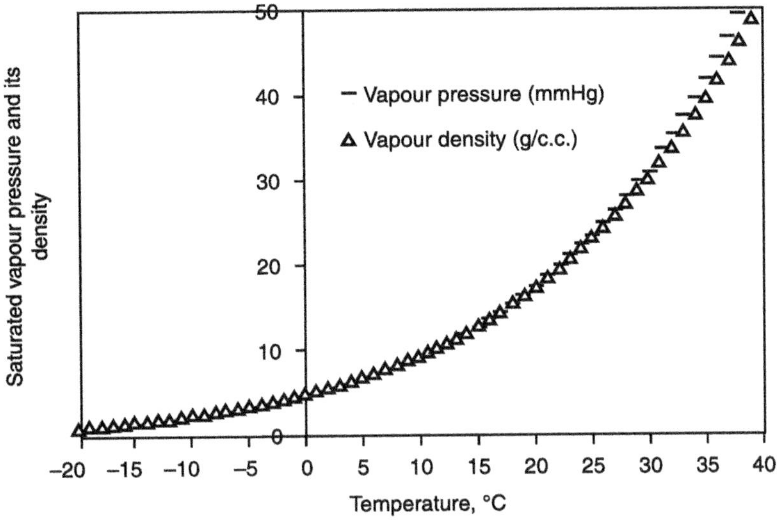

8.2 Dependence of vapour pressure and vapour concentration on temperature.

Diffusion through fibres

In general, the diffusion coefficient of fibres increases with a rise in the moisture content of the fibres. An exception to this behaviour is shown by polypropylene (Ren and Ruckman, 2003), due to its high hydrophobicity. For polypropylene fibres (which are weakly hygroscopic), the moisture sorption into the fibres can be described by single Fickian diffusion with a constant diffusion coefficient. In the case of hydrophilic fibre assemblies such as wool and cotton, vapour diffusion does not entirely obey Fick's law and is governed by a non-Fickian, anomalous diffusion. This is a two stage diffusion process. The first stage involves a quasi-equilibrium obeying Fickian diffusion with a concentration dependent diffusion coefficient, but the second stage is much slower than the first, following an exponential relationship between the concentration gradient and the vapour flux (Nordon et al., 1960; Li and Holcombe, 1992). This diffusion is time-dependent and is accompanied by the structural changes or molecular stress-relaxation process of the fibres/stiffness of the polymer chains. Under steady-state conditions, the water vapour transport is given by the relationship (Lomax, 1985):

$$wvt = DS\frac{(p_i - p_o)}{t} \qquad [8.9]$$

where p_i and p_o are the water vapour pressure at the inner and outer surfaces, t is the thickness of the medium, D is the diffusion coefficient and S is the solubility coefficient. Both the coefficients are pressure dependent, particularly at high relative humidities. The diffusion coefficient is governed by the nature of the intermolecular pore system, which is affected by the cohesive forces between polymer chains, crystallinity, density, and modifications such as cross-linking and plasticisation. The solubility coefficient depends largely on the molecular attraction between the polymer molecular chains and the water vapour, i.e. on the hydrophilicity of the medium. This can be explained by the swelling of the fibres due to absorption of moisture. Due to the affinity of the hydrophilic fibre molecules to water vapour, it is absorbed as it diffuses through the fibrous system causing the fibres to swell and reducing the size of the air spaces, thus delaying the diffusion process. Newns (1956) has explained this reduction in the diffusion rate, as caused by the stress relaxation of the fibre after swelling. Li et al. (1993) have given an account of this phenomenon; the heat of sorption produced increases the temperature of the fibrous assemblies. The moisture diffusion into the fibres is coupled with the heat transfer process, which is much slower and depends upon the ability of fibres to absorb moisture. The strength of the coupling effect is a function of a number of fibre properties, such as the moisture sorption isotherms, water diffusion coefficient, fibre diameter and heat of sorption. In the case of hygroscopic material, the later stages of the diffusion process are influenced by the absorption phenomenon. The diffusion is affected by the structural changes and the heat and moisture coupling effect which result from moisture absorption by the fibres.

Concentration-dependent diffusion

When exposed to high relative humidity or a saturated micro-climate, a hydrophilic film/fibrous assembly absorbs more moisture from the micro-climate and decreases its water vapour pressure. This increases the concentration of water in the medium, but the vapour pressure of this water remains low as few water molecules are bound to the side groups of the fibre molecule chains. The liquid water molecules which are indirectly held in the pores are not as free as those of free water. The water vapour pressure of this liquid water is low and is similar to that of ice. The diffusion rate of vapour from the micro-climate to the hydrophilic medium is controlled by the vapour pressure gradient, while the diffusion rate of vapour from the hydrophilic medium into the outer layer is controlled by the concentration gradient.

8.3.2 Moisture transfer by convection and ventilation

Forced convection is a mode of moisture transfer which takes place when air flows over a moisture layer. The mass transfer in this process is controlled by the difference in moisture concentration between the surrounding atmosphere, the moisture source and the convective mass transfer properties of the clothing. The process is governed by the equation (Incropera and DeWitt, 1996):

$$Q_m = -Ah_m(C_{vf} - C_{va}) \tag{8.10}$$

where Q_m is the mass flow by convection through area A of the fabric along the direction of the flow, C_{vf} is the vapour concentration on the fabric surface and C_{va} is the vapour concentration in the air. The flow is controlled by the concentration difference $(C_{vf} - C_{va})$ and the convective mass transfer coefficient h_m, which depends on the fluid properties as well as on its velocity. In a windy atmosphere the convection method plays a very important role in transmitting moisture from the skin to the atmosphere (Gibson et al., 1995; Gibson and Charmchi, 1997).

Apart from diffusion, absorption-desorption and convection processes, the ventilating motion of air through the fabric caused by the motion of the body relative to the surrounding environment, also plays an important role in heat and moisture transfer at transient condition. The dry and wet heat transport problem for the skin of a clothed and moving body may be described as complex. It involves several parallel/series processes of unsteady diffusion, convection, radiation, adsorption/desorption, and condensation. The complexity of heat transport is increased due to the effect of the ventilating motion of air through the fabric. The size of the air spacing between the skin and the fabric is constantly variable depending on activity level and location, thus causing a variable airflow in and out of the fabric. This induced airflow contributes to an augmentation of the rate of condensation and adsorption in the clothing system and the amount of heat and moisture loss from the body. During body motion, air must go in and out and ventilation is obtained without overall environmental air movement. Harter et al. (1981) refers to this particular aspect in clothing comfort as 'ventilation of the micro-climate within clothing'.

Clothing micro-climate ventilation is critical to the removal of sensible and latent heat from the body and consequently has a major influence on the thermal comfort and dynamic insulation values of clothing ensembles. In practical applications, ventilation of the clothing system during bodily motion occurs by the periodic movement of air in and out of the air spacing as the fabric moves outward or inward towards the skin. According to the model developed by Ghali et al. (2002) the heat loss is mainly latent during a downward movement of the fabric, and mainly sensible during the upward movement. This model has also predicted the quantities of sensible and latent heat transport at the moist to solid boundary.

Ventilation causes a dynamic change of clothing insulation due to wind penetration through the fabric or ensemble openings, wearer displacement due to motion causing a wind effect, and relative motion of the clothed limbs with respect to their clothing cover. Ventilation causes a pumping effect, causing the moisture vapour to move from the skin to the atmosphere. Clothing designed for active wear can be improved by considering the local heat loss and ventilation needs of different parts of the body (Ghali et al., 2009). The periodic ventilation effect, according to Ghali et al. (2002), causes a temperature change of about 2.5 °C in the temperature of the enclosed layer of air during one period of fabric oscillation and would not be in thermal equilibrium with the fabric yarn.

Convection takes place in the space between the incoming air and the fabric fibre, thus complicating the estimation of further heat and moisture loss through the fabric and from the clothed body. Convective heat and mass transfer in porous media is often more important than transport due to diffusion, especially if such materials are used in conditions in which there is a large pressure gradient. In the case of textile materials having high air permeability, even a small increase in the pressure gradient can produce large convective flows. The simple diffusion methods performed on these materials are often influenced by the convective flow (Gibson and Charmchi, 1997). Water vapour molecules are smaller and lighter than air molecules and the drag coefficient and viscous resistance of air molecules are higher than those of water vapour (Poynting, 1997/1925). Clothing with very fine pores will be impermeable as air cannot pass through them. In clothing with an open structure (net type), the movement of air tends to easily replace the water vapour, thus improving the water vapour transport. Moisture by convection is quite significant when the wearer is active (frequent limb movement) or in clothing with more openings. During active work/exercise, pressure in the micro-climate changes between positive and negative as the limbs move back and forth. This pumping effect increases the water vapour concentration in spaces where the pressure is low. Convective moisture vapour transfer has been related to Darcy permeability (k_D) (Gibson and Charmchi, 1997):

$$v = \frac{-k_D}{\mu}\frac{\Delta p}{\Delta x} \qquad [8.11]$$

where v = apparent gas flow velocity (m/s), k_D = permeability constant (m^2), μ = gas viscosity (17.85×10^{-6} kg/m.s for N_2 at 20 °C), Δp = pressure drop across the clothing (N/m^2 or Pa) and Δx = thickness (m). At low flow rates, k_D is constant, and with increased flow rates, k_D must be determined from plotting the pressure drop against the volumetric flow rate. Darcy permeability is found from the Darcy apparent flow resistance (R_D) as:

$$k_D = \frac{\Delta x}{R_D} \qquad [8.12]$$

Hygroscopic fibres swell at higher relative humidity and constrict the pore size, so increasing the flow resistance to gas. The relationship between RH and regain could be an indication of Darcy apparent flow resistance. Using a similarity of the flow resistance and sorption curves, flow resistance (R) as a function of relative humidity (ϕ), Gibson and Charmchi (1997) have shown that:

$$R(\phi) = R_{dry} + \left(\frac{\varepsilon_{bw}(\phi)}{\varepsilon_{bwsat}}\right)\left(R_{sat} - R_{dry}\right) \qquad [8.13]$$

where R_{dry} and R_{sat} are Darcy flow resistances in m^{-1} at $\phi = 0.0$ and 1.0 respectively, ε_{bwsat} is the volume fraction of bound water at $\phi = 1.0$, and ε_{bw} is volume fraction of water dissolved in solid phase.

Darcy flow resistance in polyester is unaffected by relative humidity; in nylon it decreases from dry to saturated condition; and fibres such as cotton, wool, silk, wool/polyester and nylon/polyester have a high flow resistance in their saturated condition.

8.3.3 Liquid moisture transport by capillaries

The transport mechanism of liquid water has a significant impact on simultaneous heat and mass transfer. Transport of liquid water across clothing increases the thermal conductivity and changes the heat transfer and moisture absorption of the fibres. Assuming interconnected capillaries and pore distributions in a fabric, Li et al. (2002) derived the diffusion coefficient of liquid water on the basis of Darcy's and Hagen Poiseuille's laws as:

$$D_l(\varepsilon_l) = \left(\frac{\gamma \cos\theta \sin^2 \beta d_e}{20\eta}\right)\left(\frac{\varepsilon_l}{\varepsilon}\right)^{1/3} \qquad [8.14]$$

where ε_l and ε are volume fraction of the liquid phase and porosity of the clothing, β is the angle of the pore with respect to the plane of the skin, and d_e is the effective pore diameter. In multi-layered clothing, there is little contact between the layers and hence the liquid transfer rate through the clothing is greatly reduced. Liquid water distribution in fabrics is not significantly affected by atmospheric pressure between 0.1MPa to 0.2MPa.

The inter-yarn pore sizes in woven fabrics are >100 μm, hence they make poor contact with the small water droplets present on the skin. However, the inter-fibre pores are small (<4 μm) and can contact the water droplets on the skin, drawing the liquid through capillary action. The inter-fibre capillaries differ in size and the average inter-fibre pore diameter is related to the porosity of the yarn and the diameter of the fibre as:

$$\bar{r} = f(\varepsilon_y, d_f) \qquad [8.15]$$

The area of the micro pores (S_p) in contact with skin is

$$S_p = A_F f_c \varepsilon_y C_f \qquad [8.16]$$

where A_F is apparent area of the fabric in contact with the skin; f_c is the fractional cover of the fabric; ε_y is the porosity of the yarn and C_f is the contact ratio determining the actual area of contact between yarn and skin. The thickness of single layer fabric is of the order of only a few millimetres, hence, the gravitational effect of the liquid mass in the capillaries may be ignored. The mass flow rate of liquid due to capillaries in the fabric on the basis of Washburn's and Hagen Poiseuille's laws (ignoring gravitational effect):

$$\frac{M}{\sqrt{t}} = S_p \sqrt{rL_p} \qquad [8.17]$$

$$L_p = \rho_l^2 \gamma \cos\theta / 2\eta \qquad [8.18]$$

The closer the weave and the more flattened the yarn, the greater will be the contact between the inter-fibre pores and the skin. The liquid parameters are: ρ_l, γ and η are the density, surface tension and viscosity of the liquid water respectively. According to Ahn and Seferis (1991), the equivalent wetted pore diameter D_e in arrays of unidirectional fibres is:

$$D_e = \frac{8r_f\varepsilon}{F(1-\varepsilon)} \qquad [8.19]$$

The form factor F assumes the value of four for flow parallel to the fibres, and the value of two for flow transverse to the fibres (Lekakou and Bader, 1998). Flow through inter-fibre pores may be written by substituting the equivalent wetted pore diameter for the mean pore diameter:

$$\frac{M}{\sqrt{t}} = A_F f_c C_f (\varepsilon_y)^{3/2} \sqrt{\frac{2L_p r_f}{(1-\varepsilon_y)}} \qquad [8.20]$$

Plain woven fabrics made from same set of warp and weft yarns, but differing only in their pick or weft density (number of weft yarns per unit length) were tested for transverse wicking with water on a Gravimetric absorption tester. The mass of liquid absorbed by the fabrics versus the square root of the time is a linear function (Fig. 8.3).

During absorbency testing, the load applied over the fabric was 21 cN/cm^2. The contact factor of the fabric over the porous plate is only about 4.1%, calculated

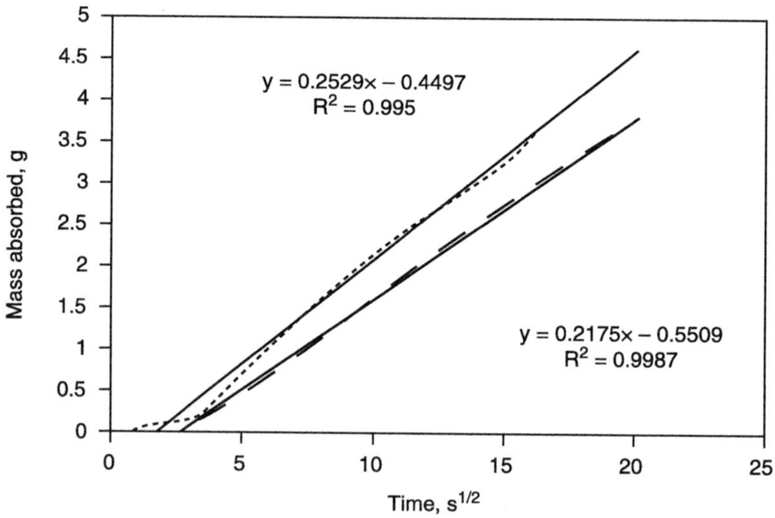

8.3 Mass of water absorbed in plain woven fabrics varying in pick density (upper: 52 picks per inch; and bottom: 40 picks per inch).

from the slope of the above equation (i.e. the initial region is excluded). This is expected as the fabrics undulate due to the interlacing yarns forming alveoli on the surface of the fabric which do not come in contact with the liquid. The curved shape of the yarns also reduces the area of contact between the inter-fibre pores and the liquid water. In the case of a clothed body, the contact factor of the fabric over the skin must remain lower than this value, especially for upper body wear. Kilinc et al. (2004) have shown that the area–fabric ratio (ratio of the actual area of contact to the apparent area of contact) depends on the woven structures and increases with applied pressure. Satin fabric showed a greater area ratio than plain fabric. The area ratio of plain fabric varies from 1 to 8% at 0.18 to 2.6 cN/cm^2 respectively. Roughly 50% of this area is occupied by the inter-yarn pores. The absorbency rate of bleached cotton fabric is 427 $kg/m^2/24h$. However, the same fabric has shown a water vapour transport rate of only 1.2 $kg/m^2/24h$. This is only the case with fabrics having high wettability. On the other hand, a nylon fabric showed zero absorbency, but had a water vapour transport rate of 1.26 $kg/m^2/24h$. This indicates that at a very high sweating rate, the liquid transport rate of the clothing is very important in reducing wetness of the skin.

Closely woven/knitted inner fabrics made from low twist yarns with coarse fibres provide improved rates of absorption of liquid water. Contact between the fabric and skin may be improved in a tight fitting garment (especially knitted) with a low percentage of elastomeric fibres. Fine capillaries will induce larger capillary pressures which tend to move liquid further away from the skin, but their liquid transport rates are lower than those with larger capillaries. For mild exercise wear in a cold climate, double knitted fabrics with a negative pore gradient from the inner to outer layers can transport liquid further away from the skin for evaporation into the atmosphere, so keeping the wetness of the skin to a minimum

8.3.4 Mechanism of coupled heat and liquid moisture transport in clothing

According to Li et al. (2002), the diffusion of water vapour in inter-fibre spaces is slow for the properties of any given material when compared to liquid water penetration of the fabrics through capillary action. The process of water vapour diffusion into fibres throughout the thickness of the fabric takes 2 to 2.5 minutes to reach a steady state, whereas liquid water diffusion takes 0.1 to 1.6 seconds to reach steady state, depending on the thickness (0.5 to 2.0 mm) of the fabric. It is not clear whether the contact factor between the skin and fabric is taken into account in simulating liquid water diffusion into the fabric. It was further observed that water vapour concentrations and temperature distributions have the same pattern of variation in different porosities of fabric and fibre types (cotton and polyester) and across the thickness of the fabrics, indicating that vapour diffusion into spaces is dependent on temperature or heat transfer. Liquid volume distributions are quite independent of temperature distributions and hence the heat transfer process does not significantly affect the liquid transfer process. The evaporation of water is influenced by the liquid transport process; when liquid water is absorbed into the fabric, evaporation is able to take place throughout the fabric. The heat transfer process has a significant impact on the evaporation process in hygroscopic fabrics but not in hydrophobic fabrics. Thicker hygroscopic fabrics with low porosity tend to raise the temperature of the middle layer due to heat of sorption. As a result of the large temperature difference between the middle and outer layers, condensation takes place on the outer layers.

The moisture sorption is largely affected both by water vapour diffusion and liquid water absorbtion, but not by heat transfer. When liquid is absorbed by the fabric, it quickly covers the fibres as its diffusion rate is faster than that of vapour and the moisture sorption is mainly influenced by the liquid transport process. Moisture sorption and evaporation significantly affect the heat transfer process, which in turn is affected by liquid and vapour diffusion.

Li et al. (2002) have given an account of coupled heat and liquid moisture transport in textiles and these are new summarised here. When dry clothing is exposed to a humid environment (micro-climate), it exhibits three stages of moisture transport behaviour. In the first stage, both water vapour diffusion into the pores by water vapour pressure gradient and liquid water diffusion by fabric capillaries, take place simultaneously. These two processes are quite fast and reach a steady state within a few minutes (for vapour) and in fractions of seconds (for liquid). The liquid transport may not occur if the skin moisture is low and the fabric has a high water vapour diffusion rate (a fabric having very high porosity). In such cases, the water vapour diffusing into the pores surrounds the fibres. In the case of both diffusions of liquid and vapour, the liquid water will surround the fibre before the vapour can do so.

The second stage of moisture transport is moisture sorption of the fibres (for hydrophilic fibres only). This is a slow process which takes between a few minutes and a few hours to complete, depending upon the fibre hydrophilicity and the fabric thickness. The moisture sorption increases the relative humidity at the surface of the fibres and the sorption process is enhanced if the fibres are surrounded by liquid moisture in the form of a thin film. At this transient stage, heat transfer is coupled with four different forms of liquid water mass transfer, due to the heat released or absorbed during sorption/desorption and evaporation/condensation. These in turn are affected by the efficiency of heat transfer. In the third and final stage, all forms of mass and heat transfer become steady and the coupling effect among them becomes less significant. The distributions of temperature, water vapour concentration, fibre water content, liquid volume fraction and evaporation rate become invariant in time.

8.3.5 Heat and moisture transfer in multi-layer clothing

Multi-layered clothing is used in cold and extreme cold weather. The thermal resistance (dry) and permeability to water vapour are important in these weather conditions. The thermal resistance (R_d) and evaporative resistance (R_{ei}) of individual layers and the gradient of temperature (T_{sk}–T_a) and vapour pressure (P_{sk}–P_a) from skin to atmosphere and heat flux (Q/A) are related as:

$$R_{di} = \frac{T_{sk} - T_a}{(Q/A)} \quad [8.21]$$

$$R_{ei} = \frac{P_{sk} - P_a}{(Q / A)} \quad [8.22]$$

The permeability index of the layer is defined as (Thwaites, 2008)

$$i_m = 60\frac{R_{di}}{R_{ei}} \quad [8.23]$$

as often the thermal resistance for multi-layer fabrics is much higher than the simple sum of individual layers; this is mainly due to the formation of air space between the fabric layers. This is especially true for a multi-layer clothing system. In designing multi-layer clothing, this factor has to be considered. The i_m value does not provide a value intrinsic to clothing as it is affected by external environmental conditions. The i_m values of 0 and 1 correspond to clothing impermeable to water vapour and air respectively. The i_m values of 0.5 and 0.4 correspond respectively to a typical nude body and to normal clothing (Parsons, 1993).

8.4 Factors influencing moisture transport

The following factors influence the moisture transport rate of clothing: porosity and thickness of clothing, complexity of pores, fibre-diameter, shape, and type (hydrophilic/hydrophobic), pore size distribution, gradients of temperature, water vapour pressure/concentration, position of layers (hydrophilic/hydrophobic), ventilation, radiative sorption constant of fibres and atmospheric temperature and pressure.

Water vapour is transported through fabric mainly by the differential pressure across the fabric layer. Yasuda and Miyama (1992) have shown that when the porosity of single layer fabrics (polyester, acrylic, cotton and wool) is above 66.4%, all the fabrics have the same water vapour pressure gradient. In fabrics with higher porosity, the characteristic permeability coefficient is nearly identical, regardless of the kinds of fibres used. In these fabrics, water permeates the fabric pores in a similar manner to inter-diffusion in air space driven by the pressure difference of water vapour. Hence, the difference in the overall water vapour transport rates is entirely due to the differential vapours that develop depending on the extent of water vapour absorption, and not to the difference of the permeability coefficients of inter-fibre pores. When the porosity of a fabric approaches ≤ 1.0, it is expected that the tortuosity factor (ratio of path length for water vapour diffusion to fabric thickness) will become close to one, for which the fabric thickness should increase. This enhances the thermal insulation essential for cold climate clothing. The competition between the favourable porosity and tortuosity against the increased fabric thickness has to be weighed in terms of the thermal insulation and vapour diffusion rate expected from the clothing.

Inner layers made of hydrophilic fibres will only exhibit the heat of sorption, which provides a sense of warmth, during the transient stage. The transient time required to achieve the maximum water concentration and rise in temperature that creates a concentration gradient across the fabric rises with increased fabric thickness and a decrease in fabric porosity. Garments with hydrophilic inner layers are useful when moving from normal temperatures to conditions of extreme cold (Li et al., 2002).

Studies on the condensation of water vapour transfer through fabrics are necessary for the development of fabrics. The liquid water permeability and water vapour hydraulic permeability are related to the convective flow of vapour or air permeability of the clothing and are important for heat and moisture regulation. The type of yarn count, twist and weave all affect hydraulic permeability (Ghali et al., 2006). The distributions of liquid water and water vapour across the clothing depend on these permeabilities. For fabrics with low vapour hydraulic permeability, the vapour transport may only take place by diffusion and the moisture distribution from the hot and moist skin to a cold boundary is found to be convex with a low vapour concentration gradient. In fabric with higher hydraulic permeability, more water vapour is transported across the fabric, leading to an increase in the amount of condensed water. The vapour concentration is high inside the clothing and the moisture content rises from the inner to outer layer, i.e. the condensation is close to the cold boundary (Cheng and Fan, 2004). It is preferable to keep condensation away from the skin and this is achieved by a high porosity fabric.

The condensed water can be rendered mobile and transported to affect both thermal comfort and heat loss from the skin. The mobility of condensate or liquid water permeability both depend on the capillary forces present in the inter-fibre pores. With an increase in the capillarity, moisture distribution from the warm and moist skin to the cold atmosphere shifts from concave to almost even (Cheng and Fan, 2004).

Waterproof breathable fabrics are normally worn during rain/wet conditions. The breathability of such fabrics is important to avoid wet conditions within the micro-climate which will cause the wearer discomfort. Experiments conducted on two-layer waterproof breathable fabrics in which the base layer is nylon fabric and the other either a polyurethane microporous membrane or hydrophilic film, have shown that higher moisture content and larger amounts of condensation in the clothing increase water vapour transport under both isothermal and non-isothermal conditions (Ren and Ruckman, 2003). If the water vapour permeability of the clothing is high, the condensation or moisture content blocks some of the pores, which in turn increases the water vapour pressure in the microclimate. This increases the water vapour pressure gradient from the microclimate to the inner surface of the clothing, enhancing the water vapour transfer through the clothing.

The water vapour permeability of multilayer clothing (Gore-Tex two-layer fabric with polyester fleece as an inner layer) is 899 $g/m^2/24h$. When the same clothing is wetted (with water content 310%), its water vapour permeability becomes 398 $g/m^2/24h$. The reduced water vapour transport of the wet clothing is due to liquid water blocking pores in the clothing. Condensation taking place below freezing point may have a marked effect on water vapour transport in waterproof breathable fabrics. Experiments using two breathable fabrics (nylon/cotton plain cloth laminated with PTFE and PET peach face cloth coated with polyurethane) at –20 °C and 20 °C, showed that condensation or moisture absorption on the covering fabric and water vapour transmission through the fabric, increase in an almost linear manner with time. Condensation on the fabric surface does not block moisture transmission. The rate of evaporation from the water surface, condensation or moisture absorption on the covering fabric and water vapour transmission through the fabric were greater in a sub-zero climate than in a conventional climate. This is believed to be due to the increased vapour pressure difference and natural convection (Zhou et al., 2006).

The radiative sorption constant of fibres β is related to fibre volume fraction, fibre emissivity (ξ_f) and radius of fibre (r_f) (Ghali et al., 2002)

$$\beta = \frac{(1-\varepsilon)\xi_f}{r_f} \quad [8.24]$$

Fibres with high radiative sorption constants reduce the percentage of condensation/water content within the fibrous batting (Li and Fan, 2006).

The air permeability of Gore-Tex-2-ply fabric is very low ($2.74\,cm^3/cm^2/s$ at 250 kPa), under the standard air permeability test method. However, this structure is reported to have high water vapour permeability. The film is adhesion bonded to the nylon fabric in a manner similar to that of mechanical hooks. The porosity of the film is about 80%. The structure consists of PTFE nodules interconnected by numerous fine fibrils (Lomax, 1985). The film has 1.4 billion pores/cm^2, with an average hydraulic pore diameter of 8.5×10^{-5}µm; much smaller than the size of a rain drop (100 µm) and larger than the size of a water vapour molecule (4×10^{-5} µm). Normally, a fabric which shows high air permeability under standard fabric test methods should also transmit water vapour satisfactorily; however, the reverse is not necessarily true (Lomax, 1985).

8.5 Improving moisture transport

Ideally, perspiration on the skin should be evaporated on the skin itself, utilising the latent heat of evaporation to cool the body. The water vapour thus generated has to be diffused quickly through the clothing. This is the most efficient way of dispersing excess body heat during intense activity. But quite often with increased perspiration, whether through increased activity, hot and humid climates or a combination of both, the surface area of the moisture on the skin available for evaporation is limited and inadequate. Some areas of the human body (such as the chest and arms) perspire more than others. The droplets coalesce, forming a film of liquid water on the skin which further reduces the surface area of liquid water for evaporation resulting in a sense of wetness of the skin and a feeling of discomfort. Properly designed clothing can draw perspiration quickly and spread it over larger areas of the outer layer for a higher evaporation rate than would take place in an unclothed body exposed to convective air currents. This will be observed by those living in tropical countries with hot and humid climates.

To avoid wetness of the skin and an associated loss of thermal insulation, liquid water must be transported as quickly as possible through the capillaries in the inner layer of clothing to the next layer away from the skin. Again, from the point of view of keeping wetness of the skin to a minimum, an inner layer made of hydrophobic fibres with the necessary capillary forces is preferable as these do not absorb moisture. The wicking rate of the inner layer may be increased by treating it with a hydrophilic finish. In the case of circular fibres, a larger pore radius increases the wicking rate, since the mass flow rate through wicking is proportional to $r_p^{2.5}$. There must be an upper limit for the pore size as small droplets will not come in contact with larger pores. Fibre surfaces with channels and grooves could also transport liquid water at a faster rate by wicking. These provide a very large number of smaller pores with minimal discontinuity along the fabric plane, and the combined effect could be as high as that provided by larger sized pores between circular fibres. Fibres with unique cross-sectional shapes such as W-profile, grooved fibres of Cool max and 3-T shaped have a large surface area which increases the evaporation rate throughout the layer.

In the case of summer clothing, the diameter of pores in the outer layer should be less than that in the inner layer, so as to exert a higher capillary pressure ($\Delta p = 2\gamma\cos\theta/r_p$) which will draw the liquid from the inner layer (i.e. one way wicking). The finer capillaries spread the liquid water over a large area for evaporation into the atmosphere. Porosities of the layers and fibre diameter are the controlling factors of pore radius.

Fleece used in the middle layers of cold weather clothing has high porosity in order to transport water vapour by diffusion. The liquid water transported to the fleece also evaporates throughout its cross-section and this provides a very large surface area for the liquid water to evaporate. The water vapour diffuses quickly through this highly porous medium and reaches the micro-porous hydrophobic membrane or the hydrophilic film of the outer layer, where it diffuses into the atmosphere by absorption-desorption (evaporation) mode. Synthetic fleece made from fine fibres or even finer wool fibres has poor wettability due to its high curvature and hydrophobicity, so if condensation occurs on these fibres, it will be in the form of droplets. Droplets of 10–100 μm diameter contribute most to the heat transfer rate. Condensation is an efficient heat transfer process and the heat released on condensation could increase the local water vapour pressure. According to the Kelvin equation defining the droplets radii r_d (resting on the fibre):

$$\frac{p_{vd}}{p_o} = e^{-\left(\frac{\gamma M}{r_d \rho RT}\right)} \qquad [8.25]$$

Due to the convex shape of the droplets, liquid water molecules at the curved interface can easily escape into the vapour phase, thus they have a higher vapour pressure (p_{vd}) than the water molecules on the flat liquid water surface (p_o). The heat transfer coefficients with drop wise condensation can be one to two orders of magnitude greater than that for film condensation. As a consequence, enhanced evaporation of water is possible with condensation taking place near the outer layer and if the outer layer is able to transfer the water vapour quickly into the atmosphere, it improves the water transport rate of the clothing.

The mass rate of change of a water droplet in a steady state vapour field is given by (Pruppacher and Klett, 1997)

$$\frac{dm}{dt} = \frac{4\pi r_d D_v M}{R}\left[\frac{p_a}{T_a} - \frac{P_{vsd}}{T_d}\right] \qquad [8.26]$$

where D_v is the effective gas phase diffusion coefficient including surface kinetics, M is the molar mass of water, P_a and T_a denote ambient pressure and temperature, and p_{vsd} and T_d are the equilibrium water vapour pressure of the droplet and its temperature. When condensation occurs close to the outer layer of the clothing, $p_{vsd} > P_a$, and $T_d \equiv T_a$; and the evaporation rate of the droplet increases. In such a situation, an outer layer with low vapour resistance will enhance the water vapour transport. The water vapour resistance of the outermost layer is more critical than that of the innermost layer in transporting water vapour through the clothing.

In the case of multi-layer clothing with Gore-Tex cover fabric (three layered: woven + membrane + warp knit) as its inner and outer layers, with the middle layers consisting of either wool or polyester, or both wool and polyester layers in combination, it was found that placing the hygroscopic wool batting in the inner region (i.e. closer to the body) and the non-hygroscopic polyester batting in the outer region (i.e. away from the body) reduces the moisture accumulation within and the total/dry heat loss through the clothing layers. This offers the potential for improving performance by optimising the positions of the batting in clothing with similar materials (Wu and Fan, 2008).

Zhou et al. (2003) have studied the water vapour transport behaviour of conventional, tightly woven, micro-porous film, and hydrophilic film-fabrics. It has been observed that the poor water vapour permeability of these fabrics increases the temperature and the water vapour pressure within the micro-climate. Condensation occurs in fabrics with low vapour transport rates (390 to 908 $g.m^{-2}day^{-1}$) with local relative humidity ranging from 91 to 99.8%. Condensation does not occur in fabrics with a higher water vapour transmission rate (>1008 $g.m^{-2}day^{-1}$) and local relative humidity ranging from 67 to 86%. These are: conventional fabrics, tightly woven fabrics and micro-porous film fabrics. PTFE laminated and tightly woven fabrics have a higher vapour transport rate than those without lamination. The influence of polypropylene and wool batting sandwiched between the inner and outer layers has been simulated for heat and moisture transfer under conditions of a sudden movement from an initial warm condition (20 °C and 80% RH) to a cold environment (–20 °C and 90% RH) (Fan et al., 2000). The water content in the polypropylene batting is almost entirely in the form of condensed liquid water or ice, whereas the wool batting has water content in both forms (regain and liquid water) for similar levels of [overall] water content. When the water vapour resistance of the inner fabric layer is increased (about three fold) and that of the outer layer is reduced tenfold, the water content (condensation) in the polypropylene batting could be reduced by 50%. Similarly, by changing the porosity of the batting by 10% in its inner quarter and by 75% in the outer quarter and maintaining linear distribution within the remaining region, the condensation could be reduced for the same water vapour transfer resistance of the inner and outer fabric.

Clothing constructed with a thin inner fabric close to the body, a thick porous fibrous batting in the middle and a thin outer fabric next to the cold environment has been simulated for heat and moisture (Fan and Wen, 2002). Batting with 50% water content exhibited a heat flux (W/cm^2) 3.5 times that of dry batting (% water content). This dramatic increase in heat flux is not solely due to the increase in conductive heat transfer. The extra heat transport is caused by the evaporation and diffusion of water vapour (Farnworth, 1986). When the porosity increases from 0.57 to 0.87, heat flux causes a reduction in effective thermal conductivity. When the porosity increases further, from 0.87, the heat flux increases due to increase in water vapour diffusion. The heat flux is mainly influenced by the initial water content, thickness of the batting and by the surrounding temperature. The water vapour resistance of the inner and outer lining fabrics, the thermal conductivity of the fibre and the porosity of the batting have only a small effect.

8.6 Clothing requirements for different environmental conditions

One of the primary purposes of clothing is to maintain a uniform body temperature under different temperature environments and to prevent the accumulation of sweat on the skin by allowing perspiration to flow to the outside environment when activity levels increase. Thus, the heat exchange between body and environment is significantly affected by the dynamic response of the clothing system and the way the clothing layers mediate the flow of heat and moisture from the skin to the environment. Under normal temperature and working conditions, the body core temperature is maintained by the transmission of heat through perspiration in vapour form, along with the regular heat transfer processes of conduction, convection and radiation. When the core temperature becomes very high, the sweat glands are activated and the body produces liquid perspiration. The intensity of sweating increases with the level of activity and the environmental temperature. This process uses latent and sensible heat to evaporate liquid sweat on the skin, thus reducing its temperature and causing the body to feel cooler. At higher atmospheric temperatures, the temperature gradient between the skin and the environment is smaller and so the convective and radiative heat losses are diminished. In tropical climatic conditions, the high relative humidity of the atmosphere reduces the vapour pressure gradient between the skin and the atmosphere, thus reducing the rate of vapour transport. The inner layer of the clothing should absorb liquid sweat by capillary sorption which enables the body to remain dry and cool. Excessive heat loss from the body should be avoided as it may lead to hypothermia, particularly if the skin is in contact with wet clothing. Therefore, in order to achieve an effective moisture transmission mechanism in a clothing system, a proper balance between the absorbent, wicking and water vapour transmission properties of the fabric has to be achieved.

Key factors controlling the thermal comfort of the clothed body are: ambient conditions (air temperature, mean radiant temperature, air velocity, and humidity); permeability (liquid and vapour); ventilation and insulation of the clothing and activity level; heat production rate, etc.

For active wear in a mild climate, water transport by convection should be considered as it is much faster than transport by simple diffusion in fabrics where air is able to circulate. A fine mesh type fabric will allow air to circulate constantly over the skin, so drawing moisture away from the skin and keeping it cool.

Clothing for extreme cold, hot and fire-fighting conditions poses a constant and major challenge to clothing designers and perhaps more importantly, calls for more effective materials. In these situations, clothing has to meet two conflicting requirements. High thermal insulation and efficient sweat transfer (Pan, 2008).

The maintenance of thermo-physiological comfort under differing environmental conditions and physical activities requires different clothing properties. During normal wear, insensible perspiration is continuously generated by the body. Steady-state heat and moisture vapour fluxes are thus created and must be gradually dissipated to maintain thermo-regulation and a feeling of thermal comfort; hence the clothing becomes a part of a steady-state thermo-regulatory system (Clark and Edholm, 1985; Fanger, 1970). In transient conditions, characterised by intermittent pulses of moderate or heavy sweating caused by strenuous activity or climatic conditions, sensible perspiration and liquid sweat occur and must be rapidly managed by the clothing. This property is important in terms of sensory comfort as well as for the thermo-regulation of the wearer (Scheurell et al., 1985). Therefore, wearer comfort requires consideration of heat and moisture transfer properties under both steady and transient conditions. Clothing requirements for some extreme conditions are discussed below.

8.6.1 Hot and dry climate

The human body tries to maintain a constant core temperature of about 37 °C (Parsons, 1993) and a skin temperature of around 30–35 °C. In hot climates, the atmospheric temperature is around 35–39 °C and may sometimes be more than 45 °C. In these conditions, heat may be gained by the body rather than lost by regular heat transfer processes, due to the negative temperature gradient. The body becomes increasingly more dependent on evaporative heat loss for maintaining its thermal balance. In such conditions, clothing, especially workwear, should allow for vapour transmission in order to reduce the thermal strain through assisting the buffering effect (Dent, 2001). Otherwise, sweat will accumulate on the skin and cause discomfort. When the exterior humidity is low, the vapour pressure gradient between the skin and environment will be high and the diffusivity of clothing is important. To be effective under these conditions, clothing must possess good properties of ventilation which permits moisture transfer by diffusion as well as by forced convection and the fibre used should retain minimal amounts of moisture if it is to provide a cool feeling to the wearer in these conditions. Hygroscopic fibres should be used and the fabric structure must be very open to facilitate ventilation.

8.6.2 Hot and humid climate

In tropical conditions, both the temperature and the relative humidity of the atmosphere are very high, around 70–75% (McNeill and Parsons, 1999). Under these conditions, the heat transfer processes are ineffective in transmitting body heat to the atmosphere and the low vapour pressure gradient between the skin and the atmosphere reduces the rate of vapour diffusion (Das et al., 2006). These conditions require sweat to be absorbed by the clothing nearest to the skin and passed through to the outer layer of the fabric by a wicking action. The outer layer, having a large surface area, facilitates the rapid evaporation of liquid sweat. If the wearer is to experience relief from the clamminess of humid weather, the fibre used next to the skin should rapidly transmit the liquid water to the outer layer, rather than retaining moisture. A smooth inner fabric is preferable for coolness. Double knitted fabric made from wicking polyester fibres (fine fibres of low modulus, with a hydrophilic finish or grafted onto a hydrophilic polymer or fibre with grooves) as an inner layer with a texturised polyester on the outside will be effective in drawing the perspiration by capillary forces and will maximise the evaporation.

When the RH of the environment is ≤70%, the body maintains a condition of reasonable thermo-physiological comfort due to the efficiency of sweating, even under high air temperatures, as for example, in deserts. When the relative humidity of the air exceeds 80–85% along with air temperature >35 °C, then a state of thermal comfort is unattainable.

8.6.3 Cold weather

The insulation of clothing is the main factor in maintaining a thermal balance in cold weather. At very low temperatures, the temperature gradient between skin and atmosphere is very high and if the insulation properties of the fabric are not high, the result will be a high loss of heat from the skin to the atmosphere, leading to a reduction in body core temperature and ultimately to hypothermia. Even in cold weather conditions, a higher rate of physical activity will increase the body's core temperature. To maintain thermal balance under these conditions, the body produces insensible perspiration to reduce its temperature by evaporative heat loss. Effective clothing should possess good water vapour permeability to transmit the vapour perspiration. Cotton should be avoided for the base layer as it becomes wet and clings to the body when the rate of sweating increases. It takes a long time to dry and to give a feeling of coolness. The sense of cold is measured by thermal absorptivity which increases with the thermal conductivity of the fibre and its density; in cotton these values are higher than in synthetic fibres. Fine merino wool and wool/polyester fibres are well suited to the base layer. These fibres can be treated so as to wick away the liquid moisture from the skin. Polyester fleece, pile fabrics and down are suitable for the middle layer and provide the best insulation. Fine fibres have less radiative loss. The thickness of the middle layer is very important in enhancing insulation. In dry, cold weather, the outer layer must be windproof for low air permeability, and waterproof in wet weather. The outer layer should also be water-repellent to minimise water pick-up. To permit evaporation of water, the outer layer should have a high water vapour permeability and low vapour resistance.

8.6.4 Heat resistant clothing

In work-wear, heat and moisture transfer properties under both steady and transient conditions must be considered for wearer comfort. The moisture vapour buffering capacities of the fabrics are important during physical activity. The buffering capacities are characterised by: the rate of increase in the relative humidity of the micro-climate, the maximum increase in micro-climate relative humidity during activity, and the time taken for the relative humidity of the micro-climate to return to a steady level following termination of activity (Yoo and Barker, 2005). Fabrics with a better regulation of moisture vapour performance should have lower values in each of the parameters mentioned above. Yoo and Barker (2005) have tested Nomex and Nomex/FR rayon work-wear fabrics, demonstrating that fabrics with fine fibres and a twill weave have a higher moisture vapour buffering capacity.

Clothing for fire-fighters requires very high thermal insulation to protect the body from burning when exposed to an external heat source. These garments are very thick and multi-layered, consisting of underwear, station uniform and a triple layered fire-fighter jacket on the outside. However, the high insulation properties of the clothing virtually shut down the channels for sweat evaporation, thereby embedding the release of heat stress. Moisture strongly affects the protection properties of these garments. The addition of moisture causes a marked change in the heat transference properties of clothing layers. The presence of water increases both the thermal conductivity and heat capacity of the fabric. Barker et al. (2006) analysed the influence of moisture on protection against low-level radiant heat exposure in a clothing system. It was found that the time before second degree burns occurred decreased the most when moisture content was low. When exposed to a sudden temperature rise, moisture in the clothing layers may evaporate and move towards the skin, causing 'steam burns' as it condenses. In this context, the distribution of moisture in the various layers and the temperatures reached within these layers during exposure to heat are critical. Keiser et al. (2008) have studied various combinations of materials in fire-fighter clothing. Depending upon the underwear, there were no significant differences in moisture accumulation in the layers of the station uniform: these appear to be similar for a given layer. In contrast, the station uniform had a significant impact on the amount of moisture stored in the underwear. The underwear behaved completely differently depending upon the station uniform layer with which it was combined. A hydrophobic finished middle layer absorbed less moisture which was unable to spread to the outer layers, thus acting as a water barrier. This resulted in the dripping of water from the underwear. The overall distribution of moisture in multi-layer protective clothing can be influenced by using defined combinations of hydrophilic and hydrophobic textile layers.

8.7 Developments in moisture management

8.7.1 Sensing and responding textiles

Temperature trigger

Fabrics made from thermo-regulating fibres (PCM fabrics) in which millions of micro-thermal spheres are incorporated, will absorb, store, distribute, and release heat in response to the body's comfort needs. The micro-thermal spheres create a dynamic thermal barrier which regulates the micro-climate next to the skin, thus maintaining warmth and comfort despite changes in the outside temperature. This technology is used in Outlast® fabrics.

A research team at North Carolina State University has demonstrated a technology which grafts thermo-responsive poly(N-isopropylacrylamide) (PNIPAM) polymer onto the surfaces of cotton, nylon, polyester and polystyrene, using atmospheric plasma treatment. This has a high potential for the development of 'smart' garments. PNIPAM-grafted surfaces undergo a reversible phase transition in response to temperature change. PNIPAM swells and the fibre diameter increases at temperatures below 32 °C, while it shrinks at temperatures above 32 °C. As a result, the pore size in the fabric decreases and increases, at low and high ambient temperatures respectively. Fabrics treated in this way undergo structural changes in terms of diameter, shape of fibres, pore sizes and fabric performance (swellability, wettability, hydrophobicity, heat transfer, and comfort) when the ambient temperature changes. Preliminary studies have demonstrated that the use of PNIPAM to treat cotton fabrics can decrease heat transfer by 8% at 10 °C, and increase it by 8% at 35 °C. The degree of change in heat transfer can be easily adjusted by selectively controlling the process parameters during surface modification.

Humidity trigger

The change of shape induced by moisture in botanical structures such as the pine cone (bio-mimetics) inspired the design of a fabric capable of adapting its air permeability in response to changes of humidity in the micro-climate of the clothing system. A UK based start-up company, MMT Textiles Ltd, has invented and patented a method of constructing bio-responsive fibres and yarns that react to humidity levels in a micro-climate. The porosity of knitted, woven or non-woven fabrics made from these fibres increases with high levels of humidity or wetness in the micro-climate, facilitating the release of moisture. In dry conditions, the structure opens up like a pine cone, reducing permeability to air and increasing the insulation properties This is referred to as 'the pine cone effect'. According to the company source, the technology works equally well with natural cotton and wool fibres and offers significantly higher levels of wearer comfort than do synthetic moisture-wicking fabrics.

8.7.2 Fabric with pore size gradient

US Patent 6427493 (2002) describes weft double-knit fabric formed from two different polyester yarns. A micro fibre (0.6 to 0.8 denier fibres) and a non-micro fibre (two denier fibres) yarn are knitted into an irregular pique knit which is a modification of the basic pique stitch. The irregular knit pique construction allows for the non-micro fibre yarn to be on the back of the fabric worn against the body, while the micro fibre yarn lies on the exposed face of the fabric. Both surfaces of the fabric are somewhat rib-like in appearance. The larger sized capillaries present in the inner layer of the fabric rapidly transport liquid moisture away from the skin to the outer layer where the smaller capillaries, due to high capillary pressure, siphon the water rapidly to the fabric outer surface for evaporation. This fabric is claimed to be suitable for athletic wear. ProwiK® is a specially engineered two sided, denier gradient knitted fabric made from nylon yarns.

8.7.3 Fibres with nano-structured surface

ProwiK® is a knitted fabric made from nylon fibres with an engineered nano-channel helix section and/or an engineered two-sided denier gradient. The fabric is suitable for the inner layer in cold weather, especially for high aerobic activities. It is claimed to be more efficient in moisture management, breathability and the transport of perspiration away from the skin to the outer layers for evaporation. Air permeability is one of the key performance tests which fabrics must pass before they may be sold as ProwiK®. The fabric dries quickly after washing, but the scientific details regarding moisture transport are not clear. The nylon fibre is partly hydrophilic with a moisture regain of 4.5% at 65% RH and it is believed that the nano-grooves on the surface of the fibre provide numerous channels for water to wick through; they improve the hydrophilic character/surface energy of the fibre surface, reducing the contact angle of water on the nylon and provide continuous capillaries for liquid water transport. The outer layer of the fabric also has a very large surface area which improves the evaporation rate of water into the atmosphere.

8.7.4 Cocona imbedded fibres and Cocona filaments

TrapTek LLC's have patented an innovative technology incorporating activated carbon derived from recycled coconut shells into nylon and polyester polymers. The technology is licensed to Burlingtone Worldwide for the manufacture of knitted cycling/active apparel from man-made fibres, cotton blends and wool blends. The hardness of the coconut shell, created by the nano-meter scale structure of the pores, gives it an advantage over softer activated carbon materials which have fewer and bigger pores. The coconut pore structure is very complex and the size of the pores is very even. These pores adsorb moisture, odours and UV rays and through the exothermic adsorption process, they cause moisture to evaporate quickly. According to the company, the activated carbon from coconut shells exhibits a greater exothermic reaction than any other known substance. Garments containing Cocona activated carbon show 45% greater wickability and significantly higher drying rates (due to 3D surface area) than garments with conventional moisture management technologies.

A Taiwanese based company, C & F, produces a double jersey fabric made from polyester and Cocona filaments consisting of 66% polyester (150D/144F) and 34% coco carbon (75D/72) and caters to the market for cycling shorts and shirts. It is reported that these fabrics have improved moisture management, protection from infra red rays and removal of odours.

8.7.5 Mixture of hydrophilic and hydrophobic fibres

Hydroweave® is a three-layer composite fabric combining hydrophilic and hydrophobic fibres into a fibrous core of batting. The batting is sandwiched between a breathable outer shell fabric and a thermally conductive, inner lining. It is recommended that the fabric be soaked in water for five minutes before wearing so that the hydrophilic fibres absorb moisture. The hydrophobic fibres which are evenly distributed in the batting core have air spaces between them which provide the right condition for the evaporation of liquid water throughout the core. When moisture in the batting evaporates, heat is removed, cooling the surrounding area while the conductive lining transfers cooling to the wearer. Contact with the wearer's skin enables heat to pass directly from the body to the fabric. The heat and moisture vapour in the inner layer are released to the outside environment through the breathable outer shell. Because of the even distribution of water-absorbent polymer throughout the batting, cooling takes place evenly throughout the entire garment which can continue to cool for up to eight hours, depending on the degree of garment contact, environmental conditions, physical activity and type of outer clothing worn.

8.7.6 100% cotton fabrics

The main problems associated with 100% cotton fabrics are their high absorbency and slowness of drying which cause a sensation of wetness. Absorbency may be reduced by treating the cotton with fluoro-polymers, silicones and waxes, etc. Most fluoro-polymers require treated fabrics to be ironed or dried at high temperatures after laundering to renew their orientation and to preserve durability. LAD fluoro-polymers, however, can be renewed through air drying. If the inside of a cotton fabric is fully treated with fluoro-polymers, it will repel water and hinder evaporation through the fabric. The printing of a fine pattern of a particular fluoro-chemical on the inner side of the fabric will reduce absorbency and the openings created by printing allows the liquid water to be wicked through the fabric (Cotton Incorporated, 2002). The time required for drying the wet fabric is not only related to the hydrophilic character of the fibres, but also to the thickness of the fabric and the drying rate may be improved by a thin cotton fabric having an open structure.

Transdry® is a double knitted 100% cotton fabric with an untreated cotton layer on the outside of the fabric and treated cotton against the skin. It claims to provide better moisture transfer than do synthetic fabrics. The cotton treated with Transdry® technology also claims to have a significantly lower (around 35%) absorbent capacity than untreated cotton.

8.8 Future trends

The next generation of textiles is likely to incorporate nanotechnology. Nano-scale structure/grooves on the hydrophilic fibres make them super-hydrophilic. Outer layers of fabric having these fibres would enhance the liquid moisture transport and spreading of liquid water over an extensive surface area for rapid evaporation into the atmosphere. Many new knitted structures with one-way wicking having a pore size gradient will be in the market in the coming years. In addition, the near future could see the development of 'smart' fabrics which are now in a developmental stage. Smart fabrics will respond to changes in the environment by adjusting their pore size or thickness at a given moment to facilitate moisture transmission.

Manufacturers will shift their attention to high-end moisture management fabrics and introduce sophisticated materials, such as carbon fibre, far-infra-red fibre, UV-cut fibre, high density titanium in the core fibre (in the outer layer for improved UV protection), anti-bacterial fibre, minus ion fibre and oil and soil repellent outer fabrics in making performance sportswear, e.g. cycling jerseys. Consumers will come to place increasing importance on the thermal comfort of garments

8.9 Sources of further information and advice

www.coconafabrics.com/
www.cottoninc.com/transdry
www.innovationintextiles.com/articles/199.php
www.tx.ncsu.edu/research_industry/

8.10 References

Ahn, K. J. and Seferis, J. C. (1991), 'Simultaneous measurements of permeability and capillary pressure of thermosetting materials in woven fabric reinforcements,' *Polymer Composites*, **12**(3), 146–152.

Barker, R. L., Guerth-Schacher, C., Grimes, R. V. and Hamouda, H. (2006), 'Effects of moisture on the thermal protective performance of firefighter protective clothing in low-level radiant heat exposures,' *Textile Research Journal*, **76**(1), 27–31.

Barnes, J. C. and Holcombe, B. V. (1996), 'Moisture sorption and transport in clothing during wear,' *Textile Research Journal*, **66**, 771–786.

Bejan, A. (2004), *Porous and Complex Flow Structures in Modern Technologies*, New York, Springer.

Cassie, A. R. D., Atkins, B. E. and King, G. (1939), 'Thermostatic action of textile fibres,' *Nature*, **143**, 162.

Chatterjee, P. K. (1985), *Absorbency*, Elsevier Science Publishing Company, New Jersey.

Cheng, X. and Fan, J. (2004), 'Simulation of heat and moisture transfer with phase change and mobile condensates in fibrous insulation,' *International Journal of Thermal Sciences*, **43**, 665–676.

Clark, R. P. and Edholm, O. G. (1985), *Man and His Thermal Environment*, Edward Arnold, London.

Cotton Incorporated (2002), '100% Cotton moisture management', *Journal of Textile and Apparel, Technology and Management*, **2**(3).

Crank, J. (1975), *The Mathematics of Diffusion*, Clarendon Press, Oxford, UK.

Das, B., Das A., Kothari V. K., Fangueiro, R. and Araujo, M. (2006), 'Study of Moisture Transmission through Fabrics for Comfort Optimisation: The Case of Work Wear for Use in Tropical Climatic Conditions,' *Autex Conference*, 11–12th June, NC, U.S.A.

Dent, R. W. (2001), 'Transient comfort phenomena due to sweating,' *Textile Research Journal*, **71**(9), 796–806.

Fan, J. and Wen, X. (2002), 'Modeling heat and moisture transfer through fibrous insulation with phase change and mobile condensates,' *International Journal Heat Mass Transfer*, **45**, 4045–4055.

Fan, J., Luo, Z. and Li, Y. (2000), 'Heat and moisture transfer with sorption and condensation in porous clothing assemblies and numerical simulation,' *International Journal Heat Mass Transfer*, **43**, 2989–3000.

Fanger, P. O. (1970), *Thermal Comfort*, McGraw Hill, New York.

Farnworth, B. (1986), 'A numerical model of the combined diffusion of heat and water through clothing,' *Textile Research Journal*, **56**(11), 653–665.

Fohr, J. P., Couton, D. and Treguier (2002), 'Dynamic heat and water transfer through layered fabrics,' *Textile Research Journal*, **72**(1), 1–12.

Fukazawa, T., Kawamura, H. and Tamura, T. (2002), 'Water vapour transfer through microporous membrane and polyester textiles at combinations of temperature and pressure that simulate elevated altitudes,' *Journal Textile Institute*, **91**(2), 434–447.

Ghali K., Othmani M., Jreije B. and Ghaddar N. (2009), 'Simplified heat transport model of a wind-permeable clothed cylinder subject to swinging motion,' *Textile Research Journal*, **79**, 1043.

Ghali, K., Ghaddar, N. and Jones, B. (2002), 'Modeling of heat and moisture transport by periodic ventilation of thin cotton fibrous media,' *International Journal Heat Mass Transfer*, **45**, 3703–3714.

Ghali, K., Ghaddar, N. and Jones, B. (2006), 'Phase change in fabrics', in *Thermal and Moisture Transport in Fibrous Materials*, edited by Pan, N. and Gipson, P., Woodhead Publishing, Cambridge, p. 416.

Gibson, P. W. and Charmchi, M. (1997), 'Modelling convection/diffusion processes in porous textiles with inclusion of humidity-dependent air permeability,' *International Communications in Heat and Mass Transfer*, **24**(5), 709–724.

Gibson, P., Kendrick, C., Rivin, D. and Sicuranza, L. (1995), 'An automated water vapour diffusion test method for fabrics, laminates, and films,' *Journal of Coated Fabrics*, **24**(4), 322–345.

Harter, K. L., Spivak, S. M. and Vigo, T. L. (1981), 'Applications of the trace gas technique in clothing comfort,' *Textile Research Journal*, **51**, 345–355.

Hong, K., Hollies, N. R. S. and Spivak, S. M. (1988), 'Dynamic moisture vapour transfer through textile, I: Clothing hygrometry and the influence of fibre type', *Textile Research Journal*, **58**, 697–706.

Incropera, F. P. and DeWitt, D. P. (1996), *Fundamentals of Heat of Mass Transfer*, 4th ed., John Wiley and Sons, New York.

Keighley, J. H. (1985), 'Breathable fabrics and comfort in clothing,' *Journal of Coated Fabrics*, **15**(2), 89–104.

Keiser, C., Becker, C. and Rossi, R. M. (2008), 'Moisture transport and absorption in multilayer protective clothing fabrics,' *Textile Research Journal*, **78**, 604.

Kilinc, F. S., Elmogahzy, Y. E., Hassan, M., Farag, R., Ebiely, R., El Dieb, A. S. and Tolba, A. (2004), 'The tactile behaviour of textile materials: New perspectives – Part I: A study on the nature of fabric handle,' *Proceedings of Cotton Beltwide Conference*, U.S. Cotton Council, January.

Lekakou, C. and Bader, M. G. (1998), 'Mathematical modelling of macro- and micro-infiltration in resin transfer moulding (RTM),' *Composites Part A*, **29A**, 29–37.

Li, Y. and Holcombe, B. V. (1992), 'A two stage sorption model of the coupled diffusion of moisture and heat in wool fabrics,' *Textile Research Journal*, **62**(4), 211–217.

Li, Y. B. and Fan, J. (2006), 'Heat and moisture transfer in fibrous clothing insulation,' in *Thermal and Moisture Transport in Fibrous Materials*, edited by Pan, N. and Gipson, P., Woodhead Publishing, Cambridge, p. 457.

Li, Y., Holcombe, B. V., Scheider, A. M. and Apcar, F. (1993), 'Mathematical modelling of the coolness to the touch of hygroscopic fabrics,' *Journal of the Textile Institute*, **84**(2), 267–273.

Li, Y., Zhu, Q. and Yeung, K. W. (2002), 'Influence of thickness and porosity on coupled heat and liquid moisture transfer in porous textiles,' *Textile Research Journal*, **72**(5), 435–446.

Lomax, G. R. (1985), 'The design of waterproof, water vapour permeable fabrics,' *Journal of Coated Fabrics*, **15**, 41–104.

McNeill, M. B. and Parsons, K. C. (1999), 'Appropriateness of international heat stress standards for use in tropical agricultural environments,' *Ergonomics*, **42**(6), 777–797.

Morton, W. E. and Hearl, W. S. (1962), *Physical Properties of Textile Fibres*, The Textile Institute, London.

Newns, A. C. (1956), 'The sorption and desorption kinetics of water in a regenerated cellulose,' *Transactions Faraday Society*, **52**, 1533–1545.

Nordon P., Mackay B. H., Downes, J. G. and McMahon, G. B. (1960), 'Sorption kinetics of water vapour in wool fibres: evaluation of diffusion coefficients and analysis of integral sorption,' *Textile Research Journal*, **9**, 761–766.

Pan, N. and Sun, Z. (2006), 'Essentials of psychrometry and capillary hydrostatics', in *Thermal and Moisture Transport in Fibrous Materials*, edited by Pan, N. and Gipson, P., Woodhead Publishing, Cambridge, p. 112.

Pan, N. (2008), 'Sweat management for military applications', in *Military Textiles*, edited by Wilusz, E., Woodhead Publishing, Cambridge.

Parsons, K. C. (1993), *Human Thermal Environments*, Taylor & Francis Publishers, United Kingdom.

Poynting, J. H. (1997), *Text book of Physics*, Charles Griffin & Co. Ltd., London, 1925.

Pruppacher, H. R. and Klett, J. D. (1997), *Microphysics of Clouds and Precipitation*, Kluwer Academic Publishers, Dordrecht, p. 506.

Ren, Y. J. and Ruckman, J. E. (2003), 'Water vapour transport in wet waterproof breathable fabrics,' *Journal of Industrial Textiles*, **32**(3), 165–175.

Scheurell, D. M., Spivak, S. M. and Hollies, N. R. S. (1985), 'Dynamic surface wetness of fabrics in relation to clothing comfort,' *Textile Research Journal*, **55**, 394–399.

Slater, K. (1999), 'Thermal comfort properties of fabrics', *Progress in Textiles: Science and Technology, Vol. I, Testing and Quality Management*, edited by Kothari, V. K., IAFL Publications, New Delhi.

Thwaites, C. (2008), 'Cold weather clothing,' in *Military Textiles*, edited by Wilusz, E., Woodhead Publishing, Cambridge, p. 158.

US Patent 6427493, 'Synthetic knit fabric having superior wicking and moisture management properties,' August, 2002.

Wehner, J. A., Miller, B. and Rebenfeld, L. (1988), 'Dynamics of water vapour transmission through fabric barriers,' *Textile Research Journal*, **10**, 581–592.

Wu, H. and Fan, J. (2008), 'Study of heat and moisture transfer within multi-layer clothing assemblies consisting of different types of battings,' *International Journal of Thermal Sciences*, **47**, 641–647.

Yasuda, T., and Miyama, M. (1992), 'Dynamic water vapour and heat transport through layered fabrics: Part II Effects of the chemical nature of fibres,' *Textile Research Journal*, **62**(4), 227–235.

Yoo, S. and Barker, R. L. (2005), 'Comfort properties of heat-resistant protective workwear in varying conditions of physical activity and environment, Part I: Thermophysical and sensorial properties of fabrics,' *Textile Research Journal*, **75**, 523–530.

Zhou, X., Zhang, J., Gao, Y., Wang, S. and Fan, J. (2006), 'A comparison of the water vapor transport through fabrics under conventional and subzero climate,' *Textile Research Journal*, **76**, 821–827.

Zhou, X., Wang, S., and Yuan, G. (2003), 'An apparatus used to investigate condensation for fabrics, laminates and films,' *Journal of Industrial Textiles*, **32**(3), 177–186.

9
Improving tactile comfort in fabrics and clothing

A. DAS and R. ALAGIRUSAMY, Indian Institute of Technology, India

Abstract: Sensorial comfort of clothing is a result of interaction between the fabric and human skin with the sensory system and the atmospheric conditions. The chapter begins by discussing the neurophysiological basis of sensory perceptions; fabric mechanical properties such as fabric prickliness, itchiness, stiffness, softness, smoothness, roughness, scratchiness, friction and tactile properties are discussed and warmth or coolness to the touch of fabrics and dampness sensations are explained. Finishing treatments for changing the textile surface for improving tactile sensation are also discussed, along with methods for improving and reducing electrostatic propensity.

Key words: sensorial comfort, human skin, tactile sensation, prickle, electrostatic propensity.

9.1 Introduction

Sensorial comfort is concerned with how a fabric or garment feels when it is worn next to skin. Smith (1984) has found that when subjects wore various fabrics next to the skin they could not detect the differences in fabric structure, drape or fabric finish but could detect differences in fabric hairiness. Some of the separate factors contributing to sensorial comfort have been identified as tickle caused by fabric hairiness and prickle caused by coarse and therefore stiff fibres protruding from the fabric surface. Matsudaria et al. (1990) found that the stiffness of protruding fibres is the dominant factor in causing prickle sensations. This is affected to a larger extent by fibre diameter and to a lesser extent by fibre length. For a fibre of given diameter, the end of a long fibre is more easily deflected by a fixed amount than the end of a short fibre, so it appears less prickly. For a fibre of a given length, a larger diameter is much stiffer by the fourth power of diameter and hence is more likely to prick.

Wet cling is associated with sweating and is caused by damp and sticky sweat residue on the skin. A factor influencing cling is the actual area of fabric in contact with skin which in turn is influenced by fabric structure. Warmth to touch is the property experienced when a garment is first picked up or put on. A fabric is usually at lower temperature than the skin and thus there will be a loss of heat from the body to the garment until the temperature of the surface in contact equalizes. The faster this heat transfer occurs, the greater is the cool feel of the fabric. The difference in the cold feel between fabrics is mainly determined by their surface structure rather than by the fibre type. The raised surface of cotton fabric feels comparatively warm on the raised surface, giving a lower contact area and hence a slower rate of change of temperature. Ironing a cotton sheet has the effect of increasing the cold feel by compacting the surface structure.

9.2 Comfort and neurophysiology

9.2.1 The neurophysiological perceptions

Skin stimuli and the skin sensory system

Coren and Ward (1989) demonstrated that human skin has very complex structure, as may be seen in Fig. 9.1 which shows the schematic of a microscopic structure in hairy skin covering most of the human body. The skin has two layers: the epidermis and dermis. The epidermis is the outer layer, consisting of several layers of dead cells on top of a single living cell and the dermis is the inner layer, containing most of the nerve endings in the skin. In addition sweat glands, hair follicles, and fine muscle filaments are housed here. Below the dermis there are layers of connective tissue and fat cells.

Figure 9.1 also illustrates some of the nerve endings on the skin. There are two type of nerve endings: corpuscular endings and noncorpuscular (or free nerve)

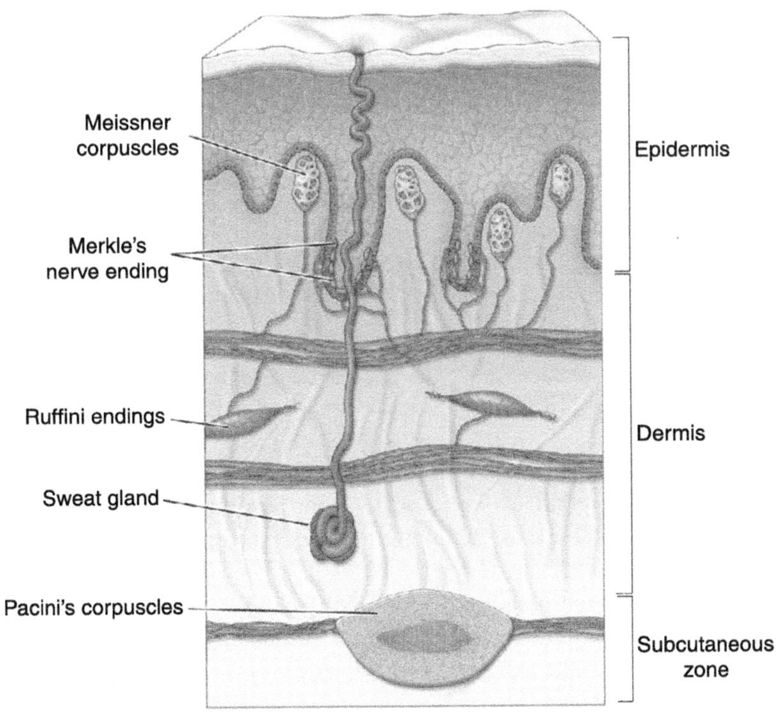

9.1 Schematic section of human skin.

endings. Corpuscular endings have small bodies or swellings on the dendrites including the Pacinian corpuscles, Meisner corpuscles, Merkle disks and Ruffini endings which are particularly responsive to touch stimuli. The free nerve endings in subcutaneous fat are associated with pain fibre, and those projecting in to the epidermis may be associated with cold fibres or pain fibres.

Sensory receptors

Skin receptors

Iggo (1988) demonstrated that human skin is the interface between the human body and its environment. It is richly innervated and contains specialized sensory receptors to direct various external stimuli. There are three major stimuli: (i) mechanical contacts with external objects, (ii) temperature changes due to heat flow to or from the body surface, and (iii) damaging traumatic and chemical insults. In responding to these stimuli, the skin receptors produce the sensation of touch, warmth or cold, and pain.

Mechanoreceptors

There are two groups of mechanoreceptors: (i) encapsulated receptors, including Pacinian corpuscles, Meissner corpuscles, Krause endings and Ruffini endings, which are all innervated by fast conducting myelinated fibres; and (ii) receptors having an organized and distinctive morphology such as the hair follicle receptors and Merkle discs. Iggo (1988) classified each mechanoreceptor with a distinctive range of properties that enable it to receive and respond to a particular parameter of a mechanical stimulus. The Pacinian corpuscles detect and respond to high frequencies of displacement up to about 1500Hz, the Meissner corpuscles and the hair follicles to middle range frequencies (20–200Hz), and the Merkle cells and Ruffini endings to steadily maintained deformation of the skin (DC to 200Hz).

Thermoreceptors

Another group of sensory receptors detects the temperature of skin. These receptors can respond to both constant and fluctuating skin temperatures. In responding to constant temperature, the receptors discharge impulses continuously to indicate temperature of skin and they are very sensitive to changes in skin temperature. There are two types of thermo receptors; cold receptors and warm receptors. The cold receptors have a peak sensitivity of around 25–30 °C and are excited by dynamic downshifts in temperature. The warm receptors have a peak sensitivity of around 39–40 °C and are sensitive to increase in skin temperature.

Nociceptors

Nociceptors are another group of sensory receptors, which respond to noxious stimuli such as heating the skin, strong pressure, or contact with sharp or damaging objects. The receptors have relatively high thresholds to function as warming devices that enable an organism to take protective action.

9.2.2 Perception and sensations related to mechanical stimuli

Dynamics of wear sensation

During wear, clothing contacts with the skin of most parts of our body, dynamically and continuously. This contact between skin and clothing has a number of features: the area of contact is large and crosses over regions with various sensitivity; the body very often changes its physiological parameters such as skin temperature, sweating rate, and the humidity at the skin surface, which generates various new thermal stimuli; and the body is often in movement causing clothing to move towards and away from the skin frequently, which often induces new mechanical stimuli. The thermal and mechanical stimuli trigger responses from various sensory receptors and formulate various perceptions such as touch, tactile, thermal, moisture, and a more complex synthetic sensation, which affect the comfort status of the wearer (Li, 2001).

Perception of touch and pressure

Any point on the surface of the human body can evoke the sensation of touch; however, the sensitivity varies from one region of the body to another. Figure 9.2 shows the average absolute thresholds for different region of female skin. The thresholds were obtained by applying a hair to the surface of the skin with different amounts of force, and are expressed as the amount of force applied to the hair. The higher the bar, the greater the force needed to trigger the sensory receptor and the lower the sensitivity. Obviously, the absolute thresholds vary considerably over the body surface. The threshold for touch sensation depends on both frequency of vibration of a stimulus and skin temperature. Each touch sensation seems to be located at a particular place on the skin and to be directly related to the amount of neural presentation at each area in the touch cortex (Coren and Ward, 1989).

In the process of fabric–skin contact and mechanical interaction during wear, clothing will exert pressure and dynamic mechanical simulation to the skin which will in turn trigger various mechanoreceptors and generate a variety of touch sensations. It is observed that the positions of a subject rather than the shape of cloth affects the amplitude of the fluctuations. It is found that an increase in pressure tends to produce stronger 'knobbiness' and 'roughness' evaluations. Makabe et al. (1993) measured clothing pressure in the covered area of the waist for a corset and

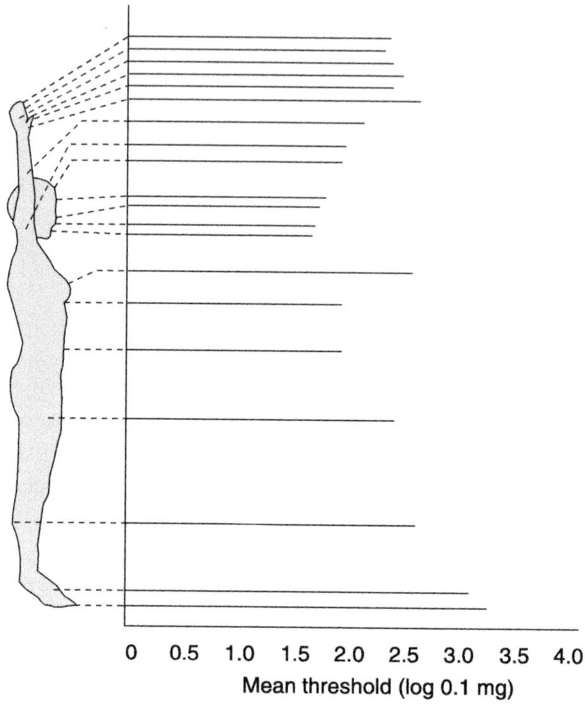

9.2 Average absolute thresholds for different regions of female skin (adapted from Y. Li., 2001).

waistband and recorded the sensory responses of the subjects on the clothing pressure. They observed that the pressure at the waist was a function of the covering area, respiration and the ability of samples to follow bodily movement.

Perception of prickle, itch and inflammation

Prickle is a sensation that is often complained of by consumers of next-to-skin garments, especially when fabric containing wool fibres is used for underwear garments. Prickle is usually described as the sensation of many gentle pinpricks. Traditionally, the prickle sensation associated with wool was considered to be associated with skin allergic response. The degree of discomfort caused by prickle varies from person to person and with the wear situations.

Westerman et al. (1984) studied the relationship between sensations of prickle and itch and human cutaneous small nerves. Skin sensations were tested on the forearms of 12 volunteers, in whom anoxia nerve blocks of the forearms were produced by inflating a blood pressure cuff to 270 mm Hg on the upper forearm. Figure 9.3 shows touch sensations were lost after about 20 minutes, but pain, temperature and fabric-evoked prickle sensation remained until about 40 minutes. This result indicated that prickle sensations are associated with small nerve fibres.

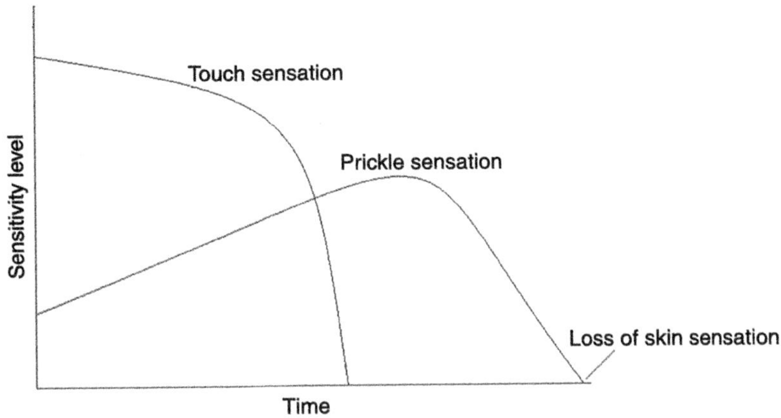

9.3 Time course of loss of prickle and touch sensations (adapted from Li, 2001).

The cause of fabric prickle as the mechanical stimulation of fabric to the skin that induces low grade activity in a group of pain nerves is shown in Fig. 9.4 (Garnsworthy et al., 1988). As a fabric begins to contact the skin, the protrouding fibres of the fabric will take all the force initially. As the body of fabric moves closer to the skin, the force increases and protruding fibres bend. When the force from individual fibres reach certain levels, large shear forces in the skin are generated and pain nerve endings in the skin are activated. Willis (1985) found that the summation of responses from the pain group of nerves is necessary for the initiation of pain sensation, and it has been demonstrated that prickles from the

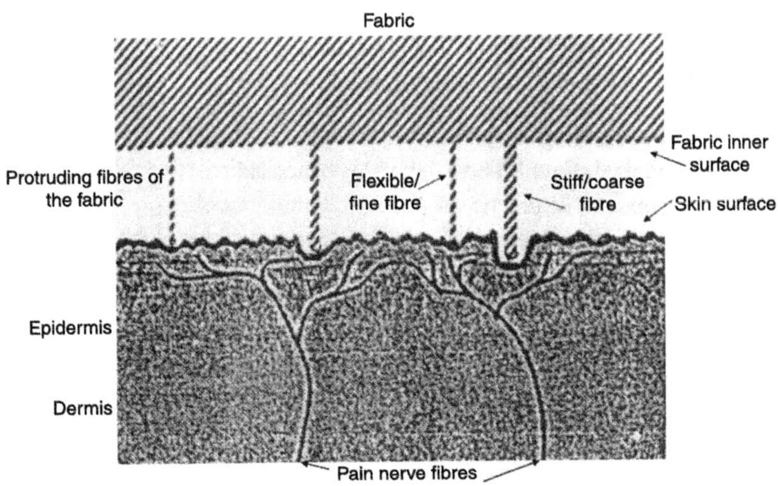

9.4 Diagramatic presentation of the mechanisms of fabric-evoked prickle sensation (adapted from Li, 2001).

fabrics could not be perceived if the density of high load bearing fibre ends is less than 3 per 10 cm^2 of the fabric, or the skin contact area is below 5 cm^2 (Garnsworthy et al., 1988).

Sensitivity to fabric prickle is influenced by a number of factors:

- Males had higher thresholds and more variations to sensitivity to prickle than females.
- Prickle sensitivity decreases progressively with age, since the skin is known to harden as age increases and prickle sensitivity decreases with hardness of the outermost skin layer.
- Pain nerve endings are very close to the surface of hairy skin, but not in glabrous skin, which explains why prickle cannot be felt with fingers.
- Prickle sensitivity is increases with moisture content of the skin, as water can soften the stratum corneum and allow the protruding fibres to penetrate more readily.
- Prickle sensitivity increases with ambient temperature in the range of 12–32 °C at constant relative humidity (60–65%) as the skin moisture content increases due to perspiration in hot and humid conditions.

Itch is a sensation that has been shown to result from the activation of some superficial skin pain receptors. The pain receptors responsible for itch may be of a different type to those responsible for prickle sensation.

Skin inflammation (reddening) occurs in a small proportion of the population, as a consequence of prickle and itch resulting from mechanical stimulation of skin pain receptors from prickly fabrics – most likely through a mechanism termed axon reflex. Inflammation may occur rapidly (in minutes) or slowly (in hours). It can be relieved quickly after the fabric is removed from the skin, unless the fabric skin contact is too long and has produced a severe reaction.

9.3 Human tactile sensation

Human skin facilitates sensation through touch via the mechanical properties of its tissues, which are responsible for conveying tactile stimuli to the dermal and subdermal receptor sites at which they are transduced into neural signals, and through the tactile receptors that perform the transduction (Visell, 2008). Human skin can be categorized as hairy (skin on most parts of the body), glabrous (the non-hairy skin on the front of the hands and bottom of the feet), or mucous (the moist skin of the mouth and other bodily openings). The skin contains a range of receptors for pain (nociceptors), temperature (thermoreceptors), chemical stimuli (chemoreceptors), limb joint and muscular states (proprioceptors), and six kinds of skin tactile mechanoreceptors (Kandel et al., 2000). Properties of five types of mechanoreceptors are present in the cutaneous and subcutaneous layers of the skin, and associated connective tissue. Details concerning the sensory correlations of such mechanoreceptors are given in Table 9.1.

Table 9.1 Skin tactile mechanoreceptors and sensory displays

Receptor	Sensory display
Pacinian corpuscle	Vibration, tickle
Meissner's corpuscle	Touch, tickle, motion, vibration, flutter
Merkel's cells	Edge pressure
Ruffini corpuscle	Stretch, shear, tension
Hair follicle receptor	Touch, vibration, proximity

In addition, many relevant psychophysical properties are known. These include the relevant ranges of frequency sensitivity, receptive field sizes, receptor densities and sensory correlates, all quantities that are highly salient to the low-level design of tactile displays. Spatial tactile resolution is among the most extensively studied features. The most frequently cited measure is the two-point threshold, which is the minimum distance between two point-like indentations applied to the skin below which only a single point of contact is detected with the senses. This value varies from 2.5 mm in the fingers, up to as much as 50 mm for other body regions (Fig. 9.5). More recently, authors have focused on inadequacies of the two-point threshold as a measure of tactile acuity. Tactile acuity exhibits high inter-individual differences and depends very much on the nature of the stimulus that is used and properties of the stimulus, such as frequency, duration, and amplitude.

Psychophysical thresholds for a particular type of stimulation indicate the minimum noticeable intensity of stimulation I_{min}, with stimulus properties (other

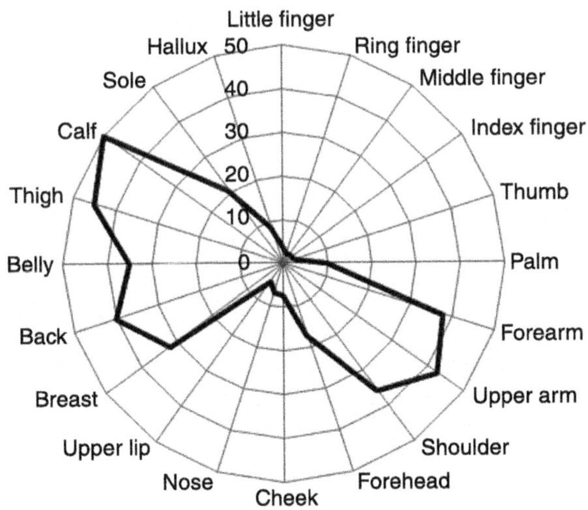

9.5 Average two-point tactile discrimination thresholds (mm) for various bodily regions (adapted from Visell, 2008).

than amplitude) held constant. Maximum stimulus intensity I_{max} is typically taken to be the threshold for pain perception. These values are significant for display engineering, as they determine the maximum dynamic range R (in dB) attainable by stimulation with a particular stimulus type, which may be expressed in decibels via equation [9.1],

$$R = 10\log_{10}(I_{max}/I_{min}) \tag{9.1}$$

In general, the maximum dynamic range will be a function of stimulus type, frequency of stimulation, location on the body, etc. Sensory transduction is not a linear, temporally or spatially independent process, and complex sensory phenomena result from tactile stimulation. Processes such as temporal and spatial integration create dependencies between the influence of stimuli located near each other in time or spatial position. Locally, sensory thresholds rise under sustained stimulation. Among other things, this makes it possible to ignore low-level tactile stimulation, such as the sensation of an object held against the skin. Adaptation is important to account for in tactile interaction design, and for sensory substitution, because such interventions may involve a continuous coupling between the user of the system and the stimulating device over an extended period of time. Tactile sensory thresholds rise by a few dB in response to only a few seconds of sustained vibrotactile stimulation, and do not attain a maximum until approximately 25 min of stimulation have passed. Full recovery requires in the order of 2 min (Kaczmarek et al., 1991). It is familiar from everyday experience that qualitative tactile sensations are wide-ranging and heterogeneous. They include pressure, texture, puncture, thermal properties, softness, wetness, friction (slip, adhesion), dynamic events (contact, release), pain, object features (like shape, edges and embossing), recessed features, and vibrotactile sensations such as tickling, itch, vibration, and buzz (Hayward et al., 2004).

9.4 Fabric mechanical properties and tactile-pressure sensations

9.4.1 Prickliness

Fabric-evoked prickle has been identified as one of the most irritating discomfort sensations for clothing wear next-to-skin. A special type of pain nerve is responsible for prickle sensation. Individual protruding fibre ends from a fabric surface are responsible for triggering the pain nerve endings, when contacting the skin. A summation of responses from a group of pain nerves seems necessary for the perception of prickle sensations.

Matsudaria et al. (1990) compared three techniques for measurement of a fabric prickle: low pressure compression testing, laser counting of protruding fibres, and a modified audio pick-up method. A KES-FB compression tester was modified to measure the relationship between pressure and fabric thickness at the initial fabric

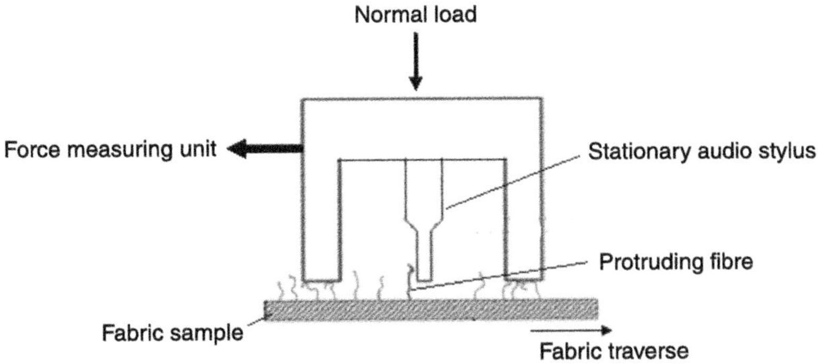

9.6 Audio pick-up device for measurement of fabric prickle (adapted from Li, 2001).

compression stage, in which protruding fibres are bent and compressed. The modified audio pick up technique shown in Fig. 9.6 was the most effective and the measured mean force per contact correlated well with the subjective perception of fabric prickle.

The fabric surface was traversed under a stationary audio stylus, from which signals were obtained from the contact between the stylus and protruding fibre. Two classical models – a loaded cantilever and an Euler column – were used to calculate the pointing force and critical buckling load as shown in Fig. 9.7.

The critical buckling load, P_E of the protruding fibre ends has been identified as the stimuli responsible for triggering the pain receptors, and can be expressed as:

$$P_E = \pi^2 (EI/4l) \qquad [9.2]$$

where E is the Young's modulus of the fibre;

I is the moment of inertia ($I = \pi d^4/64$) in case of a circular rod; and

l is the length of protruding fibre ends. The equation suggests that fibre Young's modulus, fibre diameter and fibre length are the key factors determining fabric prickliness.

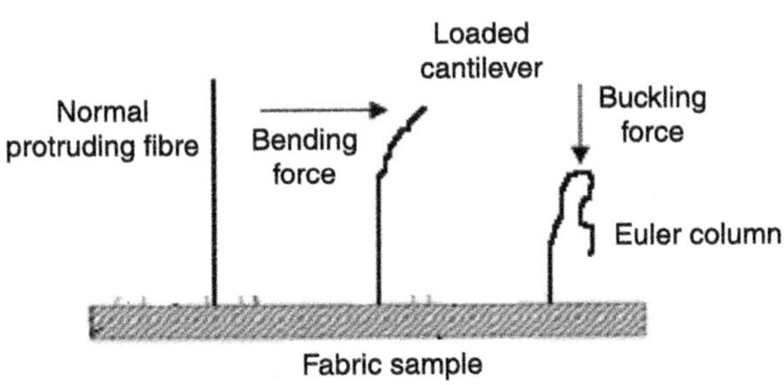

9.7 Models of a loaded cantilever, and an Euler column.

9.4.2 Itchiness

Itchiness is also found to result from activation of some superficial pain receptors. A prickling fabric usually has quality of itch sensation and perception of itchiness in clothing is highly correlated with perception of prickliness. Therefore, it could be expected that the factors influencing fabric prickle would affect fabric itchiness as well. Comparing the subjective rating of itchiness obtained from wear trials with the mechanical properties measured objectively, Li (1988) observed that the perception of itchiness correlated with fibre diameter, fabric thickness at low and high pressure, and fabric surface roughness.

9.4.3 Stiffness

Stiffness of woven and nonwoven fabric may be evaluated by using both subjective assessment and objective assessments. Magnitude estimation is generally obtained by subjective evaluation, while objective measurement of fabric stiffness is carried out by using various instruments like a Shirley Cantilever, a Cusick Drapemeter, or a Shirley Cyclic Bending Tester, etc. Elder (1984) reported a study to verify their methodology and conclusion by using woven and knitted fabrics and found that the agreement among three objective measurements – bending length, flexural rigidity and drape coefficient – was good, and that these measurements were highly correlated with subjective ratings.

Fabric stiffness has been recognized as one of the primary hand expressions used by Kawabata and Niwa (1989) in their fabric hand evaluation system, named KOSHI, carried out in Japan. Subjective ratings of garment stiffness are related to three types of mechanical properties:

1. fibre diameter and tensile breaking load;
2. fabric compression properties such as thickness at low and high pressure, the energy of compression-thickness curve, the slope of compression-thickness curve, and the resilience of compression-thickness curve; and
3. fabric frictional properties such as mean friction coefficient and mean deviation of friction coefficient.

9.4.4 Softness

Fabric softness is one of the most frequently used terms in describing clothing comfort performance by consumers. Fabric softness has multiple meanings that can be related to compression and/or to smoothness and flexibility of fabrics, depending on the fabrics being handled and end uses. The softness is opposite to stiffness that can be measured by bending length. Softness may be considered as the opposite of firmness or hardness measured by thickness tests. In Kawabata's hand evaluation system, corresponding to 'NUMERI', softness was defined as mixed feeling coming from a combination of smooth, supple and soft feeling. In the dimension of 'FUKURAMI', softness is related to the feeling from a combination of bulky, rich and well-formed impressions. The subjective perception of garment softness during wear correlated with fabric compression properties (thickness at low and high pressures, resilience and energy of compression-thickness curve), fabric tensile properties (the maximum elongation, and linearity of load elongation curve), fibre diameter and breaking load. These reflect the three aspects of fabric softness of compression, flexibility and smoothness (Li, 2001).

9.4.5 Smoothness, roughness, and scratchiness

The friction and mechanical interaction between fabric and skin during contact are the key factors determining the perception of roughness, smoothness and scratchiness. It has been identified that roughness and scratchiness are important tactile sensations determining the comfort performance of next-to-skin wear. Moisture at the skin surface can alter the intensity of fabric roughness perceived, because as moisture content increases, the friction and displacement of skin increases, which triggers more touch receptors. Behmann (1990) reported a study on the perception of roughness and and textile construction parameters. The roughness was defined as irregularities in surface that can be described geometrically as the size of roughness elements or mechanically by the friction coefficients. The study of roughness perception of woven and knitted fabrics made from nylon yarn of different diameters reveals that roughness decreases with yarn diameter logarithmically and at the same yarn diameter the knitted fabric was perceived as rougher. The perception of fabric roughness (smoothness) is associated with a number of physical properties objectively measured such as roughness, friction, prickle, shear and bending stiffness, thickness and area density.

Comparing the subjective sensory responses from wear trials with objective measured mechanical properties, the perception of roughness correlated with fabric surface roughness (maximum force, mean surface roughness coefficient, and deviation of surface roughness coefficient), compression properties (fabric thickness at low and high pressures, and energy of the compression-thickness curve), fibre diameter and fibre tensile properties (breaking load and breaking elongation), and fabric tensile properties (maximum tensile elongation, elongation recovery load).

Similarly, subjective perception of scratchiness is related to fabric tensile properties (maximum tensile elongation, energy of tensile load-elongation curve, and the slope of the tensile load-elongation curve), fabric surface roughness (maximum roughness force, mean surface roughness coefficient and deviation of surface roughness coefficient), and fabric compression properties (thickness at low and high pressure, linearity of the compression curve, energy of the compression-thickness curve and slope of the compression-thickness curve).

9.4.6 Friction and tactile properties of woven and knitted fabrics

Clothing comfort is one of the major current concerns of textile and garment manufacturers. Comfort is based on the human sensory response to clothing materials and is determined by a variety of thermal, physiological and mechanical parameters. For fabrics that come into direct contact with the skin, touch and tactile properties are especially important in connection with clothing comfort, because the sensory feel of materials is related to mechanical stimuli due to pressure and friction forces. When someone wears clothing, friction between skin and fabric can arise in many areas. This friction is the tangential force to the surface of the skin when the fabric slides over the skin. The dynamic coefficient of friction (μ) between two materials is usually defined as the ratio between the frictional force F and the applied normal load N:

$$\mu = F/N \qquad [9.3]$$

It is known that the friction of polymeric materials does not follow this law of Amonton. The frictional behavior of woven and nonwoven fabrics, the relationship between friction force and normal load, can be expressed as

$$F/A = C(N/A)^n \qquad [9.4]$$

where F is the friction force, N the normal load, A the apparent area of contact, C the friction parameter and n the friction index (non-dimensional). This deviation of Amonton's law necessitates the evaluation of friction over a range of applied normal loads. Another factor influencing the friction of textiles is the sliding velocity, as the friction of fabrics tends to increase with increasing sliding velocity (Bertaux et al., 2007).

During the evaluation of the tactile properties of textiles, contact is made through the skin where numerous receptors are located which give rise to various sensations felt by the human subject. In the epidermis, the first and very thin layer of the skin, we can find mechanoreceptor units such as the Merkel cell neurite and pain receptors. The dermis, namely the second layer of the skin, contains additional mechanoreceptive tactile units, of which three types send signals to the central nervous system: the Meissner corpuscles, the Pacinian corpuscles, and the spindle-shaped Ruffini endings. The most sensitive skin areas are the face, the torso and the hand. For instance, 17,000 mechanoreceptive tactile units innervate the hairless skin of the human hand. The Merkel cell neurites and the Ruffini endings play an important role in relaying information about surface texture and tactile and spatial form of touched objects. Due to their location, receptive field and response type, the Pacinian, the Merkel and the Meissner mechanoreceptors can characterize the roughness of fabrics, whereas the Ruffini and Meissner mechanoreceptors can characterize the friction between skin and fabrics. The prickle from the skin contact of fabrics can be detected by the pain receptors.

9.5 Warmth or coolness to the touch of fabrics

The sensation of warmth or coolness, that occurs when human skin touches another body, is a transient heat conduction phenomenon. Skin brought into contact with the surface of clothing is normally at a higher temperature than the clothing, and heat flows away from the skin. Loss of heat causes the temperature of the skin to fall, which is sensed by thermoreceptors located within the skin's dermal layer. The higher the rate of the heat flow, the more rapid the temperature drops near the thermoreceptors and the more intense the feeling of coolness (Hensel, 1981).

The intensity of the coolness sensation has been previously linked with the maximum rate of heat flow from the skin at the depth of the thermoreceptors. The simpler approach takes into account the fact that thermoreceptors detect temperature and not heat flow, and also that temperature changes are most easily perceived when they are rapid. The rate of temperature change, resulting from heat flow from the skin to a homogeneous material at a lower temperature when brought into contact with it, is determined by the thermal inertia of the material (the product of density, specific heat, and thermal conductivity). Any material that can absorb and conduct heat well will easily draw heat away from the skin and feel cool, i.e. the higher the thermal inertia the cooler it will feel to the touch. The fabric structural features, particularly surface properties, have a great influence on cool-warm feel, and a three-layered model based on inner, middle and outer layers with different thermal properties would be more appropriate.

Typical results for warm and cool fabrics are shown in Fig. 9.8. The rates of temperature change detected by the thermoreceptors that determine warm or cool sensations varied quite significantly for the two fabrics. The maximum rate of heat flow occurs shortly after an object is touched (after around 0.2 seconds), and this initial sensation is most important for the perception of coolness to the touch. The coolness sensation is estimated by measuring the initial rate of temperature change. Figure 9.8 includes the measured responses of four fabrics, i.e. cool fabric; fabric with smooth surface; fabric brushed against skin; and warm fabric. The differences in the rates of change of temperature in all the fabrics are quite substantial. This suggests that sensations of warmth or coolness to the touch are related to the surface properties of fabrics, in this case surface hairiness, in addition to the materials (Schneider and Holcombe, 1991).

9.5.1 Transient heat conduction in skin – warm/cool feeling

The warm/cool feeling felt by a human's skin contacting an object is probably related to the heat flow between the skin and the contacted object. In order to correlate the warm/cool feeling with the physical properties of the object, the relationship between heat flux and a human's warm/cool feeling must be investigated and a theoretical analysis of transient heat transfer between the skin

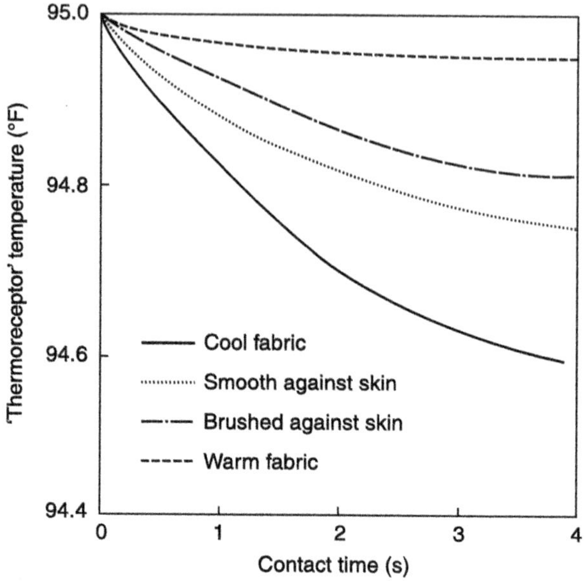

9.8 Skin temperature drop as a function of time after contact with fabric.

and the contacting object is required. Yoneda Morihiro and Kawabata Sueo (1981) explained that this analysis is also useful in evaluating the warm/cool feeling measured objectively by use of instruments.

It has been reported that the maximum heat flux which is observed shortly after the contact of the heated plate to fabric correlated well with the warm/cool feeling of human subjects to make it a convenient measure of the warm/cool feeling of fabrics. This maximum heat flux is named as q_{max}, as shown in Fig. 9.9. The q_{max} value was introduced by Kawabata and his team as a measure of predicting the warm/cool feeling of fabrics, where q_{max} is the peak value of heat flux which flows out of a copper plate having a finite amount of heat, into the surface of the fabric after the plate contacts the fabric surface. At first, a theoretical prediction of q_{max} was given as transient heat conduction within a homogeneous body and the predicted q_{max} was interpreted with respect to the effects of test conditions and the thermal properties of test specimens. It was found that the warm/cool feeling correlates well with transient heat conduction, especially with q_{max}, the peak value of heat flux which arises for a very short time after the touching of the skin to the fabric surface. The physical meaning of q_{max} is discussed on the basis of theoretical analysis of transient heat conduction from an outside object into human skin.

Figure 9.10 shows how the q_{max} of a wide range of fabrics varies with moisture content of the fabrics. It is clear that for all the fabrics the q_{max} value increases with the increase in moisture content, which means that the fabric feels cool at higher moisture content. It is also clear that the linen fabric has maximum q_{max}

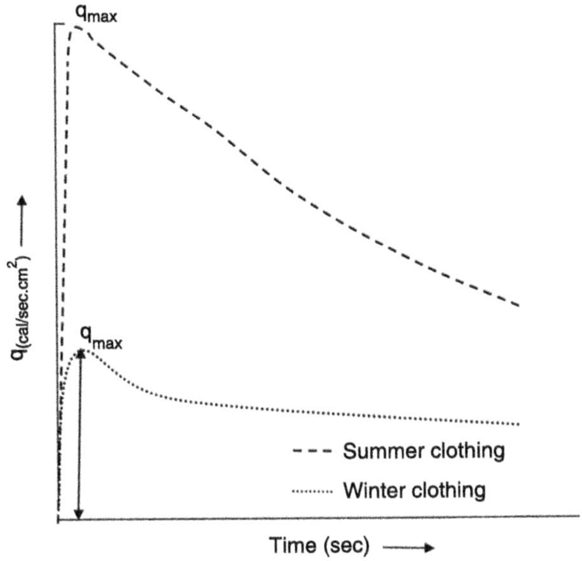

9.9 Heat flux with respect of time for winter and summer clothing.

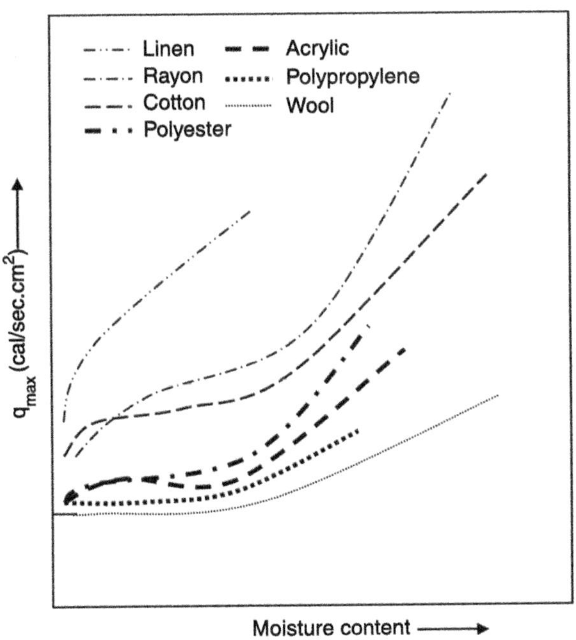

9.10 Maximum heat flux of different fabrics as a function of moisture content.

and wool fabric has minimum q_{max} up to a certain level of moisture content. This is the reason why linen feels cool and wool feels warm when they are touched.

Kawabata and Akagi (1977) reported the following experimental results in regards to the warm/cool feel of fabrics (q_{max}):

- A high correlation was obtained between physically measured q_{max} and fabric warm/cool feeing data gathered in human sensory tests.
- A high q_{max} value corresponds to a cool feeling and a low q_{max} value to a warm feeling.
- The value q_{max} depends on fabric surface condition and not on the number of fabric layers or fabric thickness.
- The value q_{max} is sensitive to fabric water content and surface geometry.

9.5.2 Dampness

Moisture in clothing has been widely recognized as one of the most important factors contributing to discomfort sensations. The skin wetness contributes to the sensation of humidity, and the sensation of dampness is related to the amount of sweat accumulated in clothing. The subjective sensations of skin and clothing wetness are considered as sensitive criteria for evaluation of the thermal function of clothing. The moisture in clothing contributes significantly to comfort perceptions during actual wear conditions (Li, 2001).

A chilling sensation is produced when damp fabrics are placed on the forearm, which is due to the temperature drop at the skin in contact with the moist fabrics. Also, the temperature drop is influenced significantly by the degree of the fabric–skin contact that is associated with fabric construction and surface hairiness. As fibre hygroscopicity is a critical factor determining the coupled heat and moisture transfer behaviour in fabric, it has a significant impact on the skin temperature drop during the contact. Comparing fabrics with different degrees of hygroscopicity, the skin temperature drop increases with the level of excess moisture as the degree of fibre hygroscopicity increases.

Ambient conditions such as temperature and relative humidity influence the skin temperature drop significantly. The skin temperature drop decreases as ambient temperature increases because of the decrease in temperature difference prior to the contact. However, ambient temperature has negligible influence on the differences of the skin temperature drop among different types of fibres, because ambient temperature mainly influences the dry heat transfer process, not the moisture exchange process. Ambient relative humidity, on the other hand, shows significant impact on both the skin temperature drop of all fibres and the differences between the fibres. When ambient relative humidity increases, the difference in moisture concentration between the fabric and the environment decreases, resulting in a smaller temperature gradient between the skin and fabric, hence a smaller skin temperature drop during the skin–fabric contact. The differences between fibres are much greater when ambient relative humidity is low, but when the relative humidity approaches saturation, the difference between fibres becomes negligible.

9.6 Improving the textile surface properties for tactile sensation

There are many new treatments for fabrics to give them a variety of looks and effects, some of which add lustre, others give muted dull effects or may make a fabric crease-resistant, crease-retentive, waterproof, etc. These 'finishing treatments' include brushing, sanding, sueding, etc. A finish often contributes much to the 'feel' and 'hand' of a fabric and it may be said that 'cloth is made in the finishing'. The treatments are performed on fabrics with the help of chemical treatments such as resin coating, silicone softener, etc., and are the last step in the textile production before the clothing operations or by mechanical treatments.

In tactile finishing treatment, three classes of products can be used: softeners; garniture/thermoplastic products; and synthetic hardenable resins. Each of them brings out tactile properties in fabrics but can induce undesirable effects. For example, typical resin treatments promote strong easy-care effects in the fabric but may induce a stiff handle. Therefore commercial finishing providers attempt to reduce the undesirable effects by using combined formulation, which allows 'one shot' treatments to give optimal results with respect to the main desired properties with minimal side-effects (Philippe et al., 2003).

For a variety of looks and effects on fabrics, there are many new finishing products and treatments. They can roughly be split into two categories: easy-care and organoleptic. Among them numerous softeners have been proposed: macro and micro silicone, fatty acid and polyethylene, etc. The results of sensory analysis of different finishes on fabrics are discussed. With silicone finishing, the 'slipping' and 'greasy' attributes clearly change along with the concentration. The result of softener macro silicone Ultratex® treatment is known to soften the fabric, and with the increase of concentration the fabric becomes more 'greasy' and 'slipping'. For the resin treatment, it is expected to have more 'nervous' and less 'crumple-like' fabrics. This is confirmed by the results obtained, since fabrics treated with a high concentration of resin finishing were significantly more 'nervous' and less 'crumple-like' than the non-treated fabric (Strazdienė et al., 2006).

Under normal wearing conditions, each of the tactile sensations (itchiness, prickliness) is important to overall comfort. To improve the comfort of garments for daily wear, one should ensure the smoothness of the fabric surface, either by using finer fibres or avoiding harsh resins, assuming these features produce the named tactile sensations. With heavy exercise and sweating, moisture related sensations (non-absorbency, clinginess, and dampness) are critical in determining overall discomfort. Sportswear should therefore have excellent water absorption properties (Lau et al., 2002).

9.6.1 Mechanical and chemical finishing

In a study (Sukigara and Ishibashi, 1994) carried out on the surface roughness of polyester crepe fabrics in the grey and finished states, subjectively and objectively, it has been observed that the finished fabrics were perceived as being more 'knobby' and 'rough' than the grey fabrics. Roughness perception increased with the pressure between the fabric surface and the subject's hand. The effects of fibre diameter and fabric process parameters on subjective prickle sensation of fabrics have been reported (Garnsworthy et al., 1988; Naylor, 1992). The fabric process parameters include silicon treatment and steam relaxation finish. The following multiple linear regression equation shows the relation between mean prickle estimate (MPE) and other parameters,

$$\text{MPE} = -3.65 + 2.83\ (\text{diameter}) - 0.60\ (\text{treatment}) - 0.25\ (\text{finish}) \quad [9.5]$$

It is evident from the above equation that the prickle sensations increase with the increase in fibre diameter and decrease with the finishing and treatment. This is due to the fact that both silicon treatments reduce fibre–fibre friction and steam finishing processes change the orientation of fibre tips. Matsudaira et al. (1990) have also reported that the successive finishing processes have appreciable impact on the fabric prickle sensations. The effect of bleaching treatment on the fabric handle characteristics has been studied by Yan et al. (2000) and they have reported that the bleached fabrics have a coarse handle and a lower fullness than those unbleached. The effect of finishing on the compression of woven fabrics has been studied by Manich et al., (2006), who reported that the finishing treatment of grey goods leads to a fuller and more compact fabric structure given the increase in fabric density (46%) and in the cover factor image analysis (9%), and the decrease in thickness (33%). Militky and Bajzik (1997) studied the effect of washing/ironing cycles on the properties of cotton fabrics. Their results show that the spread of hand rating in washed/ironed samples is higher, probably owing to changes of bulkiness and deformability. Apart from this, the hand, shrinkage and surface roughness is dependent mainly on the type of fabric. The effect of a durable flame-retardant finishing on the mechanical properties of cotton knitted fabrics has been investigated by Mamalis et al. (2001), who reported that the fabrics became hard in compressional deformation after finishing. They also concluded that the changes in the surface properties of the fabrics were small and the effect of the finishing stages seems to be weak. Strazdiene et al. (2006) reported that the finishing treatment has changed the hand of the fabric in the expected direction. Meanwhile, two experimental methods (objective and sensory approach) have shown their effectiveness to evaluate the textile touch.

9.7 Predictability of sensory comfort

The overall sensory comfort performance is very difficult to predict. Therefore it is worthwhile studying the relationship between objective fabric properties and subjective sensory perceptions and preferences, by utilizing statistical methods. With an understanding of the neurophysiological mechanisms of sensory perception of prickle and itch, the tactile comfort factor of clothing can be predicted from fibre diameter distribution and fibre end distribution densities at the fabric surface. To establish the relationship between subjective comfort perceptions and fabric physical properties Li (1988) carried out a series of psycho-physiological wear trials using T-shirts made of 8 different fibres. In the wear trials, the subjective ratings on 19 sensory descriptors were recorded under two environmental conditions, from which three fundamental sensory factors are identified as thermal – wet, tactile and pressure comfort.

The properties of the fabric measured are then classified into two categories: transport properties and mechanical properties. Transport properties mainly include thermal resistance, water vapour and air permeability, demand wetability, drop wetability and water vapour propensity; while mechanical properties include roughness and fullness, stiffness, perpendicular deformation of the fabrics, tensile stiffness of the fibres and tensile stiffness of the yarns. Friction coefficient is a measure of force generated by moving surfaces of two identical fabrics in contact at standard rate and pressure and surface roughness by force generated by moving thin wire over a surface at a standard rate and pressure.

Using correlation analysis, the predictability between the different physical factors and physiological sensory factors is assessed. The sensory factors are significantly related to the corresponding dimensions of physical properties of the fabrics, i.e. fabric roughness and fullness, fabric stiffness and wetability. Their pressure comfort, on the other hand, is correlated to fabric stiffness, fabric permeability and fabric tensile stiffness.

9.7.1 Subjective perceptions of touch comfort

To investigate fabric tactile comfort, a series of fabrics with varying constructional (weave type, blend, finish, etc) and physical (mass per unit area, thickness, thread density, etc.) parameters may be studied. In a study conducted by Hu et al. (2006), fifteen subjects (11 males and 4 females, ranging in age from 24 to 54) participated in the psychological trial and evaluated all the specimens randomly, and the evaluation process was repeated three times. Five fabric–skin contact comfort related sensory sensations, including softness, smoothness, prickliness, warmth and dampness were selected for this study into pressure (softness and smoothness), tactile (prickliness) and thermal–moisture (warmth and dampness) sensory space, respectively. The pressure and tactile-related sensations are further grouped into a mechanical sensations group. During the trial, after 20 minutes of acclimatization to the test conditions, the subjects placed one arm with the inside forearm facing upward on a small table. An operator draped a test fabric specimen across the forearm lightly. Then subjective scores on softness, smoothness, prickliness, warmth and dampness were recorded on 5-point scales, which ranged from 'soft' to 'stiff', 'smooth' to 'rough', 'soft' to 'prickly', 'cool' to 'warm' and 'wet' to 'dry' respectively, and the corresponding numerical value was recorded by the operator. All the subjective perception data were standardized in the range from 0 to 5 and the mean values of the standardized subjective sensory performances were expressed broadly in three different mechanical sensations, i.e. smoothness, softness and prickliness. It was reported that knitted fabrics have relatively higher ratings compared with woven fabrics, indicating that knitted fabrics have softer, smoother perceptions than woven fabrics.

9.7.2 Objective measurements of fabric handle and touch

The low stress mechanical properties of fabrics (e.g. bending, shear, extension) are objectively measured to assess the tactile characteristics of fabrics. The Kawabata Evaluation System for Fabrics (KESF) and Fabrics Analysis by Simple Tests (FAST) systems are available for measuring the fabric handle related characteristics. But, as far as the tactile responses are concerned, all the low stress mechanical characteristics directly or indirectly stimulate the touch, pressure, roughness and other mechanoreceptors of human skin.

The Kawabata Evaluation System for Fabrics (KESF) has the following four modules for measuring low stress and surface characteristics of fabrics:

1. KES-F1 for measurement of tensile and shearing characteristics;
2. KES-F2 for measurement of bending characteristics;
3. KES-F3 for measurement of compressional characteristics;
4. KES-F4 for measurement of surface friction and roughness.

The FAST system has been developed by CSIRO (Australia) primarily for quality control and assurance of fabrics (Minazo, 1995; Tester and Boos, 1990). It also gives the objective indication of fabric handle characteristics. It consists of a series of three instruments (i.e. FAST-1: Compression meter; FAST-2: Bending meter; and FAST-3: Extension meter) and a test method (FAST-4: Dimensional stability test) which are inexpensive, simple to use and robust in construction. It measures properties which are closely related to the ease of garment manufacturing, handle characteristics and the durability of surface finishing.

The fabric extraction principle is not a new idea at all; it has been a common practice for many years by ladies in certain parts of the world when searching for a desired scarf at a market. They would take off their rings and pull out a scarf through the ring, judging the overall quality of the scarf based on the resistance during the pull-through process (Brand, 1964). The fabric is extracted through a specially designed nozzle and the force required to extract the fabric through the nozzle is measured.

Matsudaira et al. (1990) modified the Kawabata compression tester (KESF-3) to measure the relationship between applied pressure and fabric thickness at the initial stage of fabric compression, i.e. when bending of fibres protruding from the fabric surface takes place during compression. The Wool Research Organization of New Zealand (WRONZ) developed a laser hairiness meter where the fibres protruding from the fabric surface are counted by laser beam. This gives a fairly good indication of prickliness of the fabric. The sensitivity of the instrument was found inadequate for the detection of all the fabric surface hairs, where the coarser and stiffer hairs are preferentially detected (Li, 2001).

9.7.3 Discrimination of fabrics by human haptic sensation

In a study reported by Tanaka (2001), the distinction of fabrics by human tactile perception was executed to compare the results with that of a haptic sensor. Thirty volunteers (15 female and 15 male) were asked to touch and feel small pieces of (A) figured satin, (B) crepe, (C) velvet, (D) wool, and (E) corduroy with their forefingers. All fabrics were kept out of sight in a small box, so that they differentiated the fabric only by the feeling of touch. They were informed of the five fabrics only by the letters A–E, not by their names. After getting the knowledge of the texture by repeated touch and feel, the arrangement of fabrics was changed randomly. The shuffled fabrics were renamed I–V and the discrimination was executed in the same way as for fabrics A–E. The right answer was reported in more than 70% of all fabrics tested, which means that the sensory receptors of human fingers are well functional to discriminate the fabric hand.

9.7.4 Perception testing of apparel ease variation

Understanding the relationship of apparel to the body requires an analysis of many complex factors. Because clothing can conform to the body and is our nearest environment, we expect it to fit closely yet move with us. Individuals have apparel fit preferences based upon aesthetic and functional expectations; these preferences have consequences in the volume and proportions people desire in their apparel.

Perception of fit

When a wearer judges the fit of a garment, the judgment is based on both visual and tactile information. A personal judgment of the way the garment looks on the body is made, based on visual feedback. The comfort level of the garment is judged based on both tactile and visual responses. The concept of comfort encompasses many dimensions, including physical, psychological, and social comfort. In order to learn about fit, these various tactile and visual responses need to be considered separately.

Tactile perception

Ashdown and DeLong (1995) explained the full range of tactile responses to clothing, including thermal perceptions, responses to moisture content, prickle related to allergic reactions and reactions to the surface texture of materials, abrasion of the skin surface, and pressure sensations. Two categories of tactile sensations are somesthetic sensations and kinesthetic sensations. The somesthetic sensations are touch response from the nerves in the surface of the skin. Tickle, prickle and abrasion are somesthetic responses to clothing. The second category consists of kinesthetic sensations, or the deep pressure sensations felt by the nerves in the muscles and the joints. Pressure sensations created by the resistance of the garment to movement and the weight of the garment in response to movement are kinesthetic responses to clothing fit.

9.8 Improving electrostatic propensity

9.8.1 General aspect

The phenomenon of static electricity has taken on a new importance with the advent of synthetic fibres, since these are more likely than the traditional ones to acquire an electrostatic charge of such magnitude that discomfort, or possibly even danger, is experienced. An analytical expression may be derived for the process of static charge dispersion as a textile material passes over an earthed conductive surface. Dissipation by flow only occurs at a region close to the contact position, and the residual charge is reduced, as might be expected, by increasing the radius of curvature of the contacting surface or by reducing the relative speed of movement. The degree of crystallization, orientation and crystallite size influence the charge magnitude, while high crystallinity, low orientation and large particle structure are likely to increase the charge developed (Slater, 1977).

9.8.2 Measurement of electrostatic propensity

Basic types of techniques for an estimation of static charge accumulation are direct measurement of charge or voltage acquired after frictional contact, determination of the charge imparted to a human being after walking on a carpet, fabric cling tests, corona discharge measurements, and surface-resistivity determinations.

The basic method of direct measurement depends on the fact that charges decay in a slow, mathematically predictable manner in an atmosphere of low relative humidity, so that there is time to allow the charged sample to fall in to a Faraday cage, for measurement, before charge leakage has been taken appreciably. The method is unfortunately somewhat tedious and subject to inherent-drift errors of measurement, so it is seldom used for routine testing. Two devices particularly suited for measuring charge acquisition by a carpet are noted by Albany (1972), the arrangement of each consists of a fabric holder, a rubbing mechanism and a probe near the sample surface.

The charge acquired by a human being in normal activity is of particular importance as a comfort related factor. The basic procedure described in a test method consists of monitoring the charge developed on the body of the test subject as he walks on a carpet, followed by a measurement of the rate of decay of this charge when he stops. Oxe and Keller (1972) reported the factors involved in such a measurement, and consider sole and carpet material and resistance, body weight, the electrical capacity of the condenser system, the saturation charge, moisture walking action and step frequency to be important ones. Oxe and Boschung (1975) use a shoe carpet rubbing device which they show give results comparable to those of a walking test. Ralston (1969) demonstrates reasonable correlation between rubbing and walking tests but claims that carpet and shoe materials have little effect on static build-up. Surface resistivity and volume resistivity are known to be related to electrostatic propensity, and use is made of this fact in deriving test methods. A test method is quoted for the measurement, with the provision that electrically conductive components render the method invalid. The final type of method involves the corona discharge treatment, in which the sample placed within the instrument (the Honestometer in one case) is charged by corona discharge and the charge leakage is measured and displayed on an oscilloscope. The Honestometer is cheap, reliable and easy to operate, and gives details of results obtained for several fabrics.

9.8.3 Effects of static electricity

The effect of the static electricity may be categorized very broadly into those concerning humans and those concerning the textile material. The mild discomfort experienced by a human being during the sudden discharge of a high electrostatic potential is too well known and requires no elaboration. The major safety concern is the fact that the discharge is accompanied by a spark, which may ignite a nearby hazardous substance and thus cause a fire or an explosion. Ronse (1969) discusses the industrial hazards that may arise, with particular reference to the presence of dust and vapors, which are the potential ignition sources in the textile industry. There is risk of ignition caused by static charge on the outer clothing and certain material can ignite household fuel gas at low to moderate humidity levels. A maximum resistivity of 10^{11} Ω is required between the opposite edges of a square sample for any clothing fabric for safety reasons.

In the case charging behaviour of clothing fabrics worn by an operator insulated from earth, it has been shown that the stored charge energy induced on the body can be far higher than the minimum ignition energies of various flammable gas and vapour–air mixtures. This applies to a wide range of clothing materials, including cotton. Attempts at igniting stoichiometric mixtures of coal gas and air and natural gas and air, by spark discharges from the body, indicate that the energy on the body, necessary for an ignition, is between one and two orders of magnitude higher than the minimum ignition energy of the gas concerned. To avoid dangerous levels of charge on the clothing of persons working in the presence of petrol–vapour–air mixtures with a minimum ignition energy of 0.2 mJ or higher, the resistivity of the surface material worn outermost should be less than $3 \times 10^{11}\Omega/m^2$ (Wilson, 1978).

A different aspect of static-related health problems is that under certain conditions, the spark accompanying the removal of charged clothing can bring the development of dermatitis. If a painful spark is repeatedly felt in the same skin area during clothing changes, a skin lesion may eventually develop in that area. These problems of clothing, though technical in nature, all share the fact that their presence can cause aesthetic changes in the garment and thus may affect the physiological comfort of the wearer.

9.8.4 Reduction of electrostatic propensity

Reduction of electrostatic charge may depend on the fibres in terms of charge retention and mechanisms of electron transfer during the generation and dissipation of the charge. Antistatic finishes are applied as an effective antistatic agent on the surface of the fibre and to the surface with which frictional contact is causing the potential charge accumulation.

Another approach to reduce static charge involves the creation of a conductive path for charge dissipation by other means than use of chemical finish. The most popular method of achieving this purpose appears to be the inclusion of metal, in some form, in the structure; stainless steel filament is the usual choice. Verplancke (1974) demonstrated that significant effects may be achieved by the inclusion of yarns with 12% metal content in conjunction with normal carpet yarns and an effective dissipation may be achieved with a 1% presence of metal. Metal powders are also used to coat a yarn, with or without subsequent chemical treatment, to render their attachment permanent. Layers of metal foil may be incorporated into the carpet or steel fibres may be used in constructing the backing. A sandwich technique may be used, in which a conductive viscous liquid is applied as a separate coating between the backing material and the latex, which gives better shampoo-resistance and reduced soil pick-up.

An innovation for production of conductive yarns involves the use of carbon fibres, which is used, in fact, as the core for an antistatic carpet yarn. A selected blend of two fibres, so that one always acquires a positive and the other a negative charge with respect to all shoe-sole materials, is used in making a carpet pile; the two types of charge will tend to cancel each other out, so that no net charge is produced on the body of the wearer. Gamma radiation on various synthetic fibres can give substantial improvement in antistatic behaviour, without impairment of other properties, if the optimum conditions are selected. Light radiation in an atmosphere of chlorine and oxygen with a very short dwell time, may be used for reducing static charge generation on nylon.

An electrostatic charge will inevitably be generated on garments as a result of triboelectric action. If a garment is made entirely from conducting material and provides a permanent ground connection, the charge will dissipate by conduction so rapidly that the wearer would not really be aware that it ever existed. Holdstock et al. (2004) illustrated that practical apparel fabrics, used to control static electricity, are commonly made from synthetic fibres interwoven with carbon- or metal-based fibres. The conductive or dissipative fibres typically account for a few per cent by mass of the fabric. Although much of the charge generated on the fabric can be dissipated rapidly by conduction, there will be some residual charge left on the insulating base fabric. Electrical resistance and charge decay time measurements are indicative of a fabric's ability to dissipate charge but neither of the measurements in their various forms relate to how a fabric will perform when it is made into a garment.

9.9 Future trends

Current economic pressure and high energy cost brought about by consumer demands for functional qualities from textile products they purchase have led to the manufacture of fabrics styled in a wide range of tactile and physical properties. The subjective ratings show that blend and natural fibres are preferred to man-made for all comfort attributes except smoothness, and woven was chosen over knit for smoothness, thickness, and openness. Spearman correlation between subjective rating and objective measurements showed good association for warmth and absorbency but a dearth of relationships for openness, smoothness, and thickness (Paek, 1984).

Increasingly, consumers are looking for good feeling and comfort when they buy textile goods. To respond to these demands, sensory tools are needed by industrialists to evaluate such notions. The studies have been performed and many device developments have taken place in the textile area such as mechanical, thermal and surface testing, so as to evaluate the related physical properties, but the links between measurement and the consumer feeling of comfort are still difficult to establish. Based on studies already performed in the food industry, the development of the sensory panel applied to textile goods has been implemented (Philippe et al., 2003).

9.10 Conclusions

Sensorial comfort of clothing is a result of interaction between the fabric and human skin with the sensory system and the atmospheric conditions. Human skin facilitates sensation through touch via the mechanical properties of its tissues, which are responsible for conveying tactile stimuli to the dermal and sub-dermal receptor sites. In the process of fabric–skin contact and mechanical interaction during wear, clothing will exert pressure and dynamic mechanical simulation to the skin which will in turn trigger various mechanoreceptors and generate a variety of touch sensations. An understanding of the structure of the human skin and the sensory system helps in appreciating the way in which the presence of the clothing is perceived by the human body.

Fabric mechanical properties such as fabric prickliness, itchiness, stiffness, softness, smoothness, roughness, scratchiness and friction describe the sensorial comfort of clothing. Warmth or coolness to the touch of fabrics is a function of transient heat conduction from body to fabric and vice versa. The finishing treatment, such as softeners, thermoplastic products and resins, brings out tactile properties in fabrics. The effect of the static electricity may be categorized very broadly into those concerning humans and textile material. A mild discomfort is experienced by human beings during the sudden discharge of a high electrostatic potential. The factors affecting the electrostatic propensity of the clothing, methods of characterizing this property and ways to reduce this difficulty have been suggested.

9.11 References

Albany Felt Co. (1972), B.P.1, 184,419.

Ashdown S.P. and DeLong M. (1995), 'Perception testing of apparel ease: Variation', *Applied Ergonomics*, **26**(1), 47–54.

Behmann F.W. (1990), 'Test on the roughness of textile surfaces', *Melliand Textilber*, 71,438–440 and E119–200.

Bertaux E., Lewandowski M. and Derler S. (2007), 'Relationship between friction and tactile properties for woven and knitted fabrics', *Textile Research Journal*, **77**, 387.

Brand, R.H. (1964), 'Measurement of fabric aesthetics: analysis of aesthetic components', *Textile Research Journal*, **34**, 791.

Coren S. and Ward L.M. (1989), *Sensation and Perception*, Harcourt Brace Jovanovich, New York.

Elder H.M. (1984), 'Fabric stiffness', *J. Text. Inst.*, **75**, 307–311.

Garnsworthy R.K., Gully R.L., Kandiah R.P., Kenins P., Mayfield R.J. and Westerman R.A. (1988), 'Understanding the causes of prickle and itch from skin contact of fabrics', *Australian Text.*, **8**, 26–29.

Hayward V., Astley O.R., Cruz-Hernandez M., Grant D. and Robles-De-La-Torre G. (2004), 'Haptic interfaces and devices', *Sensor Review*, **24**(1), 16–29.

Hensel H. (1981), *Themoreception and Temperature Regulation*, Monographs of the Physiological Society No. 38, Academic Press, London

Holdstock Paul, Dyer M.J.D. and Chubb J.N. (2004), 'Test procedures for predicting surface voltages on inhabited garments', *Journal of Electrostatics*, **62**(2–3), 231–239.

Hu J.Y., Hes L., Li Y., Yeung K.W. and Yao B.G. (2006), 'Fabric Touch Tester: Integrated evaluation of thermal–mechanical sensory properties of polymeric materials', *Polymer Testing*, **25**, 1081–1090.

Iggo A. (1988), 'Sensory receptors, cutaneous', in *Sensory System II: Senses other than Vision*, Pro Scientia Viva, Boston, USA, 109–110.

Kaczmarek K., Webster J.G., Bach-y-Rita P. and Tompkins W.J. (1991), 'Electrotactile and vibrotactile displays for sensory substitution systems', *IEEE Transactions on Biomedical Engineering*, **38**(1), 1–16.

Kandel E.R., Schwartz J.H. and Jessell T.M. (2000), *Principles of Neural Science*, 4th edn, McGraw-Hill, New York.

Kawabata S. and Akagi Y. (1977), *J. Text. Mach. Soc. Japan*, **30**, T13.

Kawabata S. and Niwa M. (1989), 'Fabric performance in clothing and clothing manufacture', *J. Text. Inst.*, **80**, 19.

Lau L., Fan J., Siu T. and Siu L.Y.C. (2002), 'Comfort sensations of polo shirts with and without wrinkle-free treatment', *Textile Research Journal*, **72**, 949.

Li Y. (1988), 'The Objective Assessment of Comfort of Knitted Sportswear in Relation to Psycho-Physiological Sensory Studies', Ph.D. thesis, Department of Textile Industries, The University of Leeds, Leeds, U.K.

Li Y. (2001), 'The science of clothing comfort', *Textile Progress*, **31**(1/2), 1–135.

Makabe H., Momota H, Mitsuno T. and Udea K. (1993), 'Effect of covered area at the west on clothing pressure', *Sen-iGakkaishi*, **49**, 513–521.

Mamalis P., Andreopoulos A. and Spyrellis N. (2001), 'The effect of a durable flame-retardant finishing on the mechanical properties of cotton knitted fabrics', *International Journal of Clothing Science and Technology*, **13**(2), 132–144.

Manich A.M., Martí M., Saurí R.M. and Castellar M.D. (2006), 'Effect of finishing on woven fabric structure and compressional and cyclic multiaxial strain properties', *Textile Research Journal*, **76**(1), 86–93.

Matsudaira M., Watt J.D. and Carnaby G.A. (1990), 'Measurement of the surface prickle of fabrics Part 1: The evaluation of potential objective methods', *J. Text. Inst.*, **81**, 288–299.

Militky J. and Bajzik V. (1997), 'Influence of washing/ironing cycles on selected properties of cotton type weaves', *International Journal of Clothing Science and Technology*, **9**(3), 193–199.

Minazio P.G. (1995), 'FAST – Fabric Assurance by Simple Testing', *International Journal of Clothing Science and Technology*, **7**(2/3), 43–48.

Naylor G.R.S. (1992), 'The role of coarse fibers in fabric prickle using blended acrylic fibers of different diameters', *Wool Technol Sheep Breeding*, **40**, 14–18.

Oxe J. and Boschung P. (1975), *Melliand Textilber.*, **56**, 301.

Oxe J. and Keller R. (1972), *Textilveredlung*, **7**, 417.

Paek Soae L. (1984), 'Subjective assessment of fabric comfort by sensory hand', *International Journal of Consumer Studies*, **8**(4), 339–349.

Philippe F., Schacher L., Adolphe D.C. and Dacremont C. (2003), 'The sensory panel applied to textile goods – a new marketing tool', *Journal of Fashion Marketing and Management*, **7**(3), 235–248.

Ralson R.H. (1969), 'Why the static?', *Text. Industr.*, **133**(5), 85.

Ronse W. (1969), *Industr. Text. Belge*, **11**(7/8), 59.

Schneider A.M. and Holcombe B.V. (1991), 'Properties influencing coolness to the touch of fabrics', *Textile Research Journal*, **61**, 488.

Slater K. (1977), 'Comfort properties of textiles', *Textile Progress*, **9**(4), 1–71.

Smith J. (1984), 'Comfort in casuals', *Text Horizons*, **54**, 471.

Strazdienè E., Ben Saïd S., Gutauskas M., Schacher L. and Adolphe D.C. (2006), 'The evaluation of fabric treatment by Griff tester and sensory analysis', *International Journal of Clothing Science and Technology*, **18**(5), 326–334.

Sukigarà S. and Ishibashi T. (1994) 'Analysis of frictional properties related to surface roughness of crepe fabric', *Seni Gakkaishi*, **50**, 349–356.

Tanaka M. (2001), 'Development of tactile sensor for monitoring skin conditions', *Journal of Materials Processing Technology*, **108**, 253–256.

Tester D. and De Boos A. (1990), 'Get it right FAST time', *Textile Horizons*, **10**(8), 13.

Verplancke W. (1974), *Text. Mfr.*, **101**, 42.

Visell Y. (2008), 'Tactile sensory substitution: Models for enaction in HCI', *Interact. Comput.*, doi:10.1016/j.intcom.2008.08.004.

Westerman R.A., Garnsworthy R.K., Walker A., Kenins P., Gully R.L. and Fergin P. (1984), 'Aspect of Human Cutaneous Small Nerve Function: Sensations of Prickle and Itch', paper presented at 19th IUPS Satellite Symp., Budapest, Hungary.
Willis W.D. (1985), *The Pain System: The Neural Basis of Nociceptive Transmission in the Mamallian nervous system*, Karger, Basel, Switzerland.
Wilson N. (1978), 'The risk of fire or explosion due to static charges on textile clothing', *Journal of Electrostatics*, **4**, 67–84.
Yan K., Höcker H. and Schäfer K. (2000), 'Handle of bleached knitted fabric made from fine yak hair', *Textile Research Journal*, **70**, 734.
Yoneda Morihiro and Kawabata Sueo (1981), 'Analysis of transient heat conduction and Its application, Part 2: A theoretical analysis of the relationship between warm/cool feeling and transient heat conduction in skin', *Journal of the Textile Machinery Society of Japan, Transactions*, **34**(10), T200–208.

10

Garment pattern design and comfort

P. WATKINS, London College of Fashion, UK

Abstract: To date the garment industry has focused on speeding up empirical pattern construction methods that have developed through custom and practice. Historical pattern construction and fitting alteration techniques, developed to accommodate diverse bodyshapes postures and movements, reveal techniques requiring tacit knowledge. Although 3D design using virtual mannequins is exciting lots of media attention, there are significant problems in reverse engineering the 2D pattern pieces into real world garments. Virtual fit is problematic as variables are introduced at each stage of the pattern production process. This chapter commences with identification of garment pattern construction methods, garment production modes and corresponding garment-to-body fit expectations, and goes on to review mechanical stretch testing technology and development.

Key words: stretch pattern design, close fitting garments, pressure garments, digital fashion design.

10.1 Introduction: fundamental principles of fit in apparel

Empirical pattern construction methods emerged to assist in speeding up the garment production process. This was achieved within the limitations of the available technology, but this approach to pattern design is inappropriate for today's technology. Now is the time to re-examine and try to access the theories behind these fitting methodologies so that they can be reinterpreted objectively for the ever more sophisticated technologies that are emerging.

In the garment industry there are different techniques for producing pattern profiles dependent on the industry sector. This chapter commences with identification of garment pattern construction methods, garment production modes and corresponding garment-to-body fit expectations.

Garment fit and comfort is inextricably linked and is usually viewed in direct response to the envisaged activity, culture and environment. There are difficulties in viewing these aspects in isolation; because the perception of comfort is so closely tied to subjective psychological and physiological responses, there is inevitably a degree of overlap. Physical comfort relates to the effect of the external elements, either physiological or psychological. Most individuals have a greater awareness of the negative sensation of discomfort, when the body or mind is adversely affected.

Fit can refer to either the style or the application. A style can be the fashionable silhouette of the moment, a specific garment shape, a design detail, a mode of dress adopted by a particular individual or a sector of society, or it may be a garment that fulfils a functional requirement. Fit related to the holding power of a stretch fabric in developing the pattern profile geometry is, to date, dependent on subjective expertise. A simple method for determining the degree of fabric stretch extension is illustrated which would enable a more consistent approach to the stretch pattern reduction process in the digital realm. The development of conventional approaches to pattern construction methods and their influence on stretch pattern technology when designing for mobility and comfort will focus on the shoulder and sleeve area.

The mode of garment production, ranging through mass production to couture, brings about a certain fit expectation. But generally garment design/style fit is left to the individual to interpret the acceptability of how closely the garment conforms to the body. The use of the term ‘fit’ in this chapter is in the context of pattern design development where stretch fit is a function of the proximity of the garment to the body and the fabric parameters (outlined later under distal and proximal fit).

10.1.1 Pattern construction and fit expectations

There are three traditional methods for generating patterns: drafting a basic block pattern; designing a flat pattern; and draping or modelling on the stand (a static representation of the human form). Indeed there are four, if you count the widely used method of taking a pattern from a competitor’s garment! Designers very often use a combination of all these methods. Most CAD software systems are based on computerised versions of traditional empirical haptic methods that have emerged through trial and error and, more often than not, manual intervention is needed to produce a good custom fit.

- The drafting of a block pattern follows an empirical procedure, which includes some proportional co-ordinates. The pattern has a relatively simple shape with no design embellishments and is used as a basis for style development.
- Flat pattern design involves the modification of the block pattern, which is manipulated to produce the desired design detailing.
- Modelling on the stand is the moulding or draping of cloth on a stand or a person, which is then transferred onto paper. The modelled garment can be either a basic block pattern or a design creation.

There are a number of terms used to describe modes of garment production and the expected accuracy of the fit:

- Ready-to-wear for the mass market offers a conjectural fit. Companies very often engage a fit model representative of their target market size designations for pattern design development.
- Anti fit garments are designed not to conform to a conventional fit rationale.
- Couture garments are high quality, using hand-executed techniques with extreme attention to detail and finish. They are modelled to an individual’s measurements with a number of fittings to achieve the desired fit.

- Savile Row Bespoke is high quality men's tailoring, fully hand cut and stitched to an individual's measurements with numerous fittings (Anderson, 2008).
- Bespoke tailoring is defined by the Advertising Standards Authority as being constructed from the nearest to size base pattern adjusted to an individual's measurements, which is cut and sewn by machine with a small amount of hand finishing (Anon, 2008).
- Custom fit is constructed from the nearest to size base pattern to the individual, which is altered by a small number of length and circumferential measurements.
- Made-to-measure (MTM) is constructed from the nearest to size base pattern, which is adjusted to an individual's measurements.
- 3D virtual fit takes a 2D traditional garment pattern, which is virtually sewn onto a parametric mannequin for a predictive fit. The 2D pattern is then altered in line with the perceived virtual fit.

All patterns in the clothing industry are based on garment size specifications. Despite the mass of supporting anthropometric data, traditional manufacture still relies upon measurements that have emerged through trial and error. Traditionally, garment fit is determined by the interpretation of measurement data to produce pattern-drafting co-ordinates that reflect the 'ideal' customer shape and size profile which their fit model embodies. Most block patterns used by clothing manufacturers have been developed and adapted by numerous people over many years. This means that the rationale for implementing the pattern profile, the apportionment of direct body measurements, proportional measurements and those applied for ease is often inaccessible.

Conventional non-stretch pattern construction systems have an in-built ease allowance. Ease (tolerance) is the allowance of a certain amount of fabric on a woven block pattern, which allows involuntary movement such as breathing or movements like sitting down. Because of the way pattern design systems evolve, the original construction rationale often becomes lost through successive translations in developing garment blocks for different applications. Therefore it can be extremely difficult to determine the mathematical relationship between the amount of ease applied in the pattern profile and actual body measurements.

10.2 Clothing comfort and fit

Defining comfort is almost impossible because the perception of physical comfort is subjective. Although there is not a universally accepted definition of comfort it is important to recognise the main physiological and psychological factors affecting comfort. Physical comfort relates to the effect of the external elements, either physiological or psychological. It is described in *The Concise Oxford Dictionary* (Sykes, 1980: 201) as 'freedom from pain' and general 'well being'. This definition seems to be inadequate, particularly when applied to sports participants who often expect and endure various levels of pain. Slater (1986: 158) attempts a qualitative definition in which comfort is defined as 'a pleasant state of physiological, psychological, physical harmony between a human being and the environment'. It is, therefore, a neutral sensation as one is unaware of comfort, both psychological and physiological (Smith 1986: 23). Most individuals find the positive sensation of comfort insignificant and have a greater awareness of the negative sensation of discomfort, which only becomes apparent when the body is adversely affected.

Psychological factors are inextricably linked with physical factors in determining levels of comfort: idiosyncrasies, prejudices, preferred environment, preferred temperature, posture, pain sensitivity, effects of stress, level of embarrassment, need for privacy, body consciousness, preferred garment fit, and tactile sensitivity. Harnett (1976: 8) outlines mechanical comfort which can be divided into tactile comfort and action comfort. Textile properties including thickness and weight, fibre content and the nature of fabric structure, particularly the next-to-skin surface, are obviously crucial factors for tactile perceptions of comfort. Action comfort refers to the combination of garment design and fabric properties to allow a high degree of freedom of movement without undue pressure or friction on the skin. Crowther (1985) also examined the relationship between the fabric construction and pattern design in 100% cotton denim jeans to improve comfort and fit.

10.2.1 Sizing and fit

Over the years, manufacturers have tried to develop effective garment sizing systems to improve the quality of ready-to-wear garment fit (Ashdown, 1998, 2007; Loker *et al.*, 2005). To compound this it is extremely difficult to assess the fit-quality without first defining the expected garment-to-body fit relationship. Most sizing systems use an incremental or proportional approach when grading patterns up and down to produce a range of sizes. This approach is problematic when trying to accommodate a populace with an infinite variety of body shapes and proportions. All drafting systems to a greater or lesser extent make assumptions about the body shape based on derived measurements. It is the shape proportion and posture of a person that is important, but replicating the three-dimensional body shape in a two-dimensional pattern profile can be problematic (Chen, 2007). Body shape can be described by taking the different proportions between the form, width and length of body segments. The torso can also differ in width from front to back and side to side. It becomes apparent through observation that many women who have similar measurements are vastly different in body shape, proportions and postures. Helen Douty (1954) introduced a photographic technique, which drew on Sheldon and co-workers' (1940) body shape classification system: endomorphy defined as the relative predominance of soft roundness throughout the various regions of the body; mesomorphy, predominantly muscle, bone and connective tissue, and ectomorphy have a predominance of linearity and fragility. Douty's (1968: 26–29) somatograph and the posturegraph were developed as an aid to figure analysis, aesthetics and design selection for her students. Front, back and side views of three hundred women were photographed silhouetted against a graph of six-inch squares. Her system was termed visual somatometry because the proportions, curves, irregularities, weight distribution and other characteristics were clearly visible.

There are a number of texts that address the problem of fitting and pattern alteration: Bray (1978), Liechty *et al.* (1986), Lenker (1987), Armstrong (1995), and Rasband and Liechty (2006). Fit research has also been conducted using body scan data: Ashdown *et al.* (2004), Symonds and Istook (2003), Symonds *et al.* (2004a, 2004b), and Loker *et al.* (2005). Investigating quantitative and qualitative assessment of fit is extensively covered in *Clothing Appearance and Fit: Science and Technology*, edited by Fan *et al.* (2004).

10.2.2 Fit and body cathexis

Garment sizing/fit and its infinite variables can impose a negative self-evaluation of body image (Apeagyei *et al.*, 2007; Borland and Akram, 2007). A stretch garment that conforms closely to the body contours may offer a high degree of mobility, be fashionable, aesthetically pleasing, or conform to an ideology or culture of a specific sporting activity, all of which should lead to psychological comfort. However, the contoured stretch garment can still engender some dissatisfaction. LaBat and DeLong in their study 'Body cathexis and satisfaction with fit of apparel', suggest that:

> A factor that may contribute to women's dissatisfaction with the body is that fashionable clothing reflects a standard they do not fit. When clothing does not fit, the consumer may perceive the cause as related to the body and not the clothing, with resulting negative feelings about the body.
>
> (LaBat and DeLong, 1990: 43)

The garment fit, unwittingly, can often be the root cause of this dissatisfaction and usually results in the garment constantly having to be rearranged in order to feel more comfortable. Therefore, the importance of fit to enhance comfort and mobility is crucial.

10.2.3 Technology and garment fit

Body scanning for automated measurement extraction and virtual simulation of avatars for garment design and fit and pattern technology is fast developing (D'Apuzzo, 2007; Kirstein *et al.*, 1999; Stylios, 1999; Chittaro and Corvaglia, 2003; Fontana *et al.*, 2005; Krzywinski and Rodel, 2005; Volino *et al.*, 2005; Paquet and Viktor, 2007; Decaudin, 2006; Petrac and Rogale, 2006; Petrac *et al.*, 2006; Wang *et al.*, 2007; Daanen and Hong, 2008; Wang and Tang, 2008, 2010; Wang *et al.*, 2010). Industry leaders majoring on virtual prototyping and fit, and

2D CAD made to measure (MTM) pattern design systems (PDS) include Lectra, Gerber Technologies, Browzwear, Optitex, Dressingsim, Ffitme and Assyst Bullmer. Virtual garment prototyping is highly valuable in reducing time/cost constraints and also has ecological benefits. This technology is intended to increase customer confidence in purchasing a garment appropriate for their body shape and fit preferences but Apeagyei and Otieno (2007) suggest it has some way to go before it can become an accurate fitting tool.

Although virtual avatars allow the consumer to visualise the suggested fit, fabric drape and simulated movement, it is difficult to successfully apply this technology for MTM custom fit garment design. Custom fit PDS are predominantly based on computerised hand pattern production methods. A style pattern, nearest in size to the client's own, is adjusted by substituting the client's measurements at just a few cardinal (primary) points on the pattern profile. 2D pattern pieces that have been adapted using just a few of the customer's length measurements and one-dimensional circumferential measurements (for example bust, waist and hips) are wrapped and seamed together onto a virtual parametric mannequin for fit evaluation. The resulting 3D pattern fitting process does not automatically transpose parametric variations in body shape to the 2D pattern pieces without some considerable behind-the-scenes manipulation. Without this physical intervention, garment fit will not be a true custom fit but a coincidental fit.

10.2.4 Garment pressure fit research

The ability to predict how closely stretch fabric should conform to the body for optimum performance and comfort levels is vital in stretch garment research. Harada (1982) explored the relationship between the degree of skin stretch and the degree of fabric stretch in conjunction with the proximity of the garment to the body. The study utilised Laplace's Law which relates pressure, tension and radius of curvature in the following way: $P = T/\rho$, where 'P' is the pressure exerted on the body, 'T' is the tension of the fabric, which is dependent on stretch parameters, and 'ρ' is the radius of the curved surface of the body. Assuming that the degree of fabric stretch is maintained at a constant level, the tension in the fabric will remain constant. A key variable affecting the pressure of the fabric on the body is therefore the radius of the part being covered, the smaller the curve the higher the exerted pressure. The implication of this is that the amount of pressure applied along the leg, for example, would not be linear. Parts with smaller radii (for example ankles and wrists) require less reduction in the fabric to achieve the same garment-to-body interface pressure.

The main body of research for measuring pressure garment products is for medical application. External pressure is used in the treatment of an ever-broadening range of medical conditions. Some examples are management of venous insufficiency, reduction of hypertrophic scarring, management of lymphoedema, promotion of wound healing and prevention or management of oedema after medical procedures, prevention of skin-waves after liposuction, and in conditions where sensory information is impaired such as in a stroke or sensory integration disorders.

Studies into the levels of pressure in garments of the type used to reduce the hypertrophic scarring of burns victims were undertaken by Fentem and Goddard (1979), Horner *et al.* (1980), Filatov (1985), Fentem (1986), Giele *et al.* (1995), Ng (1995), Macintyre *et al.* (2004) and Maklewska *et al.* (2006, 2007). Their research evaluated the degree and position of compression needed to attain an optimum effectiveness for compressive bandages and elastic stockings. Pratt and West (1995) suggest a mathematical formula for pattern drafting. Basically, all circumferential measurements are reduced by 20% and length measurements are reduced typically by 20%–25% of their total length. But they go on to state that applying the formula is not straightforward and needs subjective adjustment based on experience.

The starting point of most pressure garment research is Laplace's Law whereby the fabric tension and the radius of the part of the body being covered determine garment pressure. In the pattern construction, the suggested reduction of circumferences by 20% and overall length reductions of 20%–25% seem to be typical. Most research was carried out using simulated body zone circumferences. An objective approach to constructing pattern geometry that would encompass the multi-axial torso limb junctions of the shoulder and hip was not outlined. The lack of correlation between the 3D body profile and the 2D pattern geometry contouring the whole body, combined with the application of arbitrary stretch fabric parameters in the pattern reduction process, severely limits the objective evaluation of garment-to-body interface pressure variables over the whole of the body. It is difficult to evaluate and predict garment pressure consistently over the whole body contour if research is confined to a limited area only.

10.2.5 Comfort fit and pressure

Research studies, other than for medical purposes, to ascertain acceptance levels of pressure exerted on specific areas of the body, by the fit of different garments, have been conducted. Ibrahim (1968) undertook an investigation into the 'Mechanics of form-persuasive garments based on Spandex Fibers' at the Textile Research Laboratory of DuPont in America. The research was to gain an understanding of the functionality of form-persuasive garments in relation to fabric performance parameters, to provide a proper basis for design. Japanese researchers Horino *et al.* (1977), studied the simulation of garment pressure in wear. Shoh (1998) evaluated acceptable comfort pressure levels of men's socks using elastic optical fibre. In Britain, tests developed by Clulow (Sawbridge, 1989) at the Shirley Institute – now re-named the British Textile Technology Group (BTTG) – were carried out to measure acceptable levels of pressure for comfort of waistbands, sock-tops, etc. In 2001 Ian Scott, technology chief for Marks & Spencer lingerie, as part of their fit testing, introduced a bra sensor to ascertain the pressure exerted at specific sites on the body, including the shoulder area and around the rib cage. Yu (2004) outlined the development of a soft mannequin, simulating the skeletal frame, soft body tissues and skin of the lower torso of a female, to measure the contact pressure of legless pants. This enabled the correlation of garment pressure, through the use of a linear equation, to be predicted on live models.

Lindberg (1966), a Norwegian textile scientist, conducted research into how woven stretch fabrics perform. The purpose was to assess how great the stretchability of the fabric should be to provide reasonable comfort. He examined the interplay between the characteristics of the fabric and garment construction and the body. The maximum increase in fabric distortion and the distance between various restraint points (neck, shoulder, armpits, crutch, hips, seat and knees, etc.) subject to different body measurements like crouching, were recorded. He found that the fabric never stretched proportionally between two points. The grip points in a crouching position, the hips, seat and knees, form a complicated mechanical system. This was observed by drawing a series of circles with a known diameter with lines indicating the warp and the weft. When the body was mobilised the circle became elliptical, and the direction of the greatest stretch was indicated by the direction in which the ellipse had its major axis. It was possible to calculate the amount and direction of stretch at particular points on the garment where simultaneous stretch occurs.

If a non-stretch woven fabric is stretched in one diagonal direction it generally contracts almost as much in the other direction. The same applies for a stretch fabric. The stretch fabric also contracts in the opposite direction when stretched laterally. This effect is enhanced in the knit fabric because of its more malleable structure. The effect of bias stretch has significant implication for stretch contoured pattern profile geometry.

10.3 Manual and mechanical stretch testing

Extensive research has been carried out looking at fabric properties to improve comfort and fit. Examples of objective measurement systems are the Kawabata Evaluation System (KES) – which includes five highly sensitive instruments that measure fabric bending, shearing, tensile and compressive stiffness, as well as the smoothness and frictional properties of a fabric surface – and the 'Fabric assurance by simple testing' (FAST) system. However these systems are not suited to stretch garment pattern design.

Current texts on stretch pattern design are inconsistent as to the sample width, length and forces needed to quantify the degree of stretch extension (Armstrong, 1995; Cloake, 1996; Aldrich, 2004, 2007; Haggar 2004; Richardson, 2008; Shoben, 2008), which is extremely confusing for the designer. Ziegert and Keil (1988, 56) used a measurement unit of 20 cm by 20 cm with a 500 g load. The rationale for the test unit size was that it related closely to one-quarter of human body dimension of garments made with elastomers. However, Murden (1966:

356) suggested that a good approximation of the hand stretch could be achieved mechanically by taking a measurement unit of 7.5 cm wide by 25 cm long with a load approximating 1 kg/cm.

Because of this confusion I needed an understanding of fabric stretch extension characteristics. Exploratory mechanical force extension testing was undertaken using the Instron tensile testing apparatus to try to identify the forces involved in stretch fabric extension in the course, wale and bias.

10.3.1 Instron force/extension testing

The Instron tensile testing machine is used extensively to electronically calculate the extensibility of a variety of sample materials. The British Standard (BS 4952, 1992; BS EN 14704-1, 2005; ASTM D 4964–96, 1996) highlights a number of specific tests for quality assurance (QA) and quality control (QC) for stretch fabric but they are not suited to assessing the degree of fabric stretch required for garment pattern geometry.

The overall aim and objectives were: to record and plot electronically the force/extension characteristics for a range of fabric samples that had been cut in the course, wale and bias directions; to analyse the effect that fabric orientation has on the load/extension curve of a given sample; to compare the different samples for a given fabric orientation; to identify typical working ranges for the sample fabrics; and to ascertain an optimum loading for a fixed load test.

The sample fabric chosen covered a range of weights and elastane content which exhibited different bi-directional stretch characteristics and were selected because of their general suitability for a broad range of stretch performance wear. The fabric samples coded A, B, C, D and E are detailed in Table 10.1.

The test samples of the fabrics A–E were cut in the course, wale and bias (C, W and B) direction, with three sets of each orientation. The samples had a width of

Table 10.1 Fabric sampling characteristics

Code	Quality	Description	Course 0°	45° Bias	Wale 90°	135° Bias
A	21649	32gg 210g Coolmax/Lycra	36	39	28	35
B	21132	32gg 260g Animalmax	56	40	20	40
C	21132	32gg 260g Animalmax	52	47	32	48
D	22203	56gg 220g Coolmax/T902 Triskin	18	14	10	14
E	21130	32gg 180g Coolmax/Lycra	50	57	28	47

50 mm and were benchmarked with two parallel lines placed 100 mm apart. All samples were subject to specific pre-test conditioning. Following the standard Instron testing procedure, the fabric samples were clamped between the metal jaws taking care to remove excess slack material. The Instron was set up for a simple non-cyclic test and the sample was loaded until an extension of 100% was reached. The force required was recorded at 1 mm intervals for each loading and the stretch/loading characteristics were recorded using the standard Instron program. The data was then imported into a spreadsheet allowing ease of analysis.

Fabric sample orientation

The force stretch curves for samples A1, A2 and A3 and an average of sample A are illustrated in the composite Fig. 10.1. Samples B, C and D were characteristically similar.

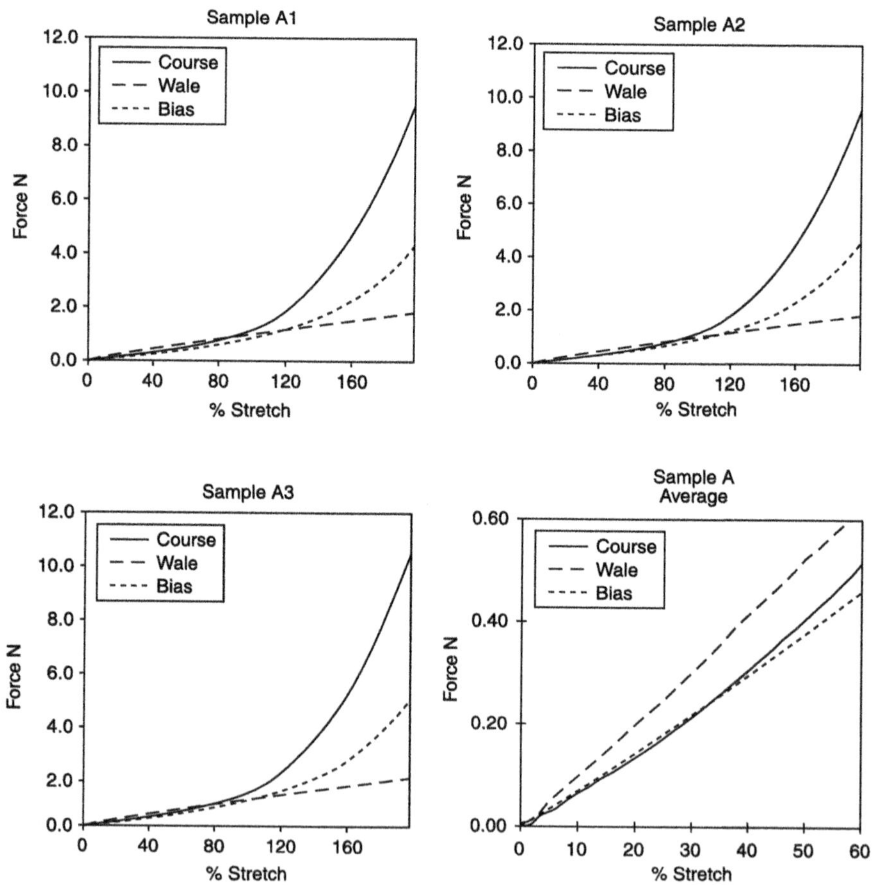

10.1 Force/stretch curve.

There is a marked difference in the extensibility between fabric orientations for a given sample. At the higher levels of stretch the general indication is that the wale offers the least resistance to stretch and the course the greatest. However for lower values of stretch, the reverse (the course offering the least resistance) is true which is more representative of the stretch extension working range of stretch garments.

Fabric sample correlation

Figure 10.2 shows the correlation between samples A to D for the course, wale and bias orientations respectively. For a given orientation there is a good correlation between samples, suggesting that fabric behaviour could be consistent within a required working range. The wale force/stretch curves, at first sight, again suggest that this orientation offers the least resistance to stretch.

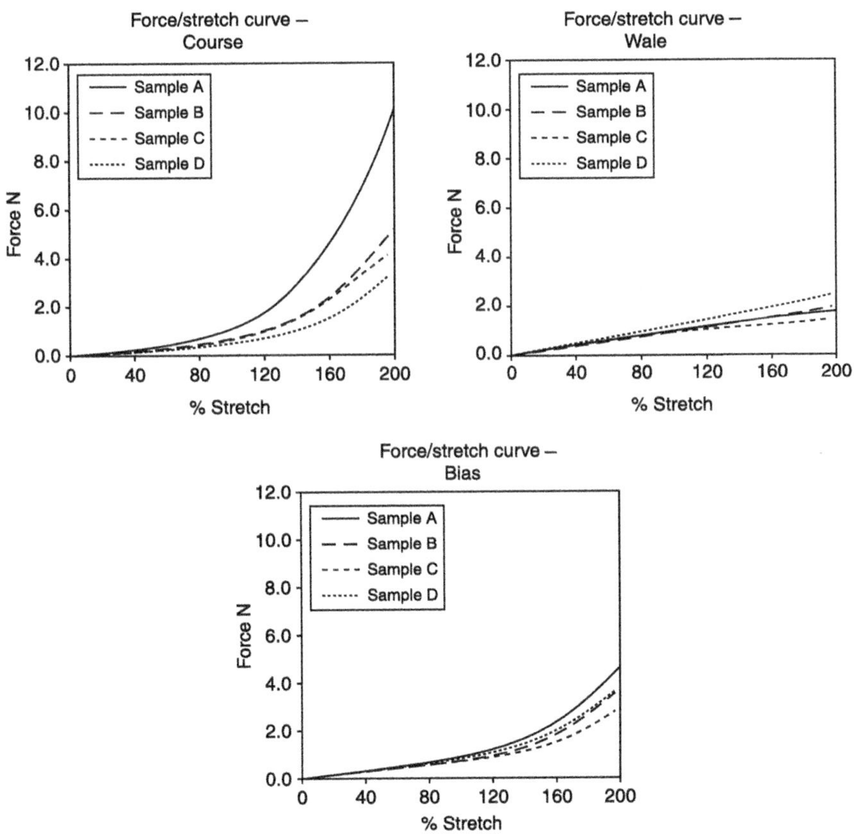

10.2 Sample orientation correlation.

Stretch extension working range

The graphs in Fig. 10.3 show the stretch extension working ranges of up to 60% stretch. Denton (1972: 16) looked at the relationship between fit, stretch, comfort and movement. It was ascertained that, in the seat area of various garments, the actual fabric stretch of the garment, in wear, was considerably less than maximum available fabric stretch percentage. The results of the Instron testing clearly illustrated that within the lower working range, the course orientation offers the least resistance. The bias orientation also requires lower forces than the wale direction, which is significant when determining the amount of the available fabric stretch to be used in the reduction algorithm applied to the pattern geometry.

Results

Analysis of the results was interesting. It was expected that the extensibility in the wale direction would be greater than the course: this was indeed the impression

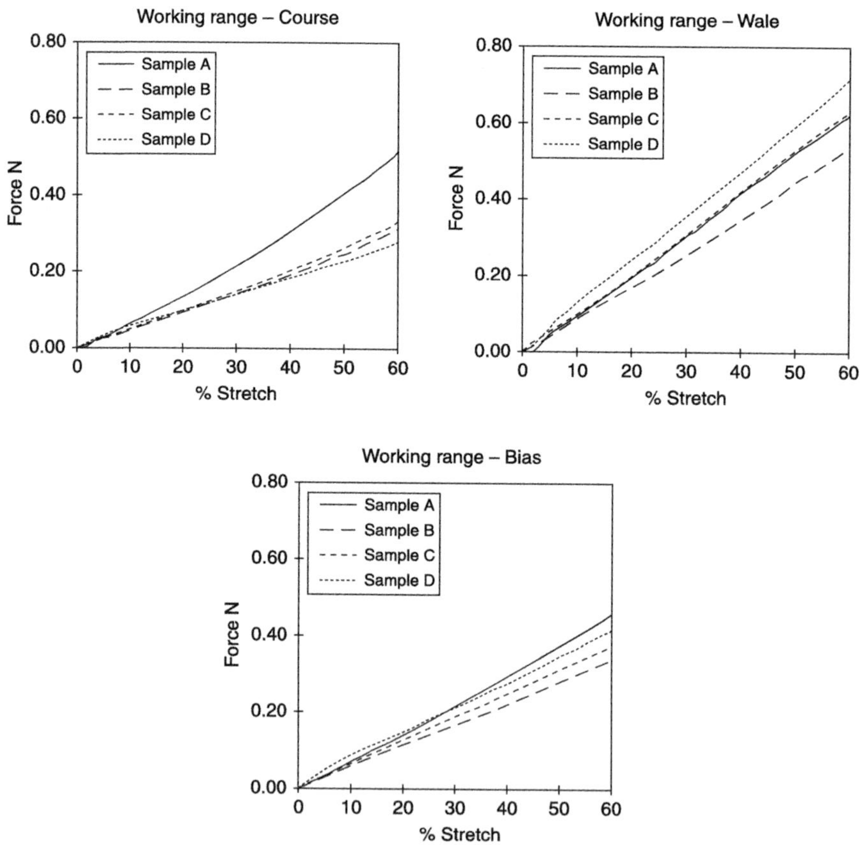

10.3 Force/stretch curves over working range.

gained from experience and clearly demonstrated by the results of the hanger load tests reported by Ziegert and Keil (1988: 56). However, although this was true when stretching each of the test fabrics up to the test limit, while observing the useful working range of up to 30–40%, it was the course direction that clearly offered the least resistance and therefore had the greatest stretch. The main observation was that the stretch characteristics were not only non-linear, as expected, but were also inverted (the course showed greater extensibility than the wale) in the crucial stretch extension working range. This has significant implications for the pattern orientation and profile geometry.

However, the designer and pattern technologist requires a more readily accessible method to estimate the degree of stretch, and the results suggested that a simple load test applying a fixed weight of 250 g to a prepared sample width of 50 mm could be employed.

10.3.2 Quad load testing

Literature on testing the degree of fabric stretch extension for garment pattern reduction is inconclusive on test fabric size, loading and application. Until an industry standard has been established, it is essential that the designer can follow a simple method to calculate the degree of stretch, which offers consistent results without requiring specially controlled conditions. These results should ideally show a breakdown of fabric extension into course, wale and bias, which can be used to calculate the relative stretch reduction factor. The author used an adapted hanger load-test, referred to as the Quad Load Test Method, designed specifically to digitally quantify fabric extension for use as part of the stretch block pattern reduction procedure. The aim and objectives were to calculate the degree of stretch extension at a specific load of 250g for sample fabrics in the four orientations of course, wale and bias 45° and 135°. The original test was conducted using the course, wale and one 45° diagonal; however, after further research this was subsequently changed to include both 45° and 135° bias orientations. Sets of four for each the five sample fabrics detailed in Table 10.1 were cut into strips measuring 50 mm × 200 mm in the course, wale and bias orientation. The test samples were identified for example as sample AC for fabric A cut in the Course direction.

Figure 10.4 shows the sample fabric pattern, illustrated as a 50 mm × 200 mm rectangle, with benchmarks on 100 mm centres between which the extended length was measured. A 25 mm fold at both ends was machined, forming slots ready for the insertion of the hanger supports. In the quad load test procedure fabric samples in the course, wale, 45° bias and 135° bias were placed on the hanger and the 250 g weight applied. After allowing one minute for the fabric to stabilise, the extended measurement between the benchmarks was recorded in Table 10.2.

The benchmark relaxed length of 100 mm was chosen because the calculation of the degree of stretch is simplified. The degree of stretch expressed as a

Table 10.2 Quad load test results

Code	Quality	Description	Polyester %	Elastane %	Colour
A	21649	32gg 210g Coolmax/Lycra	84	16	White NR5079
B	21132	32gg 260g Animalmax	88	12	White SDI10014
C	21132	32gg 260g Animalmax	88	12	White NR4888
D	22203	56gg 220g Coomax/T902 Triskin	80	20	White SDI 10515
E	21130	32gg 180g Coolmax/Lycra	84	16	White SD15243

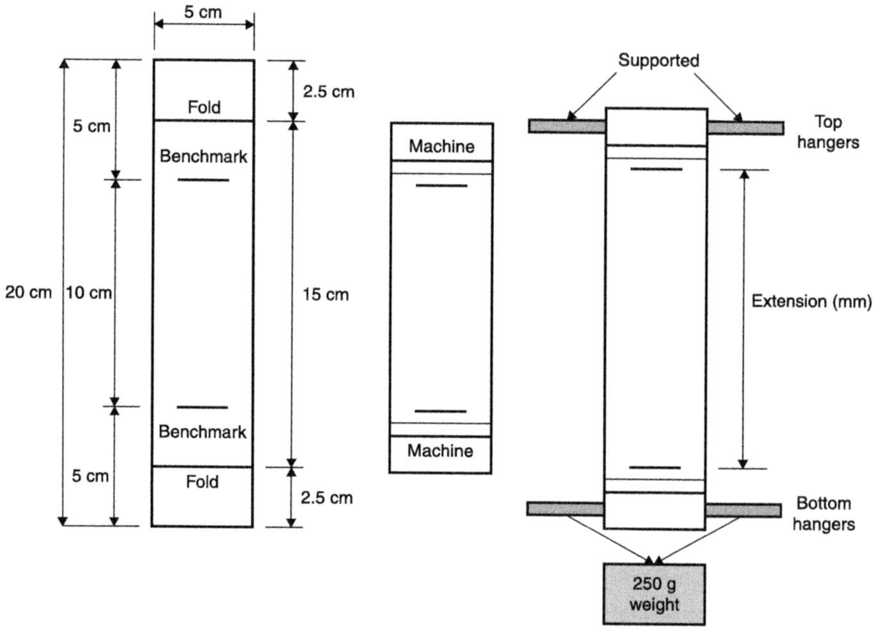

10.4 Fabric sample preparation for hanger load test.

percentage is calculated by subtracting the relaxed length from the extended length and then dividing the result by the original length or simply by subtracting 100 from the extended length.

$$\text{Degree of stretch s} = \text{extended length} - 100\% \qquad [10.1]$$

For example in the course sample fabric B, coded BC

$$\text{Degree of stretch s} = 156 - 100\%$$
$$= 56\%$$

10.3.3 Stretch distribution quad angle plots

Entering the test results into a spreadsheet enabled a graphic representation of the distribution of stretchability throughout 360° of fabric orientation to be displayed. This method was adapted from Lindberg (1966: 60) which was used to compare the bias stretch in a woven double or bi-directional stretch and a non-stretch fabric.

Although only three measurements were taken for each fabric, corresponding to 0°, 45° and 90° rotation, it was assumed that inverse symmetry would apply. However fitting experimental garments led to questioning the use of a single bias extension measurement only, because a fit disparity was observed between the right and left side of the evaluation garments. Subsequently, it was found that not all stretch knit fabrics had a corresponding degree of stretch between the bias at 45° and at 135°, as recorded in Table 10.2. The Quad Angle Plots in Fig. 10.5 compare the angular stretch distribution curves for the single 45° bias and double 45° and 135° bias measurements.

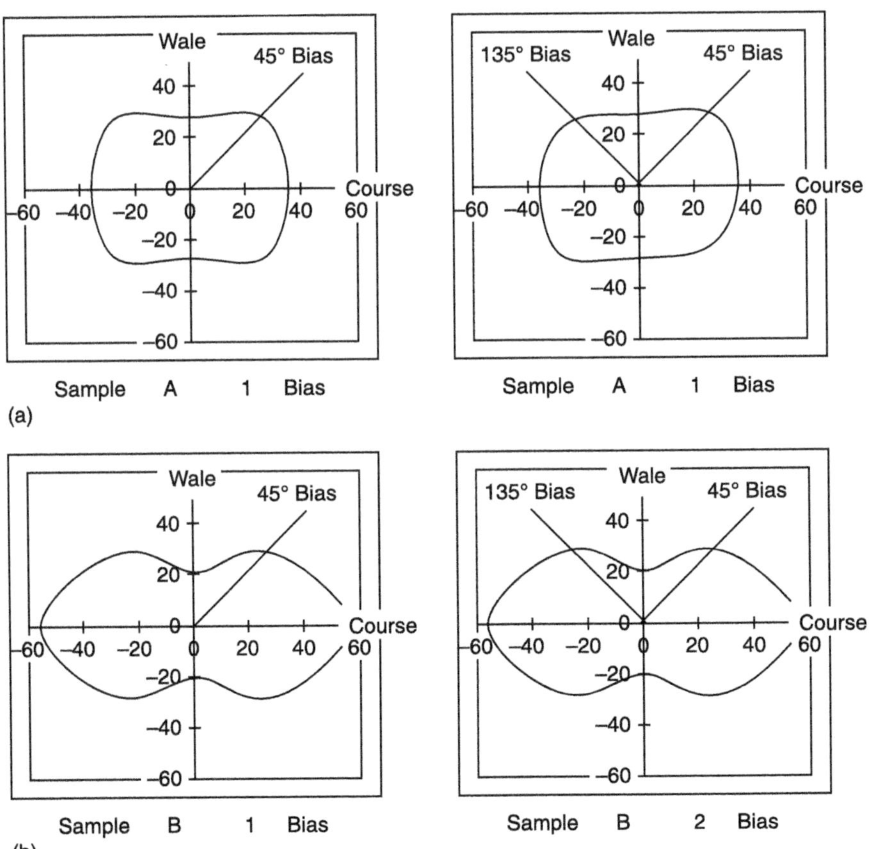

10.5 Angular stretch distribution curves. (*Continued overleaf.*)

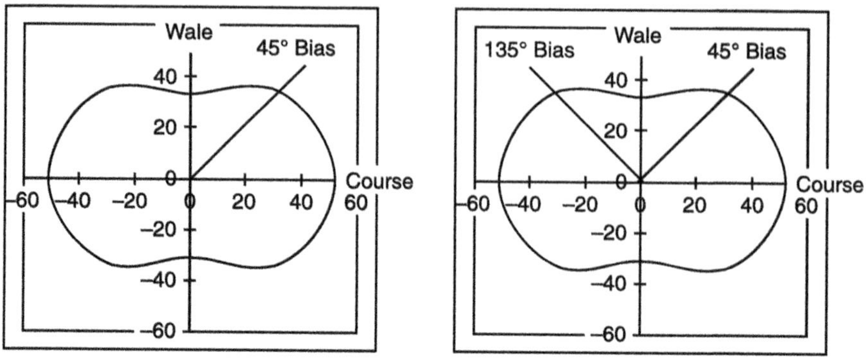

Sample C 1 Bias Sample C 2 Bias

(c)

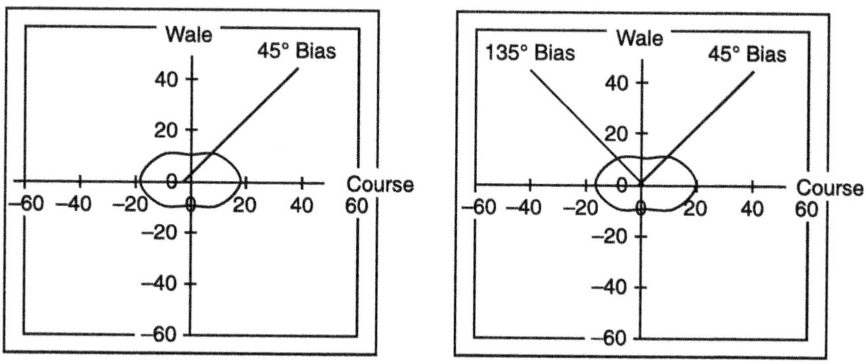

Sample D 1 Bias Sample D 2 Bias

(d)

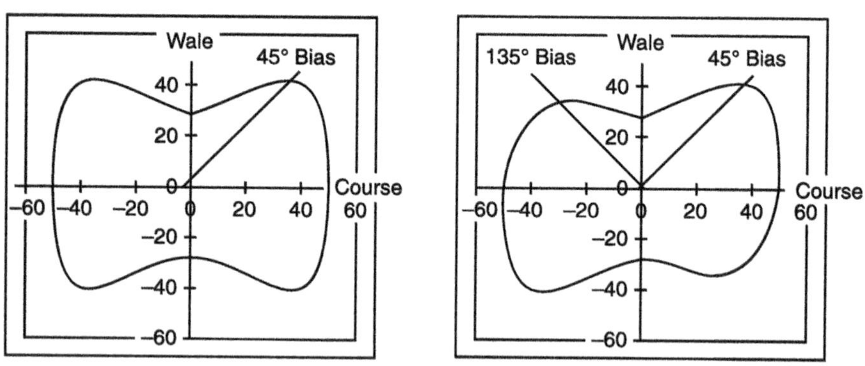

Sample E 1 Bias Sample E 2 Bias

(e)

10.5 Continued

If a fabric were to behave as a simple lattice structure that had very limited stretch in the course and wale directions, the resulting stretch distribution curve would be represented by four vectors radiating from a central point. A stretch distribution plot of a fabric that extends uniformly in all directions for a given load would be circular. The angular stretch distribution plots all clearly demonstrate that the highest stretch is in the course direction. Samples B, C and D show vertical symmetry. Samples A and E demonstrate a lack of symmetry in the bias stretch.

These plots made a significant contribution to my understanding of stretch fabric characteristics, the impact of bias stretch on pattern profile geometry and the optimal pattern orientation for dynamic fit. The results would appear to indicate that to achieve a consistent contour fit between garment right and left sides requires an equal bias measurement. Although small differences can be absorbed within the stretch fabric parameters this may not always be appropriate. In compressive garment technology, particularly in medical applications, an equal bias measurement may be crucial to obtaining an equal pressure on the body between right and left sides.

10.3.4 Digital stretch pattern technology

It is the ability of the knit stretch fabric to stretch multi-directionally that makes them suitable for form fit body profiling. The new Quad Load Test provides the input data for the fabric course, wale and 45 ° and 135 ° bias stretch extension and is readily accessible to the designer/technician because it does not rely on complicated scientific apparatus or a controlled environment. It is a convenient and simple method of quantifying stretch extension, which does not attempt to replicate British Standard test conditions in a controlled environment and therefore some inconsistencies will occur. Despite this it is possible for these inconsistencies to be accommodated within the fabric stretch reduction parameters and, therefore, should not detract from the intended purpose of the simplified test procedure. In drafting a stretch block pattern the multi-directional stretch fabric extension has to be applied using just two measurements on the x and y axis. The bias extension is the average between the course and wale becoming the course/bias and the wale/bias extension measurements referred to as bias vectors.

In its simplest form a body contouring garment could be constructed from cylindrical shapes of stretch fabric, of varying circumferences and lengths, covering the arms, legs and torso. Movement in any area of the body has to be accommodated by utilising available fabric stretch and generally must be greater than free body expansion. Therefore, the length of the body to accommodate maximum elongation will require the fabric to be reduced by a different proportion to the circumference of the body, which is not subject to the same movement excesses. This variable is referred to as the axis ratio.

Garments constructed for a variety of applications will require differing fit levels as outlined previously. The fit factor variable allows different fit level categories to be accommodated. The reduction factor takes an amount of the available stretch for the appropriate fit level. This fit factor then determines the amount of the available stretch to be applied by the axis ratio, which is the allocation of the amount of available stretch by different proportions to the vertical and horizontal pattern profile.

10.4 Stretch pattern development

Stretch garments are constructed by using a pattern that has a negative ease value. In other words, the pattern is cut to body dimensions smaller than the actual body. It is the inherent fabric stretch which ultimately determines the finished garment's size designation. Conventional pattern profiles for stretch fabrics have been developed by modifying block patterns for woven fabrics that have the ease allowance and darts removed (Haggar, 2004: 244–53). Difficulties arise in determining the amount and placement of the ease allowance to be removed. Darts are used to contour the fabric around the body form smoothly without the fabric buckling and the placement of darts and the amount of fabric suppression varies between block patterns. In a typical front bodice, the dart is suppressed (closed), removing it completely from the bust area; all or a proportion of the dart is then redistributed at the bodice shoulder or side seam. After the block pattern has had the ease allowance and darts removed, the profile is then trued into smooth lines and fluid curves. When this procedure has been completed the pattern profile is then proportionately reduced horizontally and vertically to accommodate a fabric stretch percentage. Conventionally, calculation of the stretch percentage is very subjective. Another approach to producing a stretch pattern is to model the stretch fabric directly onto a dress stand (Cloake, 1996). But this method is also subjective as it is difficult to determine how much hand stretch (force) is being used to achieve the desired pattern design. Some manufacturers just use a smaller sized pattern block in the assumption that the stretch fabric will automatically stretch in the right places to give an acceptable fit. These highly subjective approaches do not maximise the stretch fabric potential to provide a good fit-quality.

Stretch fabrics are increasingly being used across the whole gamut of clothing applications fashion sportswear, medical intimate body wear and technical garments. To date textbooks that instruct the user on how to design stretch patterns (Cloake, 1996; Armstrong, 1995; Aldrich, 2004, 2007; Haggar, 2004; Richardson, 2008) just reiterate subjective practices that date from the 1960s. Shoben (2008), in his introduction to *The Essential Guide to Stretch Pattern Cutting*, suggests pattern cutting is an art not a science, and that dealing with stretch fabrics is a minefield because the almost unlimited variations in their composition make the question of pattern size difficult. The body of the stretch block pattern has been developed using a traditional method of fabric draped onto a Kennett and Lyndsel size 12 full length dress stand. The sleeve was then constructed from the body of the block pattern using his flat pattern method. He goes on to say that his block pattern is only a starting point. Unfortunately the dependency on subjective expertise is also true for the development of pressure garment pattern design. Although extensive research has been undertaken into optimising the level of pressure, brought about by the tensioned fabric in pressure garment design, Pratt and West state in their manual:

> In some instances, measurements have to be altered 'by eye' or judgement alone. For example, when fitting a sleeve into a vest, a measurement is taken from the mid-axilla to the base of the lateral aspect of the neck. The angle at which this is drawn on the pattern will rely on your observations of the shoulder's girth, the patient's postural patterns and the site of the scarring.
>
> (Pratt and West, 1995: 24)

While this author has great respect for subjective expertise, it is difficult to translate this tacit knowledge into a digital form so that we can develop a user orientated approach to stretch pattern technology.

10.4.1 Distal and proximal fit

Garment fit expectations are not always clear, particularly in relation to stretch garments. To aid clarity this author has introduced the anatomical terms *proximal* and *distal fit*, which describe the proximity of the garment to the body on a *proximal distal fit continuum* with the body contour as the zero proximal reference point (Watkins, 2005/2006).

As one moves away from the Form Fit (zero) reference point, then the proximal (negative) value becomes greater as the garment compresses the body. Conversely, in the distal (positive) direction, the looser the garment fit becomes. For clarity, garment fit has been approximated into three values either side of the zero point along the proximal distal fit continuum. Garments along the distal continuum away from the Form Fit describe garments that are constructed from fabrics that are either non-stretch or have minimal stretch to enhance comfort. These garments are essentially an external structure ranging from 'Fitted' (D2) through 'Semi-fitted' (D4) to a 'Loose Fit' (D6).

The proximal fit describes body-contouring garments constructed in a stretch knit fabric. The increasing negative proximal fit is related to the garment pattern reduction ratio, influenced by the force exerted on the body, through the modulus or compressive retracting power of the stretch fabric. The proximal fit attributes are as follows.

- *Form Fit* (P0) describes garments that have few wrinkles and no stretch other than tare stretch (a minimal amount) in specific areas, to allow the fabric to smoothly contour the body. The stretch fabric exerts no pressure on the body and the stretch does not impede mobility. An example would be close fitting underwear with no holding power.

- *Cling Fit* (P2) includes fashion garments where the fabric stretch does not significantly compress or alter the body contour. The stretch fabric clings to the body curves accentuating the natural shape, for example stretch T-shirts.
- *Action Fit* (P4) describes garments where the retracting stretch effectively grips the body. Most stretch sportswear and exercise garments come under this heading and are produced in a diverse range of knit fabrics with differing degrees of stretch.
- *Power Fit* (P6) refers either to the garment as a whole or to specific areas where the force exerted by the stretch holds and compresses the flesh, changing the body form shape. Applications cover a wide range of sportswear, form persuasive bodywear and medical applications.

When conventional pattern co-ordinates are modified and reduced using arbitrary stretch percentage factors to develop pattern profiles in the proximal range, then an accurate fit is still expected. For a given number of measurements a better fit potential would be expected for distal fit; looser-fitting garments can accommodate a broader range of bodyshapes. Because the conventional pattern profile becomes increasingly distorted as the fabric is incrementally stretched around the body contours, even if a greater number of body measurements were taken for the proximal fit range, this would still result in a poorer fit potential. It is the inaccuracy of the garment-to-body fit relationship in combination with arbitrary stretch factors within the conventional pattern profile geometry that ultimately undermines the fit potential of custom fit stretch garments.

10.4.2 Pattern design, fit and mobility

The analysis of traditional garment pattern design and fit for non-stretch fabrics, the method and the rationale, can stimulate imaginative solutions to enhance movement in stretch garment pattern design. However, in a conventional stretch bodysuit poor fit is not always visually apparent until the garment has been worn and washed several times. The fabric may then begin to degrade at the underarm and body rise or the seam stitching bursts. Watkins (2000) outlines an objective approach to visually evaluating stretch fit. Stretch garment analysis is also interpretive as the individual's subjective assessment of comfort and fit needs to be considered. It is not only the way in which the stretch conforms to grip the body, the 'hugging power', but how the garment feels, the first impression when donned and impressions once the garment has been worn and subjected to a range of movements, that contribute to the quality of the fit analysis. Movement can be enhanced or inhibited by the garment fit and particularly problematic are the shoulder and hip areas. Joints can be classified by the extent of their range of movement. The shoulder is a multi-axial joint that has the highest degree of mobility. The body area commentaries following highlight a way in which a rigid pattern can be developed to assist the shoulder to move freely.

The bodice

The crucial areas for fit in the bodice are the shoulder angle, the breast and the armscye (armhole). The conventional bodice pattern as illustrated (see Fig. 10.6) shows the relationship between the garment pattern and the torso.

The shoulder angle

The shoulder angle is determined by posture and elevation of the shoulders and has a significant influence on the fit and comfort of a garment. Hutchinson (1977) outlines a method for determining the shoulder seam placement using a frame to measure the shoulder angle, however the technique uses complex equipment that is not readily available. Rohr (1957: 7) explains how to achieve an accurate shoulder angle by taking three simple measurements. These co-ordinates combined in the pattern draft give an accurate shoulder angle for the subject's body posture when applied to both front and back bodice constructions.

The set-in sleeve

For a conventional set-in sleeve, the head height and shape of the sleeve reflects the shape of an arm hanging in a relaxed position by the side of the body (see Fig. 10.7). The sleeve–torso angle relationship affects the degree of freedom of arm movement and the sleeve fit is at its best when the arm is fully adducted and the crown conforms smoothly around the top of the arm.

When a set-in sleeve is constructed in stretch fabric, movement is restricted as it is impossible to lift up the arm without the fabric straining. A prime example that many will be familiar with, which illustrates the point, is the cling fit stretch T-shirt with this conventional sleeve construction. When the arm is raised, the fabric adjusts to the new body position. If the underarm seam is lower than the natural armscye line, the underarm sleeve junction will automatically reposition at the anchor or grip point under the arm. Subsequently, when the arm is lowered a fold of fabric (producing the effect of an unwanted shoulder pad) appears at the apex of the sleeve crown. A fold of fabric also appears across the chest above the breasts. The T-shirt comfort/fit factor is only maintained by constant rearrangement after movement. This can lead to a negative body cathexis (LaBat and DeLong, 1990) but it is the pattern profile that is at fault and not the inadequacy of the wearer's bodyshape. Inappropriate pattern geometry in combination with the fabric stretch does not allow the crown to resume its original position when the arm is lowered.

The shirt

Conventional shirt-sleeve pattern construction allows the arms to be raised and move freely. However, it can be observed (see Fig. 10.8) that when the arm is

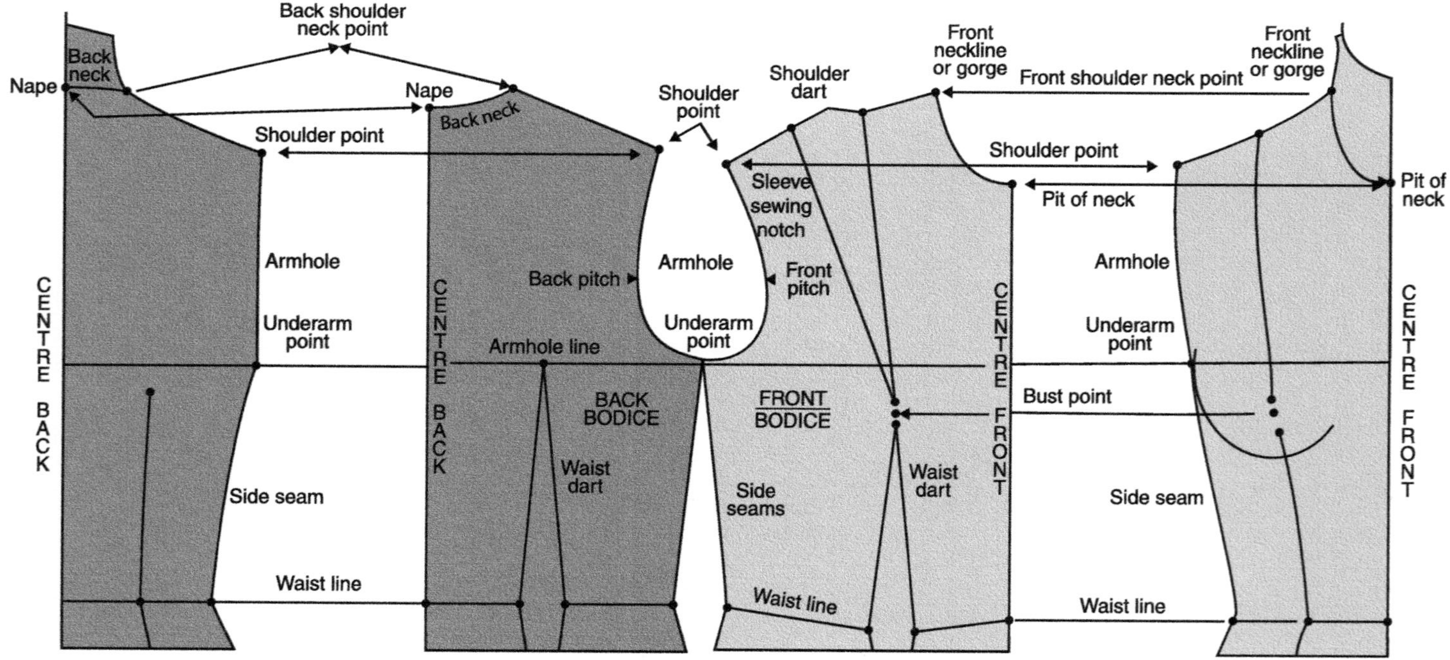

10.6 Conventional block pattern relationship to body measurements (Shoben and Ward, 1980: 39).

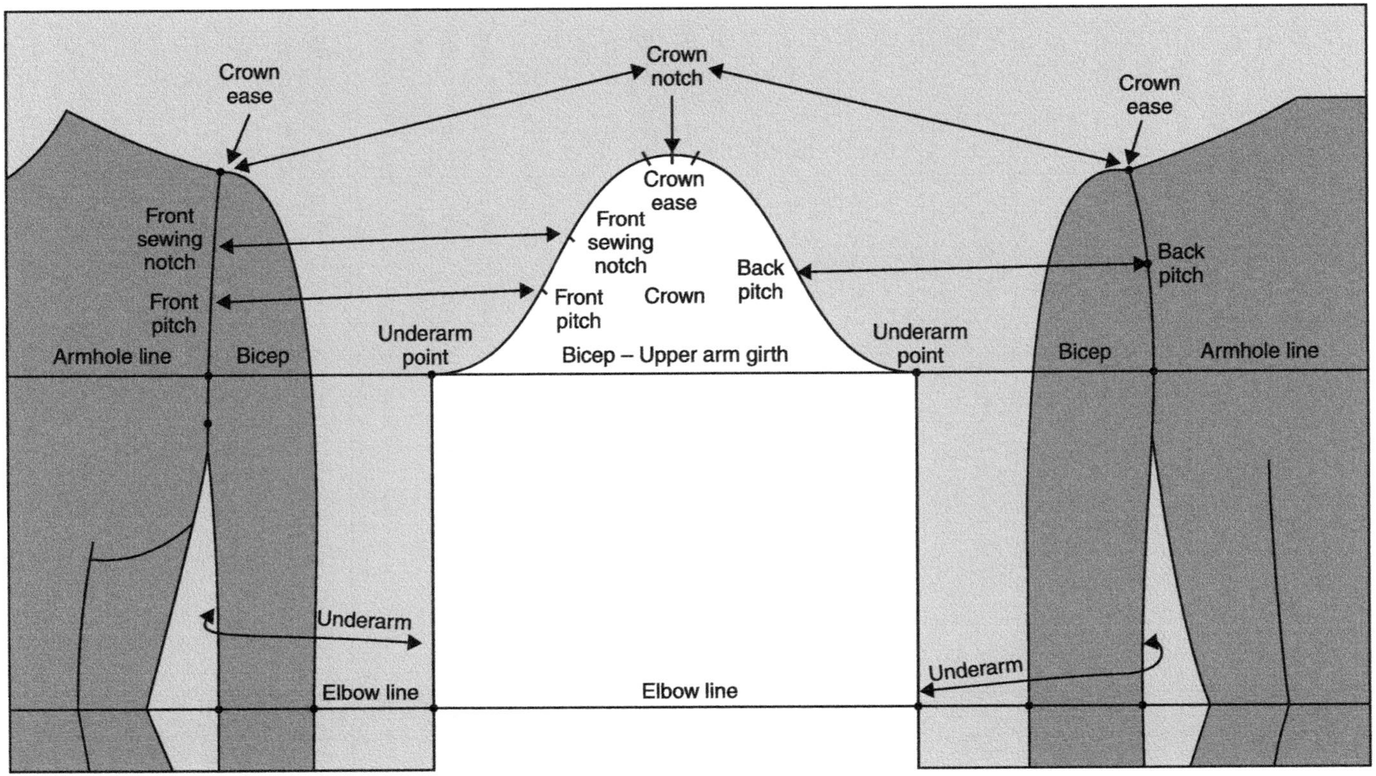

10.7 Conventional set-in sleeve pattern (Shoben and Ward, 1980: 40).

10.8 Shirtsleeve (Ladbury, 1984: 107).

lowered, diagonal wrinkles form towards the under arm. In the illustration (Fig. 10.9) the shirt-sleeve profile (solid line) is achieved by slashing and spreading the set-in sleeve pattern (dotted line). As the width of the sleeve increases, the underarm is lengthened and the crown becomes shallower, allowing the wearer to move with ease.

In a stretch pattern, if the crown pattern geometry retains a similar profile to the conventional set-in sleeve pattern, with little change in the crown depth, this impairs the quality of the garment fit. When a crown pattern profile similar to a shirt is drafted in a stretch pattern, the width of the lower sleeve may remain narrow with increased width between the underarm seam junctions. This allows the arm to move freely without fabric displacement after movement.

10.4.3 Proximal fit pattern design

The shape of the fabric affects the stretch characteristics. A visual understanding of the overall stretch curvilinear fabric distortion characteristics is essential to the process of pattern production through garment fit analysis and evaluation. Evaluation of the stretch deformation of various shapes, printed with a grid pattern and stretched, such as rectangles, trapezoids and triangles can contribute to

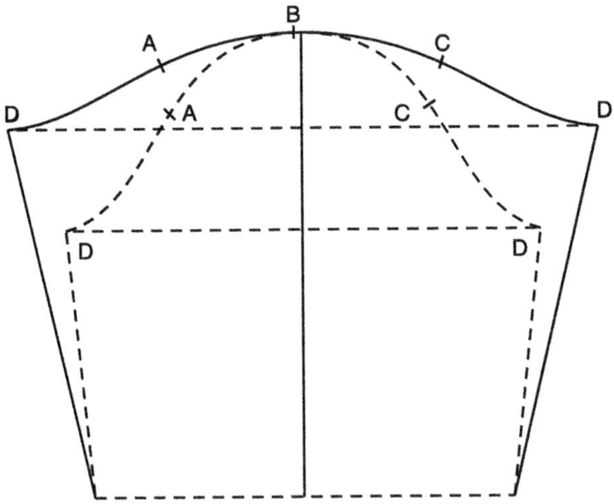

10.9 Pattern manipulation (Pivnick, 1958: 58).

maximising the stretch garment fit potential in the pattern design. The area comprising the shoulder angle, armscye, sleeve crown and the protrusion of the breasts demonstrates where directional change and protrusion need an integrational approach in balancing the pattern profile with the deformable fabric geometry for the range of movement required. The transposition of the sample shape deformation of a triangle or trapezoid is informative when applied to the garment pattern for the sleeve crown.

The dynamic crown angle

The alignment of the arm to the body determines the basic shape of the sleeve pattern and the armscye intersection of the bodice pattern. By manipulating the pattern geometry a range of movements to be performed by the arm can be accommodated.

For the proximal fit pattern profile this author has introduced 'the dynamic crown angle' that relates to the depth of the crown, which is calculated from the shoulder point at the top of the crown to the intersection between the arm and chest. This depth becomes shallower as the geometry of the pattern profile changes to utilise the fabric stretch characteristics to enhance the fit quality and accommodate a range of movements. Figures 10.10 and 10.11 illustrate the bodice to sleeve angle relationship and the shallow crown shape in the bodysuit analysis garment, which approximates a subject standing with the arms abducted at 45°.

Proximal form fit

The geometry of the stretch block pattern profile can only be developed successfully through understanding the complex relationship between the dynamic form, the

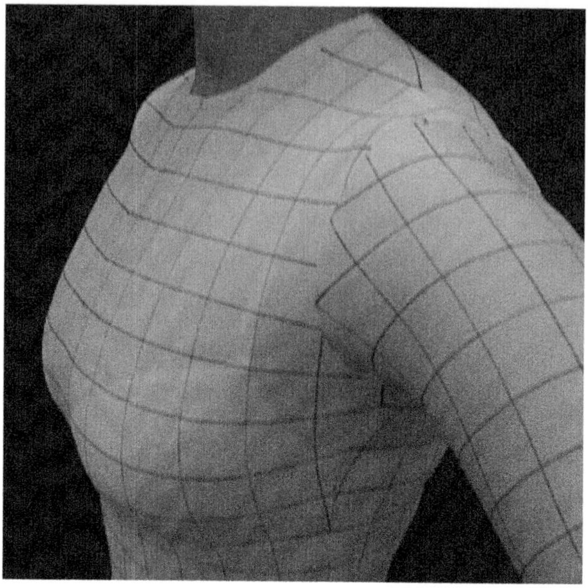

10.10 Analysis bodysuit, 25 mm grid – front.

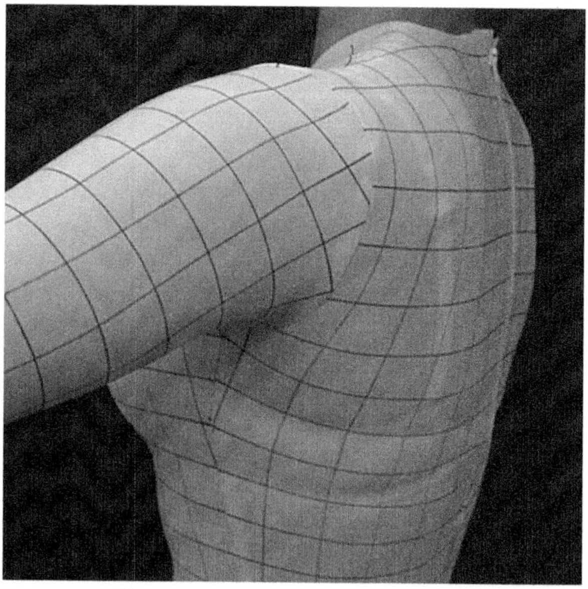

10.11 Analysis bodysuit, 25 mm grid – back.

stretch fabric behaviour and the two-dimensional pattern profile geometry. This author's approach to pattern design has been analysing traditional procedures in pattern design and garment fit to accommodate different bodyshapes, posture and movements. The 3D garment fit is evaluated then reverse engineered, using nip and tuck algorithms (haptic garment fitting experience accumulated over years) translated into digital form, and then re-applied to the evolving 2D pattern pieces. Replicating the size and shape of a person in the pattern profile is the key. Good fit is dependent on the pattern drafting co-ordinates co-operating with the stretch characteristics conforming to the shape of a person. Following research this author has developed a Form Fit block pattern using a personally extended set of traditional measurements. The new Form Fit block pattern is the basis for developing both distal and proximal garment fit. Producing a form fit flat pattern, without darts, that closely adheres to the contours of the body and without restricting movement, is complex. In woven fabric, darts and ease are used to manipulate the fabric around the form and allow movement. In a knit stretch garment without darts to contour the body, a degree of fabric stretch distortion (tare stretch) in areas of protrusion is inevitable. An optimised contour fit pattern should produce a garment that has no wrinkles, minimal stretch distortion and conforms to the body, rather like a second skin.

Proximal Action Fit

To produce the Action Fit the algorithms for the Form Fit patterns are enhanced to take into account the selected fabric stretch characteristics, the desired fit level and the radius of curvature, which can vary for adults and children or for different body zones. The resulting parametric pattern produces an Action Fit stretch bodysuit that is a true custom fit for the selected body shape size, fit level and chosen fabric.

10.4.4 Proximal fit analysis

Pressure fit encompasses a complex set of variables. It is difficult to visualise and quantify the garment-to-body stretch fabric tensional parameters when altering a garment using a manual fitting process. The quality of the fit becomes dependent on the subjective expertise of the fitter. Therefore, to objectively evaluate the proximal stretch fit a 25 mm grid system has been printed on the analysis body suit. The stretch garment-to-body pattern design fit is optimised through an iteration process and a grid system allows the designer to visualise stretch deformation over the body contours. The grid pattern deforms into different geometric shapes indicating garment-to-body alignment and the amount and direction of fabric stretch. Gridlines not only enable the observer to identify areas of unacceptable stretch, which is indicative of the pattern profile being incorrect, but also they confirm that the horizontal and vertical toile/body placement aligns as the designer intended.

10.5 Future trends

Stretch garment assessment is interpretive; the quality of the body contouring fit is inextricably linked with the stretch potential of fabric characteristics. It is imperative that the designer uses a mathematical method for quantifying the degree of fabric stretch to be applied in the pattern reduction process. The making of this tacit knowledge explicit will improve communication between industry, science, technology and practitioners to further develop emerging digital technologies in compressive stretch garment design.

10.6 Conclusions

A body contouring stretch knit garment should display no wrinkles, have minimal stretch distortion and conform to the body contours to facilitate a range of movement, without displacing or straining the fabric. The quality of the body-contouring fit is inextricably linked with the stretch potential of fabric characteristics.

Understanding the stretch behaviour, visually and mechanically, is an essential part of predicting the pattern profile geometry and the optimum orientation of the pattern placement on the fabric to improve the fit-quality, enhancing comfort and freedom of movement. This is achieved in part by maintaining the stretch extension within the lower modulus working range. The pattern orientation will affect the garment fit if the stretch fabric extension in the course and wale directions is different. Thus, if a pattern profile designed for a horizontal (course) orientation on the fabric is placed in the vertical (wale) orientation or vice versa, a garment-to-body fit disparity would occur.

Defining the fit-quality expectations and the fit level category is paramount in the assessment of the garment-to-body contouring fit relationship. A fitting scheme facilitates an integrational approach to balancing the fit requirements in the pattern profile geometry with the deformable fabric for the range of movement envisaged. Printing a 2.5 cm grid on the analysis bodysuit toile visualises the stretch fabric characteristics enabling the assessment of the interrelated factors of seam alignment placement, body landmark positioning and the amount and direction of fabric stretch in garment-to-body fit.

The purpose of the new Quad Load Test (the adapted hanger load test) is to assist the designer in quantifying stretch characteristics. The degree of stretch extension of sample fabrics in the course, wale and bias orientations results in a mathematical optimisation of the available fabric stretch to be utilised in the pattern reduction procedure. The development of QA/QC tests has contributed considerably to evolving a common language between fibre and fabric producers and garment manufacturers. The development of an industry standard to quantify the degree of stretch extension combined with a visual representation of stretch fabric parameter for stretch pattern technology would also be beneficial in engendering interdisciplinary dialogue for further research in this digital age to improve the fit-quality and comfort of stretch garments.

10.7 Sources of further information and advice

http://www.assystbullmer.co.uk/
http://www.browzwear.com/
http://www.dressingsim.com/
http://www.fitme.com/
http//www.gerbertechnology.com/
http//www.lectra.com/
http//www.optitex.com/
http://www.miralab.unige.ch/

10.8 References

Aldrich, W. (2007) *Fabric, Form and Flat Pattern Cutting*, 2nd edn, Oxford, Blackwell Science Ltd.

Aldrich, W. (2004) *Metric Pattern Cutting*, 3rd edn, Oxford, Blackwell Science Ltd.

Anderson, R. (2008) http://www.telegraph.co.uk/news/uknews/2151888/Savile-Row-tailor-Richard-bespoke-must-mean-bespoke.html (accessed 28/7/2008).

Anon. (2008) http://www.thelawyer.com/cgibin/item.cgi?id=133406&d=pndpr&h=pnhpr &f=pnfpr (accessed 28/07/2008).

Apeagyei, P. and Otieno, R. (2007) 'Usability of pattern customising technology in the achievement and testing of fit for mass customisation', *Journal of Fashion Marketing and Management*, **11**(3), 349–65.

Apeagyei, P., Otieno, R. and Tyler, D.J. (2007) 'Ethical practice and methodological considerations in researching body cathexis for fashion products', *Journal of Fashion Marketing and Management*, **11**(3), 332–48.

Armstrong, H. (1995) *Patternmaking for Fashion Design*, 2nd edn, USA, Harper Collins Publishers.

Ashdown, S.P. (1998) 'An investigation of the structure of sizing systems: A comparison of three multidimensional optimized sizing systems generated from anthropometric data with the ASTM standard D5585-94', *International Journal of Clothing Science and Technology*, **10**(5), 324–41.

Ashdown, S.P. (editor) (2007) *Sizing in Clothing: Developing Effective Sizing Systems for Ready-to-wear Clothing*, Woodhead Publishing Limited, Cambridge, England.

Ashdown, S.P., Loker, S., Schoenfelder, K. and Layman-Clarke, L. (2004) 'Using 3D scans for fit analysis', *Journal of Textile and Apparel Technology and Management*, **4**(1), 1–11.

ASTM: D 4964-96 (1996) 'Standard Test Methods for Stretch Properties of Knitted Fabrics Having Low Power', USA: *ASTM Standards*, 641–645.

Borland, H. and Akram, S. (2007) 'Age is no barrier to wanting to look good: women on body image, age and advertising', *Qualitative Market Research: an International Journal*, **10**(3), 310–33.

Bray, N. (1978) *Dress Fitting*, 2nd edn, Oxford, Blackwell Science.

BS 4952 (1992) 'Methods of Test for Elastic Fabrics', UK, *British Standards Institute*, 1–12.

BS EN 14704-1 (2005) 'Determination of the Elasticity of Fabrics', UK, *British Standards Institute*, April.

Chen, C-M, (2007) 'Fit evaluation within the made-to-measure up process', *International Journal of Clothing Science and Technology*, **19**(2), 131–44.

Chittaro, L. and Corvaglia, D. (2003) '3D virtual clothing: from garment design to web3d visualization and simulation 3D technologies for the World Wide Web' *Proceedings of the eighth International Conference on 3D Web Technology*.

Cloake D (1996) *Fashion Design on the Stand*, London, Batsford Publishing.

Crowther, E.M. (1985) 'Comfort and fit in 100% cotton-denim jeans', *Journal of the Textile Institute*, **76**(5), 323–38.

D'Apuzzo, N. (2007) '3-D body scanning technology for fashion and apparel industry', http://www.hometrica.ch/publ/2007_videometrics.pdf (accessed 11/12/2007).

Daanen, H. and Hong, S-A. (2008) 'Made-to-measure pattern development based on 3D whole body scans', *International Journal of Clothing Science and Technology*, **20**(1), 15–25.

Decaudin, P. (2006) 'Virtual garments: a fully geometric approach for clothing design', *Eurographics*, **25**(3), 625–34.

Denton, M.J. (1972) 'Fit, stretch and comfort,' *Textiles*, **1**(1), 12–17.

Douty, H.I. (1954) 'Objective figure analysis', *Journal of Home Economics*, **46**(1), 24–26.

Douty, H.I. (1968) 'Visual somatometry in health related research', *Journal of Alabama Academy of Science,* **39**(1), 21–34.

Fan, J., Yu, W. and Hunter, L. (2004) *Clothing Appearance and Fit: Science and Technology*, Woodhead Publishing, Cambridge.

Fentem, P.H. (1986). 'Advances in elastic hosiery', *Pharmacy Upate*, **5**, 200–205.

Fentem, P.H. and Goddard, M. (1979) 'Comparison of a direct and an indirect method of measuring hosiery compression', *Journal of the Textile Institute*, **70**(5), 198–209.

Filatov, V.N. (1985) 'Planning and evaluation of elastic fabrics for medical applications', *World Textile Abstracts*, **14**(1), 45–52.

Fontana, M., Rizzi, C. and Cugini, U. (2005) '3d virtual apparel design for industrial application', *Computer-Aided Design*, **37**(6), 609–22.

Giele, H.P, Liddiard, K., Currie, K. and Wood, F.M. (1995) 'Direct measurement of cutaneous pressures generated by pressure garments', *Burns in Western Australia Research Group International Symposium for Hypertrophic Scar*, Hong Kong, June, 1–5.

Haggar, A. (2004) *Pattern Cutting for Lingerie Beachwear and Leisurewear*, 2nd edn, Oxford, Blackwell Publishing.

Harada, T. (1982). 'Pursuit of comfort in sportswear', *JTN*, **334**, September, 30–33.

Harnett, P. (1976) 'Functions and properties of "thermal" underwear', *Wool Science Review*, **52–60**(1), 3–11.

Horino, T., Kawanishi, S. and Toshimi, M. (1977) 'Simulation of garment pressure in wear by strip bi-axial extension of cylindrically sewn fabrics', *Journal of the Textile Machinery Society of Japan*, **23**(2), 41–6.

Horner, J., Lowth, L.C. and Nicolaides, A.N. (1980) 'A pressure profile for elastic stockings', *British Medical Journal*, **280**, March, 818–20.

Hutchinson, R. (1977) *The Geometrical Requirements of Patterns for Women's Garments to Achieve Satisfactory Fit* (T16351), M.Phil, Department of Textile Industries University of Leeds, September.

Ibrahim, S.M. (1968) 'Mechanics of form-persuasive garments based on Spandex fibers', *Textile Research Journal*, September, 950–63.

Kirstein, T., Krzywinski, S. and Rödel, H. (1999) 'Pattern Construction for Close-Fitting Garments made of Knitted Fabrics', *Melliand Textilberichte*, vol. 80, pt 3, March, pp. E46–48/146–148.

Krzywinski, S., and Rodel, H. (2005) 'Virtual Product Development for Close-Fitting Garments of Knitwear with Elastan Yarns', *2nd International Conference of Textile Research Division NCR*, Cairo, Egypt, April, 11–13.

LaBat, K.L. and DeLong, M.R. (1990) 'Body cathexis and satisfaction with fit of apparel', *Clothing and Textiles Research Journal*, **8**(2), Winter, 43–8.

Ladbury, A. (1984) *Dressmaking with Liberty*, London, Guild Publishing.

Lenker, S. (1987) *Vogue Fitting: The Book of Fitting Techniques, Adjustments, and Alterations*, Harper & Row, New York.

Liechty, E.G, Pottberg, D.N. and Rasband, J.A. (1986) *Fitting & Pattern Alteration: A Multi-Method Approach*, USA, Fairchild Publications.

Lindberg, J. (1966) 'How stretch fabrics perform in garments', *American Fabrics and Fashions*, **72**, 58–61.

Loker, S., Ashdown, S.P., and Schoenfelder, K. (2005) 'Size-specific analysis of body scan data to improve apparel fit', *Journal of Textile and Apparel, Technology and Management*, **4**(3), 1–15.

Macintyre, L., Baird, M. and Weedall, P. (2004) 'The study of pressure delivery for hypertrophic scar treatment', *International Journal of Clothing Science and Technology*, **16**(1/2), 173–83.

Maklewska, E., Nawrocki, A., Kowalski, K. and Tarnowski, W. (2007) 'New Tool for Estimating Interface Pressure Under Burn Garments', *MEDTEX 07 Fourth International Conference and Exhibition on Healthcare and Medical Textiles. In Association with Tampere University of Technology, Finland*, Bolton, UK, July, 16–18.

Maklewska, E., Nawrocki, A., Ledwon, J. and Kowalski, K. (2006) 'Modelling and designing of knitted products used in compressive therapy', *Fibres & Textiles in Eastern Europe*, **14**(5), 111–13.

Murden, F.H. (1966) 'Elastomeric thread review (II): Elastomer and fabric test method', *Textile Institute and Industry*, **4**, 355–58.

Ng, S-F. (1995) *Design of Pressure Garments*, Ph.D, De Montfort University, Leicester, September.

Paquet, E. and Viktor, H.L. (2007) 'Adjustment of virtual mannequins through anthropometric measurements, cluster analysis, and content-based retrieval of 3-D body scans', *IEEE Transactions on Instrumentation and Measurement*, **56**(5), 1924–9.

Petrac, S. and Rogale, D. (2006) 'Systematic representation and application of a 3-D computer aided garment construction method. Part 1: 3-D garment basic cut construction on a virtual body model', *International Journal of Clothing Science and Technology*, **18**(3), 179–87.

Petrac, S., Rogale, D. and Mandekic-Botteri, V. (2006) 'Systematic representation and application of 3-D computer aided garment construction method. Part II: Spatial transformation of 3D garment cut segments', *International Journal of Clothing Science and Technology*, **18**(3), 188–99.

Pivnick, E.K. (1958) *Fundamentals of Patternmaking for Women's Apparel Part 1*, 3rd edn, New York, Pattern Publications, pp. 3–57.

Pratt, J. and West, G. (1995) *Pressure Garments, a Manual on their Design and Fabrication*, Oxford, Butterworth Heinemann Ltd, pp. 22–4 and 32–3.

Rasband, J.A. and Liechty, E.L.G. (2006) *Fabulous Fit: Speed Fitting and Alteration*, 2nd edn, USA, Fairchild Publications.

Richardson, K. (2008) *Designing and Patternmaking for Stretch Fabrics*, New York, Fairchild Books Inc.

Rohr, M. (1957) *Pattern Drafting and Grading*, 2nd edn, Eastchester, New York, M. Rohr.

Sawbridge, M. (1989) 'Comfort of clothing', *New Home Economics*, **35**(9), 5–7.

Sheldon, W.H., Stevens, S.S. and Tucker, W.B. (1940) *The Varieties of Human Physique*, New York and London, Harper & Brothers Publishers.

Shoben, M. (2008) *The Essential Guide to Stretch Pattern Cutting: Dresses, Leotards, Swimwear, Tops and More*, London, Shoben Fashion Media.

Shoben, M. and Ward, J. (1980) *Pattern Cutting and Making Up: The Professional Approach 1 Basic Techniques and Sample Development*, London, Batsford Academic and Educational Ltd.

Shoh, K. (1998) 'Comfort pressure evaluation of men's socks using an elastic optical fiber', *Textile Research Journal*, **68**(6), 435–40.

Slater, K. (1986) 'The assessment of comfort', *Journal of the Textile Institute*, **77**(3), 157–71.

Smith, J.E. (1986) 'The comfort of clothing' *Textiles*, **15**(1), 23–7.

Stylios, G. (1999) 'The next generation of garment design', http://www.iafnet.org/99conv-pres/georg-stylios.htm (accessed 20/02/2001).

Sykes, J.B. (1980) *The Concise Oxford Dictionary*, 6th edn, Oxford, Oxford University Press, p. 201.

Symonds, K.P. and Istook, C.L. (2003) 'Body measurement techniques: Comparing 3D body scanning and anthropometric methods for apparel applications', *Journal of Fashion Marketing and Management*, **7**(3), 306–32.

Symonds, K.P., Istook, C.L. and Devarajan, P. (2004a) 'Female Figure Identification Technique (FFIT) for apparel: Part I: Describing female shapes', *Journal of Textile and Apparel, Technology and Management*, **4**(1), 1–16.

Symonds, K.P., Istook, C.L. and Devarajan, P. (2004b) 'Female Figure Identification Technique (FFIT) for apparel Part II: Development of shape sorting software', *Journal of Textile and Apparel, Technology and Management*, **4**(1), 1–15.

Volino, P. and others (2005) 'From early virtual garment simulation to interactive fashion design', *Computer-Aided Design*, **37**(1), 593–608.

Wang, C.C.L. and Tang, K. (2008) 'Pattern computation for compression garment', *ACM Solid and Physical Modeling Symposium 2008*, Stony Brook, New York, USA, June 2–4, 203–11.

Wang, C.C.L. and Tang, K. (2010) 'Pattern computation for compression garment by a physical/geometric approach', *Computer-Aided Design*, **42**(2), 78–86.

Wang, C.C.L., Hui, K-C. and Tong, K.M. (2007) 'Volume parameterization for design automation of customized free-form products', *IEEE Transactions on Automation Science and Engineering*, **4**(1), 11–21.

Wang, C.C.L., Zhang, Y. and Sheung, H. (2010) 'From styling design to products fabricated by planar materials', *IEEE Computer Graphics and Application*, accepted.

Watkins, P.A. (2000) 'Analysis of Stretch Garments', *Textile Institute 80th World Conference*, Manchester, April.

Watkins, P.A. (2005) 'Custom Fit Pressure Garment Pattern Profiling', *Wearable Futures: Hybrid Culture in the Design and Development of Soft Technology Conference*, University of Wales, September.

Watkins, P.A. (2006) 'Custom Fit, Is it FIT for the Customer?', 8th *Annual IFFTI Conference. Fashion in the Digital Age*, Raleigh, North Carolina, USA, June, 20–22.

Watkins, P. A. (2007) 'Nip 'N Tuck Pattern Profiling', *MEDTEX 07 Fourth International Conference and Exhibition on Healthcare and Medical Textiles. In Association with Tampere University of Technology, Finland*, Bolton, UK, July, 16–18.

Yu, W. (2004) 'Objective evaluation of clothing fit', in Fan, J., Yu, W. and Hunter, L., *Clothing Appearance and Fit: Science and Technology*, Cambridge, Woodhead Publishing, 184–185.

Ziegert, B. and Keil, G. (1988) 'Stretch fabric interaction with action wearables: defining a body contouring pattern system', *Clothing and Textiles Research Journal*, **6**(4), 54–64.

11
Improving body movement comfort in apparel

S. P. ASHDOWN, Cornell University, USA

Abstract: Providing comfort in clothing for the moving body is a complex task. Interactions among body sizes and shapes, physiological variations, material properties, design choices, environmental challenges, and activities are exponential in their number. In this chapter we will explore some of the problems, solutions, and research methods used to decipher these issues.

Key words: dynamic anthropometry, active positions, user needs, material properties, function.

11.1 Introduction: fundamental principles of movement in apparel

Clothing comfort has been discussed and debated by researchers for decades. Though thermal comfort and tactile comfort are often the focus of discussions and research on comfort, most models of comfort include the factors of the fit of clothing, and the interaction of clothing with the activity of the person wearing the clothing, in their definition (Fourt & Hollies, 1969; Gilling, 1971; Renbourn, 1971; Sontag, 1985; Slater, 1986; Smith, 1986; Holmer, 1988; Pontrelli, 1990; Markee & Pedersen, 1991; Branson & Sweeney, 1991; Paoletti, 1991). Repeated studies identify appropriate fit as the primary factor that consumers desire in their clothing, even rating fit over thermal comfort and tactile qualities (Morris & Prato, 1981) but good fit is an elusive goal. Most clothing is designed for the 'anthropometric' position; that is, for a person standing squarely with feet slightly apart and arms at the sides. Yet we spend relatively little time in this static position; we generally are in constant motion. Renbourn (1971) discusses this issue, stating that although measurements for fitting clothing are taken in the standing position, much of our time is spent sitting, walking, and in other activities such as mounting steps, requiring the introduction of tolerances (or ease values) to allow movement in our clothing.

Well fitted and well designed clothing should not interfere with the movement appropriate to the occasion for which the garment is worn. Though we do not expect formal wear or business wear to allow the same range of movement as casual wear or sportswear, clothing that restricts the performance of ordinary movements such as those required to drive a car, reach for a subway strap, walk freely, or sit would not be consistent with comfort in clothing.

11.1.1 Human body movement

An understanding of the human anatomy and function is necessary in order to understand the body in active positions, and to design clothing that will move appropriately with the body. The movement of the human body is varied and extensive. Movement originates when signals are sent through the nervous system to muscles, which control muscle movements, connected to the joints with tendons. There are basically three types of joints: synarthroses joints which are immovable, amphiarthroses joints with limited movement, and diarthroses or freely movable joints. Movable joints come in many different shapes and configurations of ligaments that connect them, resulting in different patterns of movement. Hinge joints such as that of the elbow or the finger, and saddle joints such as that at the base of the thumb move around one plane in a folding motion. Condyloid joints such as the wrist can both rotate and bend, so they move in more than one plane, but are limited in range. Pivotal joints such as the joint at the base of the skull rotate in one dimension (head movement is further optimized due to the flexibility of the cervical vertebrae), and ball and socket joints such as that at the hip rotate in several planes and have the widest range of motion of any of the joints (see Fig. 11.1). Complex movements such as those of the hand or spinal column result from the interactions of many joints (Pansky, 1975).

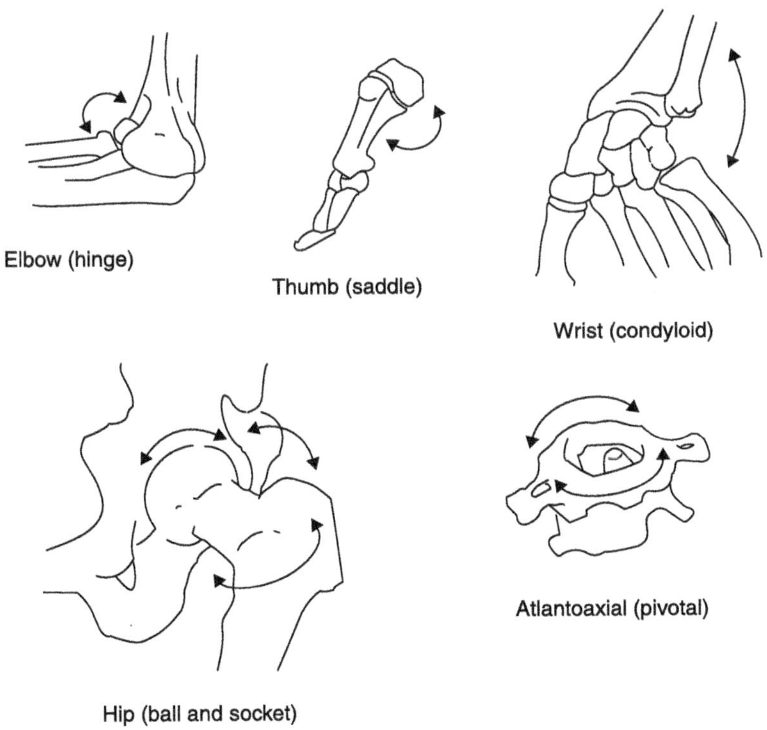

11.1 Examples of the five main joint configurations and how they move.

Connective tissues between muscle and bone and from bone to bone are also a critical part of the system. One area where the connective system is particularly extensive is the series of muscles and tendons that stabilize and allow movement in the area of the shoulder. There is not a direct bony joint connecting the shoulder blade, and through it the arm, directly to the axial skeleton. Instead, wide arrays of muscles are attached to the shoulder blade through tendons. This muscle structure connects the shoulder blade to the vertebrae, allowing the whole shoulder blade to shift over the back of the ribcage. The combination of this structure and the ball and socket joint at the shoulder results in an extensive range of movement for the arms and shoulders in all directions. This area is particularly critical in terms of clothing fit and movement, as many articles of clothing are suspended from the shoulder, yet the movements required from the shoulders, arms, and hands in our lives are varied and constant.

The study of movement, kinesiology, classifies human movement by dividing the body using three planes (the sagittal plane that divides the body from front to back, the frontal plane that divides left and right, and the transverse plane that divides the top and bottom of the body) and describing movements in relation to these planes (see Fig. 11.2). Flexion and extension occur on the sagittal plane around the transverse axis, and are forward and backward movements, as when we bend the arm at the elbow away from the anatomical position and straighten it back into the anatomical position (the anatomical position is a relaxed standing position with the arms at the side of the body and palms facing out). Generally, the range of movement is greater anterior to the frontal plane, as we move our limbs forward. Movements to the posterior of the frontal plane are much more limited, and are categorized as hyperextension when the limb is moved beyond the anatomical position; for example, the arm is hyperextended when it is straightened and moved toward the back of the body. Abduction and adduction describe movements of the limbs on the frontal plane around the medial axis. Other movements are defined in relation to the joint centers or bone axes. For example the term circumduction describes a circular motion originating at a joint, and rotation can occur around the axis of a limb. Circumduction can occur through a wide range of movement; rotation is generally somewhat more limited (Watkins, 1995).

Describing movement reliably can be difficult as the various combinations of movements of which the human body is capable are practically infinite. There are 206 bones in the adult human body, most of which are connected with movable joints, creating endless combinations of discrete positions. There is also much variation in the range of motion possible among individuals in the population related to normal variations among people: age, ethnicity, gender, health, fitness levels, body proportions, and short term effects of fatigue can all be important factors affecting movement. Cultural differences are little studied, but an important factor that can also impact the range of movement. Body movements and commonly held body positions vary among cultures and geographic areas, as differences in the way different populations sit, stand, and move are shaped by environment, culture, and

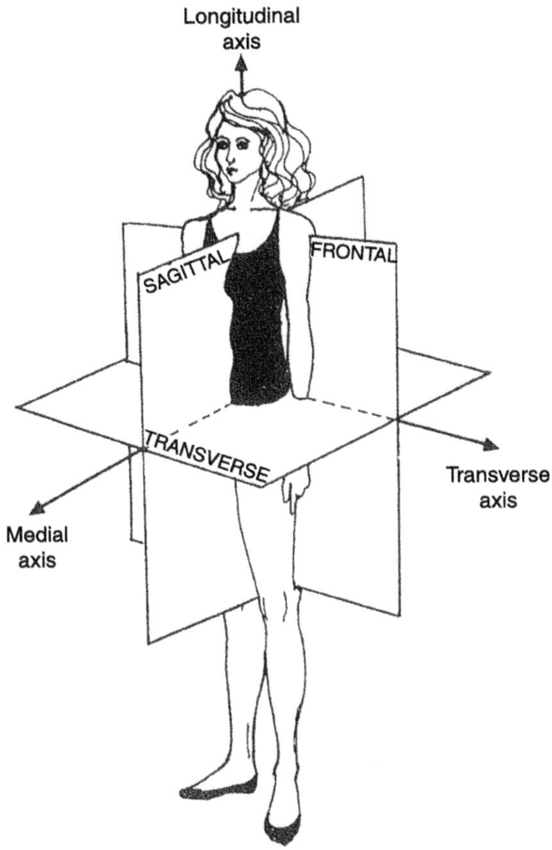

11.2 The three axes and the three planes of the body as classified in kinesiology, for defining movement (Watkins, 1995). Image courtesy of Susan Watkins.

training. Hewes (1957) developed a classification system for different standing, sitting, and kneeling positions, and created a series of world maps representing the distribution of different postures from 480 different cultures.

11.1.2 Clothing fit and movement

Discomfort in clothing related to movement is generally the result of the fact that as the body moves body dimensions change, with overall lengths increasing on one side of a bending joint and decreasing on the other side (Kirk & Ibrahim, 1966). Aldrich, Smith, and Dong (1998) describe garment distortions related to the movements of the upper body, especially the arms. Hatch (1993) quantifies body changes with movement across the shoulders at 13–16%, in buttock width at 4–6%, in arm and leg length from 35% to 45%, and in elbow and knee diameter from 12% to 22%. Lee and

Ashdown (2005) found significant variations in upper body measurements of women comparing three active postures, shoulder flexion, scapula protraction, and scapula elevation, with the greatest differences from a standard posture occurring in increased back intercye width in the scapula protraction posture, and decreased biacromion length and front interscye width in the shoulder flexion posture. Choi and Ashdown (2010) measured differences in body measurements between standing and seated measurements for young women, and found that the hip girth measurement in the sitting posture increased about 7% compared to a standing posture. Front crotch girth in the sitting posture decreased about 16%, and back crotch girth increased about 9%. The front center leg length increased about 10% and the back center leg length decreased about 19%. Lotens (1989) identified seven extreme body postures and the changes in clothing ease needed for each movement.

Clothing that does not increase in dimension over a bending joint, or that bunches and binds where body dimensions decrease, will impede movement or increase the difficulty of movement and create discomfort. This effect will be increased if the clothing is tethered to the body due to its design (i.e. a coverall that is tethered between the crotch and the shoulder and neck area) or by other equipment (see Fig. 11.3).

The level of comfort in clothing related to movement is predicated by the fit of the garment, the material properties, and the design of the garment. Movement in

11.3 This disposable protective coverall shows stress folds from the crotch to the shoulder when the wearer is in an active position. Image courtesy of the Cornell Bodyscan Research Group.

clothing can be accommodated in many ways; by loose flowing clothing (excess width, excess length, or both), by unattached areas between garments (separates), by open areas in the garment, by close fitting clothing made with elastic materials, or by design features that release as the wearer moves, such as pleats or elastic inserts (see Figs 11.4–11.7). The thickness, stiffness and frictional properties of

11.4 Depending on the activity and the environment, loose fitting garments can sometimes offer the greatest level of comfort.

11.5 Many types of protective sportswear are designed in segments to allow free movement.

11.6 Form fitting stretch materials are used in dance and sports such as gymnastics, as they provide free movement, and also show the lines of the body.

fabrics are also important properties in ensuring garment comfort with movement. A fabric that is too thick or stiff to move freely or one that sticks to the surface of the body or rubs against the skin can create discomfort.

Often the greatest demands related to movement and clothing are imposed in the process of donning and doffing the garment. Depending on the closeness of the garment form to the body, and the type and position of the fastenings of the garment, donning can require extensive and elaborate movements (Tucker & Dugas, 2008).

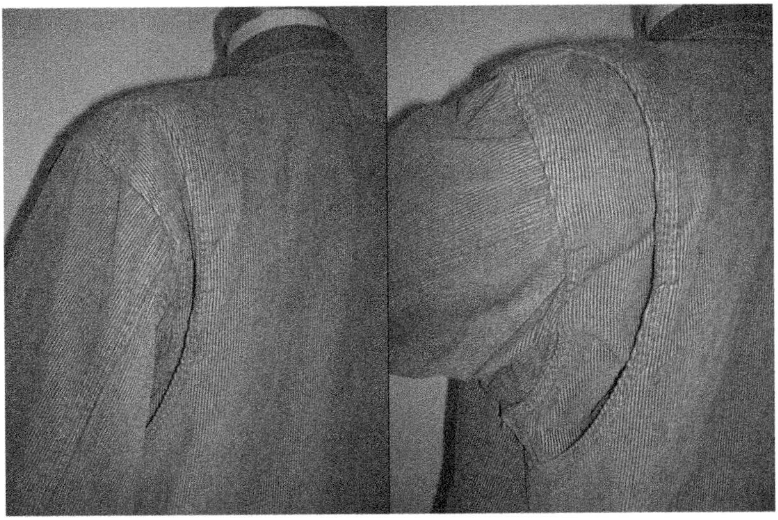

11.7 Design features can provide flexibility for ease of movement while maintaining the basic silhouette of the garment. This jacket from Duluth Trading Company has both a pleat in the shoulder area and a gusset under the arm to increase the range of movement.

11.2 Fashion and functional apparel: aesthetics, protection, performance and movement

Needs and purposes of various clothing ensembles, the actions performed by the wearer, and the environment determine the ideal relationship of the garment to the body in each case. A garment worn by a dancer may require free movement, and may also make use of the properties of the fabric to extend and exaggerate the movement of the body with flowing or fluttering extensions of light fabric. On the other hand, the worker in an industrial setting may also need free movement, but in garments close to the body so that they do not catch in the moving parts of machinery. An elite athlete needs clothing that not only accommodates movement but also extends and supports the movement of the body in specific ways, and protects and impedes the body from movements in other ways (for example, a knee brace that prevents the knee from hyperextending). Clothing designed for protection from various hazards, from thermal to chemical to projectile threats, generally presents the most challenging design problem in relation to free movement. Fabrics designed to protect the body are frequently stiff, inflexible, and can be quite thick and heavy, all properties that cause problems in designing garments that will move freely. In addition, protective garments are often designed in one piece in order to minimize or eliminate environmental challenges to any part of the body. They are frequently attached to protective gloves, boots, and helmets to create a totally encapsulating environment, further restricting movement. Finding the appropriate balance between movement free enough to perform tasks and maintain a level of comfort, and at the same time provide the needed level of protection, can be very difficult. Due to the range of variation in body size and shape in the population, and the expense of providing custom fit or a wide variety of sizes in protective clothing, users are further challenged by wearing clothing and equipment that is improperly sized for their body leading to fit issues. Women who work in fields traditionally limited to men, who need protective clothing, often must wear garments designed for male proportions that result in fit issues, and restrict or impede movement.

Movement issues in designing fashion clothing or clothing for daily wear can also differ greatly depending on the goals and needs associated with the garment. Most modern clothing is designed to fit closely to the body, and depending on the age and body type of the wearer, fit preference can dictate skin tight garments. The introduction of new fabric properties, such as the introduction of Lycra® stretch into many different types of fabrics (allowing closer fit without restricting movement) can influence the dictates of fashion. At the other end of the spectrum, garments worn in daily life can also have the purpose of hiding the shape of the body (an extreme example of this is the burka) and are therefore cut to stand away from the body. In this case, the body moves within the garment, instead of the garment moving with the body.

11.3 Materials and design strategies to provide appropriate movement performance

Fabric properties and garment design are the two primary, interacting factors that contribute to, or impede garment comfort. Material properties that have an effect on movement include weight, thickness, stiffness, stretch, and recovery. Fabrics inherently have a certain degree of flex, but the amount varies greatly. Woven fabrics are generally more rigid than knits, as knitted fabrics have inherent flexibility due to the interlaced structure of the yarns in knits. According to Hatch (1993), the appropriate range of percent elongation for textile materials for tailored clothing is between 15%–25%, for sportswear is between 20%–35%, for active wear is between 35%–50%, and for form fitting garments is between 30%–40%. Joseph (1981) identifies the amount of elongation for regular wear as between 10% and 25%, and for more active wear between 35% and 50%. It is essential that the fabrics also have a high level of recovery, generally in the range of 95% to 98%.

Fabrics used in protective clothing are often extremely rigid compared to these ranges, due to the thickness of yarns with desired properties and fabric structures or coatings designed to protect against environmental challenges. Overall, the interaction of fabric properties, garment design, and sizing and fit determine the range of movement possible in a garment (Branson & Nam, 2008).

One common solution to provide movement, when more rigid fabrics are desired for their specific properties, is to design sections in a garment that are more flexible in areas where more movement is needed. This can be done by introducing more flexible fabrics (i.e., a stretch fabric in a panel under the arm in a rain jacket), by increasing the flexibility of a fabric with design features (i.e., pleating or gathers), or by treating a fabric differently in different areas (i.e. a knitted gardening glove with rubberized areas and untreated, more flexible areas). Incorporating flexible areas in a garment design is a particularly useful strategy when a garment must be tethered to the body, and therefore cannot move freely over the surface of the body. Fourt and Hollies (1969) discuss a variety of ways that clothing can be designed with variations for the different needs within the same garment at different locations on the body.

Another design element often used to provide 'slip' of the garment over the body as the wearer moves is the use of lining fabrics with little friction. Such linings can add greatly to the comfort of a garment, reducing the 'grab' of fabrics and seam areas that have greater frictional force that can impact the skin or layers of garments. Garment layers binding together because of frictional drag forces can increase the difficulty of movement. Teitlebaum and Goldman (1972), in a study of a seven layer arctic clothing system, found that metabolic cost was approximately 16% greater than the metabolic cost when wearing a combination of underwear and one layer of clothing, and attributed this increase to either the 'friction drag' between layers or the interference with joint movement produced by the bulk of the clothing. It is important in a garment of many layers (or in ensembles made up of many layers of clothing) to consider the frictional properties of the layers, and also to carefully engineer the fit of each layer on top of the previous one, so that outer layers are not too tight and movement is not compromised (see Fig. 11.8).

The amount of frictional force of a fabric can increase as the moisture in the fabric and on the skin increases as a result of sweating, resulting in both thermal discomfort and binding of the garment with movement. Restrictions to movement from a garment also have the effect of increasing the metabolic cost of wearing the garment overall. On the other hand, in a well designed and well fitted garment the 'bellows effect' generated by movement can decrease the thermal load. Air in the garment structure can be moved through the openings of the garment in a pumping action that ventilates the body (Bittel et al., 1992; Dukes-Dobos et al., 1992).

Proper fit is also essential to create garments of any type that move and balance well on the active body. Good design of clothing requires the development of garment shapes that provide proper ease (the added circumference or length of the garment that allows the body to move) and proper set (the 'balance' of the garment that keeps it in place so that the interaction of gravity and frictional properties of the fabric do not displace the garment on the body with movement). This can be difficult to achieve, especially in a heavy garment with many layers such as a firefighter's turnout gear. Multiple prototypes and testing of the garment in active positions are necessary to create a well balanced and well fitted garment appropriate for a wide range of body sizes and proportions.

One design strategy that works well for activewear and some types of workwear is to create pattern shapes for active positions. Bike shorts that fit and move well

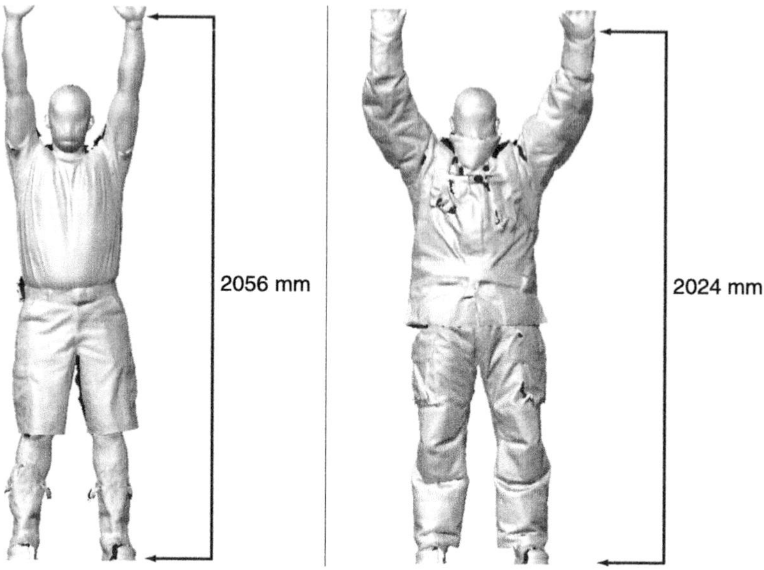

11.8 Firefighters' uniforms are made of many layers of fabric which are carefully engineered to provide thermal protection, which can impede movement. This uniform has reduced the range of movement of the wearer. Image courtesy of the Cornell Bodyscan Research Group.

when the wearer is on the bicycle, but are baggy and bind when the wearer walks are a good example of this design strategy (see Fig. 11.9). Well designed gloves will generally have fingers that are curved, as most of the movements of the fingers and hand are in a flexed position. A glove that is designed in a flat configuration will require the wearer to overcome the resistance to flex the glove to the neutral hand and finger position, before the further flexion required for most hand activities begins.

Many other creative design strategies can be applied to increase movement in clothing, such as overlapping segments or stacking segments of inflexible materials next to one another, isolating segments from one another and tethering them directly to the moving body part, or creating rigid joints that can roll, slide, rotate, or twist (Watkins, 1995).

11.4 Movement and garment stretch/pressure/compression

The use of stretch fabrics has, from the earliest times, been a primary design strategy for providing comfortable movement in garments. Denton (1971) classifies stretch garments in three categories: comfort stretch in which the garment does not conform too closely to the body, but will stretch comfortably

11.9 Patterns for bike shorts are shaped in an active position so that they fit well when the rider is on the bike. The shorts without the body inside will be the same bent shape as when they are on the body. Image courtesy of Spencer Ritenour.

with movement, stretch-to-fit garments that fit closely to body contours but that do not apply significant pressure to the body, and power stretch garments that exert compressive pressure on the body. Power stretch fabrics used for these garments are generally warp knitted fabrics with inserted elastic, and are effective under higher loads than comfort stretch fabrics (Joseph, 1981).

The knit structure in fabrics provides inherent stretch as it is a looped structure created from continuous threads that are free to shift and move under pressure, but will recover their shape when the pressure is released. Knitted garments will therefore move with body movement. Some of the earliest uses of knit garments were to provide effective movement in working garments; for example fishermen's jerseys. Modern fibers can also provide stretch in woven fabric structures, and can improve the stretch and recovery properties of knitted fabrics. Much of the movement comfort of all types of apparel, from the common t-shirt to athletic wear to protective clothing is due to the use of stretch fabrics of various types.

Fabrics and films with high modulus and good recovery are also used in modern athletic apparel to interact with the movement of the body in other ways. Braces on knees or elbows can either protect the joint from hyperextension or provide support to a joint whose function has been compromised by a previous injury. These devices provide many amateur and aging athletes as well as professional athletes with much needed support, without which their activities would be compromised.

Compression to limit body movement for elite athletes is also used in another way in modern athletic gear. Compression garments made of power stretch materials, or bands of flexible but high modulus materials incorporated into garments, provide support for working muscles, and are reputed to reduce muscle fatigue and provide warmth to the muscles, resulting in more efficient performance (see Fig. 11.10). There is evidence that use of these garments can improve the cellular processes that repair structural damage to the skeletal muscle following eccentric exercise (Trenell et al., 2006). Popular press articles claim that these garments provide muscle support and the reduction of vibration of muscles, reduction of unwanted movement of both muscle and fat tissue, a reduction of muscle damage, and increased strength and control. Articles on these garments

11.10 Many athletes feel that compressive garments improve both comfort and performance. This garment is made of power stretch material, unlike that in Fig. 11.6 which is made of comfort stretch material. Image courtesy of Skins™ USA.

also claim that they aid recovery by improving capillary circulation, wicking moisture away from the skin and improving evaporation to promote cooling, enhance body posture to optimize core strength, and provide a psychological advantage (Bernhardt & Anderson, 2005; Bringard et al., 2006; Duffield & Portus 2007; McCarthy, 2009). Though many of these performance claims are not supported in controlled studies, it is clear from their popularity with athletes that they offer increased comfort. Another mechanism that may contribute to both the comfort and performance of these athletic garments is improved proprioception of joint movement. Joint proprioception is the ability to perceive and reproduce joint movement and position in space (Peerlau et al., 1995), an important factor in maintaining the most effective form for elite athletic endeavors.

Clothing solutions that extend and improve movement capabilities are also desired by elite athletes. Swimmers now have garments with stretch properties custom tailored to their body for idealized fit to provide the best support, the most effective movement through the water, and to compress their body for a more dynamic shape. These innovations can improve performance, and have generated much controversy (Tucker & Dugas, 2008). Restrictions on the amount of body coverage and the material properties of these swimsuits have recently been introduced due to the perception that these suits give some athletes an unfair advantage in competition (France 24 International News, 2009).

11.5 Research and testing of prototype designs for comfort and movement

Little research has been done on the change in body measurements in active positions. An early study by Kirk and Ibrahim (1966) measured the skin stretch over bending joints by marking the skin above and below a joint with the joint in an extended position, and then measuring the change in the linear measure over the joint in a flexed position. The development of the 3D body scanner as a research tool has made such studies much more reliable, effective, and affordable, as many subjects can be scanned in a short span of time. Many scan studies have incorporated scans of study participants both standing and seated (Brunsman et al., 1997; Bougourd et al., 2000). However, the seated scans are often included for analysis independently of the standing scan for use by the automotive and airline industry, and by the military for cockpit design. Studies designed to compare body measurement changes require careful preparation to acquire reliable and valid measures (see Fig. 11.11) (Lee & Ashdown, 2005; Na & Ashdown, 2008; Choi & Ashdown, 2010). Accurate and appropriate landmarking is essential in these studies, and more work is needed on determining the optimal measurements that will contribute to the design of clothing. The development of methods of taking valid measurements that capture the essential information related to movement in comparative body configurations that change greatly in both dimension and shape is also a challenge.

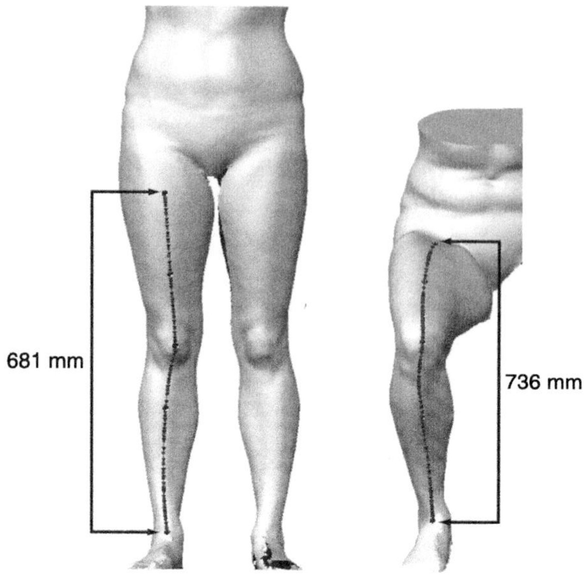

11.11 3D body scan studies are ideal for investigations of change in body surface measurements in active positions. Image courtesy of the Cornell Bodyscan Research Group.

On the other hand, methods for testing and comparing the performance of garments as the wearer moves both for understanding thermal interactions and for effective range of movement are well developed. Fit testing of clothing prototypes on human fit models is common practice in the apparel industry to refine patterns before clothing is manufactured (Bye & LaBat, 2005). Fit models sit, walk, and in the case of active wear, test the prototype by engaging in the sport for which it is designed. In the research community both controlled fit studies and wear tests that challenge the garments with actual working conditions are common practice in the design and testing of functional apparel. Studies are also common using instrumented mannequins.

Lab studies that limit the variables under consideration can be very useful. For example, Weder, Zimmerli, and Rossi (1996) used a rudimentary moving sweating arm to test heat transfer and water vapor permeability, by varying temperature, humidity, sleeve width, and whether the sleeve is open or closed at the wrist. Use of this apparatus made it possible to test for the bellows effect generated by movement in a controlled study.

Testing of full garment structures or outfits requires the use of a moving sweating thermal mannequin (Chen et al., 2004). A recent publication from the National Academies describes moving sweating thermal mannequins developed in the United States (Uncle Wiggly and Paul developed by the military, Coppelius at North Carolina State, and Newton and Huey developed by Measurement

Technology Northwest), Japan (TOM III and SAM developed by the Toyobo Corporation and JUN developed by the Bunka Women's University), Hong Kong (Walter developed by Jintu Fan at the Hong Kong Polytechnic Institute), and Switzerland (Sweating Agile Thermal Manikin [SAM] developed by the Swiss Federal Laboratories for Materials Testing and Research). The movements of which these mannequins are capable are generally limited to one plane and are externally inducted. They are designed to duplicate walking, and in some cases climbing motions, and for testing of thermal comfort only, not range of movement or for the testing of other active positions. Development of a robotic testing device that duplicates the joint motion and range of motion of a human, with the movements internally inducted, could result in a reliable lab-based test of clothing ensemble performance for range of motion (Fecht & Bennett, 1992; The National Academies, 2008). However, Goldman (2006) suggests that the effect of pumping on clo values can be reliably measured using wind speed, eliminating the need for a walking mannequin.

Testing garment prototype designs for movement with human participants has the advantage of the ability to duplicate the conditions of actual use. However, in these tests it can be difficult to control variables such as variations in the fit of the garment and in the physical capabilities of the participants (i.e. inherent range of movement, strength). Havenith and Hues (2004) describe a battery of tests for assessing the fit and performance of protective clothing, including comfort, using tests of firefighter's gear. The use of a within-subject test design, in which each study participant wears all the suits being tested and in which each participant acts as his own control is used to reduce the effects of variations in body size, strength, and metabolism.

Studies that correlate ease values in clothing both to subjective responses about the comfort and function of the garment and to functional tests are common, but are generally inconclusive because of the difficulty of defining both static (in the anthropometric position) and active fit. A study by Tremblay-Lutter and Weihrer (1996) on the optimal ease value for glove fit demonstrates some of the complexities involved.

The effects of different garments or garment assemblies on range of motion have been studied using a variety of goniometers and flexometers (see Fig. 11.12). Saul and Jaffe (1955) and Lockhart and Bensel (1977) conducted tests on military clothing for cold weather conditions to measure effects on mobility, and Huck (1988) tested different designs of firefighter turnout gear, and in another study (1991) she studied different sleeve configurations for firefighter turnout gear. Adams and Keyserling (1996) compare three different tools for testing range of motion and conclude that the universal goniometer was the most effective tool, though placing the tool reliably on the joint centers and limbs could be compromised by the clothing.

Clothing that is well designed to accommodate movement will still fail if the sizing system does not provide the range of sizes needed for the population of users, or if the user has an incorrect size. The American Society of Testing and

11.12 A goniometer can be used to measure body angles.

Materials (ASTM) developed the F-1154 standard for evaluating the comfort, fit and function of chemical protective suits (ASTM, 1988). This standard was designed to guide selection of appropriate protective equipment, and also allow consistent testing of protective clothing products. The development of these standards also gives manufacturers a way of assessing their product against their competitors, and can improve the quality and design of protective clothing overall (Abeysekera, 1992). In ASTM Standard F-1154 a set of eight standard range of movement body positions and actions (i.e. kneeling, raising arms above the body), and a set of seven work related activities (i.e. using a hand truck to move a barrel) are provided as standard motions to test the garment. A post test questionnaire is used to measure comfort, and an inspection of the garment for abrasion, punctures or tears is conducted to assess the comfort and function of the garment.

Studies of clothing comfort and function in active positions and the ASTM standard have generally focused on the development of a range of motion tests and standardized movements that are designed to explore the full range of motions of which the human body is capable. However, studies that investigate the activities in which an actual user of the clothing will engage for a specific purpose can give more valid results. In a field test evaluation of ASTM F-1154, Veghte and Storment (1992) determined that the standard was not a rigorous enough test for firefighters. The activities of different user groups can be studied to develop more focused activities for testing through observation, interviews, viewing training videotapes, and photographing the activity demonstrated by the users (Boorady et al., 2009).

Sizing standards can also educate and encourage manufacturers of functional clothing to develop effective sizing standards that will provide good fit for the majority of users. The ANSI/ISEA Standard 101 – 1996 (R2002) was developed to recommend minimum sizing for disposable protective coveralls (American National Standards Institute, 2002). However, a study of the fit of these coveralls on 166 study participants found that the smaller and larger participants were not well fitted by the size range, highlighting the difficulty of creating effective sizing systems (Keeble et al., 1992).

Creating well fitting, comfortable protective garments is a challenge for designers, as so many factors emerge in the interaction of the body, materials, the garment, and auxiliary equipment (Ashdown & Watkins, 1996; Daanen & Reffeltrath, 2008). Methods of testing design prototypes for the ability of the wearer to move comfortably can be of great value to the designer, suggesting new ways of patterning or constructing the garment. One method that has been used to test prototypes is to create a series of slashes over the garment surface and observe where these areas open when a wearer moves (Crow & Dewar, 1986; Ashdown & Watkins, 1992; Kohn & Ashdown, 1998). This is an effective way to observe even very slight interactions of the moving body and the clothing.

Traditional fit testing is an effective way both to compare the performance of designs or design prototypes, and to test an existing product to determine where there are issues that need to be addressed in a new design. Appropriate design of a fit study requires analysis of the body size of the participants in the study compared to the garments, collection of data from the user perspective, and analysis of the garment performance by fit judges (McConville, 1986).

11.6 Future trends

The development of reliable full body 3D scanning has provided a new tool that can be of great value for the study of the interaction between the body and clothing fit in active positions. Studies that use captured 3D models of the unclothed and the clothed body can show the actual relationship between body and clothing in new ways (see Fig. 11.13). Issues still remain in developing reliable methods to capture identical postures in the clothed and unclothed state, and in automating measurement analysis software so that measurements of the scans can be taken reliably and accurately, but much valuable information can be discovered for the designer of clothing using current tools (Dannen et al., 1997; Nam et al., 2005; Lee et al., 2006; Nam et al., under review).

New developments in pressure sensing equipment will also impact these studies. The development of films and fabrics with pressure sensing capability will allow analysis of the pressure of clothing where it impacts the body in movement, and of power stretch garments and the relationship between comfort and control in body shaping garments for aesthetics and athletic activities.

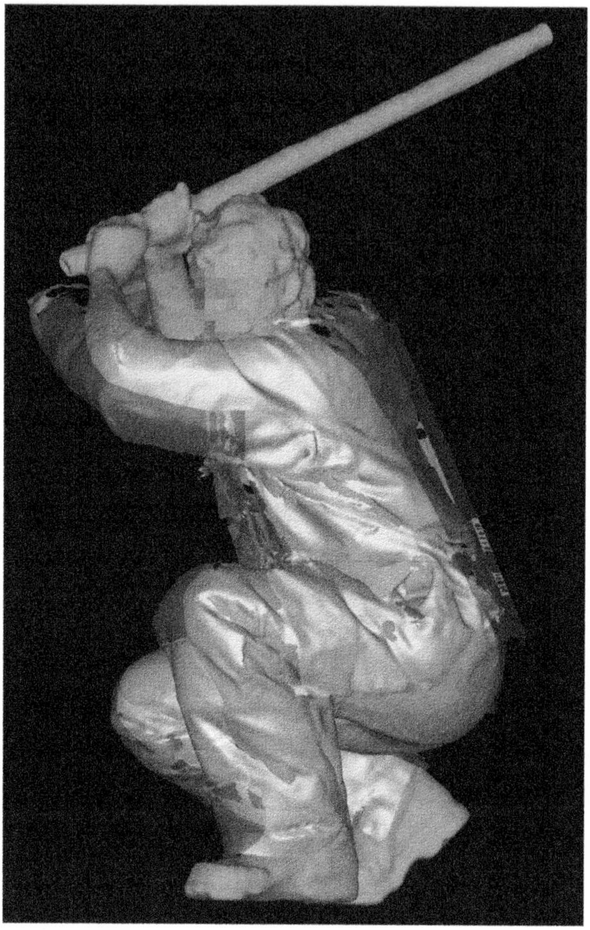

11.13 Merged 3D scans of this firefighter in minimal clothing, and in uniform, show the relationship between the garment and the body in an active position. Note the different angle of the arm. Image courtesy of the Cornell Bodyscan Research Group.

The explosive pace of development of new materials with desirable properties and new functionalities will provide designers with a whole range of solutions for designing effective protective clothing. Many new materials are under development that will provide lighter weight, and will incorporate new levels of flexibility and stretch in protective materials. New technologies under development in the apparel industry are also introducing automated affordable customized fit. This is a promising new area that could eventually eliminate the discomfort of badly fitting clothing for the great proportion of the population that cannot currently find well fitting clothing in ready-to-wear (Loker & Ashdown, 2007: Loker et al., 2008).

New development in virtual fit technology is also underway. Apparel companies are already using 3D virtual fit tools to reduce the number of prototypes needed in the design of apparel. Static 3D models are very effective at visualizing fit and the pressure of the garment on the body for ordinary ready-to-wear clothing (Hochfelder, 2009). However in order to be able to test the fit and function of protective clothing a more accurate and reliable virtual avatar is needed. Work is underway funded by the armed forces to develop such an avatar at the University of Iowa in the virtual soldier program. This research group has developed the avatar Santos, a biomechanically accurate mode and simulation of a (physics-based) human. This work has been commercialized in the creation of SantosHuman Inc. (http://www.santoshumaninc.com). Currently Santos has physics based joint movement, but work is still needed on creating the skin properties and muscle movements of an actual person. The ultimate goal of this group is to be able to accurately model clothing interactions virtually with the body avatar in motion. Work is also underway to model clothing using particle based methods, continuum finite element shell modeling, and beam grid finite element modeling in order to identify the most effective approach (Man & Swan, 2007).

Development in creation of a real time moving avatar for the apparel industry is ongoing and relatively effective garment and fabric parameters and real time generation of movement for runway depictions of fashion garments are now possible. The next important step is to create actual physics-driven animation of fabrics and garments on avatars (see Fig. 11.14). Though work on realistically creating fabric drape virtually has been underway for many years, recent developments in computing power and visualization hardware are bringing virtual fabric draping closer to reality. One example of a physics-driven virtual fabric can be seen at http://www.cs.cornell.edu/~srm/publications/SG08-knit.html (Kaldor

11.14 Physics based modeling of fabric performance and of body movement will eventually allow virtual testing of clothing to assess and improve comfort. This image shows a virtual scarf dropped on a surface, and a virtual legwarmer (Kaldor, James, & Marschner, 2008). Images courtesy of Steve Marschner.

et al., 2008). When these technical issues are resolved it will be possible to truly model clothing comfort on an active body.

11.7 Sources of further information and advice

Pansky's text, *Dynamic Anatomy and Physiology* (1975) has detailed, well written information about body movement. Appropriate sizing and fit is essential for clothing comfort, and Ashdown's (2008) book on sizing and Fan, Yu, and Hunter's (2004) book on clothing fit are good introductions to these topics. The classic textbook on functional apparel design by Susan Watkins, *Clothing: The Portable Environment* (1995) has an excellent thought provoking chapter on design for movement. Kathryn Hatch's basic textile science book published in 1993, *Textile Science*, has a chapter on comfort in clothing, for the beginning textile science student.

11.8 References

Abeysekera, J. D. A., 1992. 'Some ergonomics issues in the design of personal protective devices,' *Performance of Protective Clothing: 4th Volume*, ASTM STP 1133, James P. McBriarty and Norman W. Henry, eds, Philadelphia, PA: American Society for Testing and Materials, pp. 651–659.

Adams, P. S. & Keyserling, W. M., 1996. 'Methods for assessing protective clothing effects on worker mobility,' *Performance of Protective Clothing: 5th Volume*, ASTM STP 1237, James P. McBriarty and Norman W. Henry, eds, Philadelphia, PA: American Society for Testing and Materials, pp. 311–326.

Aldrich, W., Smith, B., & Dong, F., 1998. 'Obtaining repeatability of natural extended upper body positions,' *Journal of Fashion Marketing and Management*, **2**(4), 329–351.

American National Standards Institute, 2002. ANSI/ISEA Standard 101 – 1996 (R2002). *American National Standard for Limited-Use and Disposable Coveralls – Size and Labeling Requirements*.

Ashdown, S. P., ed., 2008. *Sizing in Clothing; Developing Effective Sizing Systems for Ready-To-Wear Clothing*, Woodhead Publishing Limited, Cambridge.

Ashdown, S. P. & Watkins, S. M., 1992. 'Movement analysis as the basis for the development and evaluation of a protective coverall design for asbestos abatement,' *Performance of Protective Clothing: 4th Volume*, ASTM STP 1133, J. P. McBriarty, & N. W. Henry, eds, Philadelphia, PA: American Society for Testing and Materials, pp. 600–674.

Ashdown, S. P. & Watkins, S. M., 1996. 'Concurrent engineering in the design of protective clothing: interfacing with equipment design,' *Performance of Protective Clothing: Fifth Volume*, American Society of Testing and Materials STP 1237, J. S. Johnson and S. Z. Mansdorf, eds., Philadelphia, PA: American Society for Testing and Materials, pp. 471–485.

ASTM, 1988. 'Standard practices for qualitatively evaluating the comfort, fit, function, and integrity of chemical-protective suit ensembles,' ASTM STP 1154 in *Performance of Protective Clothing: Second Symposium*, eds S. Z. Mansdorf, Richard Sagar and Alan P. Nielsen, Philadelphia, PA: American Society for Testing and Materials, pp. 248–252.

Bernhardt, T. & Anderson, G. S., 2005. 'Influence of moderate prophylactic compression on sport performance,' *Journal of Strength Conditioning Research*, **19**(2), 292–297.

Bittel J., Hanniquet, A. M., & Forssard, H., 1992. 'Thermal constraints related to the wearing of protective clothing: Body ventilation by fresh air,' *Performance of Protective Clothing: 4th Volume*, ASTM STP 1133, James P. McBriarty and Norman W. Henry, eds, Philadelphia, PA: American Society for Testing and Materials, pp. 597–603.

Boorady, L., Rucker, P., Haise, C., & Ashdown, S. P., 2009. 'Analysis of active positions and PPE needs of agricultural workers,' *Journal of Textile and Apparel, Technology and Management*, **6**(2). http://ojs.cnr.ncsu.edu/index.php/JTATM/article/viewFile/646/455.

Bougourd, J. P., Dekker, L., Ross, P. G., & Ward, J. P., 2000. 'A comparison of women's sizing by 3D electronic scanning and traditional anthropometry,' *Journal of Textile Institute*, **91**, Part 2(2), 163–173.

Branson, D. & Nam, J., 2008. 'Materials and sizing,' *Sizing in Clothing; Developing Effective Sizing Systems for Ready-To-Wear Clothing*, S. P. Ashdown, ed., Woodhead Publishing Limited, Cambridge.

Branson, D. H. & Sweeney, M., 1991. 'Conceptualization and measurement of clothing comfort: Toward a metatheory,' *Critical Linkages in Textiles and Clothing Subject Matter: Theory, Method, and Practice*, ITAA Special Publication #4. Monument CO: International Textile and Apparel Association, pp. 94–105.

Bringard, A., Perrey, S., & Belluye, N., 2006. 'Aerobic energy cost and sensation during submaximal running: Positive effects of wearing compression tights,' *International Journal of Sports Medicine*, **27**, 373–378.

Brunsman, M. A., Daanen, H., & Robinette, K. M., 1997. 'Optimal postures and positioning for human body scanning,' *Proceeding of International Conference on Recent Advances in 3-D Digital Imaging and Modeling*, IEEE Computer Society Press, Los Alamitos, CA.

Bye, E. & LaBat, K., 2005. 'An analysis of apparel industry fit sessions,' *Journal of Textile and Apparel, Technology and Management*, **4**(3). Available at http://www.tx.ncsu.edu/jtatm/volume4issue3/articles/Bye/Bye_full_129_05.pdf

Chen, Y. S., Fan, J., Qian, X., & Zhang, W., 2004. 'Effect of garment fit on thermal insulation and evaporative resistance,' *Textile Research Journal*, **74**(8), 742–748.

Choi, S. Y. & Ashdown, S. P., 2010. '3D body scan analysis of dimensional change in lower body measurements for active body positions.' *Textile Research Journal* Online First, August 13, 2010, http://trj.sagepub.com/. Forthcoming in print issue, *Textile Research Journal*.

Crow, R. M. & Dewar, M. M., 1986. 'Stresses in clothing as related to seam strength,' *Textile Research Journal*, August, pp. 467–473.

Daanen, H. A., Brunsman, M. A., & Robinette, K. M., 1997. 'Reducing movement artifacts in whole body scanning,' *Proceedings of International Conference on Recent Advances in 3D Digital Imaging and Modeling*, Ottowa, Ontario, Canada, pp. 262–265.

Daanen, H. A. M. & Reffeltrath, P. A., 2008. 'Function, fit and sizing,' *Sizing in Clothing*, S. P. Ashdown, ed., Woodhead Publications, Oxford, pp. 202–219.

Denton, M. J., 1971. 'Fit, stretch, and comfort,' paper presented at the Third Shirley International Seminar 'Textiles for Comfort', The Cotton Silk and Man-made Fibres Research Association, Shirley Institute, Manchester, England, pp. 1–11.

Duffield, R. & Portus M., 2007. 'Comparison of three types of full-body compression garments on throwing and repeat-sprint performance in cricket players,' *British Journal of Sports Medicine*, **41**, 409–414.

Dukes-Dobos, F. N., Reischl, U., Buller, K., Thomas, N. T., & Bernard, T. E., 1992. 'Assessment of ventilation of firefighter protective clothing,' *Performance of Protective*

Clothing: 4th Volume, ASTM STP 1133, James P. McBriarty and Norman W. Henry, eds, Philadelphia, PA: American Society for Testing and Materials, pp. 629–633.

Fan, J., Yu, W., & Hunter, L., eds., 2004. *Clothing Appearance and Fit: Science and Technology*, Woodhead Textiles Series No. 33, Woodhead Publishing Limited, Cambridge.

Fecht, B. A., Bennett, D. W., 1992. 'Robotic mannequin technology for enhanced product testing,' *Performance of Protective Clothing: 4th Volume*, ASTM STP 1133, James P. McBriarty and Norman W. Henry, eds, Philadelphia, PA: American Society for Testing and Materials, pp. 734–741.

Fourt, L. & Hollies, N. R. S., 1969. 'The comfort and function of clothing,' *Technical Report 69-74-CE*, Clothing and Personal Life Support Equipment Laboratory, Harris Research Laboratories Division of Gillette Research Institute, Rockville, MD, pp. 68–72.

France 24 International News, 2009. 'No more polyurethane swimsuits from 2010,' http://www.france24.com/en/20090724-ban-polyurethane-swimsuits-swimming-world-record-bernard-phelps (accessed 29 July, 2009).

Gilling, D. R., 1971. 'Assessment of the comfort properties of military clothing,' paper presented at the Third Shirley International Seminar 'Textiles for Comfort', The Cotton Silk and Man-made Fibres Research Association, Shirley Institute, Manchester, England.

Goldman, R. F., 2006. 'Thermal manikins, their origins and role,' *Thermal Manikins and Modeling*, edited by J. Fan, Hong Kong Polytechnic University, Hong Kong, pp. 3–18.

Hatch, K., 1993. *Textile Science*, Minneapolis: West Publishing.

Havenith, G. & Hues, R., 2004. 'A test battery related to ergonomics of protective clothing,' *Applied Ergonomics*, **35**, 3–20.

Hewes, G. W., 1957. 'The anthropology of posture,' *Scientific American*, **196**(2), 123–132.

Hochfelder, B., 2009. 'Stylin' and profilin' at very high speeds,' *Advanced Imaging*, June 2009, pp. 21–22.

Holmer, I., 1988. 'Thermal properties of protective clothing and prediction of physiological strain,' *Performance of Protective Clothing, Second Symposium*, ASTM STP 989, S. Z. Mansdorf, R. Sager, and A. P. Nielsen, eds, Philadelphia, PA: American Society for Testing and Materials, pp. 82–86.

Huck, J., 1988. 'Protective clothing systems: A technique for evaluating restriction of wearer mobility', *Applied Ergonomics*, **19**(3), 185–190.

Huck, J., 1991. 'Restriction to movement in fire-fighter protective clothing: evaluation of alternative sleeves and liners,' *Applied Ergonomics*, **22**(2), 91–100.

Joseph, M. L., 1981. *Introductory Textile Science*, Holt, Rinehart, & Winston, New York.

Kaldor, J., James, D. L. & Marschner, S. 2008. 'Simulating knitted cloth at the yarn level,' *Proceedings of SIGGRAPH 2008*, Los Angeles, California, August. http://www.cs.cornell.edu/~srm/publications/SG08-knit.pdf

Keeble, V. B., Prevatt, M. B., & Mellian, S. A., 1992. 'An evaluation of fit of protective coveralls manufactured to a proposed revision of ANSI/ISEA 101,' *Performance of Protective Clothing: 4th Volume*, ASTM STP 1133, James P. McBriarty and Norman W. Henry, eds, Philadelphia, PA: American Society for Testing and Materials, pp. 675–691.

Kirk, W. & Ibrahim, S. I., 1966. 'Fundamental relationship of fabric extensibility to anthropometric requirements and garment performance,' *Textile Research Journal*, **36**, 37–47.

Kohn, I. & Ashdown, S. P., 1998. 'Use of video capture and image analysis to quantify the fit of apparel,' *Textile Research Journal*, **68**(1), 17–26.

Lee, J. & Ashdown, S. P., 2005. 'Upper body change analysis using 3-D body scanner,' *Journal of the Korean Society of Clothing and Textiles* (English Edition), **29**(12), 1595–1607.

Lee, Y. A., Ashdown, S. P., & Slocum, A. C., 2006. 'Measurement of surface area of 3-D body scans to assess the effectiveness of hats for sun protection,' *Family and Consumer Sciences Research Journal* **34**(4), 366–385. Available at http://fcs.sagepub.com/cgi/reprint/34/4/366

Lockhart, J. M. & Bensel, C. K., 1977. *The Effects of Layers of Cold Weather Clothing and Type of Liner on the Psychomotor Performance of Men*, Report No. NATICK/TR-77/018, US Army Natick Research & Development Command, Natick, MA.

Loker, S. & Ashdown. S.P., 2007. 'Virtual sensation: Dress online,' in *Dress Sense: Emotional and Sensory Experiences of the Body and Clothes*, Donald Clay Johnson and Helen Bradley Foster, eds., Berg Publishers, NY.

Loker, S., Ashdown. S. P., & Carnrite, E. (2008) 'Dress in the third dimension: On-line interactivity and its new horizons,' *Clothing and Textiles Research Journal*, **26**(2), 164–176.

Lotens, W. A., 1989. 'Optimal design principles for clothing systems,' in *Handbook on Clothing*, Research Study Group on Biomedical Research Aspects of Military Protective Clothing, eds, NATO, Brussels, pp. 1701–1715.

Man, X. & Swan, C. C., 2007. 'A mathematical modeling framework for analysis of functional clothing,' *Journal of Engineered Fibers and Fabrics*, **2**(3), 10–28.

Markee, N. L. & Pedersen, E. L., 1991. 'The conceptualization of comfort with regard to clothing,' *Critical Linkages in Textiles and Clothing Subject Matter: Theory, Method, and Practice*, ITAA Special Publication #4. Monument CO: International Textile and Apparel Association, pp. 81–93.

McCarthy, M., 2009. 'What are you wearing? Base layers go high tech,' *32 Degrees: The Journal of Professional Snowsports Instruction*. Spring issue, pp. 46–50.

McConville, J. T., 1986. 'Anthropometric fit testing and evaluation,' *Performance of Protective Clothing*, R. L. Barker & G. C. Coletta, eds, Philadelphia, PA: American Society for Testing and Materials.

Morris, M. A. & Prato, H. H., 1981. 'Consumer perception of comfort, fit and tactile characteristics of denim jeans,' *Textile Chemist and Colorist*, **13**(3), 60–66.

Na, H. & Ashdown, S. P., 2008. 'Comparison of 3-D body scan data to quantify upper body postural variation in older and younger women,' *Clothing and Textiles Research Journal*, **26**(4) 292–307.

Nam, J. H., Branson, D. H., Ashdown, S. P., Cao, H., and Carnrite, E. (forthcoming). 'Analysis of ease values and fit of liquid cooled vests from 3D body scan data taken in working positions,' *International Journal of Clothing Research and Technology*.

Nam, J., Branson, D. H., Ashdown, S. P., Cao, H., Jin, B., & Peksoz, S., 2005. 'Fit analysis of liquid cooled vest prototypes using 3D body scanning technology,' *Journal of Textile and Apparel, Technology and Management*, **4**(3), available at http://www.tx.ncsu.edu/jtatm/volume4issue3/articles/Nam/Nam_full_138_05.pdf

The National Academies, 2008. *Soldier Protective Clothing and Equipment: Feasibility of Chemical Testing Using a Fully Articulated Robotic Mannequin*. Authored by the Board on Chemical Sciences and Technology (BCST) and the Earth and Life Studies (DELS), The National Academies Press, Washington, DC, available at http://www.nap.edu/catalog.php?record_id=11959#toc

Pansky, B., 1975. *Dynamic Anatomy and Physiology*, Macmillian Publishing Company, New York.

Paoletti, Jo B., 1991. 'Wanted: An interdisciplinary definition of clothing comfort,' *Critical Linkages in Textiles and Clothing Subject Matter: Theory, Method, and Practice*, ITAA

Special Publication #4. Monument CO: International Textile and Apparel Association, pp. 106–108.

Perlau, R., Frank, C., & Fick, G., 1995. 'The effect of elastic bandages on human knee proprioception in the uninjured population,' *The American Journal of Sports Medicine*, **23**(2), 251.

Pontrelli, G. J., 1990. 'Comfort by Design,' *Textile Asia*, **21**(1), 52–61.

Renbourn, E. T., 1971. 'The psychology of clothing with materials in mind,' paper presented at the Third Shirley International Seminar 'Textiles for Comfort', The Cotton Silk and Man-made Fibres Research Association, Shirley Institute, Manchester, England.

Saul, E. V. & Jaffe, J., 1955. *The Effects of Clothing on Gross Motor Performance*, Headquarters Quartermaster Research & Development Command, Technical Report EP-12.

Slater, K., 1986. 'Comfort or protection: The clothing dilemma,' *Performance of Protective Clothing: 5th Volume*, ASTM STP 1237, James S. Johnson and S. Z. Mansdorf, eds, Philadelphia, PA: American Society for Testing and Materials, pp. 486–497.

Smith, J. E., 1986. 'The comfort of clothing,' *Textiles*, **151**, 23–27.

Sontag, M. S., 1985. 'Comfort dimensions of actual and ideal insulative clothing for older women,' *Clothing and Textiles Research Journal*, **4**(1), 9–17.

Teitlebaum & Goldman, 1972. 'Increased energy costs with multiple clothing layers,' *Journal of Applied Physiology*, **32**(6), 743.

Tremblay-Lutter, J. F. & Weihrer, S. J., 1996. 'Functional fit evaluation to determine optimal ease requirements in chemical protective gloves,' *Performance of Protective Clothing: 5th Volume*, ASTM STP 1237, James S. Johnson and S. Z. Mansdorf, eds, Philadelphia, PA: American Society for Testing and Materials, pp. 367–383.

Trenell, M. I., Rooney, K. B., Sue, C. M., & Thompson, C. H., 2006. 'Compression garments and recovery from eccentric exercise: A 31P-MRS study,' *Journal of Sports Science and Medicine*, **5**, 106–114. Available at http://i.skins.net/UserFiles/File/trenell%20rooney%20published%20paper.pdf

Tucker, R. & Dugas, J., 2008. 'The Speedo LZR racer: feedback on a developing debate, and the value of technique,' *The Science of Sport*, available at: http://www.sportsscientists.com/2008/03/speedos-lzr-swimsuit.html, March 25, 2008 (Accessed 29 July, 2009).

Veghte, J. H. & S. Storment, 1992. 'Field test evaluation of ASTM Standard F-1154 with chemical protective suit ensembles,' *Performance of Protective Clothing: 4th Volume*, ASTM STP 1133, James P. McBriarty and Norman W. Henry, eds, Philadelphia, PA: American Society for Testing and Materials, pp. 311–321.

Watkins, S., 1995. *Clothing: The Portable Environment*, 2nd edn, Ames, IA: Iowa State University Press.

Weder, M. S., Zimmerli, T., & Rossi, R. M., 1996. 'A sweating and moving arm for the measurement of thermal insulation and water vapour resistance of clothing,' *Performance of Protective Clothing: 5th Volume*, ASTM STP 1237, James S. Johnson and S. Z. Mansdorf, eds, Philadelphia, PA: American Society for Testing and Materials, pp. 257–268.

Part III

Improving comfort in particular types of clothing

12

Evaluating the heat stress and comfort of firefighter and emergency responder protective clothing*

R. BARKER, North Carolina State University, USA

Abstract: In order to include heat stress and comfort as design parameters in the development of advanced protective ensembles for firefighters and emergency responders, it is necessary to be able to measure and evaluate these factors. This chapter will focus on describing the laboratory test methods that are currently available for this purpose, and on identifying research needed to advance the state of these testing technologies and protocols.

Key words: firefighter, emergency responder, protective clothing, sweating, manikin, hot plate.

12.1 Introduction

Many of the inadequacies of current protective gear for firefighters and emergency responders can be associated with heat stress and discomfort, resulting from hot, bulky protective garments. The additional need for barriers to protect against chemical and biological hazards increases the challenge of creating comfortable, functional protective clothing. Advanced protective ensembles must minimize heat stress while providing protection. The impact of protective clothing on heat stress depends on the extent to which the clothing affects the heat transfer between the responder and the environment. The breathability, or moisture vapor permeability of the clothing, can affect the evaporation of moisture from the body and heat exchange. Clothing weight, stiffness, and bulkiness add additional burden that can increase metabolic heat production in the stressful conditions typically associated with firefighting and emergency response.

Protective clothing systems for firefighters and other emergency responders impose, by their unique and often contradictory sets of properties, a metabolic and sensory burden that in many cases impedes performance and safety. Heat stress is a significant risk for first responders involved in strenuous activity, especially in hot and humid environments. The primary goal of testing and evaluation is to provide a systematic way to assess the impact of protective clothing on heat stress.

12.2 Background

A state-of-the-art testing approach for heat stress and comfort must be based on a multilevel concept, advancing from the measurement of constituent fabric heat/moisture transfer properties to, ultimately, the analysis of complete garment properties. Consequently, the following levels of testing and evaluation must be considered:

- Heat and moisture transport properties of fabrics.
- Heat and moisture transport of garments and predicted heat stress limits using a sweating instrumented manikin.
- Controlled, human wear trials in an environmental chamber.
- Field tests, conducted in actual use conditions.

These elements must be interwoven in stages to produce a database on the protective clothing system properties that translate to performance in the field.

12.3 Laboratory tests for clothing heat stress

Several testing technologies that have emerged can be used to critically assess materials. The most significant of these are guarded sweating plates for small samples and sweating thermal manikins for complete ensembles.

12.3.1 Sweating hot plate tests

Sweating hot plate tests are the most widely used instrumental means of measuring the heat and moisture vapor transfer properties of materials used in protective garments. These tests measure the dry thermal resistance (insulation) and the evaporative resistance of fabrics. Sweating hot plate test methodologies have been available for many years and are standardized by the American Society for Testing and Materials (ASTM) and the International Standards Organization (ISO). The most comprehensive standardized method is ASTM F1868, Standard Test Method for Thermal and Evaporative Resistance of Clothing Materials Using a Sweating Hot Plate.[4,17,22]

As a result of incorporation into NFPA standards, sweating hot plate methods are increasingly used in the United States to quantify the heat stress potential of materials used in protective clothing for firefighters and other emergency responders. ASTM F1868, Part C (Procedure for Total Heat Loss in a Standard Environment) is referenced in several NFPA standards, including NFPA 1951, 1971, 1977 and 1999, as a basis for specifying heat stress performance in protective clothing used by firefighters.[4,26–30]

12.1 Guarded sweating hot plate in environmental chamber.

A sweating plate test apparatus consists of a guarded flat plate housed in an environmental chamber. The plate is electrically heated to skin temperature (35 °C) and covered by the test material, with the side of the fabric that normally faces the human body in wear facing the plate. A guarded ring, heated to the same temperature as the test plate, prevents lateral heat loss. Water is fed to the surface of the test apparatus that is covered by a moisture vapor permeable cellophane sheet in order to shield the fabric from liquid water.

The entire test apparatus is housed in an environmental chamber to provide for testing in a controlled ambient condition. ASTM F1858, Part C, specifies ambient air temperature and humidity at 25 °C, 65% RH. These conditions create thermal heat loss through the test fabric, or heat loss that is influenced by both the dry thermal insulation and evaporative resistance of the test material. Measurements are made under both dry and simulated sweat wetted skin conditions: dry tests are conducted to determine conductive thermal resistance, while wet tests are conducted to determine apparent evaporative thermal resistance. The total heat loss of the test material is calculated using an equation that combines both conductive and evaporative heat transfers as:

$$Qt = \frac{10\,°C}{R_{ef} + 0.04} + \frac{3.57 kPa}{R_{ef}^{A} + 0.0035}$$

where Qt = total heat loss (W/m^2), R_{cf} = average intrinsic thermal resistance of the laboratory sample (km^2/W), and R_{ef}^{A} = average apparent intrinsic evaporative resistance of the laboratory sample (kPam^2/W).

The sweating plate method specified by NFPA standards is a non-isothermal test. Other sweating plate protocols, including ASTM F1868, Part B and ISO

11092,[4,22] measure evaporative resistance using isothermal conditions (the sweating plate and ambient air temperature are both set at 35 °C). The rationale for use of isothermal testing conditions is that it eliminates complications that can be involved in measuring evaporative resistance across a fabric due to moisture condensation in the fabric layers. However, the non-isothermal procedure, specified in the NFPA standards, has the advantage of being a more realistic simulation of total heat loss from sweating skin into a cooler environment.

12.3.2 Sweating manikin tests

Sweating hot plate tests, made on flat fabric samples, cannot provide information critical to garment design and cannot validate the effects of garment fit, seaming, joining and pumping effects from articulation and movement. This is an important consideration since the overall heat stress burden of protective clothing is determined, not only by the breathability of constituent fabric components, but also by air layers trapped inside the clothing. The air volume in the protective ensemble is determined by the garment design and fit. In addition, sweating plate tests do not account for the effects of additional layers, such as reinforcements, padding, trim

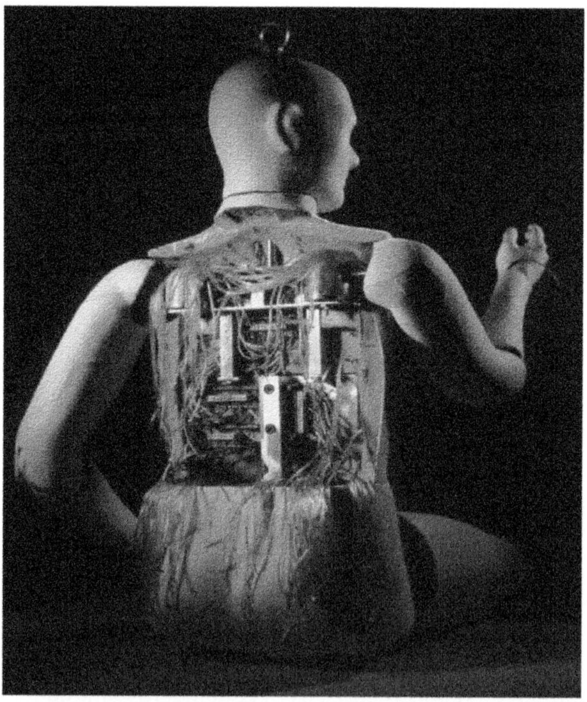

12.2 Sweating manikin.

or the like. Therefore, a major gap exists in available data for realistic research on the heat stress associated with protective clothing systems. Sweating manikin tests provide opportunity to address this shortcoming in instrumented approaches.

The 'Coppelius' type sweating manikin is an example of a sweating manikin technology.[24] Its main features are:

- A computer controlled heating system with 18 individually controlled body sections.
- A computer controlled sweating system with 187 individually controlled sweating glands; sweating over the whole body with the exception of head, hands, and feet.
- Anatomical body dimensions, size 40.
- Prosthetic joints to permit movements and different postures.

The manikin is housed in a climatic chamber and water is supplied from a reservoir, placed on a balance near the ceiling in the chamber. A micro-valve system in the manikin distributes the water to the 187 sweat glands, and the computer system allows individual control of each sweat gland. The condensed water on the dressed manikin is recorded by measuring the change in the weight of the clothed manikin during the test. This measurement is made from the output of the sensitive balance from which the manikin is suspended. Test garments are weighed before and immediately after the test. This is done to estimate the amount of moisture condensation in the individual clothing layers. Moisture condensation in the skin material of the manikin is calculated as the total weight change subtracted by the moisture condensed in the clothing.

Because of the complex nature of sweating manikins, successful installation and operation of these facilities requires a high level of specialized expertise and sophisticated laboratory support infrastructure. Nevertheless, laboratory facilities throughout the world currently use instrumented sweating manikins to evaluate the heat stress potential of clothing. Existing sweating manikin technologies are different with respect to specific details of the manikin test apparatus. They may also use different measurements and environmental control systems and employ different testing protocols. For these reasons, a standardized sweating manikin test procedure has been badly needed.

The ASTM sweating manikin test method provides standardized means of calculating the total evaporative resistance of clothing ensembles. By adapting this method, and combining with an existing ASTM standard procedure for measuring clothing insulation using a dry manikin (ASTM F1291),[1] a garment level equivalent of a total heat loss test is possible.[2]

Studies have shown the potential value of using sweating manikins as tools for assessing the heat stress and comfort of clothing.[13,14] However, a systematic validation of manikin generated total heat loss data in conjunction with human physiological testing is yet to be conducted for firefighting/emergency responder protective gear.

12.3.3 Physiological validation for instrument tests

The utility of sweating hot plate and sweating manikin test methods is limited by the lack of a sufficient number of well-qualified studies that establish correlations with the heat stress associated with wearing emergency responder protective gear. This situation is partly a result of the complexity of interrelated physical and physiological factors that combine to determine the impact of protective clothing on human heat stress response. Researchers have long recognized that the heat stress burden placed on a clothed individual is influenced by many interrelated environmental, physiological, and clothing material variables. These include ambient temperature, thermal radiation, relative humidity; wind velocity, heat and moisture transfer properties of clothing materials, clothing fit, stiffness, design, and tactile sensation. Some measured indictors are skin temperature, body core temperature, heart rate, metabolic rate, peripheral blood flow volume, sweat rate, and wetted area of skin. Psychological preference also plays a critical part. Perceived expectations, peer pressure and habit may determine the stress response regardless of measured physical and physiological parameters.

Because many variables can influence the specific translation between instrument readings and heat stress, performance criteria based on test measurements of the clothing material properties must be based on a comprehensive understanding of environmental and use conditions. Understanding these translations is an important part of an overall assessment of the gaps and limitations in laboratory tests in heat stress associated with protective clothing.

Surveys of physiological studies on the heat stress in firefighter gear underscores the diversity of the physiological experimental designs that have been used. Investigations run the gamut from field trials to controlled laboratory studies. They have incorporated a wide range of different physiological stressing factors, and used different environmental conditions, exercise regimens and clothing configurations.[5,8,18–21,23,25,31–34] There is a lack of a common basis of comparison, which therefore points up a need to establish standard protocols for the physiological testing.

Many studies have been aimed at evaluating the heat stress impact of the breathability of the moistures barrier component used in firefighter turnouts. Studies have specifically focused on defining the correlation between the total heat loss (THL), measured using a sweating hot plate method, and heat stress in firefighter turnouts.[8,25,31] These investigations have provided technical information to NFPA committees interested in establishing minimum performance requirements for heat stress. NFPA 1971 Standard for Structural Firefighting Protective Clothing and Equipment currently requires a minimum sweating plate heat loss of 205 W/m^2 for materials, used in turnout components. NFPA 1999 (EMS), NFPA 1951 (Urban Search and Rescue) and NFPA 1977 (Wildlands Protective Ensembles) standards set the minimum acceptable total heat loss requirements at 650 W/m^2.

In summary, setting a minimum performance for emergency responder protective clothing based on sweating plate, or THL values, is difficult because of the complicating ensemble, environmental and use factors in these translations. Sweating manikin test methods provide opportunity to gain additional information since added clothing layers, as well as garment fit and design can be tested. In any case, more well-designed physiological studies are needed to quantify the translation between these tests (sweating plate and sweating manikin) and the heat stress of emergency responder ensembles. These correlations need to be developed for a broader range of environmental and use conditions. In this way, minimum performance levels for total heat loss can be established for a wide variety of mission related conditions of emergency response.

12.4 Laboratory tests for clothing comfort

Although physiological criteria are central to heat stress tolerance, the perceived comfort of protective clothing and acceptability of the clothing system must also be evaluated. One reason for this is that uncomfortable protective clothing might be removed or incorrectly worn, thus increasing the potential for hazardous environmental exposures and injury. Emergency response conditions that produce clothing related discomfort could occur long before disabling heat stress jeopardizes the responder. Discomfort responses caused by wearing protective clothing are associated with sensory perceptions of thermal and tactile sensations. Because human reaction to physical stimuli is subjective in nature, objective tests for these qualities have often proved to be challenging.

Comfort researchers recognize that clothing comfort has two main aspects that combine to create a subjective perception of satisfactory performance. These are thermophysiological and sensorial comfort. The first relates to the way clothing buffers and dissipates metabolic heat and moisture. The latter relates to the interaction of the clothing with the senses of the wearer, particularly with the tactile response of the skin.

Thermophysiological comfort has two distinct phases. During normal wear, insensible perspiration is continuously generated by the body. Steady state heat and moisture vapor fluxes are thus created and must be gradually dissipated to maintain thermoregulation and a feeling of thermal comfort, thus the clothing becomes a part of the steady state thermoregulatory system. In transient wear conditions, characterized by intermittent pulses of moderate or heavy sweating caused by strenuous activity or climatic conditions, sensible perspiration and liquid sweat occur and must be rapidly managed by the clothing in order to maintain thermal regulation. The behavior of clothing in these two different domains may be predicted by certain measurable fabric properties, including thermal insulation and water vapor permeation resistance.[9]

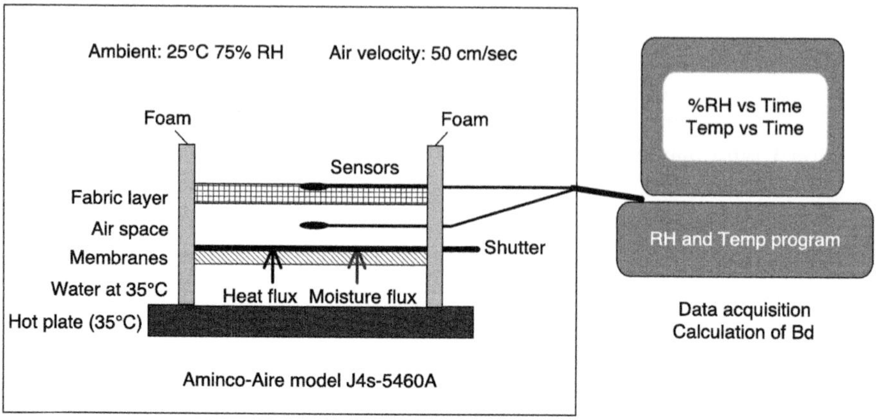

12.3 Dynamic heat and moisture measurement system.

12.4.1 Measures of steady state vapor and heat transmission

Dry and evaporative heat transfer can be measured using sweating hot plates allowing calculation of various indices of thermal comfort including insulation (clo) and moisture vapor permeability.

12.4.2 Measurement of pulsed vapor and heat transmission

Measurement of fabric and microclimate response to pulsed moisture loads has been performed using the setup illustrated in Fig. 12.3.[6] A momentary vapor pressure gradient is created using a diffusion column with a shuttering device housed in an environmental chamber. Strategically placed high sensitivity/rapid response probes track the moisture and temperature pulse history in the microclimate and across the fabric layers. A consecutive series of moisture pulse can also be created and tracked, allowing simulation of a variety of expected end use scenarios.

12.4.3 Comfort prediction models

A theoretical model developed by Woo and Barker has been used to integrate the various measured comfort related physical properties of the fabrics into a prediction of human comfort limits for given climatic and metabolic work load conditions.[10,35] The model is based on rates of heat loss and storage and their effect on body core temperature. Sensible and evaporative heat loss, as well as percent of skin area wetted by sweat, is considered. The model predicts the range of body activity within which an individual wearing a clothing system is thermophysiologically comfortable. Above these limits, heat stress is likely and below them hypothermia may occur.

The model is based on Woodcock's equation[36] for energy dissipation from the body. The total energy dissipated from a sweating human through clothing layers into an ambient environment, assuming no internal reactions (condensation, absorption or re-evaporation) within fabric components, can be expressed as:

Total energy dissipation – Dry heat transfer + Evaporative heat transfer, or

$$Q = Mn = (1/I)(T_s - T_a) + (1/R_v)(P_s - P_a)$$

where:

Q = total energy dissipation, w/m^2
Mn = net metabolic rate, w/sq m (usually, external work efficiency = 0 so Q = Mn)
I = thermal resistance of fabric to convective and radiant transfer, m^2°C/w
T_s = skin temperature, °C
T_a = ambient temperature, °C
R_v = water vapor resistance of fabric, m^{2}kPa/w
P_s = saturated vapor pressure at skin temperature, kPa
P_a = vapor pressure of ambient, kPa.

It is useful to express fabric water vapor and thermal resistance as a ratio of those of free air. This ratio, known as the permeability index (i_m), ranges from 0 for impermeable to 1 for materials as permeable as free air. Also, the evaporative cooling term assumes the body is completely wetted by sweat. But in fact, beyond 20% sweat wetted area (SWA), and in the terms of the more commonly used clo (I) units and introducing i_m and SWA and the relevant conversion constants, we obtain the formula for determining the upper and lower comfort limits:

$$(6.46/I)(T_s - T_a) < Mn < (6.46/I)[(T_s - T_a) + 3.3\, i_m(P_s - P_a)]$$
$$< (6.46/I)[(T_s - T_a) + 16.5\, i_m(P_s - P_a)$$

This model assumes that evaporative heat transfer in addition to dry heat transfer can extend the thermal comfort zone. The model contains three functional parameters: those that are a function of fabric type (I, i_m), those that are a function of environmental conditions (T_a, P_a, air velocity) and a parameter that is a function of the amount of metabolic heat generated (Mn).

Resultant comfort ranges can be plotted for given environmental conditions. Since the plots represent the functional range of the fabric or fabric combinations in terms of allowable exertion, they are key tools in the analytical determination of improved comfort performance.

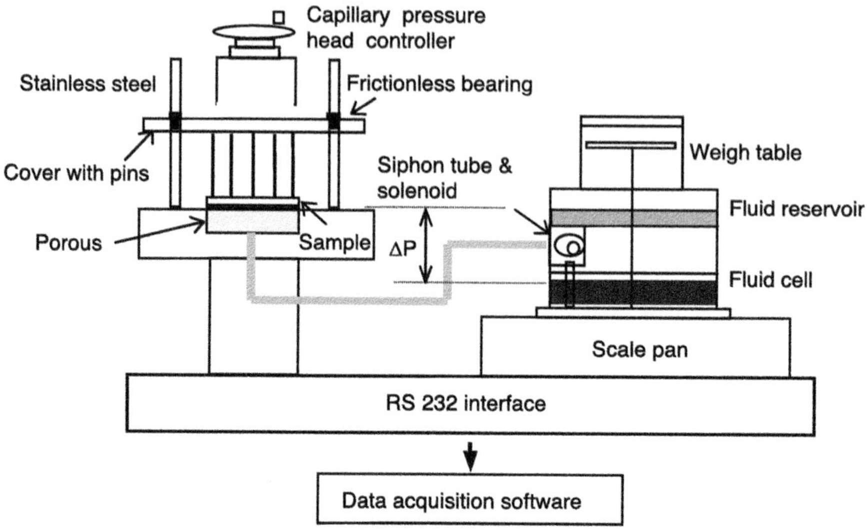

12.4 Gravimetric Absorbency Testing System (GATS).

12.4.4 Measurement of liquid moisture absorption

The ability of a clothing material to transport moisture from sweat-wetted skin is crucial to wear comfort. Laboratory testing technologies have been developed to characterize the ability of a fabric to wick liquid moisture from sweating skin. One such test system is the Gravimetric Absorbency Testing System or GATS. The GATS procedure measures demand wettability. The test indicates the lateral wicking ability of the fabric, or the ability of the material to take up liquid in a direction perpendicular to the fabric surface. The GATS apparatus incorporates a special test cell and cover to assess absorption behavior in the presence of evaporation (Fig. 12.4). In this arrangement, liquid is drawn from a fluid reservoir by the capillary action of the fabric and the hydrostatic pressure of the fluid delivery system is adjusted by controlling the position of the sample platform. Liquid is delivered to the test material placed on a porous plate. Numerous pins, uniformly distributed over the area of the test surface, restrain the test fabric. The amount (grams) of liquid siphoned from the reservoir is recorded as a function of time and these data are used to calculate absorption capacities and rates, and the percentage of moisture evaporated by the fabric. Applications of this device for protective clothing are discussed in the references.[6,9–10]

12.4.5 Measurement of fabric mechanical properties

Fabric weight and mechanical properties related to stiffness can be important determinants of comfort and ergonomics factors in protective clothing. Available laboratory tests for measuring these properties of clothing materials have not been widely used for protective clothing applications. The Kawabata Evaluation System (KES) is the most advanced laboratory testing system for measuring the surface and mechanical properties of fabrics. The KES instruments measure mechanical properties that correspond to the fundamental deformation of fabrics in manipulation that occurs in garment wear. Five different tests are performed using KES, including compression, bending, shear, tensile and surface properties, generating eighteen different mechanical characteristics.

Data on the mechanical properties of fabrics can be used to identify the contribution of tactile stiffness of material components to the flexibility and comfort of protective clothing designed for emergency responders.

12.4.6 Measurement of ergonomic factors

The ergonomic functionality of emergency responder ensembles is an important performance characteristic that requires evaluation.[5] Evaluation of ergonomic performance involves characterizing the effects of ensembles and components on dexterity, range of motion, and the ease with which protective suits can be donned and doffed. Ergonomic tests have not been widely applied for PPE, partly because required protocols generally can involve elaborate human subject requirements and use subjective methods of assessment. ASTM F1154 is an example of an available standardized test[3] which describes standard practices for qualitatively evaluating the comfort, fit, function, and integrity of chemical protective ensembles. Exercise requirements in ASTM F1154 are used in conjunction with liquid- or gas-tight integrity testing of chemical protective ensembles. Procedural options are also described in this method to evaluate the effect of a protective ensemble on the ability of a test subject to perform routine work tasks. These protocols require adaption to be suitable for different types of protective ensembles and functionalities.

An additional need exists to develop improved test methods for evaluating the impact of protective gloves on manual dexterity. Glove hand function tests, such as those described in NFPA 1971 for gloves used by structural firefighters, typically assess the effect of the glove on a prescribed exercise, such as placing pegs into a peg board. Performance is evaluated based on comparison with a bare hand control.[28] Investigation of a new hand function tested for assessing multi-layer glove dexterity is discussed in the references.[15]

12.5 Research needs

Due to their primary function, protective materials often trap heat and moisture, creating a negative impact on comfort, health, safety, and efficiency. Traditional assessment techniques that rely on physiological protocols, while necessary for validation, are time consuming and costly. Currently, only a few studies support physiological testing capabilities. New materials and material combinations are continuously being introduced and considerable merchandising in the protective clothing market has served to confuse the issue of heat stress impact on protective gear. There is a critical need for reliable instrumented test methods that can be used to evaluate the heat stress burden of protective clothing for emergency responders. This review has identified the following specific areas where research is needed:

1. *Development of an improved scientific validation for performance criteria based on instrument test for heat stress.*
Several testing technologies that have emerged are used to critically assess the heat stress potential of materials. Most significant of these are guarded sweating plates for small samples and sweating thermal manikins for complete ensembles. These methods provide consistent, reproducible measures. However, there is a continuing need to correlate and validate these measurement techniques with human physiological measures of heat stress tolerance for specific environments and protective clothing articles.

2. *Development of an improved basis for testing standards based on sweating manikins.*
Much of the emphasis for firefighting and emergency responder clothing in standards is focused on material performance. This approach fails to consider garment design features as they impact thermal comfort. Manikin testing of full ensembles will provide better laboratory prediction of garment field performance and fit. It will also address the effects of design features, which currently are not evaluated. Correlation of bench level tests results with more representative full ensemble testing is needed to provide developers with the input needed to design better protective ensembles.

3. *Development of an application basis for laboratory measurements of non-steady state tests for heat stress and comfort.*
Steady-state heat transfer measurements made on sweating hot plates or sweating manikins do not provide information on dynamic changes in sweating environmental conditions. Laboratory testing procedures are available which could be applied to measure these phenomena.

4. *Development of an application basis for laboratory procedures for measuring sensorial comfort.*
Perceived comfort has been largely overlooked in testing standards for firefighter protective clothing. Laboratory test methods measure material properties associated with sensorial comfort including predictive tests for sweat absorption and tactile factors such as fabric stiffness and softness. A qualified basis is needed for applying these test methods to materials used in emergency responder protective clothing.

5. *Development of a set of practical guidelines that would inform emergency responders how their selection of a protective ensemble would impact them to function in specific situations.*
The impact of protective ensembles on heat stress tolerance is ultimately determined by a complex set of variables involving not only elements of the protective clothing system, but use of variables controlled by environmental and other factors. Although integrating models have been developed, more research is needed to advance the basis upon which physical parameters are used to predict heat stress tolerance.

Test methods, and associated performance criteria, developed through research in these areas will advance the development of protective ensembles for emergency responders that will provide for reduced heat stress, as well as improved comfort and ergonomic functionality. Research is needed to develop and adopt test methods for evaluating the effect of protective ensembles and components on dexterity, range of motion, and ease of donning and doffing.

12.6 References

1. American Society for Testing and Materials (2010), 'ASTM F1291, Standard Test Method for Measuring the Thermal Insulation of Clothing Using a Heated Manikin,' ASTM International, West Conshohocken, PA.
2. American Society for Testing and Materials (2010), 'ASTM F2370, Standard Test Method for Measuring Evaporative Resistance of Clothing Using a Sweating Manikin,' ASTM International, West Conshohocken, PA.
3. American Society for Testing and Materials (2010), 'ASTM F1154, Standard Practices for Qualitatively Evaluating the Comfort, Fit, Function, and Integrity of Chemical-Protective Suit Ensembles,' ASTM International, West Conshohocken, PA.
4. American Society for Testing and Materials (2009), 'ASTM F1868, Standard Test Method for Thermal and Evaporative Resistance of Clothing Materials Using a Sweating Hot Plate,' ASTM International, West Conshohocken, PA.
5. Barker, R.I., Deaton, A.S., and Liston, G. (2009), 'Human Factors Performance of a Prototype Firefighter Suit with Deployable Features', presented at 13th International Conference on Environmental Ergonomics, Boston, MA, August 2–7.
6. Barker, R.L. (2002), 'From fabric hand to thermal comfort: the evolving role of objective measurement in explaining human comfort response to textiles,' *International Journal of Clothing Science and Technology*, **4**, 181–200.
7. Barker, R.L., Guerth-Schacher, C., Hamouda, H., and Grimes, R. (2002), 'Heat transfer in moist thermal protective fabric systems,' *Proceedings of 2nd International Conference on Advanced Fiber and Textile Materials, Ueda, Japan*, November 11–12.
8. Barker, R., Myhre, L., Scruggs, B., Shalev, I., Prahsarn, C., and Miszko, T. (2000), 'Effect of measured heat loss through turnout materials in firefighter comfort and heat stress. Part I: Performance in a mild environment,' in *Performance of Protective Clothing, Seventh Volume*, ASTM STP 1386, Cherilyn N. Nelson and Norman W. Henry III, eds., American Society of Testing and Materials, Philadelphia, PA, pp. 519–534.

9. Barker, R.L., Yoo, S.C., Shalev, I., and Scruggs, B.J. (1999), 'Thermophysical and sensorial properties of heat resistant workwear materials. Part I: Instrument analysis of fabric properties,' published in *Proceedings of the IFAI Textile Technology Forum, San Diego, CA*, October.
10. Barker, R.L., Yoo, S.C., Shalev, I., and Scruggs, B.J. (1999), 'Thermophysical and sensorial properties of heat resistant workwear materials. Part II: Perceived comfort response to garments,' published in *Proceedings of the IFAI Textile Technology Forum, San Diego, CA*, October.
11. Barker, R.L., An, S.K., Shalev, I., and Scruggs, B.J. (1992), 'Comfort properties of single layer aramid workwear fabrics,' *Proceedings of the Fifth Conference on Environmental Ergonomics*, W.A. Lotens and G. Havenith, eds., TNO-Institute for Perception, Soesterberg, The Netherlands, pp. 26–27.
12. Barker, R.L., Woo, S.S., Radhakrishnaiah, P., Hatch, K.L., Markee, N.L., and Maibach, H.I. (1990), 'In vivo cutaneous and perceived comfort response to fabric. Part II: Mechanical and surface related comfort property determinations for three experimental knit fabrics,' *Textile Research Journal*, **60**(8), 490–494, August 1990.
13. Deaton, A.S. and Barker, R. (2003), 'Heat loss through flame resistant protective clothing systems,' *Proceedings of the 2nd European Conference on Protective Clothing, Montreaux, Switzerland, May 21–24.*
14. Deaton, A.S., Barker, R., and Thompson, D. (2002), 'An Advanced sweating manikin for measuring the heat stress of protective clothing,' *Proceedings of the Second IFAI Symposium on Safety Fabrics, Charlotte, NC*, October 23.
15. Dodgen, C.R., Golhke, D.J., Stull, J.O., and Williams, M. (2000), 'Investigation of a new hand function test aimed at discriminating multi-layer glove dexterity,' in *Performance of Protective Clothing: Issues and Priorities for the 21st Century: Seventh Volume*, ASTM STP, 1386, C.N. Nelson and N.W. Henry, eds., American Society for Testing and Materials, West Conshohocken, PA, pp. 162–178.
16. Frim, J. and Romet, T.T. (1988), 'The Role of the Moisture/Vapour Barrier in the Retention of Metabolic Heat During Fire Fighting,' Report DCIEM No. 88-RR-40, prepared for Department of National Defence – Canada, Defence and Civil Institute of Environmental Medicine, Downsview, Ontario, October.
17. Gohlke, D.J. (1997), 'Total heat loss test method,' in *Performance of Protective Clothing: 6th Volume*, ASTM STP 1273, J.O. Stull and A.D. Schwope, eds., American Society for Testing and Materials, West Conshohocken, PA, pp. 190–206.
18. Goldman, R.F. (1974), 'Clothing design for comfort and work performance in extreme thermal environments,' *Transactions of the New York Academy of Sciences*, Series II, **36**(6), 531–544.
19. Goldman, R.F. (1990), 'Heat stress in firefighting: the relation between work, clothing, and environment,' *Fire Engineering*, May, 47–52.
20. Huck, J. and McCullough, E.A. (1988), 'Fire fighter turnout clothing: physiological and subjective evaluation,' in *Performance of Protective Clothing: Second Symposium*, ASTM STP 989, S.Z. Mansdorf, R. Sager and A.P. Nielson eds., American Society for Testing and Materials, West Conshohocken, PA, pp. 439–451.
21. Huck, J. and McCullough, E.A. (1987), 'Firefighter Turnout Clothing: Physiological/Subjective Evaluation,' report prepared by the Institute for Environmental Research, Manhatton, Kansas.
22. ISO 11092 (1995), 'Textiles-Physiological Effects – Measurement of Thermal and Water Vapour Resistance Under Steady-State Conditions (Sweating Guarded Hot Plate Test),' International Organization for Standardization, Geneva.

23. Mäkinen, H., Ilmarinen, R., Griefahn, B. and Künemud, C. (1996), 'Physiological comparison of fire fighter turnout suits with and without a microporous membrane in the heat,' *Performance of Protective Clothing, Fifth Volume*, ASTM STP 1237, S.S. Johnson and S.Z. Mansdorf, eds., American Society for Testing and Materials, West Conshohocken, PA, pp. 396–407.
24. Meinander, H. (1997), 'Experience with a Sweating Manikin-Ready for Standard Use?', European Seminar on Thermal Manikin Testing at the National Institute for Working Life.
25. Myhre, L.G., Barker, R.L., Scruggs, B.J., Shalev, I., Prahsarn, C., and Miszko, T. (2000), 'Effect of measured heat loss through turnout materials of firefighter comfort and heat stress. Part II: Performance in a warm environment,' in *Performance of Protective Clothing: Issues and Priorities for the 21st Century: Seventh Volume* ASTM STP 1386, C.N. Nelson and N.W. Henry, eds., American Society for Testing and Materials, West Conshohocken, PA, pp. 535–545.
26. National Fire Protection Association (2008), 'NFPA 1999 Standard on Protective Clothing for Emergency Medical Operations,' National Fire Protection Association, Quincy, MA.
27. National Fire Protection Association (2007), 'NFPA 1951 Standard on Protective Ensembles for USAR Operations,' National Fire Protection Association, Quincy, MA.
28. National Fire Protection Association (2007), 'NFPA 1971 Standard on Protective Ensemble for Structural Fire Fighting,' National Fire Protection Association, Quincy, MA.
29. National Fire Protection Association (2005), 'NFPA 1991 Standard on Vapor-Protective Ensembles for Hazardous Materials Emergencies,' National Fire Protection Association, Quincy, MA.
30. National Fire Protection Association (2005), 'NFPA 1977 Standard on Protective Ensembles for Wildland Fire Fighting,' National Fire Protection Association, Quincy, MA.
31. Stull, J.O. and Duffy, R.M. (2000), 'Field evaluation of protective clothing effects on fire fighter physiology: predictive capability of total heat loss test,' in *Performance of Protective Clothing: Issues and Priorities for the 21st Century: Seventh Volume*, ASTM STP 1386, C.N. Nelson and N.W. Henry, eds., American Society for Testing and Materials, West Conshohocken, PA, pp. 481–503.
32. Umbach, K.H. (1988), 'Physiological tests and evaluation models for the optimization of the performance of protective clothing,' in *Environmental Ergonomics*, I.B. Mekjavic et al., eds., Taylor & Francis, New York/London, pp. 139–161.
33. Veghte, J.H. (1983), 'Field Physiological Evaluation of Gore-Tex and Neoprene Vapor Barriers in Fire Fighters Turnout Clothing,' report prepared by W.L. Gore & Associates, Biotherm, Inc., Dayton, OH.
34. White, M.K. and Hodous, T.K., 'Physiological responses to the wearing of fire fighter's turnout gear with Neoprene and Gore-Tex barrier liners,' *American Industrial Hygiene Association Journal*, **49**(10), 523–530.
35. Woo, S. and Barker, R.L. (1988), 'Comfort Properties of Nonwoven Barrier Fabrics,' *TAPPI Proceedings 1988 Nonwovens Conference, TAPPI, Allenton, GA*, pp. 167–181.
36. Woodcock, A.H. (1962), 'Moisture transfer in textile systems, Part I,' *Textile Research Journal*, August, pp. 628–633.

13 Improving comfort in military protective clothing

S. DUNCAN, DRDC Suffield, Canada, T. MCLELLAN, DRDC Toronto, Canada and E. G. DICKSON, Royal Military College of Canada, Canada

Abstract: The military has unique requirements for protecting individuals in combat roles where it is expected that they may encounter toxic materials. Accordingly, the military has developed its own specialised protective clothing and associated equipment to prevent incapacitating physiological effects and possible death when combat operations must be conducted in a contaminated environment. The performance of military individual protective equipment (IPE) will be discussed from a Canadian perspective. Specifically, the focus will be on systems developed for the Cold War, systems developed following the Gulf War of 1990–91, and a new conceptual approach to IPE that is being considered to address a much more unconventional threat state which has manifested itself over the past decade. Knowledge of the threat and mission times are key to establishing comprehensive performance specifications for a protective material that will meet critical toxicological protection requirements and at the same time delay the onset of severe heat strain sufficiently to ensure missions can be completed successfully. Thus comfort, functionality and lower burden must be balanced with the level of protection that is appropriate for the most probable current threats. Designing IPE for very specific threat states, and to protect against realistic hazards for only as long as operational requirements demand, should lead to a new generation of IPE for the asymmetric threat environment which offers significant capability enhancements.

Key words: individual protective equipment (IPE), protective clothing, military, protection, heat strain, burden, Cold War, Gulf War, asymmetric threat.

13.1 Introduction

The military has unique requirements for protecting individuals in combat roles where it is expected that they may encounter toxic materials. Of specific concern are the military engineered and highly toxic chemical and biological warfare (CBW) agents. Accordingly, the military has developed its own specialised protective clothing and associated equipment (boots, gloves and full face air purifying respirator), collectively referred to here as individual protective equipment (IPE), to prevent incapacitating physiological effects and possible death when combat operations must be conducted in a contaminated environment. For the most part, the clothing and equipment are worn as an integrated protective ensemble.

Protection has and continues to be the most important component of IPE. This is hardly surprising because without some form of protection against the highly incapacitating and potentially lethal military agents, limited or no combat related activities are possible as long as levels of contamination within the operational zone will cause immediate casualties. Equipment used by soldiers in combat is meant to enhance the probability of mission success, however this has not necessarily been achieved for military IPE. It restricts mobility and functionality, hinders integration with other equipment, and imposes unavoidable physical, physiological and psychological burdens on the users of this equipment. Consequently, although soldiers may be protected, their task performance effectiveness generally is dramatically diminished when wearing IPE, and when exposed to high heat and or humidity, long work durations or high metabolic activity, they may suffer heat strain and associated degradation in physical and cognitive functions.

In this chapter the performance of military IPE will be discussed from a Canadian perspective. Defence research institutions in Canada have a long history of developing protective clothing and equipment to protect the military from potentially devastating casualties should they encounter these CBW agents on the battlefield, and also to provide them with an operational capability in contaminated environments. Specifically, the focus will be on systems developed for the Cold War, systems developed following the Gulf War of 1990–91, which will be referred to hereafter as 'post-Gulf War IPE', and a new conceptual approach to IPE that is being considered to address a much more unconventional threat state which has manifested itself over the past decade that involves an ill-defined enemy with unknown capabilities, who may strike without warning against military or civilian targets.

Canada's developmental path for IPE is not unique by any means. Many other countries have retired their Cold War IPE and brought into service new systems following the Gulf War of 1990–91. Only a few have embarked on a new developmental course in light of the current threat state. Following the use of chemical agents during World War I and the ensuing threat of encountering vast quantities of CBW agents in a Cold War conflict, Western military organisations typically have opted to design their IPE to protect against the highest battlefield exposure concentrations with less regard given to equipment comfort and functionality, to ensure that deployed forces could survive to maintain their positions and operational tempo, and conduct counter-offensive manoeuvres in a contaminated environment. While comfort, functionality and heat strain have long been recognised as a deficiency in IPE design, the Gulf War of 1990–91 brought the limitations of IPE to the forefront as the impact on combat operations became immediately evident. Unfortunately, although new material technologies permitted more comfortable and functional protective garment designs in the period following, protection requirements remained the same and there was little gain in reducing burden. A significant enhancement in operational capabilities, including greater mobility, ergonomic functionality, integration with other equipment platforms and reduction in burden requires alignment of the operational strategies and mission activities with the true threat state. To achieve more comfort, functionality and reduction in burden will clearly require the use of lighter weight materials, but paradoxically, this also may result in lower overall levels of protection. Thus the path to greater comfort, functionality and lower burden must be balanced with the level of protection that is appropriate for the most probable current threats. Accordingly, there is a clear dichotomy around the expectations of military IPE, requirements typically demand maximum protection with minimal degradation in human performance. Without exception, protection requirements negatively impact comfort, burden and functionality. The converse is equally true as well. Integration of IPE into combat roles is a significant challenge when it can so adversely affect the nature of the operation, and in the past it often has been given a lower priority as a result. Optimising comfort, burden and functionality along with protection has now become a key goal for the military in order to increase the probability of mission success when wearing IPE. All training for operations is based on a mission specific risk and threat analysis. However, IPE can only be effective in use when it is perceived by the end user to provide net benefit in practice and not just in theory, so that it is appropriately integrated into techniques, tactics and procedures doctrine.

It is the intent of this chapter to discuss military IPE from a Canadian defence research perspective as it was developed and intended to be used in the Cold War era, the post-Gulf War era, and now into new and less well-defined threat states involving, for example, rapid intervention forces in failing states and domestic counter-terrorism operations. The dichotomy of protection versus burden is at the heart of all issues the military face when using IPE, directly impacting comfort, functionality, duration of wear and integration with other equipment. Ultimately, protection and burden must be finely tuned to the anticipated threat, expected hazard and desired operational capabilities.

13.2 Historical perspective

Most of the significant changes in military CB protective clothing in recent history have evolved in response to two specific military events, the Cold War and the Gulf War of 1990–91. Concomitant with the first use of chemical agents in World War I, chemical protection was designed primarily for the respiratory tract and the eyes, these being the primary targets of the agents used, which included phosgene, chlorine gas, and sulphur mustard (HD). Better polymers and manufacturing techniques led to some general improvements in respirator technology during World War II, but the end products bear witness to the fact that little if any consideration was given to equipment integration issues and more importantly, to the effect that these systems would have on the task performance of the soldiers. HD, being a vesicant, also presented a dermal hazard. Some of the earliest dermal protection involved capes of light oilskin cloth. A considerable amount of effort was directed at the development of clothing impregnants based on active-chlorine-containing compounds and other chloroimides for anti-vesicant garments in World War II, but production limitations, logistic problems and the fact that such impregnants tended to be skin irritants, particularly in hot, humid climatic conditions, led to most of this protective clothing being held in reserve in each theatre of operation as an insurance against gas warfare starting. Some work was undertaken to look at the efficacy of clothing impregnated with carbon adsorbent, but this technology remained poorly developed even through World War II, and comparable effort was expended determining the injuries that individuals would sustain if wearing regular combat clothing.

With the advent of Cold War politics, the associated arms race, and build-up of chemical and biological agents for offensive purposes, it was realised that having to fight in a contaminated theatre of operation was a distinct possibility. This led to the development of dedicated protective suits (with hoods), to be worn over existing combat clothing, with polymeric full face-piece respirators, protective gloves and overboots (an example is illustrated in Fig. 13.1). Operationally, a minimum of 24 hours protection against direct liquid contamination was considered essential for Western countries. The protective suits were typically constructed from a material system that included a carbon adsorbent liner, an inner comfort layer and an outer shell material that either readily wicked liquids laterally along the yarns to increase the evaporative surface area, or was coated with a liquid repellent finish to prevent liquid penetration. The carbon adsorbent liner typically was made using open-cell foams containing carbon powder, or non-wovens impregnated with granular activated carbon. For 24 hours of protection, a substantial amount of carbon was necessary, making the materials thick and heavy. Accordingly, the protective suits of this period were stiff, bulky, cumbersome and ill-fitting. Because it was recognised that thermal burden could be a debilitating side effect when wearing IPE, the material systems were intentionally designed to be air permeable to permit some degree of evaporative cooling to address heat strain, which it was hoped, in the cooler continental climate of central Europe, would minimise heat strain casualties. On the other hand, eastern-bloc countries tended to favour protective suits constructed from impermeable rubbers and polymer coated fabrics, largely dictated by strategies linked to their offensive weapons programmes. These no doubt would have been effective at preventing the penetration of agents, however they would have imposed significant physiological burden due to minimal evaporative heat transfer through such materials.

As the Gulf War of 1990–91 began to unfold, participating countries were confronted with the dilemma that protective systems designed for the central European theatre would, if worn in the hot desert climate using standard operating procedures at that time, lead to significant heat strain casualties. Following the Gulf War, improved carbon adsorbent fabric technologies were developed that

13.1 Canadian Cold War design IPE.

resulted in thinner materials with slightly better hand properties still capable of providing similar levels of protection. While the post-Gulf War protective systems were lighter, more functional and of similar air permeability to their Cold War predecessors, the concept of wear remained largely the same in that they were used as protective suits, to be worn over, or to replace, a combat uniform. While some improvement in heat tolerance was observed with these materials, heat strain remains a significant concern when these systems are worn in hot climates or during very high work rates.

13.3 Threat level and concept of operations

For the military community, it is the inherently high toxicity of the nerve agents that makes the assurance of protection so critical. Two considerations figure prominently in establishing the protection performance levels for military IPE for a given agent: (i) knowledge of the dosage at which the resulting physiological effects are sufficiently benign to have a negligible impact on military task performance; and (ii) the highest probable dosage to which individuals may be exposed while conducting operations. In the case of chemical vapour exposure, the ratio between these for any given agent is referred to as the Target Protection Factor (PF_T), i.e. the highest probable external dosage divided by the highest allowable dosage within the IPE. The PF_T will be set high if high exposure concentrations of chemicals are anticipated, for very toxic compounds, or if operations in a contaminated area will be of long duration.

Historically, the threat level has defined the protection requirement, which then limits the amount by which protective materials and clothing designs can be adapted to address physiological thermal imbalance associated with wear in hot climates or during high work rate activities. During the Cold War it was understood that if CW agents were used, the attacks would involve prolonged barrages using large quantities, and would cause extensive contamination within the theatre of operation. Protective materials designed for the Cold War had to protect against direct liquid chemical agent contamination to the material for upwards of 24 hours and against continuous chemical agent vapour challenges for in excess of six hours. Operations were expected to continue despite the contamination – 'fighting dirty' – and it was anticipated that this could keep soldiers fully suited in their chemical protective gear for numerous days at a time. Re-use also implied the possibility of laundering up to ten times while maintaining protection. The materials required to meet these various protective requirements were incapable of providing the level of evaporative heat transfer necessary to prevent the onset of heat strain at higher metabolic work rates and higher environmental temperatures. Thus it became necessary to develop work/rest tables to establish maximum times personnel could work in full IPE at different metabolic work rates in order to prevent their core temperature and/or heart rate exceeding a medically defined physiologically safe threshold. This translated into safe wear times for different military tasks depending on the environmental conditions. Commanders then were faced with the difficult task of balancing wear times with mission requirements in order to minimise self-inflicted physiological heat strain casualties.

With the protection levels being a rigidly ensconced requirement, a process for varying the level of protective dress was implemented during the Cold War, which was based on defensive strategies developed in the 1960s with less modern IPE, to address the ongoing deleterious impact that IPE had on operations. Protective posture levels were established which allowed components of the IPE system to be introduced in a tiered approach, as warranted by the threat level and mission. In the absence of an impending threat, the preference was to leave the protective gear in a depot well back from the combat zone. If the threat was determined to be low, personnel were instructed to carry their IPE with them. As the threat level escalated, the protective garment would be donned and the boots, gloves and respirator carried. In this dress state individuals were considered to be at low protective posture. More imminent threats would demand that the boots be worn next, followed by the gloves and finally the respirator. Once the closures at the respirator-hood interface, and torso, ankles and wrists were fully sealed the individual was at full protective posture. A good deal of logistical effort was required to monitor the changing threat, inform the personnel of the correct protective posture required, and finally to ensure compliance. The pace of the operation had to adjust to the donning of protective equipment. Commanders were expected to select the level of protective posture most appropriate for the mission and attempt to address heat strain by mixing the levels of protective posture among their troops. They did not want their troops in a higher state of protective posture than was necessary because of the desire to maintain the highest levels of operational readiness, effectiveness and assurance of mission success.

It is now accepted that the threat to western military organisations has shifted away from involvement in Cold War battlefield scenarios (where there was a risk that the warfighter would face the large-scale use of chemical and biological agents and the necessity of fighting on contaminated ground in contaminated clothing for extended periods of time). The current operational construct consists of NATO- and UN-led task forces involved in counter-insurgency operations in support of failed states, as well as possible urban counter-terrorism operations on home soil. The resulting asymmetric battlespace includes an ill-defined enemy that typically does not have the military infrastructure to commit to a conventional confrontation, but is well adapted to hit and run guerrilla tactics. Where CBW agent use may be involved, it is anticipated that the area that may be contaminated by the most hazardous militarised agents will be much lower, primarily because the adversary will not have access to the quantities of agent available to the former Soviet Union, and will likely lack sophisticated technical means of delivering the agent over large areas. While very localised exposure doses and area contamination densities may still be large, those who are not the direct target of such an attack are unlikely to see exposures anywhere near comparable to Cold War scenarios even if they were to remain in place – unlikely in a mobile battlespace where rapid completion of the mission and departure from the contaminated area is preferred in order to further reduce the potential exposure.

Such a change in threat state combined with a highly mobile battlespace has led the protection science community to question the need to maintain the requirement to protect an individual against agent exposure for the significant duration of time considered necessary during the Cold War. Most, if not all, chemical protective garments now in service around the world continue to be designed to meet or exceed former Cold War agent challenge requirements. This has resulted in chemical protective suits in use today that are little different from a system design perspective than those that were in use 30 years ago. Post-Gulf War era design changes have had a minimal impact on the deleterious effects on soldier's physiological and psychological performance levels resulting from the use of CB protective IPE. In fact, recent world events have shown that operational decisions involving troop deployments where there is a suspected CB threat are being made based on minimising the impact that chemical protective suits have on the ability of soldiers to carry out their duties. For example, the decision made by the US to enter Iraq in the early spring of 2003 probably took into consideration the fact that to delay further would have resulted in soldiers fighting in CB protective equipment in the hot Middle East summer, which may have increased the risk of heat strain casualties.

In order to make appreciable gains in the reduction of the burden imposed by CB protective suits, and similarly improve ergonomic functionality, it is necessary to design protective materials and suits specifically to meet the operational requirements of the asymmetric threat state. One approach that we have pursued is to engineer the performance of materials to protect for very precise exposure times against realistic challenge levels expected in the most likely types of incidents. In other words, materials are not designed to protect against all possible worst case exposures, rather, they are meant to provide a toxicologically appropriate and physiologically relevant level of protection only for a short period of time, beyond which it is assumed that other operational strategies exist which may be implemented to mitigate exposure. At the simplest level, reducing the length of time over which protection is required, means that protective garments can be constructed from thinner, lighter-weight materials with better hand properties. Accordingly, these garments should be less physiologically burdensome, resulting in reduced risk of heat strain, and have a higher level of comfort and improved ergonomic functionality. The focus of our work has become the integration of CB agent protection into a standard combat uniform, also referred to as the duty uniform, worn by land, sea and air forces. This will provide the user with a minimum level of protection available all of the time rather than having to choose between no protection, or an overgarment that offers too much protection that is not needed most of the time. It specifically addresses the military's capability goals for the asymmetric counter-terrorism and counter-insurgency environments, which consist of rapid response, high in-theatre mobility, technical superiority and maintenance of operational tempo.

13.4 Understanding system level whole-body protection: baseline performance

A brief overview is warranted on the system level protection performance of IPE. Whole-body system tests quantify the protection performance of protective suits and equipment at different regions of the body, using a variety of possible hazard types (e.g. C or B, vapour, liquid, or aerosol) by measuring the amount of test material that penetrates through the protective suit system to the skin surface. A detailed discussion of the development of a system level analysis methodology to assess the chemical agent vapour protective performance of IPE on a per body region basis under different environmental and chemical vapour challenge conditions has been reported by Duncan and Gudgin Dickson.[1] This methodology uses a non-toxic chemical, typically methyl salicylate, disseminated in vapour form, to simulate a vapour exposure from chemical agents having a moderate vapour pressure, particularly dermally active sulphur mustard. Standard test procedures and conditions include a vapour exposure in a chamber at some predetermined challenge concentration, an exposure time of 30–120 minutes, temperature of 20–30 °C, relative humidity of ~50%, and wind speed of ~2 m s^{-1}. The concentration of the challenge chemical, C_{Si}, which penetrates through the protective suit system to the skin at different body regions, is measured using calibrated dosimeters from the following relationship

$$C_{Si} = \frac{m_i}{tAv}$$

where m_i is the mass measured on the dosimeter at body region i, t is the duration of the vapour exposure, A is the sampling area of the dosimeter and v is the uptake rate of the dosimeter. For a given chamber vapour challenge to the protective suit, protection factors can be determined at each body region where a dosimeter was located according to

$$PF_i = \frac{C_X}{C_{Si}}$$

where PF_i is the body region protection factor at body region i, C_X is the concentration of vapour in the chamber to which the protective suit system is exposed, and C_{Si} is the concentration of vapour measured inside the protective suit at body region i. The beneath the suit vapour concentration-time dosage, which can be related to a toxic exposure and effect, is determined by multiplying C_{Si} by the duration of the exposure t. Values of $C_{Si}t$ that exceed a critical effects threshold may represent a toxic exposure to the individual. Accordingly, higher values of $C_{Si}t$ measured beneath the IPE are necessarily more problematic from a health perspective than lower values.

Often, marked variations in $C_{Si}t$ beneath the suit are measured at different regions of the body using this method. These may be caused by materials within the IPE having different air permeability, or being exposed to different impinging air flows because of changes in the relative orientation to the wind, or activity level of the individual wearing the IPE. More commonly, the variation is due to the effect of non air-tight closures on the protective suit, such as at the ankles, wrists, front torso, neck, and hood where it integrates with a respirator facepiece. Typical protection performance results for the air permeable protective suit with non air-tight closures illustrated in Fig. 13.1 are provided in Fig. 13.2. The variation in the performance of the protective suit system is highly evident across the body from head to feet (left to right in the graph). In particular, we observed almost an order of magnitude increase in the beneath-the-suit dosage at the hood area, ankles

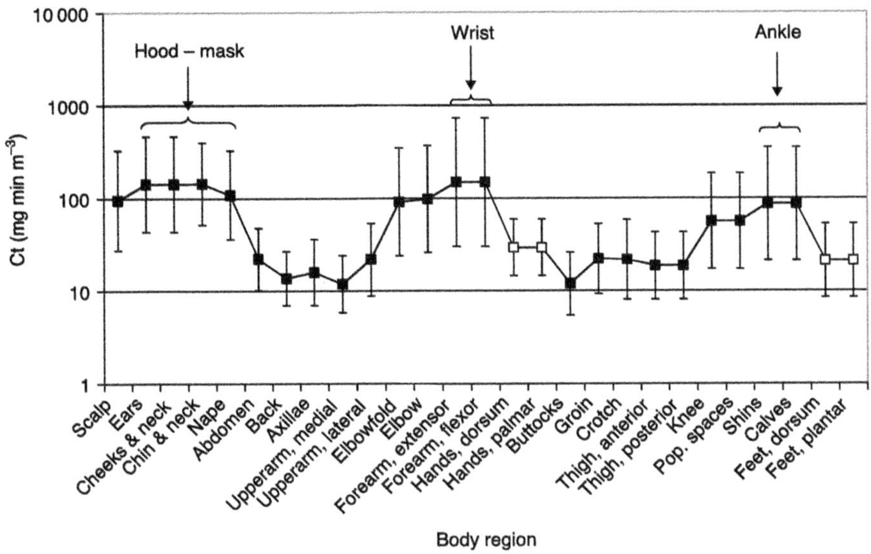

13.2 Variation in vapour dosage measured across the body beneath the air permeable Canadian Cold War IPE with non air-tight closures. The white squares represent data obtained beneath the gloves and boots. The geometric mean vapour dosage (Ct) is shown with the error bars representing the geometric standard deviation. N = 26 tests using 13 subjects.

and wrists compared to the upper and lower torso region. As seen here, it is not uncommon to observe a geometric standard deviation of a factor of 2–3, partly as a result of the differences between individuals, but also as a result of differences each time the IPE is worn by a given individual. This is largely due to the inherent variation coming from numerous factors associated with donning a garment and carrying out physical activities in it; while every effort is made to standardise and control test parameters, disparity between closure tightness and activity intensity, and differences in garment bellowing, may have the most significant impact on performance variability. None of these factors will be a constant during a military operation. Accordingly, a protective suit will not provide the identical level of protection to all individuals, and the difference between the individual obtaining the highest level of protection and the individual with the lowest level of protection at a given body region may be an order of magnitude. This inherent statistical variability in the performance associated with wearing IPE must be taken into account when setting requirements and attempting to improve design.

13.5 Civilian style protective systems

It is useful to illustrate the results for protective systems across the spectrum of performance so that the reader can better situate the typical performance of military IPE. At the highest end of the whole-body protection scale is the fully encapsulated, air-tight suit worn with self-contained breathing equipment. These systems, typically sold for emergency response to releases of hazardous materials, incorporate materials that protect the user from volatile organic vapours, toxic/corrosive liquids, and toxic/pathogenic aerosols. They are most commonly referred to as Level A protective suits based on their original description and classification by the US Environmental Protection Agency (EPA), and appropriate design and requirements for such totally encapsulated systems are described in standards by the National Fire Protection Association (NFPA).[2] At the other end of the spectrum are non-air-tight suits worn with conventional air purifying respirators, commonly referred to as Level C protective ensembles again based on the EPA description; Class 3 ensembles (also referred to as Level C) meeting NFPA 1994 standards[3] also fall within this category. Figure 13.3 illustrates examples of two such systems. In either case, the materials used to construct these suit systems are fairly impermeable and provide protection against volatile organic vapours, toxic/corrosive liquids, and toxic/pathogenic aerosols; the list of chemicals against which Level C suits are evaluated may be fewer than for Level A suits. The most significant difference between Level C suits compared with Level A suits is the presence of closures at the ankles, wrists, front torso, and the

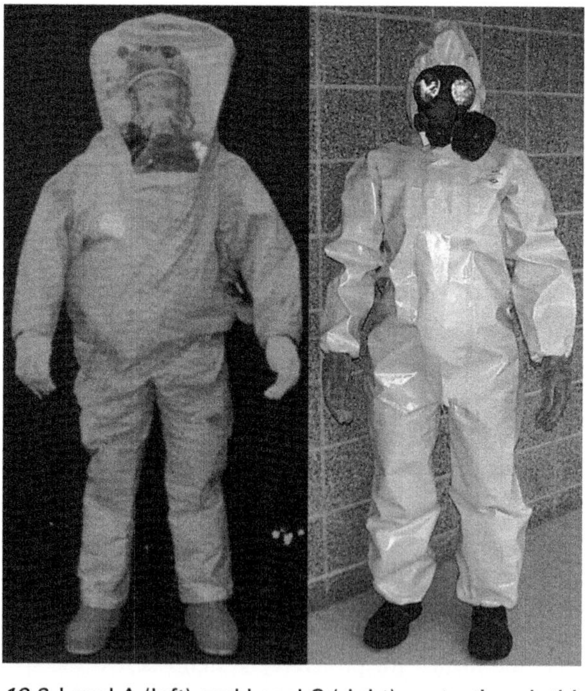

13.3 Level A (left) and Level C (right) protective clothing systems.

hood where it integrates with the respirator. None of the closures in a Level C suit provide an air-tight seal, and while they do restrict air flow in and out of the garment at these locations, they do not prevent it. Depending on design, vapour and aerosols may readily penetrate into the ensemble at these closure locations.

Because closures are the recognised weakness of Level C protective systems, one of our studies compared the whole-body protective performance of a Level A and a Level C suit. The latter is shown worn with the Canadian C4 military air purifying respirator (see Fig. 13.3). The aim was to isolate and quantify the effect that closures had on whole-body protection, since in both systems the suit materials were impermeable to air and to the chemical agent simulant vapour used in the test. Only shorts and a T-shirt were worn under both protective suit configurations. The system level protection performance as a function of body region for these systems is provided in Fig. 13.4.

To facilitate the interpretation, the data beginning at the left of Fig. 13.4 show results for the head, neck, then torso and progressing to the feet at the right of the figure. As expected, the Level A protective suit gave very low and uniform beneath-the-suit vapour dosages over the entire body. The value of $C_St = 3$ reflects the limit of detection for this experiment. The system level protection performance for the Level C protective suit is markedly different as evidenced by the data shown with the square symbols. Beneath-the-suit vapour dosages range from above 1000 mg min m^{-3} at the lower arms and lower legs to 100–200 in the torso

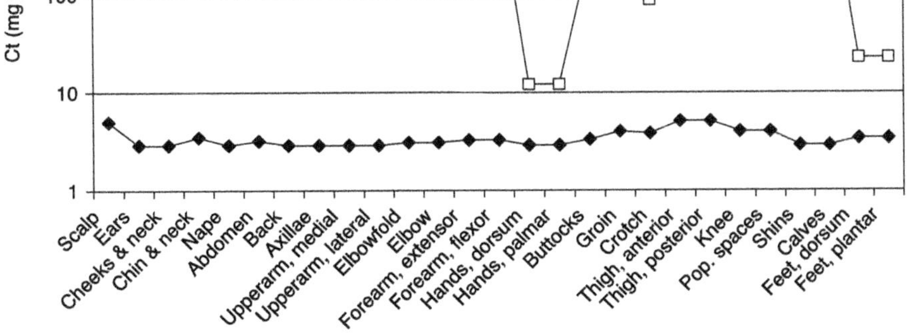

13.4 Vapour dosages measured at different body regions beneath a Level A (diamonds) and Level C (squares) protective suit (both N = 1). Limit of detection equals a Ct dosage of 3.

region. The distinctly lower values on the hands and feet reflect the fact that the skin was covered at this region by tight fitting impermeable gloves and boots. Of particular note is the trend for highest dosages near the non-air-tight closures at the ankle, wrist and neck. As illustrated previously in Fig. 13.2, the hood, ankle and wrist closures tend to be the locations of leakage, with ingress of chemical vapour likely facilitated by the bellows effect due to movement,[4,5] With time, vapour inside the suit disperses by diffusion, convective flow and gross pumping of air throughout the interior cavity. This type of IPE can be modelled as a leaky system in steady state,[6] and the protection performance shown to be governed by the leakage into the suit at the closures, with absorption of the chemical agent into the skin or back into the adsorbent layer being the only removal mechanism. Accordingly, the overall protection performance of an air impermeable suit is governed by the presence of closures and their effectiveness (Fig. 13.4).

13.6 Adsorptive undergarments

Layering a close-fitting carbon adsorbent undergarment (see Fig. 13.5A) underneath a Level C protective suit had a significant impact on the whole-body protection performance of the system, reducing the beneath-the-suit vapour dosage over the body by several orders of magnitude (Fig. 13.5B, open triangle symbols). While the carbon adsorbent layer did exactly what it was designed to do, the significant extent to which this occurred was somewhat surprising. However, the effect of the closures at the wrist and head location remains evident – with higher dosages being observed at these sites and at adjacent body regions. The substantially higher result for the hands reflects that this was the only location where skin was not covered by carbon adsorbent. It may be concluded that the incorporation of a carbon adsorbent layer under a Level C protective suit enhances the whole-body protection level to that approaching a Level A fully encapsulated protective suit. The tight fitting nature of the carbon adsorbent-containing undergarment certainly plays a role in the observed increase in protection, primarily by limiting the ingress of air next to the skin due to the bellows effect, minimising under-the-suit airflow and reducing the diffusion path length, which contributes to a more effective adsorptive system. When combined with the air impermeable over-layer, this is an extremely effective protective concept.

Even using air permeable layers such as street clothes worn over the adsorptive undergarment, the protection was still significantly increased (Fig. 13.5B), resulting in lower beneath-the-suit vapour dosages at most body regions than the Cold War IPE (Fig. 13.2). Duncan et al. (2001) have shown this for other protective suit concepts as well.[7]

In summary, as these examples illustrate, a protective suit system such as an uncompromised Level A fully encapsulated suit, provided near perfect whole-body protection, showing uniformly low beneath-the-suit vapour dosages (<5 mg min m^{-3}) over all body regions (see Fig. 13.4). At the other extreme, a protective suit system

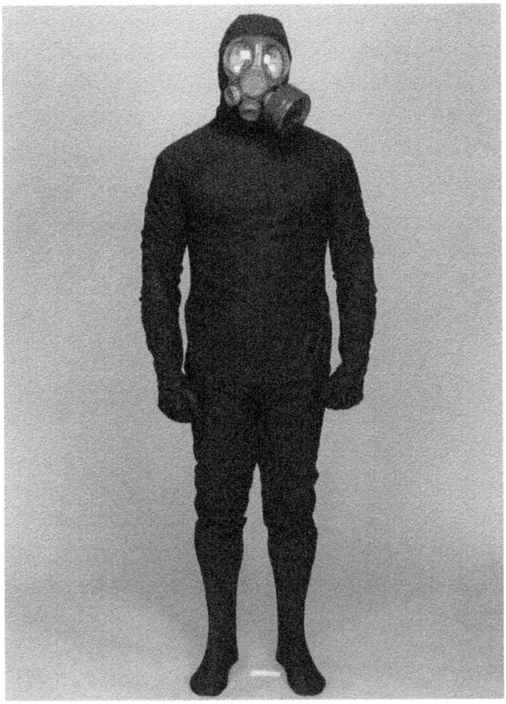

(a)

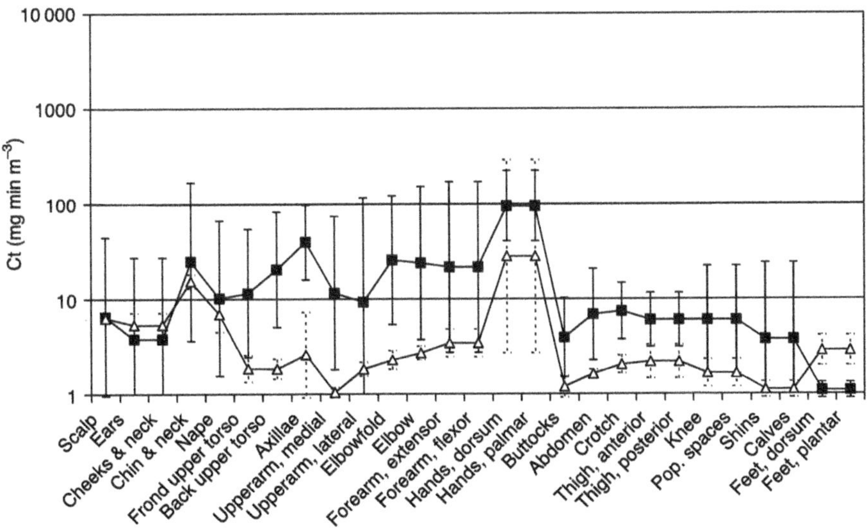

(b)

13.5 (a) Illustration of tight-fitting carbon adsorbent undergarment that was worn under Level C suit or street clothes; (b) geometric mean beneath-the-suit vapour dosages measured at different body regions for these configurations; solid squares, street clothes; open triangles, level C suit (both N = 4). Limit of detection equals a Ct dosage of ~1.

with very poor closures, such as was the case for the Level C system, was characterised by beneath-the-suit vapour dosages typically between 100 and 1000 mg min m^{-3}, with the highest values at body regions adjacent to closures. Protective suit systems using adsorptive carbon fabrics can provide better performance (Fig. 13.2) while permitting some evaporative cooling, and when the adsorbent layer is tight fitting to the skin, dramatic improvements in protection are seen (Fig. 13.5), although due to additional trapped air layers physiological burden might be expected to increase. This will be discussed in more detail later. Coupling a next-to-skin adsorbent layer with an air impermeable outer layer gives system level protection against dermally active CW agents nearly equivalent to a Level A system.

13.7 Cold War individual protective equipment

Protection requirements for Cold War theatres of operation were based on exposure hazard levels developed from various threat assessments for the offensive use of CBW agents. Lengthy wear and re-use were required as described previously. Accordingly, material systems had to have a substantial activated carbon loading to provide and maintain the necessary level of protection, even after carbon loss that resulted during laundering. They also had to have physical properties that made them durable enough to meet the extended wear requirements. These factors resulted in material systems that tended to be heavy, thick and stiff (poor hand properties). It was recognised early on in the development of the Cold War material systems that with protection required for the most probable expected threat, and the operational requirement to fight in a contaminated environment for extended periods of time, the debilitating problem of heat strain due to the thermal burden of the materials would need to be addressed. Air permeable fabric-based materials continued to be considered the most viable option and, except on the hands and feet, replaced earlier impermeable oilskin materials and previous air permeable cloths impregnated with active chlorine compounds and designs coated with carbon granules.

The Canadian Cold War protective garment, referred to as the Chemical Warfare (CW) Protective Coverall, was designed as a one-piece coverall with integrated hood (see Fig. 13.1). It has a front zipper opening extending from the lower abdomen to the neck. The closures at the wrists and ankles utilise a material tab with a hook and loop fastener (Velcro™) and the opening at the hood is elasticised to integrate tightly with the respirator. Standard operational protocol to achieve the highest level of protective posture in the shortest transition time from the unprotected state was to wear the protective garment over the combat uniform. Some key properties of the CW Protective Coverall material system are provided in Table 13.1. It contains approximately 115 g m^{-2} carbon adsorbent to achieve the desired level of protection.

The whole-body system level protection performance of the Canadian Cold War integrated protective ensemble was depicted in Fig. 13.2. Part of the reason

Table 13.1 Characteristics of the material system used in the Canadian CW Protective Coverall

	Material system	Mass ($g\ m^{-2}$)	Thickness (mm)	Air permeability ($cm^3\ cm^{-2}\ s^{-1}$)
Cold War	Outer shell: 50/50 nylon cotton Adsorbent: open cell carbon impregnated foam Comfort layer: nylon tricot	500	2.32	15

for the significantly higher beneath-the-suit vapour dosage at the closure locations is due to the stiff, bulky nature of the material, which results in more folds, wrinkles and buckling of the material at the closure site, allowing paths for vapour to penetrate inside the suit, particularly as a result of the bellows effect.

McLellan[8] investigated the heat strain associated with wearing the Canadian CW Protective Coverall (Cold War IPE) in its standard configuration, as an overgarment, over top of a combat uniform, and compared with several post-Gulf War era lighter weight protective garments worn over shorts and a T-shirt. Items common to the IPE configurations were combat boots, impermeable overboots, impermeable gloves and the Canadian C4 air purifying respirator. The thermal exposure regime involved subjects dressed in IPE (at full protective posture) in a climate controlled chamber set at 40 °C, 30% RH and wind speed of 0.1 m s^{-1}, walking at 4.0 km h^{-1} for 45 min followed by a period of rest in a seated position for 15 min of each hour. The changes in heart rate and rectal temperature are provided in Fig. 13.6 and Fig. 13.7, respectively. The maximum heart rate measured during the heat exposure for the subjects wearing the Cold War IPE was approximately 10% higher than for those wearing the lighter weight IPE. While there was little difference in the final rectal temperature between groups, the rate of change in rectal temperature over the course of the experiment was 35% higher in those wearing the Cold War IPE. Accordingly, heat strain impacting the performance of individuals wearing Cold War IPE resulted in tolerance times, defined as the difference in time between entry into and removal from the environmental chamber, that were ~29% lower than the tolerance times observed for the lighter weight IPE, significant at $P = 0.05$. These results are consistent with other studies of the Canadian CW Protective Coverall, which show high heat strain and lower tolerance times for this system.[9–12] The rate of sweat production was statistically similar for both groups, 0.448 kg $m^2\ h^{-1}$ (SD = 0.0.157) and 0.425 kg $m^2\ h^{-1}$ (SD = 0.127). However, the rate of sweat evaporation, defined as the difference in pre- and post-trial corrected dressed weight, was different, measuring 0.150 kg $m^2\ h^{-1}$ (SD = 0.016) for the Cold War IPE compared to 0.195 kg $m^2\ h^{-1}$ (SD = 0.038) for the lighter weight IPE. Thus, evaporative

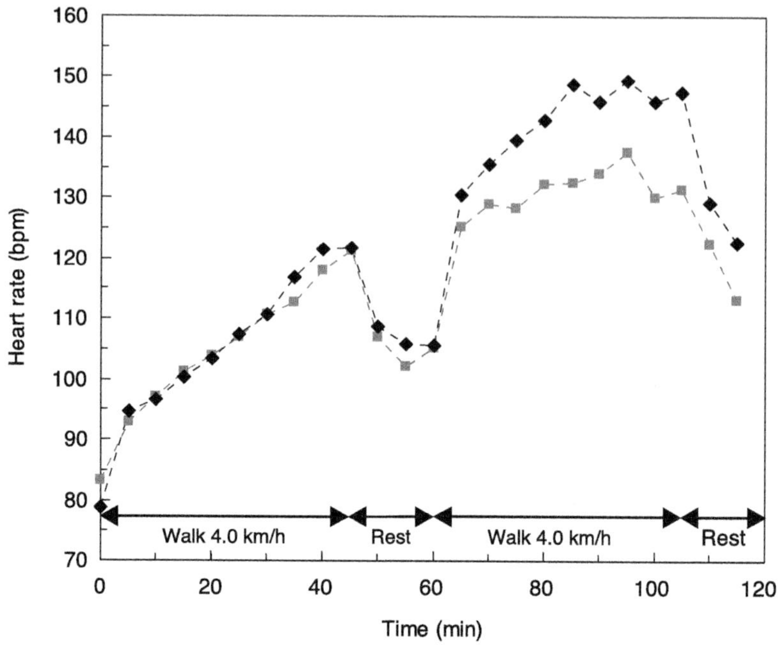

13.6 Mean change in heart rate for subjects wearing Cold War IPE – Canadian CW Protective Coverall (diamond symbols) and lighter weight IPE (square symbols) during a walk/rest regime at 40 °C, 30% RH and wind of 0.1 m s^{-1}. Data derived from McLellan.[8]

efficiency, defined as the evaporative sweat loss expressed as a percentage relative to the total sweat produced, was significantly lower for the Cold War IPE, 36% versus 47%, reflecting the greater resistance to evaporative heat transfer with the use of the overgarment concept. Notably, wearing the Cold War IPE (Canadian CW Protective Coverall) without the combat uniform, in a similar concept configuration to the lighter weight IPE worn only over shorts and a T-shirt, improved the rate of sweat evaporation (0.196 kg m^2 h^{-1}; SD = 0.052), evaporative efficiency (48.4%) and tolerance times to values similar to the lighter weight IPE. These findings highlighted the additional thermal strain associated with wearing the combat clothing when the CB protection was not integrated into the uniform.

In another study on the Canadian CW Protective Coverall, McLellan[13] reported on the relationship between tolerance time and metabolic rate associated with wearing IPE at different levels of protective posture while exercising at different intensities and exposed to 40 °C and 50% relative humidity. Indices of thermal strain, such as the rate of increase in rectal temperature, and cardiovascular strain, such as the rise in heart rate, reflected the increase in thermal and evaporative resistance in the progression from low to high states of protective posture (and concomitant encapsulation of the body). It was also evident that differences

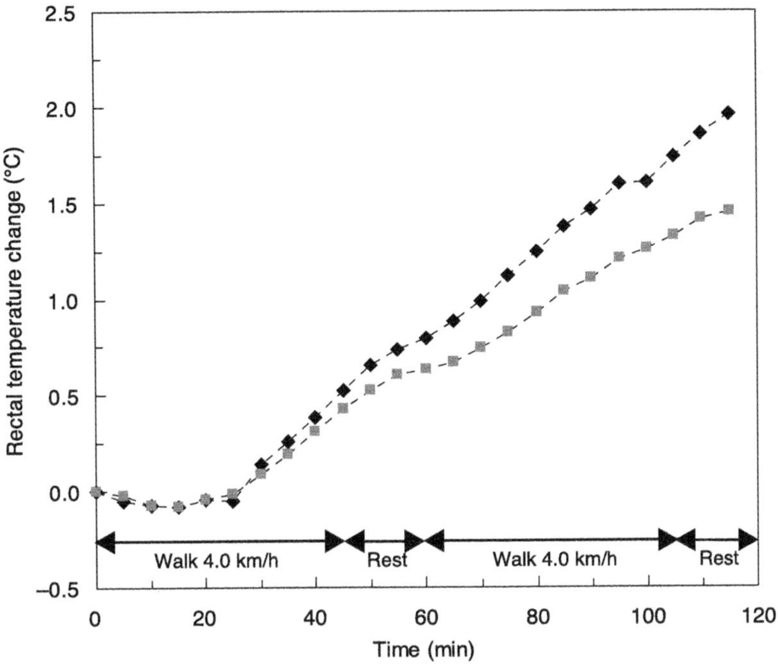

13.7 Mean change in rectal temperature for subjects wearing Cold War IPE – Canadian CW Protective Coverall (diamond symbols) and lighter weight IPE (square symbols) during a walk/rest regime at 40 °C, 30% RH and wind of 0.1 m s^{-1}. There was a significant difference between the time for a 1.5 °C change in rectal temperature for the Cold War IPE (88 min (SD = 10.4)) and the lighter weight IPE (111.7 min (SD = 22.3)). Data derived from McLellan.[8]

among the levels of protective posture became less evident as the rate of heat production, or metabolic rate, increased. The kinetics of evaporative heat loss are largely established by the characteristics of the clothing.[12,14] High rates of activity while wearing protective clothing may result in situations where the rate of heat storage and heat tolerance are governed primarily by the rate of heat production and not by environmental conditions.[13] In these situations, variations in ambient water vapour pressure, which help to establish the gradient for evaporative heat loss, may have little impact on heat tolerance.[13] Craig et al.[15] and Schvartz and Benor[16] first reported a hyperbolic relationship between voluntary tolerance time and the rate of heat storage. The curvilinear relationship between tolerance time and metabolic rate for different ambient temperatures and water vapour pressures while wearing NBC protective clothing is shown in Fig. 13.8. Each of these curves is accurately defined by a hyperbolic function where the vertical asymptote of the equation defines the metabolic rate that delineates compensable and uncompensable heat stress.[9,12,13,17] By definition, uncompensable heat stress

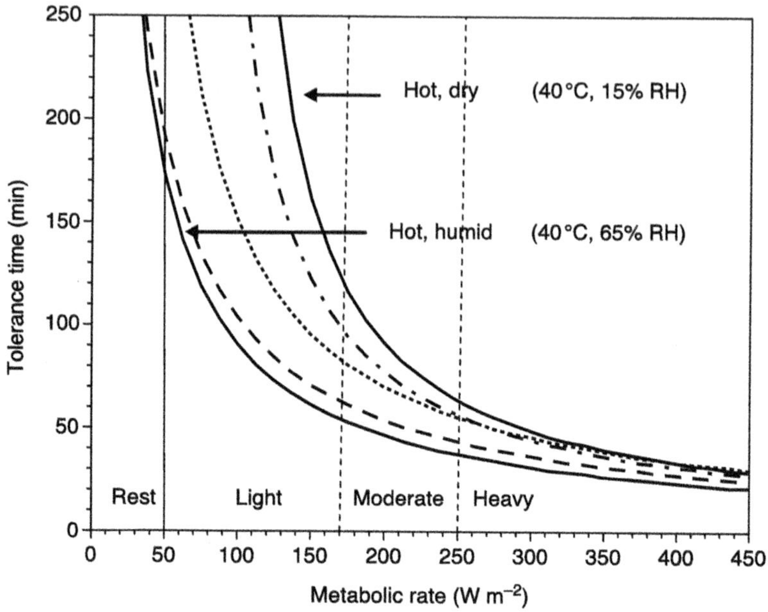

13.8 Relationship between tolerance time and average oxygen consumption (VO_2) for the Canadian Cold War IPE – CW Protective Coverall, worn at full protective posture. Data for the various environmental conditions were derived from McLellan[13] and McLellan et al.[9,17,12]

(UHS) represents a condition where the evaporative heat loss required to maintain a thermal steady-state exceeds the maximum evaporative capacity of the ambient environment.[18] As such, during UHS there is a continued increase in body heat storage and tolerance time is finite. Conversely, during compensable heat stress, where a thermal steady-state can be achieved, increases in body heat content, in theory, do not limit performance and tolerance times are infinite.

Figure 13.8 is further divided into work intensities classified as light (less than 170 W m^{-2}), moderate (between 170 and 250 W m^{-2}) and heavy (greater than 250 W m^{-2}) according to guidelines adopted for the US military.[19] As the rate of heat production increases, it is apparent from Fig. 13.8 that variations in temperature and water vapour pressure have less of an impact on tolerance time. At work intensities above 250 W m^{-2}, tolerance times converge at approximately 50 min. Differences in tolerance time among different ambient conditions are more noticeable at lighter metabolic rates. This is because time is required for the sweat produced at the skin surface to evaporate and condense as it passes through the various clothing layers until it finally evaporates to the environment,[12,20] where the ambient vapour pressure establishes the gradient that drives evaporative heat transfer.[21] Clearly, in the dry desert conditions (40 °C and 15% relative humidity) this vapour pressure gradient and resultant evaporative heat loss (and resultant cooling) will be much greater than in the humid environment of 40 °C with 65% relative humidity. With lower rates of heat production these differences in the rate of evaporative cooling translate into marked differences in tolerance time. Conversely, at higher rates of heat production there is insufficient time for effective vapour pressure gradients to be established between the person inside the protective ensemble and the environment.[12] As a result, the rate of heat production becomes the primary factor determining the rate of heat storage and tolerance time. The curves shown in Fig. 13.8 can also be used to help explain whether implementing work and rest schedules while wearing this IPE is the correct strategy.[22] For example, the vertical asymptote for the desert condition (40 °C and 15% relative humidity) occurs at a metabolic rate of 100 W m^{-2}. At metabolic rates greater than this value there will be continued heat storage since the conditions represent UHS, whereas at metabolic rates below this asymptotic value cooling will occur since these conditions represent compensable heat stress. The reader should note that the metabolic heat production at rest is about 50 W m^{-2}. If the soldier was performing a reconnaissance patrol moving at a slow speed the rate of heat production might approximate 175 W m^{-2}. After 30 minutes of patrol core temperature might increase from 37.0 °C to 38.0 °C in the UHS conditions. If this were then followed by 15 minutes of rest, which would now represent compensable heat stress, core temperature might decrease to 37.5 °C. The next 30 minutes of patrol would increase core temperature to 38.5 °C followed again by another rest period which would lower core temperature again. A final 30-minute patrol might then increase core temperature to 39.0 °C where the soldier might begin to experience symptoms of exertional heat illness. Thus, in contrast to the 60 minutes of continuous work that would lead to core temperatures around 39 °C, 90 minutes of total work are performed with this work and rest schedule. Therefore, in these dry desert conditions, implementing work and rest schedules is the correct strategy for increasing the total work output. However, the metabolic rate that defines the vertical asymptote for the hot and humid environment (40 °C and 65% relative humidity) occurs at a non-physiological value of 10 W m^{-2} implying that even under resting conditions there will be continued body heat storage. Thus, during rest periods the soldier's core temperature will continue to increase and the next work bout will begin at an even higher core temperature than at the beginning of the previous rest period. As a result, less total work will be performed before symptoms of exertional heat illness are seen. In these very hot and humid environments, therefore, implementing work and rest schedules may not be the most prudent strategy if the objective is to move to a safer area where the protective clothing can be removed and the soldier can then cool more effectively.

To summarise, wearing the combat uniform under the protective garment in the standard Cold War IPE dress configuration was considered an operational necessity driven by the tactics of the Cold War battlefield. Commanders anticipated little, if any, warning that CBW agents would be used by the opposition forces and if they were indeed encountered, troops would be required to transition as quickly as possible from combat dress to full protective posture, which meant that the IPE would be donned directly over the combat clothing. The cost of getting personnel into IPE quickly so that they could continue as an effective fighting force was a more burdensome protective clothing system with higher thermal resistance due to the combat clothing and additional air layer. The outcome, later confirmed by researchers as noted above, impacted individuals on a physiological level, reducing the evaporative heat loss and increasing heat strain. Making the protective materials air permeable was a mechanism to address heat strain by affecting the thermal resistance and the water vapour permeability of the materials. It was successful in relative terms, given the alternative of constructing garments from impermeable materials.

13.8 Post-Gulf War individual protective equipment

The demise of the Cold War impacted military organisations in many ways, not the least of which for Canada was the withdrawal of Canadian soldiers from Western Europe. However, one area which saw little change in doctrinal strategies in Canada and other western countries was chemical biological defence. In particular, despite the disappearance of the Soviet Union as a direct military threat, most western military organisations continued to consider large-scale attacks with agent the most probable threat and accordingly, maintained and even updated their inventories of Cold War IPE. This would come back to haunt them following Iraq's occupation and annexation of Kuwait in August 1990. The conflict that followed, which came to be known as the Gulf War, was a United Nations-authorised military conflict between Iraq and a coalition force from 34 nations tasked with removing Iraqi forces from Kuwait. With Iraq's known use of CW agents against Iran during the Persian Gulf War (1980–1988),[23] coalition forces were forced to deploy to Kuwait and Iraq with Cold War legacy IPE designed for an Eastern European operational theatre. It is not uncommon for temperatures in the Middle East to routinely exceed 40 °C during the summer months. It soon became evident that under such climatic constraints soldiers dressed in full protective posture quickly succumbed to heat strain, and were largely incapable of carrying out their tasks. Work began almost immediately in defence R&D laboratories to develop new materials and protective garment designs that would reduce physiological heat strain when wearing IPE in hot climatic conditions.

While a variety of carbon adsorbent technologies were pursued during the early development of post-Gulf War IPE, including more advanced granular carbon slurry impregnation and printing of wovens and knits, two technologies have come to dominate the protective materials used, woven carbon cloth[24] and carbon beads[25] (see Fig. 13.9a,b). The former is made from a woven polyacrylonitrile cloth that has been pyrolised to carbon maintaining the woven fabric structure intact, and the latter are discrete beads of carbon typically <1.5 mm in diameter bonded to a supporting layer. Both forms of adsorbents contain meso- and micro-pores, which gives these media a large functional surface area and the capacity for high physical

(a)

(b)

13.9 Most common post-Gulf War style carbon adsorbent barrier layers. (a) carbon beads; (b) woven carbon cloth. Note: removal of the non-woven abrasion layer laminated to the carbon beads plucks some beads away giving the impression in (a) that the loading of the carbon beads is lower than is actually the case.

adsorption of organic compounds. Post-Gulf War protective material systems typically have the same general configuration as earlier configurations, which includes an external outershell fabric and the carbon adsorbent layer laminated to a comfort layer that goes next to the skin. The outershell may be constructed from a material with 'wicking' properties designed to physically spread the liquid drops that come in contact with the material over a large area to increase the rate of evaporation and reduce the mass challenge to the material per unit area, or it may be coated with a liquid repellent fluoro-polymer which limits wetting by causing liquids that come in contact with the surface to bead and roll off. The carbon beads may be pre-supported on a thin non-woven and both adsorbent layers may be covered with an additional thin non-woven layer to increase the robustness of the layer and prevent abrasion against the outershell layer.

As the post-Gulf War era became a reality, the Canadian military began to consider that protection requirements might have to be reduced in order to address the pressing and urgent necessity to field lower burden IPE for operations in hotter climatic regions. However, with new threats poorly defined and still evolving, most of the classical threat assessments that were conducted still considered the large-scale offensive use of CBW agents to be a possible exposure scenario, which continued to drive material and protective suit specifications within the military community at large. Consequently Cold War protection levels continued to be a requirement for IPE being developed for use in the post-Gulf War, and this remained the case throughout the 1990s even as countries fielded more functional protective garments. The new carbon cloth and carbon bead adsorbent technologies were very effective adsorbents, were inherently thinner, and the carbon loading was readily manipulated to realise some advantage in hand properties and weight and thickness reduction whilst maintaining the protection performance similar to Cold War protective materials. One slight change in approach involved a reduction, in some cases, in the extended wear requirement in an attempt to glean the most physiological benefit from material and garment design. Some key physical properties of the Canadian post-Gulf War protective material system are provided in Table 13.2. The carbon cloth adsorbent layer has a mass of approximately 100 g m^{-2}.

Carbon bead and carbon cloth adsorbent technology are characterised by a fundamentally different permeation response during the initial period of exposure to liquid contamination. This is often missed by researchers who measure permeation incrementally at only a few time points over the course of a permeation experiment, rather than collecting cumulative data in real-time. Typical agent permeation curves are shown in Fig. 13.10 for two carbon bead protective fabric systems and one carbon cloth material system that were contaminated with 5 g m^{-2} of the chemical warfare agent sulphur mustard. All systems utilised an outershell with a liquid repellent coating, which prevented the agent from wicking into the material, thus the results are for vapour penetration only. Note the dramatic difference between the carbon bead material (curve A) and the carbon cloth material (curve B), both

Table 13.2 Typical physical properties of the Canadian post-Gulf War protective material

	Material system	Mass ($g\ m^{-2}$)	Thickness (mm)	Air permeability ($cm^3\ cm^{-2}\ s^{-1}$)
Gulf War	Outer shell: 50/50 nylon cotton Adsorbent: activated carbon woven cloth (pyrolised polyacrylonitrile woven fabric) Comfort layer: nylon tricot	445	1.23	13

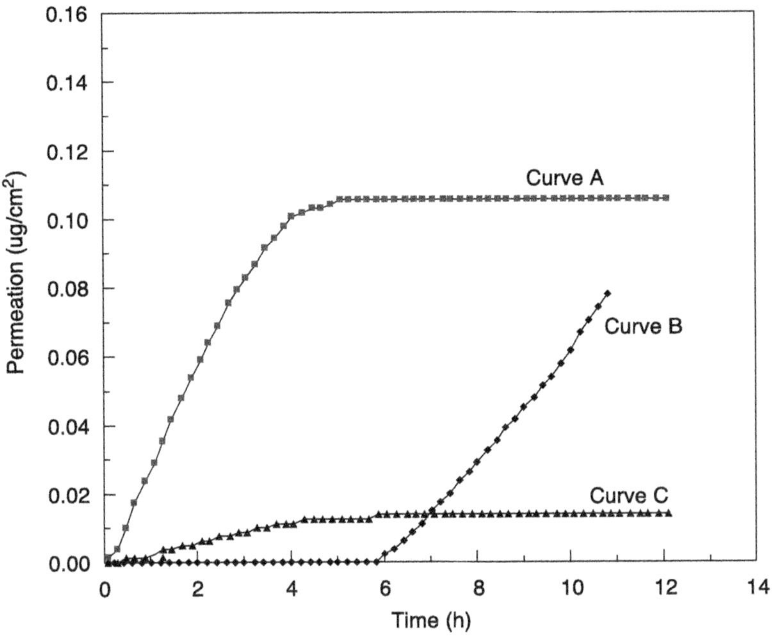

13.10 Cumulative permeation results for post-Gulf War style carbon adsorbent protective materials that were contaminated with $5\,g\,m^{-2}$ sulphur mustard. Air flow across the material surface was $0.5\,m\,s^{-1}$. Curve A is a cabon bead protective fabric with an air permeability of 48 $cm^3\ cm^{-2}\ s^{-1}$; curve B is a carbon cloth protective fabric with an air permeability of 17 $cm^3\ cm^{-2}\ s^{-1}$; curve C is a carbon bead protective fabric with an air permeability of 5 $cm^3\ cm^{-2}\ s^{-1}$.

of which had a carbon loading of approximately $100\,g\,m^{-2}$. Within minutes of the liquid agent contacting the surface of the carbon bead material system, vapour had permeated through. In contrast, it took approximately five hours for the agent to breakthrough the material system with the carbon cloth. The experiments were completed at 30°C with a $0.5\,m\,s^{-1}$ air flow across the surface of the material which resulted in the agent completely evaporating from the surface within four hours. This is supported by the results from the carbon bead material which show no further agent permeation after five hours. The results for the carbon cloth material are distinctly different. The agent was not observed to breakthrough the material system for five hours, approximately an hour after the agent had evaporated from the surface of the material. Thus, agent diffusing into the material system was completely adsorbed by the carbon cloth layer. The near steady-state permeation rate evident following breakthrough suggests that the delayed breakthrough was due to residual agent desorbing from the carbon cloth. The distinctly different response for the two protective materials is a function of the morphology and structure of the carbon adsorbent layer in each fabric system, which in turn may be related to air permeability. The air permeability of the carbon bead adsorbent layer, which included the tricot comfort layer that it was laminated to, was 105 cm^3 cm^{-2} s^{-1} while the air permeability of the carbon cloth adsorbent layer (including tricot comfort layer) was 22 cm^3 cm^{-2} s^{-1}. With the addition of the outershell the air permeabilities of the complete material system were 48 cm^3 cm^{-2} s^{-1} and 17 cm^3 cm^{-2} s^{-1} respectively. As is evident from Fig. 13.9, while the mass density of the adsorbent layer in each material system may be similar, the inherent difference in the morphology and structure of the technologies gives rise to very different coverage. The monolayer packing arrangement of the carbon beads results in an unavoidable inter-bead space where there is no adsorbent. Consequently, some of the diffusing agent may channel between the beads, penetrating through the adsorbent layer and the material system itself.

The carbon cloth on the other hand, is derived from a woven precursor fabric material, and the multi-fibre yarn and fabric weave (see Fig. 13.9), gives it a three dimensional overlapping structure with little opportunity for through-going channels, increasing the probability of diffusing agent being adsorbed. Curve C shows the permeation response for a carbon bead protective material with a carbon load of ~140 $g\,m^{-2}$, and overall air permeability of ~5 cm^3 $cm^{-2} s^{-1}$, a factor of 10 lower than the carbon bead material depicted by curve A and more than a factor of 3 lower than the cloth system (curve B). Even given the higher carbon loading and lower air permeability, agent was detected penetrating the material system within minutes of liquid contamination on the surface of the material. However, the combination of lower air permeability and higher carbon loading both contribute to substantially lower total cumulative permeation for this carbon bead based material system. Obviously, a protective material with any range of physical properties may be developed to achieve any level of protective performance. Neither adsorbent technology is inherently superior to the other. These results illustrate that the penetration performance against agent may be a function of the type of carbon adsorbent, the carbon loading and the air permeability. It is the latter parameter that has important implications regarding thermal burden.

Figure 13.11 illustrates how much more effective the Canadian post-Gulf War carbon cloth material system is at adsorbing sulphur mustard vapour than is the carbon impregnated foam material used in the Canadian Cold War IPE. When challenged at the same constant vapour concentration at a wind speed of $5\,m\,s^{-1}$, catastrophic breakthrough begins at seven hours for the carbon foam material due the agent saturating the adsorbent. Only a low level constant leakage of agent is observed to occur through the carbon cloth material. In this case the protection performance is primarily a function of the efficiency of the carbon adsorbent layer, as the air permeability of the two material systems is similar, 15 $cm^3\,cm^{-2}\,s^{-1}$ and 13 $cm^3\,cm^{-2}\,s^{-1}$.

The Canadian post-Gulf War protective garment now in-service is a one-piece coverall with integrated hood (see Fig. 13.12). The closures at the ankles and wrists, as well as the front torso closure and the hood/respirator interface, are similar in design to the Cold War garment. However, the body of the garment is more tailored and includes an elasticised waist band. Canada continues to be the only country that offers a one-piece protective garment. This was a user driven requirement carried over from the Cold War era largely because soldiers continued

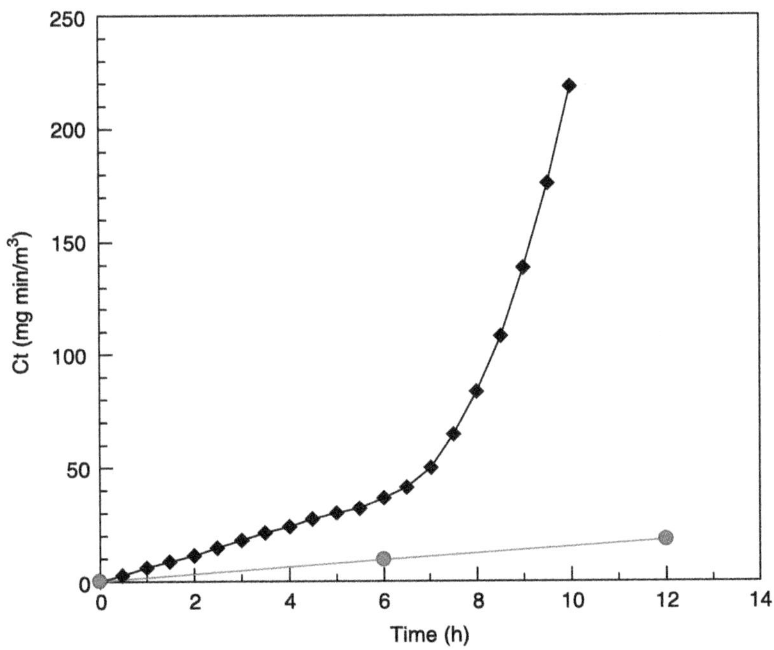

13.11 The concentration-time dosage penetrating through two protective material systems with different carbon adsorbents. Diamond symbols refer to the Canadian Cold War carbon impregnated foam material containing 115 g m^{-2} carbon. Square symbols refer to the Canadian post-Gulf War carbon cloth material system containing 100 g m^{-2}. Air permeability of each material system is similar (see text for details).

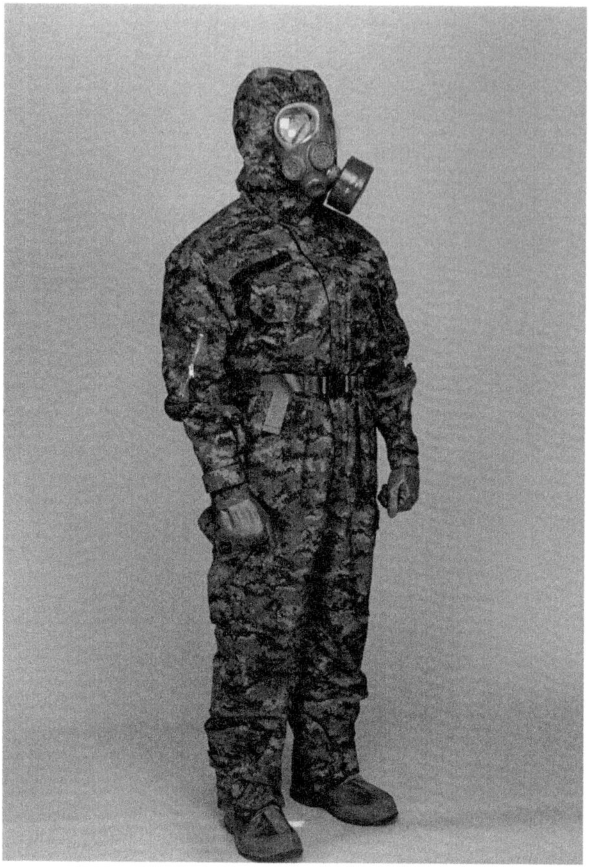

13.12 Current Canadian in-service IPE (post-Gulf War).

to insist that IPE consisting of a full one-piece coverall must somehow provide more effective protection than a two-piece system comprised of pants and a jacket. In fact, our research has shown that a two-piece protective garment can provide equivalent protection provided that there is sufficient overlap at the waist between the pants and jacket. The post-Gulf War protective garment was designed to be a stand-alone system, and standard operating procedures for use in a combat role dictated that it would be worn in place of (and not over) the combat uniform. This implied that soldiers would dress in the IPE before combat, as the threat level required. Of course individuals had the option to wear the Gulf War garment over regular combat clothing in the event the threat level changed suddenly while in the midst of conducting an operation.

The whole-body vapour protection performance of the Canadian post-Gulf War IPE is provided in Fig. 13.13. Both the magnitude and trend of the beneath-the-suit vapour dosages across the 27 body regions is similar to Cold War IPE

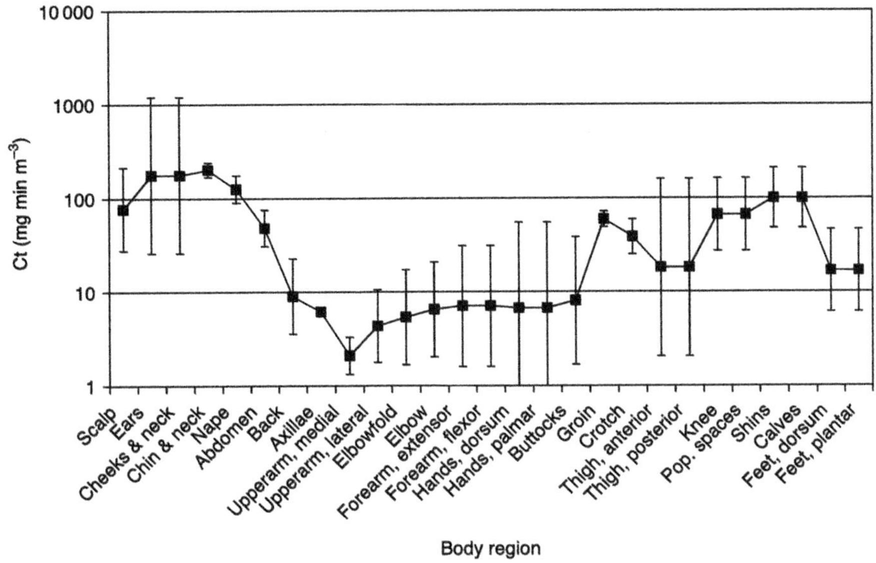

13.13 Vapour dosages measured beneath the post-Gulf War chemical protective suit (N =3 tests). The geometric mean Ct is shown with the error bars representing the geometric standard deviation.

performance when measured under the same conditions. The highest dosages are found in the head region due to the hood interface with the respirator, and dosages typically increase nearer to garment closures.

Increasing the air permeability of a protective material system to augment evaporative cooling and lessen physiological heat strain is not without concerns regarding its impact on protection. This is particularly the case for exposure to aerosols <10 μm in diameter. From a military CBW agent perspective, aerosols include those of biological origin (anthrax), inert solid particulates supporting toxic chemical coatings, and toxic chemical liquid aerosols. Many would argue that aerosols are primarily a respiratory hazard and that the use of self-contained breathing apparatus (SCBA) or air purifying respirators (APR) with filtration canisters that incorporate high efficiency particulate filters (HEPA) largely addresses this issue. However, contamination of IPE is an obvious outcome resulting from an exposure to an aerosol source. Particles <10 μm have little difficulty in penetrating air permeable materials and those <3 μm may readily penetrate completely through material layers and deposit on the underlying skin. For a BW aerosol, unless there is an open trauma injury to the skin, there is minimal risk from percutaneous infection. Other aerosols such as those impregnated with toxic compounds, and neat liquid aerosols, do present a percutaneous hazard and for this reason aerosol penetration of IPE should be a concern. Further, what often is not considered when it comes to aerosol contamination of IPE is both the secondary re-aerosolisation of particulates from the fabric layers during the removal of the garment, and the logistics of decontaminating individuals who have been contaminated to the skin level. Secondary re-aerosolisation from outer layers during their removal may re-distribute the contamination on the individual, and more specifically, may lead to contamination of areas well beyond the original exposure zone, and force additional precautionary measures to ensure that the contaminated individuals remain in the highest level of respiratory protection until the end of the decontamination process, with a means of verifying the integrity of the respiratory protection.

Figure 13.14 shows the aerosol deposition velocity at nine body regions for two post-Gulf War IPE systems. One was constructed from a carbon cloth material system (Fig. 13.9b) with an air permeability of 17 $cm^3\ cm^{-2}\ s^{-1}$ while the other was constructed from a carbon bead material system (Fig. 13.9a) with an air permeability of 48 $cm^3\ cm^{-2}\ s^{-1}$. The general equation for determining aerosol deposition velocity (DV_A, m min^{-1}) is

$$DV_A = \frac{(m_{DA} - m_B)}{AC_m t}$$

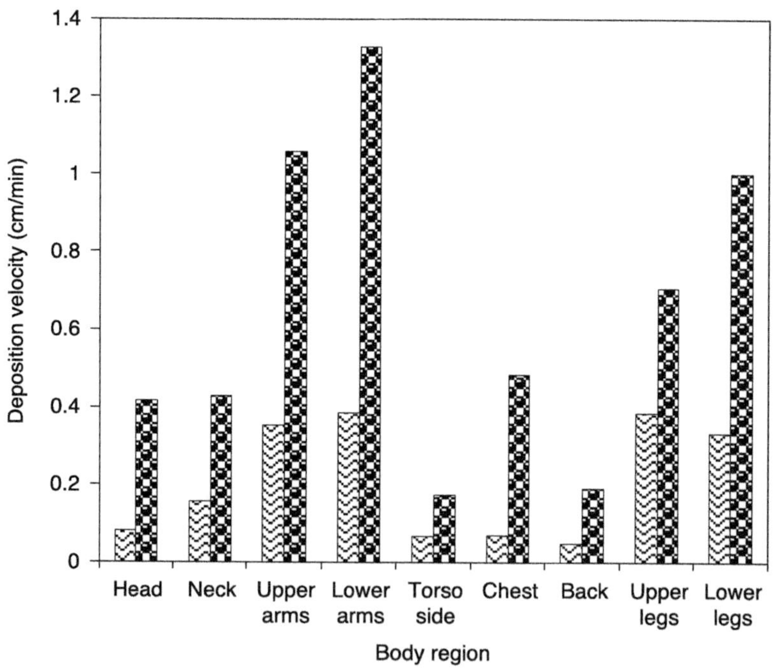

13.14 Aerosol deposition velocity at nine body regions for two post-Gulf War IPE systems based on a carbon cloth (hatched) protective material with an air permeability of 17 $cm^3\ cm^{-2}\ s^{-1}$, and a carbon bead (circles) protective material with an air permeability of 48 $cm^3\ cm^{-2}\ s^{-1}$.

where m_{DA} is the mass of deposited sample, m_B is the background mass, A is the sample area, C_m is the external aerosol mass concentration (mg m^{-3}), and t is time (min). Measuring the fluorescence of tagged aerosols can also be used to determine the aerosol deposition velocity. Not unexpectedly, we found that the aerosol deposition velocity on the skin of subjects wearing IPE constructed from the lower air permeability protective material was two to three times less than when wearing IPE of the same design made from the higher air permeability material system.

Air permeability is only one factor that affects aerosol penetration. Just as important is the gross fabric structure, interstitial pore size and the tortuosity of the void space. Figure 13.15 shows aerosol penetration through fabric swatch samples as a function of different particle diameters: curves 2 and 3 depict the penetration for the post-Gulf War carbon bead and carbon cloth materials systems respectively described above. Also shown is the Cold War material with an air permeability of 15 $cm^3 cm^{-2} s^{-1}$ utilising a carbon impregnated open-cell foam adsorbent layer. A comparison of the Cold War material (curve 1) and the Gulf

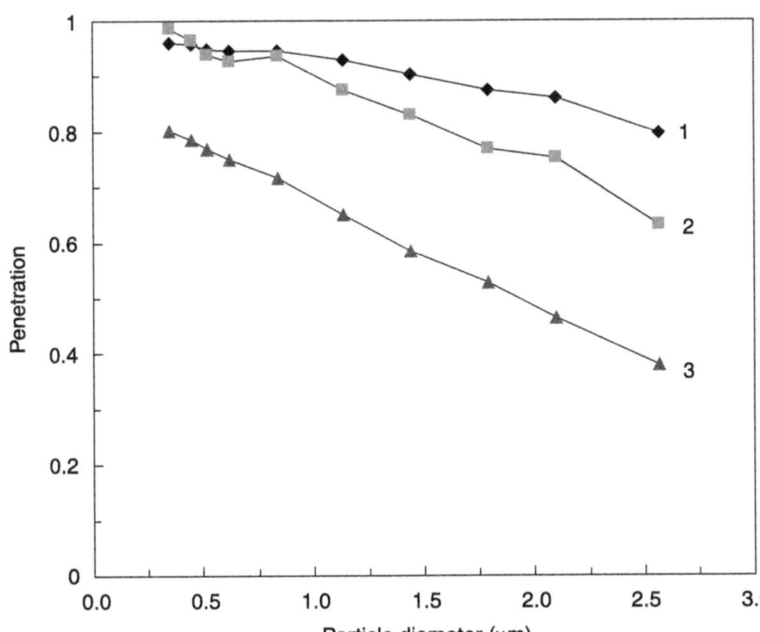

13.15 Aerosol penetration through fabric swatches as a function of particle diameter. Curve 1 (diamond symbols) is a Cold War material based on a carbon impregnated open-cell foam adsorbent layer with an overall air permeability of 15 $cm^3 cm^{-2} s^{-1}$; Curve 2 (square symbols) is a post-Gulf War material based on a carbon bead adsorbent layer with an overall air permeability of 48 $cm^3 cm^{-2} s^{-1}$; Curve 3 (triangle symbols) is a post-Gulf War material based on a carbon cloth adsorbent layer with an overall air permeability of 17 $cm^3 cm^{-2} s^{-1}$.

War material with the higher air permeability (curve 2) shows that despite there being a three-fold difference in the air permeability, aerosol penetration is the same for particle diameters from 0.3 to 1.0 μm, and, in fact, penetration is higher for the lower air permeable Cold War material considering particles up to 2.5 μm in diameter. On the other hand, the Cold War material and the Gulf War material with the tightly woven carbon cloth adsorbent layer (curve 3), have similar air permeability but the latter a much lower aerosol penetration profile across the full range of particle diameters. Accordingly, these differences can only be attributed to the very different fabric structures between these material systems. A clear trade-off manifests itself from a material and garment design perspective: attempting to address physiological comfort by increasing air permeability of materials to foster improved evaporative cooling may come with a loss of protection against aerosols. However, maintenance of high aerosol protection requires an awareness of the potential impact of fabric structure.

The study of McLellan et al.,[8] which included two post-Gulf War IPE of the same design but one constructed from a protective material system with an air permeability of 17 $cm^3 cm^{-2} s^{-1}$ and the other from the system with an air permeability of 48 $cm^3 cm^{-2} s^{-1}$, found no significant differences in any of the key indices of heat strain between the two IPE systems (see Table 13.3) under these conditions of UHS. Although increasing the air permeability of a protective

Table 13.3 Key indices of heat strain while wearing post-Gulf War IPE constructed from protective material systems with different air permeability

Heat strain indices	Post-Gulf War IPE	
	Material system air permeability 17 $cm^3 cm^{-2} s^{-1}$	Material system air permeability 48 $cm^3 cm^{-2} s^{-1}$
Rate of sweat production (kg $m^2 h^{-1}$)	0.425; (0.127)	0.399; (0.118)
Rate of sweat evaporation (kg $m^2 h^{-1}$)	0.195; (0.043)	0.210; (0.054)
Evaporative efficiency (%)	47.2; (7.8)	51.8; (10.1)
Time for 1 °C increase in rectal temperature (min)	82.6; (9.4)	87.5; (19.0)
Time for 1.5 °C increase in rectal temperature (min)	111.7; (22.3)	130.5; (38.3)
Final rectal temperature (°C)	38.9; (0.27)	38.9; (0.31)
Tolerance time (min)	157.1; (50.1)	169.6; (54.5)

Note: Heat strain exposure involved alternating periods of 45 minutes of treadmill walking and 15 minutes of seated rest at 40 °C and 30% relative humidity. Values are means and (standard deviations); N = 8. IPE was worn over boxer shorts and a T-shirt. Data derived from McLellan et al.[8]

ensemble will, in theory, promote greater heat loss to the environment, it must be remembered that these materials will promote greater heat gain through mechanisms of convective heat transfer when environmental temperatures exceed skin temperature (approximately 36–38 °C during exercise in hot environments). Nevertheless, under environmental conditions where the only mechanism of heat loss available to the body is through the evaporation of sweat (and respiratory heat loss) increasing air permeability of the fabric should reduce indices of thermal and cardiovascular strain. In this study by McLellan et al.,[8] although all indices of thermal and cardiovascular strain were reduced for the more air-permeable ensemble, these differences did not attain statistical significance when compared with the other ensemble. In summary, although it has obvious advantages, increased air permeability does not *ensure* better performance under all conditions.

McLellan[11] investigated the heat strain associated with wearing the Canadian Cold War IPE when worn with and without standard combat clothing underneath, and a new French designed hot-weather (post-Gulf War) IPE[26] worn in place of a combat uniform (over top of boxer shorts and T-shirt) in time of threat. In this study it was determined that the mean tolerance time, the rate of sweat evaporation and the evaporative efficiency of sweat from the clothing were significantly higher for the French designed post-Gulf War IPE, and sweat rates were significantly decreased, compared to the Canadian Cold War IPE when worn with a combat clothing layer beneath. The study also found that removal of the combat clothing layer from under the Canadian Cold War IPE gave results that were significantly different than when worn in the conventional configuration with combat clothing. Thus, it was shown that there were no differences for the mean tolerance time, the rate of sweat evaporation, the evaporative efficiency and sweat rates between the French post-Gulf War IPE and the Canadian Cold War IPE when worn in a stand-alone configuration without combat clothing. Removal of the combat clothing layer from under the Canadian Cold War IPE lowered the thermal insulation from 2.35 clo to 1.86, and the ratio of water vapour permeability to thermal insulation ($i_m\,clo^{-1}$) increased to 0.17 from 0.14.[19] In environments where external temperatures are below skin temperature, these changes in clo and $i_m\,clo^{-1}$ due to removal of the combat clothing layer would suggest an increase in both convective and evaporative heat loss.[18] For the French post-Gulf War design, some indices of heat strain such as rectal and skin temperature, and skin and garment vapour pressure, were improved when compared to the Canadian Cold War IPE worn without the combat clothing, implying a more effective evaporation from the skin and increased permeability of water vapour through the French garment but again these improvements did not translate into greater heat tolerance. As McLellan[11] points out, heart rates were not significantly different between these IPE, which suggests that the overall reduction in the physiological strain was minimal. Clearly, changing the dress protocol for Canadian Cold War IPE by using it as a stand-alone system worn over only boxer shorts and T-shirt, similar to the French post-Gulf War IPE, has a much greater effect on reducing heat strain. The removal of the additional clothing layer from the protective system would decrease thermal and evaporative resistance and contribute to an increase in the total heat loss.

Both Canada and the United States (US) have undertaken research on protective undergarment concepts designed to be worn underneath combat clothing. However, McLellan et al.[8] found that undergarments were associated with greater heat strain than post-Gulf War style IPE. This has detracted from them being given serious consideration as a protective system for general infantry operations, despite the high level of whole body protection that they afford the user (see Fig. 13.5). Levine et al.[27] have also reported that heat strain associated with wearing a US undergarment design was similar to a US IPE system worn over only underwear. One area where protective undergarments have found a niche is in special operations where mission durations may be shorter and therefore heat strain may be less of a factor depending on the rate of heat production; in addition, other specialised equipment may need to be worn and mobility is particularly critical, making additional bulky overlayers associated with conventional IPE undesirable.

Australian defence researchers designed and developed a post-Gulf War era chemical protective garment for use in hot climatic regions in the mid 1990s. The concept of use is similar to the French IPE, which is a stand-alone protective garment to be worn over minimal underwear. Amos and Hansen[28] investigated the physiological strain induced by the Australian low burden post-Gulf War IPE. Overall, they found that subjects experienced considerably less physiological strain when wearing this IPE compared to the Australian Cold War IPE, which, similar to Canadian Cold War IPE, also is worn over combat clothing. The protective materials used in the construction of the post-Gulf War protective garment had higher air permeability and lower thermal insulation than their existing in-service Cold War protective garment. This resulted in higher sweat evaporation efficiency (enhancing evaporative cooling), lower skin temperatures, lower rate of increase in rectal temperature and heart rate, and significant improvement in tolerance time under the conditions used, from 59 ± 7 minutes for the Cold War IPE increasing to 93 ± 9 minutes for the new Gulf War IPE.

The differences in the heat strain indices that Amos and Hansen[28] measured between the post-Gulf War era IPE worn with the additional protective equipment and when it was worn only as a combat uniform clearly showed that wearing impermeable overboots, impermeable gloves, full face APR and hood contributed to substantial heat strain. They found that wearing this equipment resulted in significantly lower sweat evaporation efficiency, higher skin temperatures, increased heart rate and larger increases in rectal temperature. McLellan et al.[17] observed similar differences in the rate of increase in rectal temperature in troops during exercise at 30°C and 50% RH in protective clothing with and without APR and hoods. There are two obvious effects of wearing mask, hood and gloves. The first is that the protective posture is such that garment closures located at the wrist and at the interface between the hood and APR must be sealed tightly to prevent the ingress of contaminated air inside the clothing by bellows action. Accordingly, this would prevent under the clothing ventilation which might otherwise have provided some additional cooling. The second effect is the restriction of evaporative cooling from the hands, face and neck region. The hands represent a relatively small area of the body, however the head region has been shown to be an important site for heat transfer.[29]

While intrasubject variability and different end-point criteria may lead to ambiguous interpretation of how long individuals can tolerate wearing IPE under heat stress conditions, it is worthwhile to consider the impact that wearing post-Gulf War style IPE may have on walking distances and reconnaissance area coverage of soldiers in the field. The study by McLellan et al.[8] comparing heat strain in the Canadian Cold War IPE with several protective systems designed for use in hot weather climates such as the middle east, found tolerance times increased by as much as 60 minutes compared to the Cold War IPE, for the post-Gulf War IPE system constructed from the highest air permeable protective material. Since subjects were walking at 4 km h^{-1} for 45 minutes of every hour, under the metabolic work conditions of this study, these gains in tolerance time translate into walking distances of 3 km. The potential advantage to the military commander in terms of additional area covered during reconnaissance patrol with a group of soldiers wearing IPE in full protective posture, assuming the distance is the radius of a circle, is 28 km^2.

In summary, despite some moderate gains in reducing heat strain with the development of post-Gulf War IPE, it was generally not much more so than that achievable simply by wearing Cold War IPE as a stand-alone system without the combat clothing underneath. A significant component of the heat strain imposed by all protective IPE, regardless of the material system or garment design, would appear to be the additional CBW protective equipment that the soldier is required to wear, such as impermeable overboots, gloves, APR and hood. The properties of these items that govern heat strain have changed little, if at all, in design between the Cold War era and the post-Gulf War period. It is nevertheless true that post-Gulf War protective fabrics are superior in their hand properties (feel, stiffness, drape and bulk), providing a lighter garment with a more tailored and comfortable fit, which would contribute to higher garment functionality and improved ergonomics. Accordingly, with the same protection level requirements, post-Gulf War IPE is essentially only a lighter weight, more functional version of Cold War IPE. While this is partway to the target performance, it does not deliver the efficiency gain required by commanders to maintain the operational tempo at the highest level and ensure mission success.

13.9 Asymmetric operations (individual protective equipment)

The asymmetric threat state is wide ranging in its definition and may include, for example, such scenarios as counter-terrorism operations, counter-insurgency operations in support of failed states, and security at VIP meetings. Consistent with an ill-defined enemy with unknown capabilities, who may strike without warning against military or civilian targets, there is no 'typical' asymmetric mission profile. If there is a thread of commonality, it is that mission planning for asymmetric threats purposely defines them to be much shorter in duration than Cold War or Gulf War operational scenarios. In Canada the military has defined an operational window for an asymmetric event involving chemical and/or biological agents of between two and six hours. This is the planned duration of time for successful mission completion in a contaminated environment, or extrication to positions of safety. Acknowledgement of shorter mission times and a lower probability of encountering CB agents, particularly in large-scale quantities, recognition that other equipment such as ballistic vests or helmets may enhance protection, and alteration of tactical procedures, are all key to a new approach to the development of protective clothing for the soldier operating in an asymmetric threat environment. It is important to be aware that such considerations need not involve increasing exposure to toxic materials. An alternate approach is to reduce the duration of time that protective materials are required to maintain exposures below established critical end-point toxicological levels; essentially to match the performance of protective materials to the operational window for an asymmetric event. Accordingly, it may be possible to achieve the reduced time requirement for protection with thinner, lighter protective materials, perhaps even with different technologies than carbon adsorbents, compared to materials that were necessary to meet the 24 hour protection requirements of the Cold War and post-Gulf War era. Thus it becomes feasible to consider incorporating CBW agent protection into a true combat uniform concept that could be worn daily during a combat operation.[31,32] The most obvious advantage with a protective combat uniform concept is that the soldier has protection on all of the time, providing them with additional confidence and allowing them to focus on completing the mission. The logistical burden of having to carry a separate garment for protection into a theatre of operation is eliminated and transition time to full protective posture is minimised as it is only necessary to don protective gloves, hood and respirator. More to the point, it gives commanders additional operational scope, battlefield mobility and decision time, knowing that their soldiers are protected, to consider various options and responses as the mission unfolds. The expected positive impact of a protective combat uniform for the soldier include lower burden and concomitant reduction in the incidence of debilitating heat strain, improvement in ergonomic functionality, including ability to complete tasks more effectively with less effort and improved equipment integration. Whilst our R&D programme is ongoing, we have to date developed several new protective material systems, and we have incorporated design features into a uniform concept targeted specifically to reduce thermal burden when the IPE is worn.

The approach taken in Canada to date is to refocus material system protection requirements from 24 hours to 2–6 hours protection, in line with the operational window for an asymmetric event. Reducing thickness, weight and lowering stiffness should impact positively on comfort, thermal resistance, evaporative resistance and total heat loss. The protective materials that have been developed permit agent to permeate through at a rate that would not exceed a toxicologically safe limit within the constraints of the asymmetric operational window. We refer to this as 'controlled breakthrough' and it is believed to be the first time that such a concept to intentionally allow agent through a protective material has been used in the design of protective materials. This approach ties protection performance directly to mission operations.

Some key physical properties of an asymmetric operational protective material system that we have developed are provided in Table 13.4. The adsorbent layer contains approximately 60 g m^{-2} carbon to achieve the desired duration of protection described above. It can be seen that its mass is ~33% lower than the typical Canadian post-Gulf War protective material and ~29% thinner (compare Tables 13.1 and 13.2).

Figure 13.16 presents the normalised permeation through one of our asymmetric operational protective materials exposed to 5 g m^{-2} HD with evaporation allowed. Curve A shows the results for a single continuous agent challenge applied at t=0 min. The drops, 1 μl in volume, typically evaporate between four and five hours after application. Curve B shows the cumulative permeation that resulted when the protective material was contaminated five separate times over the course of 27 hours. The same amount of agent was applied for 30 minutes in each application and then removed. Monitoring was continuous during and between agent applications. Agent application times are indicated on the x-axis with the open square symbols (t=0–30 min; t=2.5–3 h; t=5–5.5 h; t=21.5–22 h; t=24.25–24.75 h). The normalised permeation refers to the ratio of HD permeated through the material to a critical toxicological allowable limit based on the use of a rate limiting film with an uptake rate of 1.3 cm min^{-1}, similar to skin. It can be seen that for a single continuous challenge of agent, the maximum allowable amount of agent through the protective material occurred at approximately ten hours after the agent was applied. When the material was contaminated multiple times but for short, discrete durations, it was capable of protecting for in excess of 26 hours,

Table 13.4 Properties of asymmetric operations protective material

	Material system	Mass (g m^{-2})	Thickness (mm)	Air permeability (cm^3 cm^{-2} s^{-1})
Asymmetric ops	Outer shell: 50/50 nylon cotton Adsorbent: activated carbon woven cloth (pyrolised polyacrylonitrile woven fabric) Comfort layer: nylon tricot	297	0.87	31

during which the maximum allowable amount was not exceeded. The two different agent challenges are simplistic representations of possible challenges that might be encountered in the field. As the data in Fig. 13.16 illustrate, the implications of considering the nature of the threat and the appropriate challenge scenario are significant. When the material was exposed to agent multiple times but for short durations it provided protection for 150% longer than when exposed to a single continuous agent challenge. Cumulative permeation was <30% of the maximum allowable (curve B). Less agent adsorbs into the carbon layer in the protective material during a short duration exposure, thus the residual capacity remains high and the materials can provide effective protection against multiple exposures over many hours. It should not be lost on the reader that this material exceeded the aforementioned military requirement of six hours protection by four hours and >22 hours for exposure scenarios A and B respectively. Thus it may be feasible to

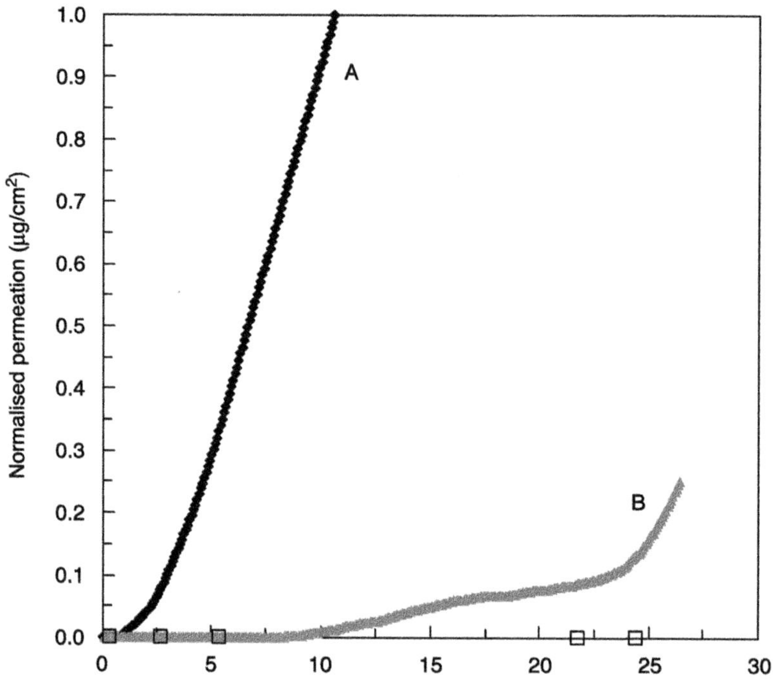

13.16 Normalised permeation through Canadian asymmetric threat operational protective material when exposed to 5 g m^{-2} HD with evaporation allowed. Curve A shows the results for a single, continuous contamination challenge applied at t=0 min. Curve B shows the cumulative permeation that resulted from five separate 30 min contamination events (at times indicated by the squares on the time axis) over the course of 27 hours. See text for further details.

scale-back the protection performance of the material closer to the allowable limit in favour of achieving further gains in reducing physiological burden. These results also illustrate the importance of characterising the real-time permeation performance of materials developed for use in asymmetric combat uniforms. Much more information can be obtained, particularly in terms of agent breakthrough time and flux rate, than is possible by measuring cumulative permeation at only two or three time points during an experiment.

The cumulative penetrated concentration time dosage through two asymmetric operations material systems is shown in Fig. 13.17. The material depicted by the circle markers has the properties listed in Table 13.4. Data shown by the square markers was obtained from a material with a similar carbon adsorbent layer only covered with a fine non-woven that reduced the overall air permeability of the material system to $<10\,cm^3\,cm^{-2}\,s^{-1}$. When challenged at a constant vapour concentration at a wind speed of $1\,m\,s^{-1}$, both materials show a nearly linear increase in the penetrated dosage over the six hour exposure. Notably, for the conditions measured, the dosages are within the asymmetric threat operational

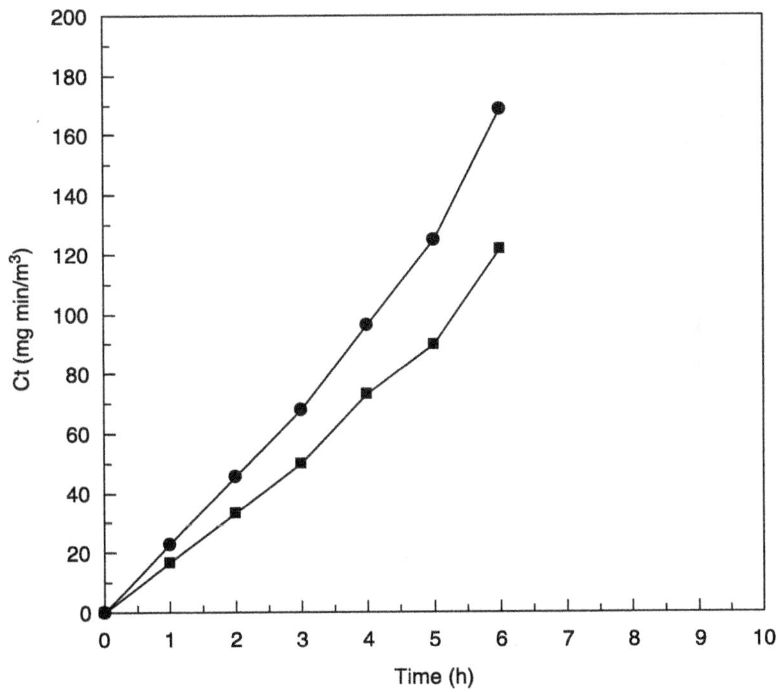

13.17 The concentration-time penetrated dosage through two asymmetric operations protective material systems with different air permeabilities. Circles: air permeability = 31 $cm^3\,cm^{-2}\,s^{-1}$; squares: air permeability ~ 5 $cm^3\,cm^{-2}\,s^{-1}$. All values are at the detection limit and reflect the maximum penetrated dose possible.

window. The reduction in the penetrated dose observed is consistent with a lower air permeability. Full consideration of the static liquid challenge presented in (Fig. 13.16) and the dynamic vapour exposure (Fig. 13.17) is necessary when designing the protection performance of asymmetric operations materials. Performance can be affected simply by reducing the air permeability of the material, in addition to more obvious approaches like varying the carbon adsorbent loading. The impact that such changes have on thermal must also be considered.

Amos and Hansen[28] and McLellan[33] have both shown that a certain amount of heat strain occurs even wearing conventional military combat uniforms (non-protective). In the study by Amos and Hansen,[28] subjects walked for 100 minutes at 5 km h^{-1} and 3% gradient under environmental conditions of 30 °C, 60% RH and impinging wind speed of 1.1 m s^{-1}. They observed a change in rectal temperature for the combat uniform of 0.7 ± 0.3 °C, a change in heart rate of 34 ± 13 beats min^{-1}, and a change in skin temperature of 0.6 ± 0.4 °C. The sweat evaporation efficiency for the combat uniform was 83%. The subjects wearing only a combat uniform in McLellan's study[33] walked for 90 minutes at 4.5 km h^{-1} (no gradient) under environmental conditions of 35 °C, 50% RH and impinging wind speed of 1.0 m s^{-1}. For the combat uniform McLellan observed a change in rectal temperature of 0.85 ± 0.3 °C, a change in heart rate of 42 ± 17 beats min^{-1}, and a change in skin temperature of 2.5 ± 0.9 °C. Sweat evaporation efficiency for the combat uniform was not reported.

From these studies, it is apparent that even if it were technically possible to develop protective materials that had exactly the same physical properties as the fabric used in a combat uniform, heat strain would not be eliminated. At best one would achieve thermo-neutrality relative to a combat uniform. Amos and Hansen[28] were able to demonstrate this for the light weight protective garment developed by Australian defence researchers. Part of their study investigated the heat strain associated with wearing the light weight protective garment worn in low threat posture (without impermeable overboots, impermeable gloves, full face APR and hood), to that of a standard combat uniform worn only with leather combat boots. They observed no statistically significant differences in any of the heat strain indices (sweat evaporation efficiency, heart rate, rectal temperature and skin temperature) between systems, suggesting that subjects were able to achieve a similar thermal steady-state wearing either garment system. For a given wind speed, evaporative heat loss at the skin surface is determined by the ratio of the water vapour permeability coefficient to the thermal resistance of the clothing ensemble together with the vapour pressure gradient between the skin and the environment.[21] Amos and Hansen's results have shown that light weight protective garments with relatively high air permeability have greater evaporative efficiency, which enables a high degree of evaporative cooling, reducing all the key heat strain indices. It should be pointed out that heat transfer from the skin through clothing to the environment also is affected by the thermal properties of each layer of clothing covering the skin.[21] As discussed earlier, when environmental temperature exceeds skin temperature the lower thermal resistance clothing can actually promote greater heat gain.

There is a significant issue related to the protection performance of garments constructed from materials with high air permeability. The open structure of the fabric may lower evaporative resistance and increase total heat loss through evaporative cooling, but as with aerosol penetration, it also allows volatile organic chemicals in the vapour phase that are not adsorbed effectively by activated carbon to penetrate more readily through the material system to the skin, resulting in higher beneath-the-suit vapour dosages. Whilst it is not necessarily the case that higher air permeability protective materials have lower carbon loading, generally the residence time of the air in the carbon layer as it flows through the material will be less, and therefore there will be less time for adsorption to occur. This is particularly pronounced at higher environmental wind speeds as well. Figure 13.18 shows the dramatic reduction in whole body protection that occurs due to exposures at higher wind speeds. A three-fold increase in the wind speed ($1.6\,m\,s^{-1}$ to $5\,m\,s^{-1}$) led to an order of magnitude increase in the beneath-the-suit vapour dosage across the torso, arms, groin area and thighs. Similar dosage values at the head and shins/ankles are due to a common problem of ingress of vapour through the garment closures at these

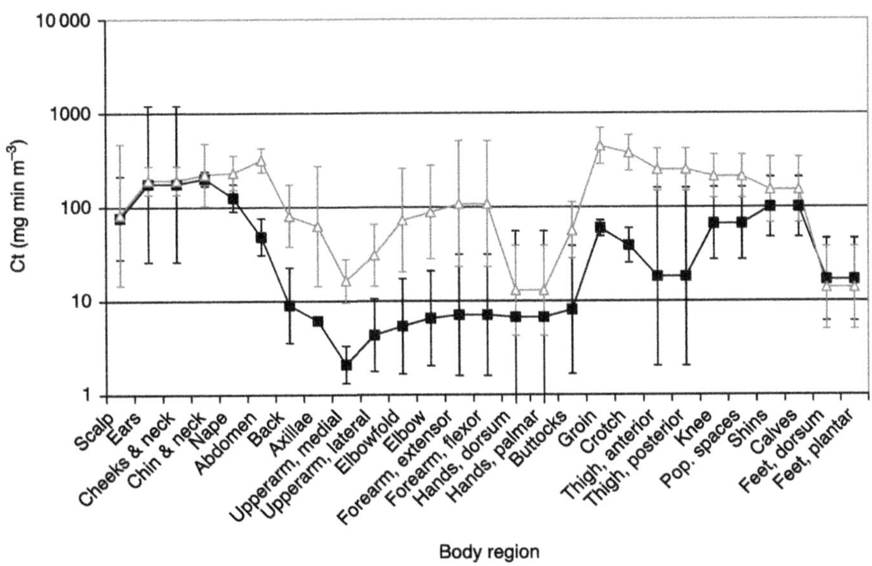

13.18 Effect of wind speed on vapour dosages measured under post-Gulf War IPE systems of the same design and material air permeability ($14\,cm^3\,cm^{-2}\,s^{-1}$). Data indicated by square symbols for IPE exposed at a wind speed of $1.6\,m\,s^{-1}$; data indicated by triangle symbols for IPE exposed at a wind speed of $5.0\,m\,s^{-1}$.

locations. In situations where low beneath-the-suit vapour dosages are required at high wind speeds, we see that the physiological advantage of high air permeable IPE becomes problematic and simply may not be compatible with sufficient protection, recognising however that the hazard may dissipate more quickly at higher wind speeds.

We have also considered design changes to the protective combat uniform as a mechanism to reduce thermal burden. By incorporating features into the design of the uniform that passively or actively aid in the reduction of thermal burden there will be less of a necessity for the evaporative heat transfer requirement to be accommodated in its entirety through the protective material. One of our interests has been the incorporation of passive venting on the garment to assist in evaporative cooling at key locations on the body. Up to now, research based on experimentation with human subjects to support this idea is limited. However, the biophysics of heat transfer[34] and documented effects from bellowing of clothing due to physical movement[35,36] appear to have convinced many manufacturers of outdoor athletic apparel to include vents in some of their clothing to address thermal burden and/or moisture issues. Several studies have reported that vents reduce vapour condensation inside clothing[37] and significantly improve the transfer of vapour from the skin to the environment.[38] Goldman[39] showed that inserting a vent across the back reduced heat storage in subjects wearing impermeable rain wear walking at 35 °C. Interestingly, Lotens and Wammes[38] showed that heat transfer was more effective for garments with lower air permeability. Thus the advantages of incorporating vents in clothing may be superseded by high air permeability materials.

McLellan[33] investigated the impact of the thermoregulatory strain that occurred when wearing a protective uniform with and without vents worn in low protective posture (no overboots, gloves, APR or hood), on heat strain indices including tolerance time after transitioning to full protective posture with the vents completely sealed (overboots, gloves, APR and hood worn). This methodology differed from many earlier studies in that it demonstrated the importance of considering the physiological strain associated with wearing protective stand-alone uniforms during low protective posture on determining the level of heat strain that evolved after the transition to full protective posture. This is noteworthy since if the stand-alone uniform is intended to replace the overgarment concept of protection, comparisons must be made throughout the continuum of transition from low through high dress states. McLellan[33] observed a significant difference in the cardiovascular strain between the uniforms with and without vents after 45 minutes exercise during low protective posture. After the transition to full protective posture, heart rates for the uniform with vents (sealed) remained significantly reduced compared to the uniform with no vents, despite the IPE being otherwise identical (see Fig. 13.19). While the average skin temperature across the body was unaffected by the vents in low protective posture, the skin temperature at the specific vent locations (arms and legs) was significantly different than at the same body regions under the uniform

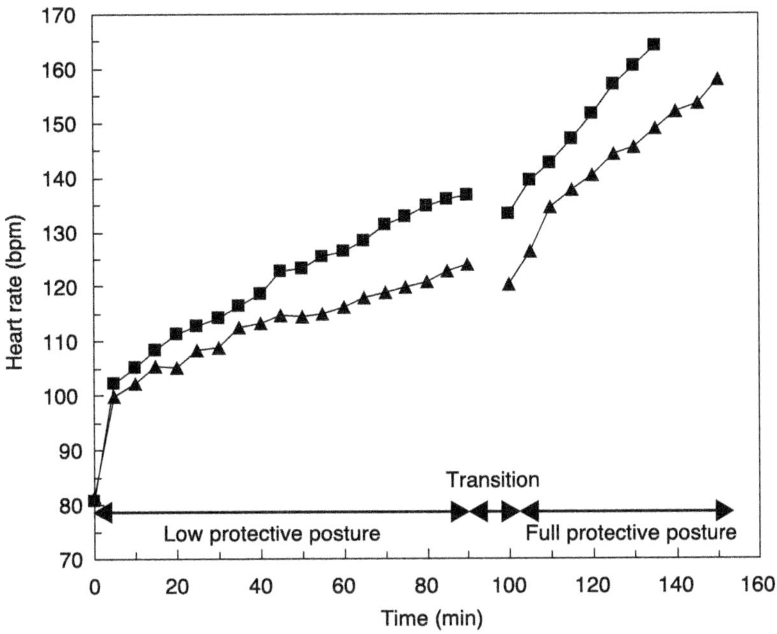

13.19 Variation in mean heart rate while wearing protective combat uniform at low protective posture (overboots, gloves, APR and hood not worn) and subsequent effect on heart rate after transition to full protective posture. Protective combat uniform with no vents indicated by square symbols and protective combat uniform with vents indicated by triangle symbols. Vents were sealed during full protective posture. Heat strain activity was walking at 4.5 km h^{-1} and exposure to 35 °C and 50% RH. Data as presented in McLellan.[33]

without the vents. This affected the average skin temperature of the uniform systems following transition to full protective posture; the average skin temperature for the uniform with vents (sealed) remained significantly reduced compared to the uniform with no vents. Differences in core body (rectal) temperature changes were not as marked between the vented and non-vented garments during low protective posture. The increase in the rectal temperature change was greater for the uniform without vents after 75 minutes of heat stress (see Fig. 13.20). After the transition to full protective posture the difference in the rectal temperature change between the two systems remained evident for the duration of the additional heat stress. The significantly higher tolerance time observed for the uniform with the vents may be attributed to the difference in the rectal temperature at the beginning of the heat stress in full protective posture.

McLellan's[33] study has provided supporting evidence that changes to uniform design, such as incorporating vents, are an additional avenue to be considered along with the development of new protective material systems, which can impact

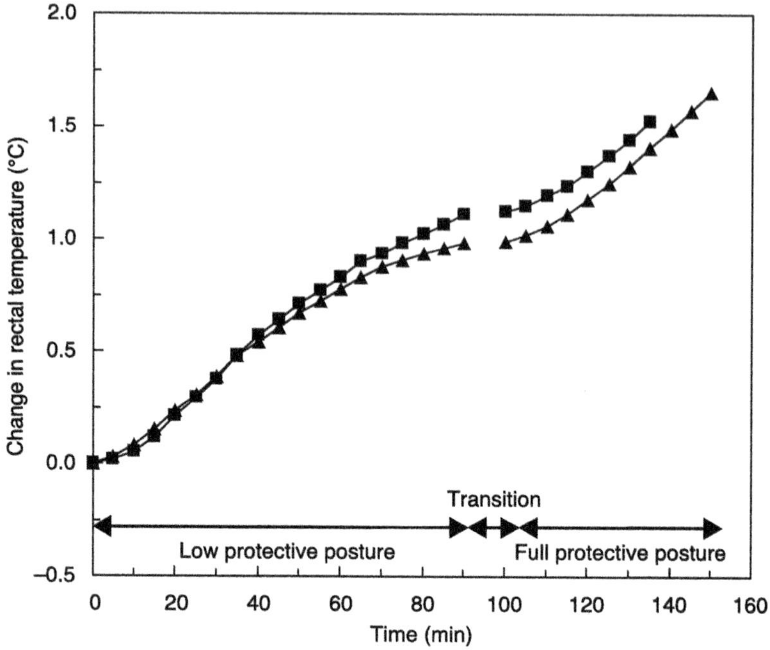

13.20 Change in rectal temperature while wearing protective combat uniform at low protective posture (overboots, gloves, APR and hood not worn) and subsequent effect on change in rectal temperature after transition to full protective posture. Protective combat uniform with no vents indicated by square symbols and protective combat uniform with vents indicated by triangle symbols. Vents were sealed during full protective posture. Heat strain activity was walking at 4.5 km h^{-1} and exposure to 35 °C and 50% RH. Data as presented in McLellan.[33]

physiological burden in a positive way. The subjects in this study wore a fragmentation vest (without the armour plates) and tactical assault vest (with simulated ammunition magazines) on the torso over the protective uniform. These additional items of equipment completely covered other vents that were located on the front torso of the garment. The results of his study clearly revealed that incorporating vents into the uniform at locations that are not impeded by other equipment assisted with heat transfer in low protective posture and extended the tolerance time after transitioning to full protective posture with sealed vents. It seems consistent to consider that when it is not necessary to wear the fragmentation vest, opening vents on the torso may further aid in the transfer of heat away from the body during heat stress.

We note significant progress has been realised to reduce weight and thickness, and optimise air permeability and evaporative and thermal resistance, to increase the total heat loss through our asymmetric operations CB protective materials (see Table 13.5). Of course, an adequate protective capability must be maintained

Table 13.5 Thermal and evaporative resistance of various IPE protective materials at 40 °C, 20% RH, wind speed 3 m s^{-1} with 6.4 mm spacer between hotplate skin and fabric system

	IPE protective material			
	Cold War	Post-Gulf War	Asymmetric	Reference – combat uniform (single layer)
Weight (g m^{-2})	500	445	297	170
Thickness (mm)	2.32	1.23	0.87	0.31
Air permeability (cm^3 cm^{-2} s^{-1})	15	13	31	65
Thermal resistance (R_{ct} – m^2 °C W^{-1})	0.25	0.21	0.24	0.20
Evaporative resistance (R_{et} – Pa m^2 W^{-1})	39.58	26.29	22.28	20.40
Total heat loss (THL – W m^{-2})	117.6	159.9	174.9	190.6
Permeability index (I_m)	0.38	0.49	0.65	0.60

in the quest to reduce physiological burden. It is evident that current adsorbent technology, whether it is carbon based or not, when incorporated as a layer into a protective material system, will add weight and thickness and reduce the air permeability of the composite system. Accordingly, we may be able to approach the physical characteristics of single layer combat uniform material, but we will never achieve it with such technology. Currently we have attained as much as a 50% improvement in performance in the evaporative resistance and total heat loss for asymmetric materials compared to the Cold War material (Table 13.5). However, the gains realised with the asymmetric threat materials have only been made with current carbon adsorbent technology by considering a new operational construct for the asymmetric threat environment which acknowledges shorter mission times, use of other equipment and tactical procedures as part of an extended protection umbrella, and a lower probability of encountering CB agents. These new materials are being designed specifically to bring their protection performance in-line with the asymmetric CB operational window of between two and six hours. Demands for protection beyond these times could not be managed using these materials.

Further changes to the design of material systems and protective garments to achieve balanced protection and lower physiological burden are being explored, but there will be situations where the rate of heat storage and heat tolerance are governed primarily by the rate of heat production and not by environmental conditions. Thus, there always will be scenarios of use that, no matter how well designed the protective system is, will result in undue thermal strain on the user.

13.10 Conclusions

It is recognised that military requirements for protection are outside the general norms for most work professions. When the military use IPE it always will be under intense, physically demanding conditions. In most operational settings some level of heat strain will be inevitable; when it will occur and how much it will degrade task efficiency, will depend on factors such as climate, metabolic work rate, psychological stress and opportunities for rest. The use of air permeable protective materials in IPE is essential to afford the users some reasonable amount of work time to complete tasks effectively and ultimately to limit the severity of heat strain. However, if the air permeability of the materials is too high, it reduces the protection performance of the IPE system; the consequence of this may be fatal. There is no single, ideally correct value for the air permeability of a protective material. Knowledge of the threat and mission times are key to establishing comprehensive performance specifications for a protective material that will meet critical toxicological protection requirements and at the same time delay the onset of severe heat strain sufficiently to ensure missions can be completed successfully. Less severe threats, more locally confined hazards and lower probabilities of occurrence may allow the down-scaling of protection duration requirements for materials that will translate into gains in the reduction of heat strain. By incorporating features into the design of the uniform that passively or actively aid in the reduction of thermal burden there will be less of a necessity for the evaporative heat loss requirement to be accommodated in its entirety through the protective material. Recent results of work with vents highlights the potential for such approaches. Other ideas that we are currently exploring for asymmetric threat IPE systems include multiple, air permeable materials, high moisture permeable and agent resistant membranes, single layer protective undergarments and designing the protection into combat uniforms around other equipment that is worn, such as ballistic protection and rain gear. Comfort is highly desirable, but in a military context it must be achieved as an outcome from a more intensive process to finely tune protection and burden to the anticipated threat, expected hazard and desired operational capabilities. It is not possible to provide perfect protection nor is it possible to eliminate heat strain – the aim should be to manage these effectively within the context of the operational requirements to achieve a level of comfort that will not impede mission success.

In summary, IPE offers an obvious capability enhancement, permitting safer operations in a contaminated environment by reducing the likelihood of casualties from exposure to toxic materials. At the same time, other mission activities potentially may be impaired due to an increase in physiological burden and the reduction in functionality that result from wearing IPE, and further, affected individuals may be at greater risk of becoming a casualty as a result of lowered performance. Higher levels of target protection necessarily penalise design choices for lower burden and improved ergonomic functionality, and the opposite is true as well. Obviously, it is desirable to find the optimum trade-off, where the likelihood of becoming a casualty while operating in a contaminated – or potentially contaminated – environment is at a minimum, and the opportunity to achieve mission success is at its highest. Designing IPE for very specific threat states, and to protect against realistic hazards for only as long as operational requirements demand, should lead to a new generation of IPE for the asymmetric threat environment which offers significant capability enhancements. Canadian defence research for protection against toxic materials has clearly demonstrated the potential gains to be realised within the asymmetric threat context pursuing technologies for protective combat duty uniforms.

13.11 Future trends

Where does the future lie for military IPE? Certainly, if we want to achieve protective materials that are single layer systems that can be used in place of standard combat clothing, the current separate adsorbent layer must be re-invented in some new manifestation, perhaps as an adsorptive coating or other constituents functionalised or embedded on yarns in fabric. The difficulty in attaining a single layer protective material becomes more evident when one considers that such a material also may require hydro/oleophobic finishes for liquid repellency and anti-fouling, infra-red absorbance, unique dyes for disruptive camouflage, be non-irritating and non-toxic to skin contact, have supple hand properties akin to existing combat cloth, good abrasion resistance, and be able to maintain its protective efficacy over an extended wear/wash life-span. This is a tall order for a single layer protective material to meet.

The area of science being investigated most actively in terms of the development of new materials is nanotechnology. A key area of interest from a defence perspective relates to the human platform, and much of the research and development effort has focused on the protection, performance and survivability of the future soldier. In particular, there has been considerable effort directed at reactive constituents that could be added to materials to provide self-decontamination (active coatings).[40,41] Other areas of interest include lightweight nanocomposites, selectively permeable membranes, biomimetic materials, polymer membranes with switchable porosity to augment evaporative heat transfer, super-repellent and self-cleaning surfaces, photo-adaptive camouflage, closures with 'gecko-feet' nano-adhesion, and miniaturised, wearable and network-linked sensors for physiological, environmental and situational awareness.[42–52] Individually, these functionalities may offer little gain, the real benefit will come at the system level when the technologies can be harmonised into a human centric, fully integrated protective system (smart uniform) offering higher protection, lower burden, longer endurance and enhanced self-supporting capabilities to optimise human performance and maintain well-being under a myriad of operating conditions. While any practical and serious application of nanotechnology in IPE is years away, a glimpse into a nano-crystal ball might reveal smart combat uniforms with superamphiphobic coatings, capture and self-decontaminating finishes, activated carbon nanofibres, selective nanopore membranes, imbedded CBW agent nanosensing arrays, environmental and physiological monitoring, active wound treatment, adaptive insulation and ventilation, adaptive camouflage and thin, flexible antiballistic armour.

13.12 Acknowledgements

The authors would like to thank the scientists and technicians at Defence R&D Canada and The Royal Military College of Canada who either provided direct technical support, or participated in thoughtful discussions, which made it possible to acquire the data and complete the many studies reported and discussed in this chapter. The majority of the work was funded through the Applied Research Programme of Defence R&D Canada, an agency within the Department of National Defence, unless otherwise indicated in the acknowledgements of cited references. Special thanks are due to the Technical Editor, Dr Guowen Song, and Woodhead Publishing for the invitation to contribute to this book.

13.13 References

1. Duncan EJS, Gudgin Dickson EF (2003). 'A new whole-body vapour exposure chamber for protection performance research on chemical protective ensembles', *American Industrial Hygiene Association Journal*, **64**: 212–221.
2. NFPA 1991-05 Standard on Vapour Protective Ensembles for Hazardous Materials Emergencies, NFPA 1981-07 Standard on Open-Circuit Self-Contained Breathing Apparatus (SCBA) for Emergency Services.
3. NFPA 1994-07 Standard on Protective Ensembles for Chemical/Biological Terrorism Incidents.
4. Nielsen R, Olesen BW, Fanger PO (1985). 'Effect of physical activity and air velocity on the thermal insulation of clothing', *Ergonomics* **28**: 1617–1631.
5. Vogt JJ, Meyer JP, Candas V, Libert JP, Sagot JC (1983). 'Pumping effects on thermal insulation of clothing worn by subjects'. *Ergonomics* **26**: 963–974.
6. Dickson, EG, Bodurtha P, Fedele P (2005). 'Predicting protection performance behaviour of protective ensembles against airborne contaminants: modeling and measurement', *Proceedings of the First International Conference on NBC Individual Protective Equipment, Bath, UK*, September, 6 pp.
7. Duncan S, Tremblay-Lutter J, Grant T, Gudgin Dickson E (2001). 'System protection performance of a next-to-skin chemical vapour protective suit', *Proceedings of the Seventh International Symposium on Protection Against Chemical and Biological Warfare Agents, Stockholm, Sweden*, June, 7 pp.

8. McLellan TM, Bell DG, Smith IF, Morris AC, Miyazaki M (1997). 'Heat strain in the current Canadian Forces NBC ensemble compared with new hot-weather NBC garments.' Defence and Civil Institute of Environmental Medicine, Technical Report DCIEM No. 97-R-25, 1997.
9. McLellan TM, Meunier P, Livingstone S (1992). 'Influence of a new vapour protective clothing layer on physical work tolerance times at 40 °C', *Aviat Space Environ Med* **63**: 107–113.
10. McLellan TM, Bell DG, Dix JK (1994). 'Heat strain with combat clothing worn over a chemical defence (CD) vapour protective layer', *Aviat Space Environ Med* **65**: 757–763.
11. McLellan TM (1996). 'Heat strain while wearing the current Canadian or a new hot-weather French NBC protective clothing ensemble', *Aviat Space Environ Med* **67**: 1057–1062.
12. McLellan TM, Pope JI, Cain JB, Cheung SS (1996). 'Effects of metabolic rate and ambient vapour pressure on heat strain in protective clothing', *Eur J Appl Physiol* **74**: 518–527.
13. McLellan TM (1993). 'Work performance at 40 °C with Canadian Forces biological and chemical protective clothing', *Aviat Space Environ Med* **64**: 1094–1100.
14. Craig FN and Moffitt JT (1974). 'Efficiency of evaporative cooling from wet clothing', *J Appl Physiol* **36**: 313–316.
15. Craig FN, Garren HW, Frankel H, Blevins WV (1954). 'Heat load and voluntary tolerance time', *J Appl Physiol* **6**: 663–644.
16. Schvartz E and Benor D (1972). 'Heat strain in hot and humid environments', *Aerosp Med* **43**: 852–855.
17. McLellan TM, Jacobs I, Bain JB (1993). 'Influence of temperature and metabolic rate on work performance with Canadian Forces NBC clothing', *Aviat Space Environ Med* **64**: 587–594.
18. Givoni B, Goldman RF (1972). 'Predicting rectal temperature response to work, environment and clothing', *J Appl Physiol* **32**: 812–822.
19. Gonzalez RR, Levell CA, Stroschein LA, Gonzalez JA, Pandolf KB (1993). 'Copper manikin and heat strain model evaluations of chemical protective ensembles for The Technical Cooperation Program (TTCP).' Natick, MA: US Army Research Institute of Environmental Medicine, Technical Report No. 94-4.
20. Cain B and McLellan TM (1998). 'Model of evaporation from the skin while wearing protective clothing', *Int J Biometeorol* **41**: 183–193.
21. Gonzalez RR, McLellan TM, Withey WR, Chang SK, Pandolf KB (1997). 'Heat strain models applicable for protective clothing systems: comparison of core temperature response', *J Appl Physiol* **83**: 1017–1032.
22. McLellan, TM (1994). 'Tolerance times for continuous work tasks while wearing NBC protective clothing in warm and hot environments and the strategy of implementing rest schedules'. Defence and Civil Institute of Environmental Medicine, Technical Report DCIEM No. 14.
23. AJ Mauroni (2003). *Chemical and Biological Warfare: A Reference Handbook.* Santa Barbara, California, ABC-CLIO, Inc.
24. Cal, MP and Rood MJ (1997). 'Gas phase adsorption of volatile organic compounds and water vapor of activated carbon cloth', *Energy and Fuels* **11**: 311–315.
25. Yamamoto T, Endo A, Eiad-ua A, Ohmori T, Nakaiwa M, Soottitantawat A (2007). 'Synthesis of monodisperse carbon beads with developed mesoporosity', *AIChE Journal* **53**: 746–749.

26. Etienne S, Melin B, Pelicand JY, Charpenet A, Warme-Janville B (1994). 'Physiological effects of wearing light weight NBC battle dress in hot environments', in: Frim J, Ducharme MB, Tikuisis P, eds. *Proceeding of the Sixth International Conference on Environmental Ergonomics, Montebello, Canada*, pp. 30–31.
27. Levine L, Quigley MD, Latzka WA, Cadarette BS, Kolka MA. 'Thermal strain in soldiers wearing a chemical protective undergarment: results from a laboratory study and field study.' Natick, MA: U.S. Army Research Institute of Environmental Medicine, Technical Report No. 2-93.
28. Amos D, Hansen R (1997). 'The physiological strain induced by a new low burden chemical protective ensemble', *Aviation Space Environmental Medicine* **68**: 126–131.
29. Caputa M, Cabanac M (1988). 'Precedence of head homoeothermia over trunk homoeothermia in dehydrated men', *Eur J Appl Physiol* **57**: 611–615.
30. McCaffrey TV, Geis GS, Chung JM, Wurster RD (1975). 'Effect of isolated head heating and cooling on sweating in man', *Aviat Space Environ Med* **46**: 1353–1357.
31. Duncan EJS, Tremblay-Lutter J, Caron P (2004). *CB^{plus} Material and Combat Uniform Protection and Performance Requirements*, DRDC Suffield TM 2004-222, December, 41 pages.
32. Tremblay-Lutter J, Caron P, Duncan S, Claesson O, van Houwelingen T, Stromqvist M, Gudgin Dickson E (2004). 'CB^{plus} combat operations uniform innovation: innovation in protection and benefits of international collaboration', *Proceedings of the Eighth International Symposium on Protection against Chemical and Biological Warfare Agents, Gothenburg, Sweden*, June, 7 pp.
33. McLellan TM (2008). 'Chemical-biological protective clothing: effects of design and initial state on physiological strain', *Aviat Space Environ Med* **79**: 500–508.
34. Gonzalez RR (1987). 'Biophysical and physiological integration of proper clothing for exercise', *Exerc Sport Sci Rev* **15**: 261–296.
35. Nielsen R, Olesen BW, Fanger PO (1985). 'Effect of physical activity and air velocity on the thermal insulation of clothing', *Ergonomics* **28**: 1617–1631.
36. Vogt JJ, Meyer JP, Candas V, Libert JP, Sagot JC (1983). 'Pumping effects on thermal insulation of clothing worn by subjects', *Ergonomics* **26**: 963–974.
37. Lotens WA, Havenith G (1988). 'Ventilation of rainwear determined by a trace gas method', in: Mekjavic IB, Banister EW, Morrison JB, eds, *Environmental Ergonomics Sustaining Human Performance in Harsh Environments*, London, UK: Taylor and Francis, pp. 162–176.
38. Lotens WA, Wammes LJA (1993). 'Vapour transfer in two-layer clothing due to diffusion and ventilation', *Ergonomics* **36**: 1223–1240.
39. Goldman RF (1974). 'Clothing design of comfort and work performance in extreme thermal environments', *Trans NY Acad Sci* **36**: 531–544.
40. Walker J, Schreuder-Gibson H, Yeomans W, Ball D, Hoskin F, Hill C (2003). *Development of Self-Detoxifying Materials for Chemical Protective Clothing*, U.S. Army Natick Soldier Center, Natick, MA, July, 8 pages.
41. Nagarajan R, Zukas W, Hatton TA, Lee S (2010). *Nanoscience and Nanotechnology for Chemical and Biological Defense*, (ACS Symposium Series No. 1016). Oxford University Press, 384 pages.
42. Stevens CL, Gnatowski MJ, Duncan S (2007). 'Nylon-12 nanocomposite thin films as protective barriers', Rapra International Conference on Polymers in Defence and Aerospace 2007, Smithers Rapra Technology Ltd., Toulouse, France. Paper 4, pages 1–12.

43. Steeves DM, Farrell R, Ratto JA (2007). 'Investigation of polybutylene succinate-*co*-adipate (PBSA)/montmorillonite layered silicate (MLS) melt-processed nano-composites', *J Biobased Mat Bioenergy* **1**: 94–108.
44. Koziol K, Viltela J, Moisala A, Motta M, Cunniff P, Sennett M, Windle A (2007). 'High-performance carbon nanotube fiber', *Science* **318**: 1892–1895.
45. Chiang C-Y, Mello CM, Gu J, Silva ECCM, Van Vliet KJ, Belcher AM (2007). 'Weaving genetically engineered functionality into mechanically robust virus fibers', *Adv Mater* **19**: 826–832
46. Sitti M, Fearing RS (2003). 'Synthetic gecko foot-hair micro/nano-structures as dry adhesives', *J Adhes Sci Tech* **17**: 1055–1073.
47. Lee HJ, Michielsen S (2007). 'Preparation of a superhydrophobic rough surface', *J Polym Sci Part B: Polym Phys* **45**: 253–261.
48. Brewer SA, Apperley DC, Dennis M, Duncan S, Stone CA, Willis CR (2009). 'The influence of humidity on the protective performance of a membrane based on poly(vinyl alcohol)', *J Mater Chem* **19**: 7897–7903.
49. Klossner RR, Queen HA, Coughlin AJ, Krause WE (2008). 'Correlation of chitosan's rheological properties and its ability to electrospin', *Biomacromolecules* **9**: 2947–2953.
50. Achmann S, Hagen G, Kita J, Malkowsky IM, Kiener C, Moos R (2009). 'Metal-organic frameworks for sensing applications in the gas phase', *Sensors* **9**: 1574–1589.
51. Thandavamoorthy S, Gopinath N, Ramkumar SS (2006). 'Self-assembled honeycomb polyurethane nanofibers', *J Appl Poly Sci* **101**: 3121–3124.
52. De Vrieze S, Westbroek P, Van Camp T, De Clerck K (2010). 'Solvent system for steady state electrospinning of polyamide 6.6', *J Appl Poly Sci* **115**: 837–842.

14
Balancing comfort and function in textiles worn by medical personnel

W. CAO, California State University – Northridge, USA and
R. M. CLOUD, Baylor University, USA

Abstract: Wear comfort, including psychological and physiological comfort, is of particular concern to the surgical personnel, whose efficiency needs to be supported rather than impaired. Improvements in protective performance of medical apparel products often are achieved at the expense of comfort, making wearers less likely to comply with universal precaution guidelines. This chapter focuses on issues involved in balancing the comfort and function of textile products worn by medical personnel. The discussion will identify factors which influence the comfort and function of surgical gowns, nursing uniforms, gloves, and masks and consider the needs/concerns of government agencies, researchers, manufacturers and users in guiding comfort improvements.

Key words: surgical gowns, surgical gloves, surgical mask, thermal comfort, sensorial comfort, fit/mobility.

14.1 Introduction

Medical textiles, an emerging type of industrial textiles, have played an increasingly important protection role in the healthcare industry. The term 'medical textiles' encompasses a wide range of soft goods used for medical and hygiene applications, including those for surgical, orthopedic, and dental uses. As a niche market in the global economy, the medical/hygiene textiles sector has grown steadily and is expected to continue experiencing strong, long-term demand based on the aging consumer and healthy lifestyle proponents. Part of an estimated \$3.5–4.0 trillion a year healthcare industry, hygiene products are valued at approximately \$73 billion. Medical textiles accounted for 9% of global consumption of technical textiles in 1990 and consumption is predicted to have annual growth at 4.5% through 2010 (Sparrow, 2009).

Medical textiles are classified as non-implantable (e.g., wound dressing, bandages, gauzes), implantable (e.g., artificial arteries, sutures, vascular grafts), extracorporeal devices (e.g., artificial organs) and healthcare and hygiene products (Rajendran & Anand, 2003; Horrocks, & Anand, 2000). Surgical textiles for medical personnel fall within the classification of healthcare and hygiene products. Fibers used in medical textiles must be non-toxic, non-allergic, non-carcinogenic and be able to be sterilized without imparting any change in physical or chemical characteristics. Strength, flexibility, absorbency or biodegradability could be expected for particular products.

A primary function of surgical hygiene products is to minimize the cross-infection between healthcare workers and patients. Most disposable hygiene products and a significant proportion of medical textiles are made of nonwoven textiles, which are valued at $14.5 billion a year with an average annual sales growth rate of 7% expected over each of the next five years (Nonwoven Industry Report, 2006). Several types of natural and synthetic fibers/materials are used in healthcare/hygiene products. The fibers used for specific healthcare/hygiene products, primary fabric structures, ranges of textile materials employed within each classification and the requirements are provided in Table 14.1 (Rajendran & Anand, 2003; Horrocks & Anand, 2000; Williamson, 2003; Cao & Cloud, 2010).

Surgical textiles, such as surgical gowns, gloves and mask, are a familiar sight in operating rooms for protection against the spread of hazardous microorganisms/viruses for both surgical teams and patients. Wear comfort, including psychological and physiological comfort, is of particular concern to the surgical personnel, whose efficiency needs to be supported rather than impaired. The near environment created by wearing gowns, gloves and mask can increase the potential for heat fatigue, which has been shown to contribute to increased mistakes, impaired

Table 14.1 Healthcare/hygiene products

Product (surgical clothing)	Function	Requirement	Fiber type	Fabric structure
Gowns	Barrier for contaminated body fluid from patients. Barrier to prevent the release of pollutant particle from the surgeon	Liquid penetration Comfort Sterile Flexible	Cotton Polyester Polyethylene Polypropylene	Woven Nonwoven Composite materials Membrane
Gloves		Softness Pliability Comfort Easy donning Tactile sensitivity	Natural latex rubber Synthetic materials	Film Knit
Masks		High filter capacity High air permeability Lightweight, non-allergenic	Viscose Glass Polyester	Nonwoven

performance, and less efficient work (Rutala & Weber, 2001). In addition, providing adequate physiological comfort for healthcare workers can have a positive psychological influence. Comfort of medical apparel, like any other apparel, is affected by thermal/moisture comfort, sensorial comfort, and fit/mobility issues.

14.2 Surgical gowns

Surgical gowns were originated to protect patients from bacterial infections. In recent decades, their protective functions have extended to healthcare workers due to increased occupational exposure to blood-borne pathogens such as human immunodeficiency virus (HIV) and hepatitis B and C viruses (Laufman et al., 2000). As medical personnel recognized the risks in performing the medical tasks (particularly surgeries, implants and heart by-pass operations), more attention was given to the level of the protection gowns can provide for surgical staff and emergency personnel. A recent standard, developed by the American Society for Testing Materials (ASTM), ASTM F2407-06, Standard Specification for Surgical Gowns Intended for Use in Healthcare Facilities, classifies a barrier material's performance into Levels 1, 2, 3 and 4.

The standard classifies barrier levels (Table 14.2) based on the results of the following tests: AATCC 42 (Water Resistance: Impact Penetration Test); AATCC 127 (Water Resistance: Hydrostatic Pressure Test); and ASTM 1671 (Standard Test Method for Resistance of Materials Used in Protective Clothing to Penetration by Blood-Borne Pathogens Using Phi-X174 Bacteriophage Penetration). The water impact penetration test (AATCC 42) determines the ability of a material to resist distilled water penetration under one time spray contact. The hydrostatic pressure test (AATCC 127) method was originally designed to determine a

Table 14.2 Classification of surgical gown barrier performance

Level	Test	Challenged with	Result
1	AATCC 42[1]	Water	<4.5 g
2	AATCC 42	Water	<1.0g
	AATCC 127[2]	Water	>20cm
3	AATCC 42	Water	<1.0g
	AATCC 127	Water	>50cm
4	ASTM 1671	Surrogate blood bacteriophage Phi-X174	Pass

Note: Information in this table is from ASTM F2407, 'Specification for Surgical Gowns Intended for Use in Healthcare Facilities'.
[1] Criterion is based on change in weight gain of blotting paper.
[2] 1 cm water pressure = 1 mbar pressure. Criterion is based on pressure at which three drops of water penetrate fabric.

rainwear fabric's water repellency capability. It tests the resistance of fabric to liquids under hydrostatic pressure. ASTM 1671 is designed to determine the ability of a material to resist the penetration of microorganisms under constant contact.

Surgical gowns may be constructed of single layer, single-use (nonwoven fabrics) or reusable materials (woven fabrics), or gowns reinforced with a second layer of material (films, coatings, plastic, etc.) in the front panel and in the forearm area. A case study by Wong et al. (1998) demonstrated that wearing appropriate protective barriers can significantly improve protection for healthcare personnel. Because surgical procedures involve varying levels of blood exposure, surgical gown fabrics offering different levels of liquid protection have been made available by manufacturers. In addition to the barrier effect, surgical gowns need to demonstrate good wear comfort because surgeons have to wear gowns for several hours while doing strenuous work (Aibibu et al., 2003a). ASTM F2407-06 standard provides an elaborate indicator in evaluating the protective performance of gown materials, but to compare comfort levels of various surgical apparel is even more challenging (Akridge, 2004).

14.2.1 Thermal and moisture comfort

Even though it is crucial to be protected by surgical gowns, surgeons are not likely to sacrifice comfort and fit during wearing (Belkin, 1993; Stanley, 1994). A survey investigating wearers' and purchasers' satisfaction for surgical gowns reported continuing issues with thermal discomfort of gowns (Brandt, 1993). Thermal comfort exists when a balance occurs between the heat the human body loses and the heat the body generates. It is a condition of mind which describes satisfaction within a hot/cold environment, which relies on both internal and external temperature sensitivity.

Thermal and moisture comfort of gowns are highly correlated with the conditions of surgical environments, length of procedure and amount of exposure. Brandt (1993) indicates that operating room temperatures range from 15.6 °C/60 °F to 25.6 °C/78 °F, and Mangram and co-workers (1999) report relative humidity measurements of 30–60%. The radiant heat from overhead lights also increases the temperature of the surgical theatre. During surgical procedures, nurses and doctors perform in a high-stress environment under strong lighting causing their bodies to release heat (Cao & Cloud, 2010). If internal body heat can not be dissipated appropriately, heat fatigue will occur, which may impair performance. Research indicates that a sweating surgeon is significantly more likely to contaminate the surgical field than a non-sweating surgeon (Mills et al., 2000). Medical personnel vary in their ability to tolerate heat and cold. If health care workers wear the same gowns/scrubs but perform in a higher temperature or higher humidity environment, they will feel exhausted easily. Surgical personnel may feel more comfortable wearing a one layer disposable gown instead of a film-reinforced gown, but the latter garment would be required for adequate protection when performing a lengthy open-heart surgery.

Surgical gowns play a critical role in modifying the medical personnel's response to temperatures in the operating room; they should assist in maintaining an isothermal environment for health care workers (Recommended Practice, 2003). The results from a series of studies to investigate the perceived comfort response to fabric indicated that thermophysiological comfort parameters related more to fabric structures than to fiber content (Hatch et al., 1990). Pissiotis et al. (1997) surveyed surgeons to compare the comfort of four types of disposable single layer gown, a reusable cloth gown and four types of disposable reinforced gowns used during 250 major operations. It was concluded that only four percent of the surgeons felt comfortable in the reusable cotton gown, and more than 61% of surgeons prefer to wear either disposable single layer or reinforced gowns. Surgeons also felt less comfortable when wearing the reinforced gowns compared with non-reinforced disposable gowns from the same materials due to the inhibition of evaporation from the skin. Membranes and coatings tend to impair the wear comfort (Aibibu et al., 2003b).

A study to evaluate thermal comfort properties of surgical gowns made of dual functional finish cotton and nonwoven fabrics showed that finished surgical gowns allowed more heat to be dissipated from the skin of the subject than the equivalent untreated gowns. The nonwoven surgical gown showed no difference in comfort properties from the cotton gown (Cho et al., 1997). Theoretically, thermal conduction/insulation properties of materials and air permeability of fabrics are factors that impact thermal comfort of the gown materials. Intuitively we understand that tightly woven materials or those with impervious lamination will increase the heat stress due to lack of ventilation. Lightweight or thin fabric is lower in heat insulation, while, conversely, thick fabric contributes to heat retention/storage. McCullough (1992) measured the thermal comfort and sensorial comfort of six surgical gowns; she found that the materials differed in flexibility, weight, softness, scratchiness, slipperiness, papery feel, clamminess and stickiness, which impact the comfort evaluation of wearers.

Advanced technologies in fabric production and innovative seaming techniques are the approaches manufacturers have taken for providing enhanced protective surgical gowns without sacrificing comfort/breathability. One approach is to make smart materials treated with phase change materials to regulate temperatures within the comfort range (Mondal, 2008). The brand, Smart Gowns®, is claimed by manufacturers to employ a breathable yet impervious gown using a microfiber composite and a monolithic film which reacts to increasing temperature by increasing the moisture transfer rate from the body. The moisture vapour transfer rate of the material is expected to increase exponentially as the moisture level rises beneath the garment (Akridge, 2004). Another approach is to use layered, breathable materials with an active cooling mechanism. A good example is Acturel® brand fabric. This breathable, impervious barrier material is said to maximize comfort by adjusting to the wearer's body temperature over time. Acturel® is made of three layers – polyester nonwoven as an inner layer, Hytrel® as a breathable membrane sandwich [or middle] layer and spunbonded polypropylene as the outer layer. The combination of these three layers is claimed to cool the body down by removing more moisture from the skin than typical gowns, allowing the wearer to stay fresher longer (The Miracles of Science™, 2008).

Surgeons have specific requests regarding surgical gowns in terms of their ideal comfortable fabric. They prefer softer, pliable and more comfortable products which can maximize the balance between protection and greater moisture dissipation (Akridge, 2004). Cotton products are typically expected to be more comfortable due to the absorbency properties of cotton. However, a complex network of ultra thin polyester and polyamide fibers was found to provide water-repellency and moisture transference more effectively than cotton (Suprun et al., 2003).

Ideal surgical apparel should withstand liquid penetration, but allow excess body heat to be dispersed by means of air and moisture transfer. Comfort is dependent on the permeability and flexibility of the fabric (T-PACC, 1997). One of the unsatisfactory properties of the medical clothes which inhibit its use is low air permeability of the clothing (Suprun et al., 2003). The level of comfort could be influenced by the fabrics' construction, such as the fineness and shape of the filaments, weaving structure, and fabric density, which will affect the pore structure of the fabric and lead to the effect on comfort (Aibibu et al., 2003b). Manufacturers have taken many approaches to improve the permeability of the materials through the structural design. One example is MicroCool® brand fabric, which features an impervious plastic film sandwiched between two layers of spunbonded nonwoven. The structure is designed to allow moisture vapor to pass through the fabric from the inside of the garment, but prevent the passage of fluids or microorganism from outside (Akridge, 2004).

14.2.2 Sensorial comfort

Surgical gowns should drape easily. They must allow necessary mobility without rubbing and chafing and should feel soft and smooth (May-Plumlee & Pittman, 2002). Medical clothes with rigid materials restrict the using from medical personnel due to the discomfort touch (Suprun et al., 2003). The stiffness of gown materials may affect perspiration and movement (Abreu et al., 2003).

In 1995, 80% of U.S. surgical gowns were made by wood pulp/polyester spunlaced fabric which was very soft and breathable. However, layers of materials need to be used to maintain the protection level needed for surgical gowns, which increased the thickness and weight of the gown materials (Akridge, 2004). In 2004, many surgical apparel fabrics were made using synthetic fibers which can provide higher level of protection with thinner fabrics (based on the same weight of the material) than the products made by natural fibers (Akridge, 2004).

Astound® surgical apparel uses microfiber technology in a spunlaced basis structure. This upgraded fabric was advertised to be 35% lighter and offers 25% more air permeability than the previous spunlaced product by the same manufacturer (Williamson, 2003). Suprel® is a combination of polyester (for strength) and polyethylene (for comfort and softness). The fabric features a very smooth texture with less surface friction than competitive products (The Miracles of Science™, 2008). A study investigating the influence of the sterilization process on the tactile feeling of three types of surgical gowns (outer layer and inner layer were made by different combination of non-woven, PE and laminate) found no difference in the products after sterilization (Philippe et al., 2003).

14.2.3 Fit/mobility comfort

Neckline, sleeves, cuffs and overall fit are important attributes to be considered for both comfort and protection in gown selection (Ellis, 2005). Surgical apparel should allow the wearer a proper range of movement when performing work-related tasks. Inappropriate fit of gowns can affect the protection levels negatively (Stanley, 1994). Excess fabric is common because the gowns must be designed to fit a diversity of body shapes and sizes within a limited range of sizes (May-Plumlee & Pittman, 2002). The survey investigating wearers' and purchasers' satisfaction with surgical gowns reported that surgeons and nurses didn't feel the gowns comfortable due to the size (too large). This discomfort further brought the safety concern for patient blood and body fluid contamination from the surgeons and nurses. Survey data suggested desired changes and improvements in gown fit and sizing related to comfort (Brandt, 1993).

Gowns should be easy to don and doff without cross-contamination, which requires the gown to have appropriate openings (Stanley, 1994). The design influence in comfort of surgical gown through a case study by May-Plumlee and Pittman (2002) has drawn the following conclusions. The way to close the neckline of the gowns, distance between snaps arranged in the neckline and hook and loop tape closure utilization all contribute some adjustability for the ease and flexibility in fit. The ease of comfortable movement in sleeves depends on the style of sleeve (raglan style versus set in sleeves), excess fabric in the shoulder and upper chest area, sleeve length and width and cuffs fit. Reusable gowns have better fit than disposable gowns due to the ease in adjusting neckline size and closer fit in the torso (May-Plumlee & Pittman, 2002). Proper fit of surgical gowns can even ensure an optimal glove/gown interface (Ellis, 2005)

In summary, factors which could influence the fit/mobility issue of gowns are structural design (neckline, closure, tie, cuff, etc.), size analysis (circumferences, limiting excess fabric in the shoulder and upper chest area), material influence (stretch, lightweight, etc.) and fit evaluation in cuff, forearm, thigh, and chest and abdomen area (May-Plumlee & Pittman, 2002). One of the innovative technologies adopted by manufacturers was using stretch/smooth materials, like Sorona™ and

Suprel®, to facilitate mobility of gowns during wearing. Sorona™ can provide comfort stretch and recovery due to the differentiated polymer ingredient by 1,3 propanediol (PDO) and other monomers and additives. Suprel® glides easily without catching or grabbing, allowing for freedom of movement (The Miracles of Science™, 2008).

14.2.4 Comfort issues in nursing uniform

Dressing for comfort and practicality has always been important for nurses (Kelly, 1957). Changes in nurses' uniforms during their history have included length of skirt, length of the sleeve, color of uniform, number of pieces, presence or absence of bib or apron, and the position of waistline and neckline (Kelly, 1957). Despite the revolution in appearance, cost and suitability, whether it is comfortable (soft, cool, lightweight, absorbent, antistatic, elastic, etc.) is always a primary criterion to be considered by the nurses (Kelly, 1958). However, limited research work has investigated the thermal/moisture and sensorial issues for nurses' uniform.

Fit/mobility of the uniform is crucial to nurses in the practice (Kelly, 1957). However, manufacturers and distributors have given more attention to style and material quality than to function (freedom of movement) of the uniform. Some nurses have suggested design changes such as a gored skirt or added reinforcement under the arms (Kelly, 1957). Janowski (1983) mentioned that high hemlines on nursing uniforms in 1975 restricted movement/posture during work for activities such as stooping to check urine outputs and bending the knee to grab life-saving equipment. Comparatively, pant suits with long tunics and close-fitting, stretchable pants provide more flexibility/ease for bending and reaching. Comfort can be improved even further if the suits are made using ventilated fabrics (Janowski, 1983). Another concern is the limited range of sizes of the uniform, requiring each size to fit a variety of body shapes (tall, short, slender, overweight). Research is needed to determine the optimal sizing scheme for nurses.

14.3 Surgical gloves

Gloves are worn during surgery to minimize the cross-infection between patients and the surgical team. Protection and comfort performance are key factors in glove effectiveness and depend on materials used in the glove, the methods used to construct the glove, glove layers, the surgical procedure involved and the hazards/risks experienced. Most medical personnel are careful in selecting gloves, because they have increasing expectations of the softness, pliability, comfort, ease of donning and tactile sensitivity (Williamson, 2003). One goal of product developers has been to minimizes the possibility of allergic reactions by users and maintain a soft touch to the wearer (The Miracles of Science, 2008).

Most disposable gloves are made from two types of materials: natural rubber latex and synthetic materials (Tanner, 2008), either is permeable. When the level of protection becomes a concern for health care workers who wear synthetic gloves instead of latex ones due to allergies or personal preference, double-gloving is used. However, this practice can cause problems. Tanner and Parkinson (2007) conducted a study to detect potential areas of penetration in gloves. During a perforation, they observed that the moisture from the surgical site seeps into the layer between the two pairs of gloves worn, as a clear bright spot.

Previous research (Tiefenthaler et al., 2006) indicated that wearing either a single pair of surgical gloves or a double layer of surgical gloves could impair normal touch sensitivity significantly. Loss of sensitivity may bring difficulties in inserting needles successfully, leading to dissatisfied patients and physician resistance to wearing glove. More recently, extra-thin surgical gloves have been produced which could improve touch sensitivity (Kopka et al., 2005).

In addition, manufacturers of glove materials face challenges related to fit and skin sensitivity. Surgical gloves made from natural latex rubber are soft, fit well, and are more comfortable to wear than gloves made from other materials. However, the latex material and the chemicals used in the manufacture of synthetic materials could cause other adverse reactions during use, such as skin irritation and allergies (Gritter, 1998). Allergies related to wearing latex gloves were identified as a health problem for healthcare workers as early as 1980 (Gritter, 1998). The increase in these associated health problems inspired many manufacturers/suppliers to seek alternative materials to face this challenge. Materials that have been tested or proposed include synthetic material, latex-free material and powder-free material. Comparatively, synthetic substitutes could be competitive at this weak point, but are inferior to natural latex rubber in dexterity, tactile sensitivity, comfort, flexibility and durability (Tiefenthaler et al., 2006). The most common latex-free surgical gloves used in the United Kingdom are made from nitrile, neoprene or polyisoprene (Tanner, 2008). A study by Bowler (2006) indicated that gloves made of neoprene and polyisoprene were rated poorly for fit, feel and comfort due to poorer elastic quality of the material. One of the best gloves available is a knitted glove worn for high-risk surgery and even these are evaluated by wearers as reducing dexterity and comfort (Tanner, 2008). Another approach from manufacturers is to use thinner material to lower the skin irritation. For example, the Aloetouch® exam glove was purported to have improved comfort and reduced skin irritation achieved in part by using a 20% thinner synthetic glove material (Williamson, 2003).

The improper sizing of gloves leads to less tactile sensitivity and limited range of motion. These problems impair the healthcare worker's ability to stay focused and could lead to compromises in medical techniques, posing a potential risk for the wearer and the patient (Tanner, 2008).

Another concern is the exposed area in glove/gown interface caused by either glove slippage or gown sleeve length (Ellis, 2005). During surgical procedures, this problem compromises the protective function of gowns and gloves. The technology is available on selected surgical gowns to secure the interface between gowns and gloves while still providing comfort and barrier properties from the manufacture's statement. Secure-Fit® is claimed to be a proprietary, slip resistant coating of deep blue coloration, which can be thermally embedded into the fabric of surgical gowns (Kimberly-Clark Innovations, 2007).

14.4 Surgical masks

Surgical face masks are an important component of surgical apparel (Leonas & Jones, 2003). Masks function as filters for the mouth and the nose area, usually covering most of this area, and are expected to block the dual contamination and protect both the patients and healthcare workers. Requirements for heathcare workers to wear surgical masks for protection from potential airborne hazards have been set forth in Association of Operating Room Nurses (AORN, 1998) and Centers for Disease Control and Prevention (CDC) guidelines (Garner, 1986). Despite the excellent protective efficiencies that are achievable with surgical masks, comfort issues are still under consideration for improvement. Healthcare workers can encounter significant breathing difficulties if they wear an inappropriate or poorly adjusted mask for long periods in the operation room (Adams et al., 1959). Proper adjustment is needed to allow the wearer to breathe appropriately.

Surgical masks were first advised and designed by Henrot and Sold in 1868 (Mellinger, 1931). After being adopted by prominent surgeons, various types of masks became prevalent in the market. In 1918, laws in the United States were passed requiring gauze masks to be worn. Early product developers realized that fine gauze in multiple layers gave excellent protection; however, wearers could only tolerate the masks for a few minutes before suffering from lack of air to breathe. In 1929, there were twenty-two types of masks available varying in type of filter fabric and number of layers (Mellinger, 1931). Sizes were not standardized and performance varied considerably as well. Manufacturers faced continuing challenges to develop permeable materials which could be constructed into a properly fitted mask with high bacterial filtering efficiency and low air flow resistance. Cellu-cotton, a nonwoven material, replaced gauze in some masks; however the short fibers provided difficulty in maintaining the form of the mask and its sterilization (Betzold, 1943). Another early version of surgical mask was made from 'plastacele' and was quite uncomfortable (Betzold, 1943). The 'vegetable fiber' mask was rated high in comfort in terms of low weight, some resistance to air flow and well fitted, but no study confirmed the bacterial filtering efficiency (Betzold, 1943).

Soon after the introduction of the surgical mask, disputes arose with regard to the comparative benefits of high filtering efficiency versus task effectiveness. These debates continue today. Compliance of healthcare workers with recommended infection control practices has been shown to be compromised due to the discomfort issues with masks (AORN, 1998). Mellinger (1931) devised a light, cool, comfortable and attractive mask which claimed more protection than gauze masks. The doctor/inventor explained that most moisture in the breath could evaporate, but a small amount might stick in between the upper edge of the mask and the adjacent cheek during a long surgery, bringing slight discomfort to surgeons. A wear study in 1943 to compare the effectiveness and comfort of six commercial masks pointed out that wearers were not active when they wore the masks, even though the testing area was dry and the ambient temperature was around 70 °F (Betzold, 1943).

Enerson, Eisenfeld and Kajikuri (1967) conducted a study investigating the influence of heat and moisture trapped in masks on surgeons' discomfort and performance. They determined that with temperature increases in the operating room, there was a significant rise in temperature and humidity in the mask wearer's facial environment, which led to the likelihood of mental and physical performance decreasing (Romney, 2001). Nielsen et al. (1987) found that facemask air temperature could impact thermal sensations of the whole body significantly. Preventing the increase of microclimate temperature and humidity inside the facemask could reduce heat stress on the body (Hayashi & Tokura, 2004). Li et al. (2005) investigated the effects of wearing N95 masks and surgical facemasks with and without nano-functional treatments on thermophysiological responses and the subjective perceptions of discomfort. They measured participants' heart rates, microclimate temperature, humidity and skin temperature inside the facemask, together with perceived humidity, heat and breathing resistance in the facemask. The results indicated that both surgical facemasks were significantly lower for perception of humidity, heat, breath resistance and overall discomfort than N95 facemasks. Participants felt surgical masks were more comfortable than N95 in terms of less heat stress, fit, tight, itchy, fatigued, odorous and salty. Nano-treated facemasks and N95 mask were highly preferred.

Over the past decade, the effectiveness of masks in trapping bacteria has become increasingly important and the challenges in achieving comfort have remained. Thin muslin or gauze masks are rated low in protection for microbes, but they are perceived as comfortable due to the air leakage around the sides and the opening in the weaving structure, even for the most finely woven fabric. The commonly-used masks made of solid substances such as plastic, cardboard or other impervious materials perform poorly in moisture transmission (Adams et al., 1959). A properly fitted rubber filter mask of sufficient filtration potential used in 1958 was unattractive and uncomfortable after wearing for a couple of hours in the operating room (Adams, Fahlman & Lord, 1959). The mask generated an unpleasant feeling from the contact between bare rubber and the skin, and there was an unpleasant odor that developed when autoclaving the rubber. Adams, Fahlman and Lord's study (1959) examined masks made from finely spun, layered glass fibers. When they placed the mask directly against the wearer's nose and mouth, the wearer was uncomfortable due to the sucking in and billowing out of the material with each breath. The researchers concluded that with enough surface area and with room air at 60% RH or less the mask could provide good evaporation and a comfortable feeling (Adams et al., 1959).

A properly fitted mask involves several aspects. The first aspect is the form of the mask. The shape of mask depends on the surfaces of the two sides of the nose, the cheeks, the under surface of the chin and the confluence of these three surfaces. Some products are constructed in a round-shape; some are in an egg-shape. Fiberglass, nylon, heat-stable moldable plastic and resin-impregnated cotton have been used as the frame of masks to support the curved shape of the mask and maintain reasonable weight and durability (Adams et al., 1959). The second aspect of mask fitting is the adaptability of the mask to the irregular and various contours of faces. Adams, Fahlman and Lord's study (1959) found four shapes and eighteen sizes of human faces based on the measurements and analyses of hundreds of faces. The third aspect is the need to restrict air leakage around the mask when inhaling and exhaling. Silk, cotton, nylon, polyethylene and polystyrene complexes, various synthetic esters, rubber in different forms (solid, foam and inflatable) and polyurethane foam have been studied for their effectiveness in the adapter ring (Adams et al., 1959).

14.5 Future trends

In addition to performing as an effective barrier against liquids/particles/bacteria/germs, and maintaining high mechanical stability, surgical textiles need to function as a thermal and moisture regulator, to maximize the comfort zone for medical personnel. The advances in fibers, polymers, chemical technology and fabric/web forming technologies, such as nanotechnology, innovative composite materials and new nonwoven technologies are being incorporated into medical textiles. For example, phase change material microcapsules could be embedded in the surgical apparel, patient bedding materials, bandages and products to regulate patient temperatures in intensive care units (Mondal, 2008), keeping the skin temperature within the comfort range. Nanocapsules could be integrated into the medical textiles to collect and distribute medical parameters of the patient by means of sensors or drug (Hofer & Swerev, 2003). Anti-allergen finishing treated fabrics can help relieve patients suffering from asthma and allergies caused by dust mites (Fisher, 2006).

Concomitantly, there must be a continuing effort to develop appropriate standards/test systems to verify the function of the innovative technologies. The standards need to be established in the situation which could simulate the real wearing situation in the large degree. The optimizing of new techniques and materials needs to be evaluated both in the lab and during wear study (Cao & Cloud, 2010). The driving technology force in materials development for medical textiles will lead to the products meeting the stringent performance specifications and market expectation in conjunction with environmental concern, personal safety and comfort requirements. This long term quest for researchers who work in this field needs the attention of government, academia, business and manufacturers.

14.6 References

Abreu, M.J., Silver, M.E., Schacher, L. & Adolphe, D. (2003). Designing surgical clothing and drapes according to the new technical standards. *International Journal of Clothing Science and Technology*, **15**(1), 69–74.

Adams, R., Fahlman, B. & Lord J. (1959). New fashions in surgical attire. *The American Journal of Nursing*, **59**(8), 1102–1107.

Aibibu, D., Lehmann, B. & Offermann, P. (2003). Barrier effect of woven fabrics used for surgical gowns. *AUTEX Research Journal*, **3**(4), 186–193.

Aibibu, D., Lehmann, B. & Offermann, P. (2003). Image analysis for testing and evaluation of the barrier effect of surgical gowns. *Journal of Textile and Apparel Technology and Management*, **3**(2), 1–8.

Akridge, J. (2004). Raising the bar on surgical gowns and drapes. *Healthcare Purchasing News*, September, 21–29.

Association of Operating Room Nurses (AORN), Recommended Practices Committee (1998). Recommended practices for surgical attire. *AORN Journal*, **68**, 1048–1052.

Belkin, N. (1993). The challenge of defining the effectiveness of protective aseptic barrier. *Technical Textiles International*, October, 22–24.

Betzold, K.V. (1943). How safe is your mask? *The American Journal of Nursing*, **43**, 59–60.

Bowler, G. (2006). Safer surgical gloves: evaluation and implementation. *J Perioper Prac*, **16**(2), 67–70.

Brandt, B. (1993). Surgical gowns: a survey of wearer and purchaser satisfaction with protection. Nonwovens Industry, September 1. Available from http://www.thefreelibrary.com/Surgical+gowns:+a+survey+of+wearer+and+purchaser+satisfaction+with+...-a014428963

Cao, W. & Cloud, R. (2010). Effects of temperature on liquid penetration performance of surgical gowns. *International Journal of Clothing Science and Technology*, **22**(5), 319–332.

Cho, J., Tanabe, S. and Cho, G. (1997). Thermal comfort properties of cotton and nonwoven surgical gowns with dual functional finish. *Journal of Physiological Anthropology*, **16**(3), 87–95.

Ellis, K. (2005). Evaluating and Purchasing Surgical Apparel and Sharps. Available from http://www.infectioncontroltoday.com

Enerson, D.M., Eisenfeld, L.I. & Kajikuri, H. (1967). Heat and moisture trapping beneath surgical face masks: A consideration of factors affecting the surgeon's discomfort and performance, *Surg.*, **67**, 1007–1016.

Fisher, G. (2006). Development trends in medical textiles. *Journal for Asia on Textile & Apparel*. Available from http://www.adsaleata.com/Publicity/ePub/lang-eng/article-40/asid-76/Article.aspx

Garner, J.S. (1986), CDC guidelines for prevention of surgical wound infections. *American Journal of Infection Control*, **14**, 71–80.

Gritter, M. (1998). The latex threat. *The American Journal of Nursing*, **98**(9), 26–33.

Hatch et al. (1990). Invivo cutaneous and perceived comfort response to fabric 1.

Thermophysiological comfort determinations for 3 experimental knit fabrics. *Textile Research Journal*, **60**(7), 405–412.

Hayashi C. & Tokura, H. (2004). The effects of two kinds of masks on clothing microclimate inside the mask in participants wearing protective clothing for spraying pesticides. *International Archives of Occupational and Environmental Health*, **77**, 73–78.

Hofer, D. & Swerev, M. (2003). The future of medical textiles: high-tech for the well-being of the patient. *Journal of Textile and Apparel Technology and Management*, **3**(2), 1–3.

Horrocks, A.R. & Anand, S.C. (2000). *Handbook of Technical Textile*. Abington, Cambridge, UK: Woodhead Publishing Limited in association with The Textile Institute.

Janowski, M.J. (1983). My love affair with uniforms. *American Journal of Nursing*, pp. 1241–1244.

Kelly, C.W. (1957). What nurses want in a uniform. *American Journal of Nursing*, **57**(10), 1282–1284.

Kelly, C.W. (1958). The fabric makes the uniform. *American Journal of Nursing*, **58**(10), 1384–1388.

Kimberly-Clark Innovations (2007). Retrived online on October 16 from http://www.kimberly-clark.com/pdfs/Health%20Care.pdf

Kopka, A., Crawford, J.M. & Broome, I.J. (2005). Anaesthetists should wear gloves-touch sensitivity is improved with a new type of thin glove. *Acta Anaesthesiologica Scandinavica*, **49**, 459–462.

Laufman, H., Belkin, N.L. & Meyer, K.K. (2000), A critical review of a century's progress in surgical apparel – how far have we come? *Journal of American College of Surgeons*, **191**, 554–68.

Leonas, K.K. & Jones, C. (2003). The relationship of fabric properties and bacterial filtration efficiency for selected surgical face masks. *Journal of Textile and Apparel Technology and Management*, **3**(2), 1–8.

Li, Y., Tokura, H., Guo, Y.P., Wong, A.S.W., Wong, T., Chung, J. & Newton, E. (2005). Effects of wearing N95 and surgical facemasks on heart rate, thermal stress and subjective sensations. *International Archives of Occupational and Environmental Health*, **78**, 501–509.

Mangram, A.J., Horan, T.C., Pearson, M.L., Silver, L.C. & Jarvis, W.R. (1999), Guideline for prevention of surgical site infection. *American Journal of Infection Control*, **27**, 97–132.

May-Plumlee, T. & Pittman, A. (2002). Surgical gown requirements capture: A design analysis case study. *Journal of Textile and Apparel Technology and Management*, **2**(2), 1–10.

McCullough, Elizabeth A. (1992). 'Liquid barrier and thermal comfort properties of surgical gowns,' in *Pacific Polymer Science* 2, Yukio Imanishi (ed.), Berlin/Heidelberg: Springer Verlag, pp. 316–322.

Mellinger, H. (1931). A face mask, an efficient and new-type protection. *The American Journal of Nursing*, August, 913–915.

Mills, S.J.C., Holland, D.J. & Hardy, A.E. (2000). Operating field contamination by the sweating surgeon. *Australian and New Zealand Journal of Surgery*, **70**, 837–839.

Mondal, S. (2008). Phase change materials for smart textiles – an overview. *Applied Thermal Engineering*, **28**, 1536–1550.

Nielsen, R., Berglund, L.G., Gwosdow, A.R. & Dubois, A.B. (1987). Thermal sensation of the body as influenced by the thermal microclimate in a face mask. *Ergonomics*, **30**, 1689–1703.

Nonwoven Industry Report (2006). Available from http://www.nonwovens-industry.com/news/2006/04/06/medical_and_hygiene_study_available

Philippe, F., Abreu, M., Adolphe, L. & Silva, M. (2003). Influencing of the sterilization process on the tactile feeling of surgical gowns. *Clothing Science and Technology*, **15**(3/4), 268–275.

Pissiotis, C.A., Komborozos, V., Papoutsi, C. & Skrekas, G. (1997). Factors that influence the effectiveness of surgical gowns in the operating theatre. *The European Journal of Surgery*, **163**, 597–604.

Rajendran, S. & Anand, S.C. (2003). *Development in Medical Textile*. Oxford Street, Manchester, UK: The Textile Institute.

Recommended practice for selection and use of surgical gowns and drapes (2003). *AORN Journal*, **1**, 206–210.

Romney, M.G. (2001). Surgical face masks in the operating theatre: re-examine the evidence. *Journal of Hospital Infection*, **47**, 251–256.

Rutala, W.A. & Weber, D. (2001). A review of single-use and reusable gowns and drapes in health care. *Infection Control and Hospital Epidemiology*, **4**, 248–257.

Sparrow, N. (2009). Breakthrough research in advanced wound-care products and the development of a nonwoven material for OR use that combines comfort and durability are keeping the textiles sector humming. Available from http://www.devicelink.com/emdm/archive/06/03/018.html

Stanley, L. (1994). OSHA ruling still causing shifts in surgical gown marketplace. *Health Industry Today*, November.

Suprun, N., Vlasenko, V. & Ostrovetchkhaya, Y. (2003). Some aspects of medical clothing manufacturing. *International Journal of Clothing Science and Technology*, **15**(3/4), 224–230.

Tanner, J. & Parkinson, H. (2007). Surgical glove practice: the evidence. *Journal of Perioperative Practice*, **17**(5), 216–225.

Tanner, J. (2008). Choosing the right surgical glove: an overview and update. *British Journal of Nursing*, **17**(12), 740–744.

The Miracles of Science™ (2008). DuPont. Available from http://www2.dupont.com/DuPont_Home/en_US/index.html

Tiefenthaler, W., Gimpl, S., Wechselberger, G. & Benzer, A. (2006). Touch sensitivity with sterile standard surgical gloves and single-use protective gloves. *Anaesthesia*, **61**, 959–961.

T-PACC (1997). *Surgical Gown Wear Trials*. Textile Protection and Comfort Center, North Carolina State University.

Williamson, J. (2003). OR apparel makers weave comfort, value into innovative products. *Healthcare Purchasing News*, September, 28–31.

Wong, D.A., Alexander, J.A. & Kay, L. (1998), Risk of blood contamination of health care workers in spine surgery: a study of 324 cases. *Spine*, **23**, 1261–1266.

15
Improving comfort in sports and leisure wear

V. T. BARTELS, Bartels Scientific Consulting GmbH, Germany

Abstract: Sports, everyday and leisure wear share a huge portion of the clothing market. Due to the differing activities undertaken by the wearers, and consequently of varied sweat production, sportswear has to fulfil different physiological demands from everyday and leisure wear. Whereas the latter should concentrate on the management of sweat in vapour form, sportswear needs to transport liquid sweat and must dry quickly. There is, however, a mutual influence between sports and everyday clothing which is mainly driven by comfort: sportswear such as fleece has become everyday clothing and outdoor garments have become more fashionable.

Key words: physiological comfort, physiological function, comfort testing, sportswear, leisure wear.

15.1 Introduction

A study on improvement in the comfort of clothing requires a chapter on sports, everyday and leisure wear. These categories of clothing are of great economic importance, because of their huge market share. There is a universal need for everyday wear and most people will also want sports and leisure wear. Wearer comfort in these categories is important in today's textile markets. Due to its superior comfort characteristics, clothing originally designed for sportswear has become widespread in everyday and leisure wear.

In the next sections, these facts are considered in more detail and the specific physiological (i.e. comfort) needs which must be met by sports, everyday and leisure wear are discussed. These criteria can be tested either by wearer trials or by means of textile physiological laboratory apparatus. Based on these investigations, the comfort of the garments can be optimised and a number of examples are given. Finally, some recent and future trends are discussed.

15.2 Market share of sports and leisure wear and affected group of users

Everyday and sportswear takes by far the largest market share of all types of clothing when unit numbers and metres of textiles are taken into account. Hence, the related annual turnover is enormous. According to Umbro (2007), the international sportswear market in 2006 had a total value of approximately £45–50 billion. The main reason for the huge market share of these categories of garments is the group of users concerned. We all need everyday clothing, most of us have sportswear and many people own special leisure wear.

15.3 Definition of sports and leisure wear

There are no common definitions for sports, everyday or leisure wear, respectively. In this chapter, the following general differentiation is used (see also Fig. 15.1):

- Sportswear: clothing intended for use in sports which usually have high activity levels, leading to the increased production of sweat.
- Everyday wear: clothing for everyday and 'normal' situations which produce only a small amount of perspiration. Everyday wear also comprises business and fashionable clothes.
- Leisure wear: clothing which is worn during leisure time, frequently at home. It often has connections with a sense of well-being.

(a)

15.1 Comparison of sports, everyday and leisure wear. (a) Sportswear, intended to be used while heavy sweating. (b) Everyday wear, underwear to be used in normal wearing situations. (c) Leisure wear, home-suit worn while relaxing. (All photos: Triumph International.)

(*Continued opposite*)

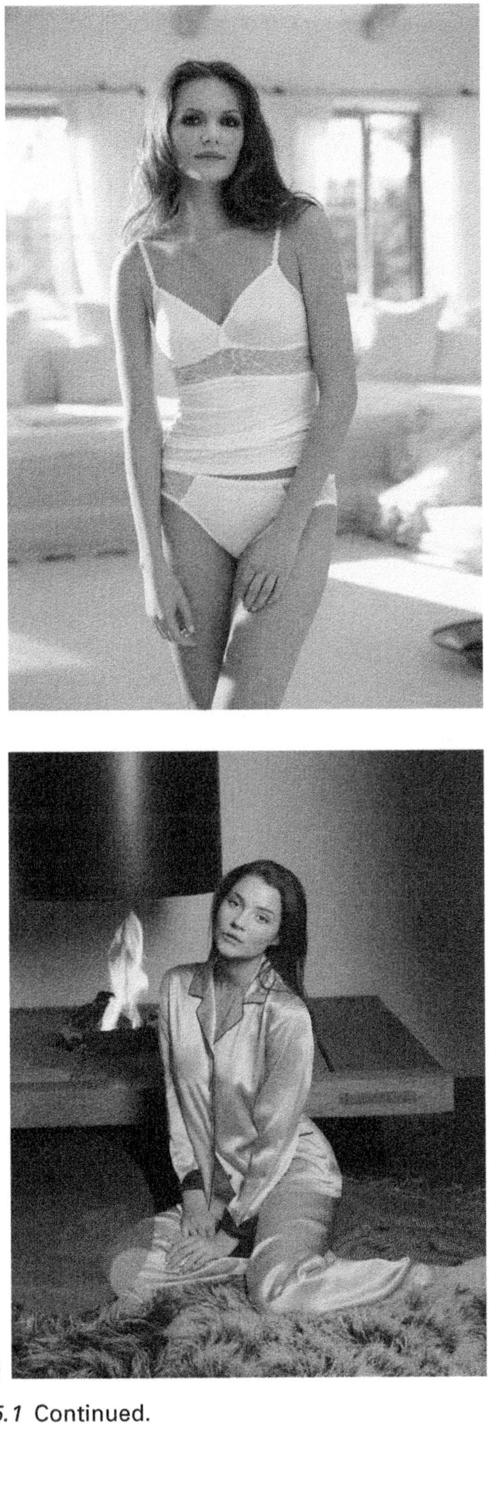

15.1 Continued.

15.4 Influence of sportswear on everyday and leisure wear fashion

Sportswear has a considerable influence on everyday and leisure wear fashion. Most people regularly wear clothes originally designed for sports applications such as:

- boxer-shorts
- polo-shirts
- jogging suits
- sneakers.

In addition to these 'classics', outdoor clothing particularly influences everyday and leisure wear fashion, e.g.:

- functional shirts or blouses
- hiking trousers
- fleeces
- soft-shells
- foul weather protective jackets.

Comfort is the main reason for the influence of sportswear on everyday and leisure wear. Consumers who have experienced the superior comfort of sportswear look for the same advantages in other garments. This is a significant example of the importance of clothing comfort. Most of us have a few favourite pieces of clothing and these are usually the most comfortable. Less comfortable clothes tend to remain in the wardrobe.

It should not being forgotten that as well as the influence of sportswear on everyday and leisure wear, the reverse is also true: the interaction between different clothing groups is mutual. For example, clothing designed for sailing or use in gyms has had a 'fashion function' for many years. Looking at the example of outdoor clothes, it is obvious that they have become more fashionable and thus acceptable to a larger group of consumers. It is certainly not by accident that designers of outdoor fashion are increasingly involved in women's tailoring. With this growing market for women's outdoor clothing, a paradigm change from mainly technically oriented men's wear to more fashionable women's clothing has occurred. An example is shown in Fig. 15.2.

15.5 Physiological demands on sports, everyday and leisure wear

Garments for sport, everyday and leisure must fulfil the needs created by their intended use. These needs do not only comprise technical or fashion aspects, but also the physiological function of comfort. It is important to recognise that the physiological demands vary across the different types of clothing. This is discussed in the following sections.

15.2 Today, outdoor clothing needs to be both functional and fashionable (photo: Mike Arzt/Helly Hansen).

15.5.1 Aspects of comfort

All clothes have to meet demands concerning four different aspects of comfort (Mecheels, 1998):

- Thermo-physiological comfort deals with the interaction between the body's thermo-regulation and its clothing. It comprises heat and moisture transport from the body through the clothing into the atmosphere and the consequent sensations of coolness, warmth, chilling and sweating.
- Skin sensory comfort describes the contact and interaction between textile and skin. It includes pleasant sensations like softness and smoothness and unpleasant feelings such as scratching, stiffness or the clinging of fabric to sweat dampened skin.

- Ergonomic comfort is concerned with the fit of the garment, the ease of movement within it and how it is donned. The decisive factors are the pattern of the clothing and the elasticity of the materials.
- Psychological comfort influences the subjective comfort sensation. This depends on fashion and on the attitude of the wearer towards brand image, colour, pattern, textile material and production.

Detailed discussions on the different aspects of comfort are given in the other chapters of this book.

Thermo-physiological comfort

Sports, everyday and leisure wear has to fulfil the following demands for thermo-physiological comfort:

- Support the wearer's thermo-regulation.
- Keep the wearer at a comfortable body temperature.
- Keep the micro-climate between skin and textile as dry as possible.

For all types of clothing, the heat balance of the wearer must be maintained. i.e. the body's heat production and heat loss should be equal (see Fig. 15.3). However, this does not mean that the different types of clothing have to fulfil the same demands, because they may be worn for different activities and under varying climatic conditions.

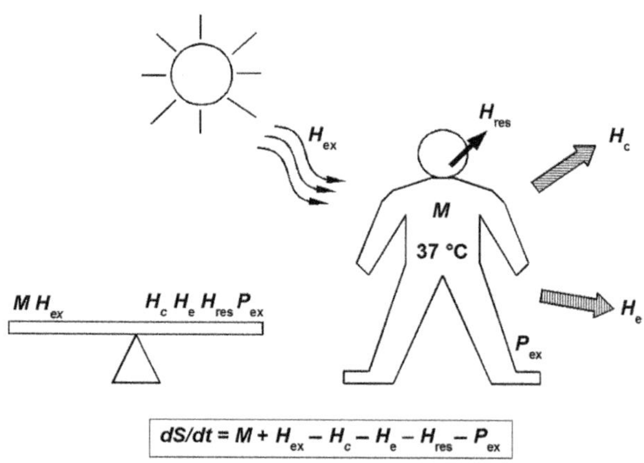

15.3 The prerequisite for a good thermo-physiological comfort is the fulfilling of the heat balance. Heat production (metabolic rate *M* and external heat gain H_{ex}) and heat loss (dry heat flux H_c, evaporative heat flow H_e and respiratory heat loss H_{res}) of the body plus external working power P_{ex} should be of the same amount, the change of body heat storage dS/dt being zero.

In looking at the requirement to support the wearer's thermo-regulation, the difference in metabolic heat production (M) between sports and everyday wear must be taken into account. Everyday or leisure wear is typically used at lower rates of activity such as sitting, standing or normal walking (at about 4 km/h), which would produce approximately $M = 90$, 115 or 280 W, respectively. On the other hand, sportswear is often worn for activities with elevated metabolic rates. Rapid walking (at 5 km/h) gives a metabolic heat production of $M = 350$ W and this will be increased to values of 1000 W or more by activities such as running or biking.

These differences in metabolic heat production have to be taken into consideration if wearer comfort in different situations is to be achieved. It is particularly important for sportswear to dissipate body heat in order to prevent overheating. On the other hand, everyday and leisure wear may need higher thermal insulation if the wearer is to be kept comfortably warm.

Keeping the micro-climate dry in both sportswear and everyday wear cannot be achieved in the same way. Sportswear, everyday and leisure wear all need a high water vapour transport rate (i.e. good breathability). The water vapour resistance should be as low as possible so moisture can escape from the clothing and sweat can evaporate to cool the body.

The amount of sweat on the skin, and consequently in the textile, is typically quite different. In sportswear liquid sweat is the norm; in everyday and leisure wear it is exceptional and is most likely to occur as a vapour. The management of liquid and vaporous sweat follows different physical principles (Mecheels, 1998):

- Vaporous sweat is mainly transported via diffusion; a smaller but sometimes significant portion may be moved through convection (movement of air inside the clothing) and ventilation (exchange with the atmosphere).
- The absorption of vaporous sweat is controlled by the hygroscopic properties of the textile and the chemical nature of the fibres.
- Liquid sweat is transported by the effects of (a) adsorption and migration along fibres and yarns and (b) capillary action. Both effects are strongly related to the hydrophilicity of the fibre surfaces.
- Liquid sweat may be stored in the free space between the fibres as well as being absorbed. Fibres should be hydrophilic to improve liquid storage.

The drying time of textiles is one of the most important physiological parameters for sportswear comfort. As sports clothes are more likely to be subject to liquid sweat, they need to dry quickly, otherwise, post-exercise chill may occur, which is very unpleasant and may lead to a cold or stiffness. However, everyday and leisure wear garments are less likely to become soaked with sweat and do not need to dry so quickly. Consequently, drying time is less significant for the comfort of everyday and leisure clothing.

The efficiency of sportswear has a significant influence on the performance of the wearer, so support of the body's thermo-regulation is of great importance. Leisure-time joggers as well as world champions want clothing that will improve

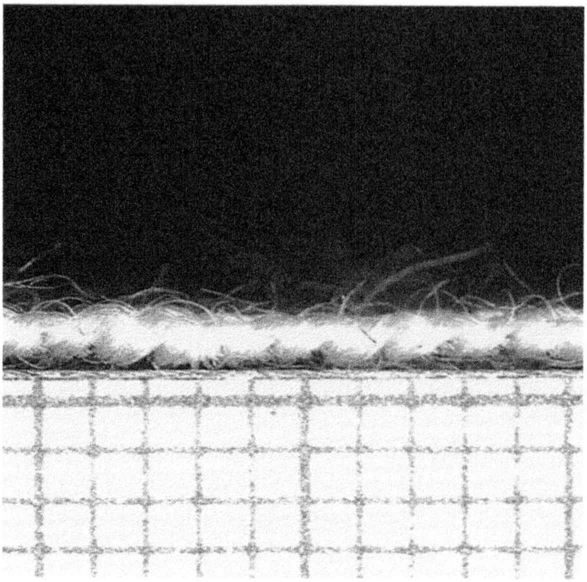

15.4 A slightly hairy inner surface improves the skin sensorial comfort of a textile (photo from Bartels, 1999).

their performance. For extreme sport applications such as marathon running, mountain climbing or open sea sailing, clothing can play an essential part in keeping the wearer competitive.

Skin sensory comfort

When worn directly in contact with the body, sports, everyday and leisure wear should feel soft and smooth on the skin. The textiles must not scratch or feel stiff. A pleasant sensation of friction between skin and textiles is often obtained by means of a slightly hairy inner clothing surface (see Fig. 15.4).

For sportswear, the sensory aspects are of even greater importance than for everyday and leisure wear. The garment should not cling to skin moistened by the higher levels of sweat produced during sport activities. As wet skin is much more readily irritated than dry skin, particular care should be taken to achieve a good level of skin comfort. Unfortunately this is not always taken into consideration and as a consequence, user satisfaction may be low.

Ergonomic comfort

Ergonomic comfort is one of the main characteristics of many leisure wear garments, following the principles of well-being (see Fig. 15.1(c)). Examples for constructions which offer this superior ergonomic comfort are:

- loose patterns, as in bathrobes
- soft materials, such as loop plush
- elastic materials, as used in leggings.

Due to its ergonomic comfort, this special leisure wear is also attractive to consumers as everyday wear. However, for everyday clothing, ergonomic comfort will be in competition with price and fashion.

For sportswear, the situation is more complex as ergonomic comfort must be balanced with the specific needs of a particular sport. For example, a loose fit is not efficient for cycling garments which have to be streamlined. Often sportswear is tight fitting aiming to enhance the transport of liquid sweat. To achieve a good degree of ergonomic comfort under these circumstances, elastic materials, such as elastane which contains knitted material, are frequently employed.

Psychological comfort

Although difficult to quantify, the psychological aspect of comfort is important in sports and everyday clothing. In addition to their high level of comfort, sports clothes are often worn in everyday situations due to the status which may be conferred by their image. Examples could be a jacket from a well known brand or a football shirt with the name of a certain player on it. Wearing sporty clothes may also be perceived as conferring a sporty image.

Theorem: A textile construction that is comfortable for sportswear is not automatically suitable for everyday or leisure wear and vice versa.

15.6 Testing sports, everyday and leisure wear comfort

The physiological demands on sports, everyday and leisure wear described in this section have been identified. The next step is to assess and optimise comfort properties. This requires testing procedures for different aspects of comfort. As Part II of this book deals with comfort testing in detail, this section will discuss only the issues concerned specifically with sports, everyday and leisure wear.

There has been little testing of comfort for everyday and leisure wear and few results have been published. Sports clothes in particular are often assessed by different types of comfort tests (Bartels, 2005a). European standards for protective clothing include physiological demands, e.g. on thermal insulation (EN 342), or water-vapour resistance (breathability) (EN 343). Such standards do not exist for everyday, leisure or sportswear, thus testing is optional.

15.6.1 Field trials

Field trials are widely used to test comfort. A group will be allocated sample pieces of clothing to wear during their normal activity. Subsequently, the test subjects are asked for their experiences. Well known sportsmen or journalists frequently serve as test subjects and the trial results are published in advertisements or sports journals.

The great advantage of field trials is the testing of samples under real-life conditions. For example, with regard to ergonomic comfort, the optimum placement of zips may be assessed, or judgement made on the ease of specific movements (such as raising the arms in mountaineering).

15.6.2 Wearer trials with human test subjects in climatic chambers

For controllable test conditions, wearer trials in a climatic chamber can be employed. The climatic chamber keeps the ambient conditions of temperature, humidity and wind speed constant and replicable. The activity can be also be kept constant. For sportswear tests, treadmills or bicycle ergometers are frequently used (Bartels, 2005a). By means of probes attached to the body of the test subject, a large variety of physiological data may be obtained, such as heart rate, skin temperature and humidity in the micro-climate between skin and textile.

Wearer trials in a climatic chamber are time- and work-intensive. To obtain reliable statistics, a number of tests must be performed and a large amount of data is produced, which needs statistical analysis. Moreover, as test subjects cannot always be in exactly the same condition (fitness, weight, hydration status, etc.), the reproducibility of wearer trials will be limited by reproducibility of the test subjects.

15.6.3 Laboratory test methods, benefits and limitations

The main aim of laboratory test methods is to overcome the limitations of wearer trials. Results on testing apparatus can be more easily reproduced than those on human beings. Laboratory tests also have the advantages of speed and lower costs over wearer trials which have to be carefully designed.

However, physiological laboratory tests do not directly measure comfort. They test certain parameters, of which the significance for comfort has to be demonstrated. This physiological validation is crucial (5I3M, 2003; ICEE XI, 2005), because only then will the meaning of a test result for wearer comfort be known. There must be a mathematical correlation between the wearer's perception of comfort and results from the laboratory apparatus (Bartels and Umbach, 2003). Otherwise, the test apparatus is useless from a comfort point of view.

15.6.4 Laboratory test methods for thermo-physiological comfort of textiles

Testing apparatus exist to measure thermo-physiological comfort for both textile materials and ready-made clothing. It is therefore important to distinguish between material tests and clothing tests. Frequently used test apparatus for the breathability of sportswear materials, i.e. two-dimensional textile samples, are:

- Cup methods such as ASTM E96 or BS 7209.
- Tests with the so-called 'skin model', more exactly named as sweating guarded hotplate, according to EN 31092, ISO 11092 or ASTM F1868 (McCullough et al., 2004).
- The Permetest apparatus designed by Hes.

Cup methods are low cost and easy to use. As a consequence, they are frequently employed by sportswear producers within their quality management programmes and by testing firms. Unfortunately, the repeatability of cup methods and the analysis of their test results for the assessment of human comfort are limited. Their utility for comfort testing is therefore questionable (Gibson, 1999).

The sweating guarded hotplate is detailed, very accurate and displays a good correlation with human comfort sensations. It is therefore included in European standards for the testing of protective clothing (EN 342, EN 343, EN 471). Many sportswear companies also test their materials with this 'skin model'. The disadvantages are the high price of the apparatus and the relatively long test duration of approximately 45 minutes for a single sample.

The Permetest apparatus is described in detail in Chapter 5 of this book.

The guarded hotplate has been used for the measurement of heavy perspiration (liquid sweat transport and uptake, drying time) since the 1980s (Umbach, 1987). It offers a detailed and physiologically validated method for these tests.

The wicking test (NCSU, 2009) is simple and inexpensive and is therefore it is frequently employed, although its physiological results are questionable.

15.6.5 Laboratory test methods for thermo-physiological comfort of clothing

Ready-made garments and complete clothing systems are tested by means of manikins. These simulate the body's heat dissipation and some are also able to simulate the evaporative heat flow (so-called 'sweating manikins'). A discussion on the methodology of testing with thermal manikins is given in Anttonen et al. (2004) and Goldman (2007) and in the proceedings of the International Manikin Meetings (e.g. 5I3M, 2003).

A moveable manikin capable of simulating wearer activity is preferable for testing sportswear. Movement causes air inside the clothing system to be disturbed and exchanged with the atmosphere. These processes are called convection and ventilation, respectively, and in well-designed clothing are able to dissipate about 30% of the body's heat (Mecheels, 1998). Moveable manikins are able to quantify these effects.

Manikins with the capacity to simulate sweating are being developed. The difficulties involved in building a manikin capable of producing sweat in a way comparable to human physiology is approached in varied manners by different institutes and manufacturers, but the differing concepts do not always lead to the same results (McCullough, 2001). However, the development of such a manikin is generally held to be desirable for effective sportswear assessment.

15.6.6 Laboratory test methods for skin sensory comfort

Sports, everyday and leisure wear garments are often worn next to the skin, for example underwear, T-shirts and bathrobes. The sensory properties are important for comfort, particularly in sportswear as increased sweating may cause skin irritation.

The Hohenstein Institutes in Germany have developed and improved a complex system of skin sensory test apparatus and assessment formulae over several decades (Mecheels, 1982; Bartels and Umbach, 2002b). It comprises parameters which describe the inner textile surface, such as textile hairiness, number of contact points with the skin, stiffness of the material (see Fig. 15.5), the tendency

15.5 Laser based measurement of the fabric's bending angle for the characterisation of textile stiffness (Bartels and Umbach, 2005; photo: Hohenstein Institutes).

to cling to sweat moistened skin, and the parameters which describe the sweat uptake from the skin.

A laser-based method of characterising the roughness and hairiness of a textile surface was developed by Renner and co-workers (Tourlonias et al., 2006). This group also worked on the mechanical characterisation of fabric contact (Breugnot et al., 2006). Another apparatus, the 'friction meter', measures the force between a rotating sample and an artificial test indicator (Tronnier et al., 2004). Developed at a medical institute, it is utilised in textiles for patients with atopic eczema.

15.7 Textile constructions for sports, everyday and leisure wear

15.7.1 Textile constructions for everyday and leisure wear

As discussed in section 15.2, everyday wear, leisure wear and sportswear have to be constructed differently in order to fulfil the different requirements made of them. For everyday and leisure wear, the management of liquid perspiration is less important and the concentration is on the management of vaporous sweat. Hygroscopic fibres are favourable for the buffering of vaporous sweat impulses. Cellulose fibres such as cotton or viscose, other man-made regenerated fibres and also protein fibres such as wool, can absorb a considerable amount of moisture within their molecular structure. Humidity is drawn away from the micro-climate, which stays drier and thus more comfortable. Non-hygroscopic synthetic fibres like polyester, polyamide, polypropylene, etc. cannot absorb moisture within their molecular structure and are therefore less comfortable for use in everyday and leisure wear.

But the chemical nature of a fibre is only one of the textile construction parameters affecting comfort and is not the most important. Fibre type, weave or knit construction and finishing are also critical in determining the comfort of a textile. The design of clothing also has a significant influence.

Yarn type and weave or knit construction should be carefully chosen with respect to skin comfort. The inner surface of a textile worn next to the skin should be somewhat hairy, as shown in Fig. 15.4. The hairs act as spacers between skin and textile, increasing the friction between skin and textile in a pleasant way. Such a structure may be obtained by using spun yarns instead of filaments. If filaments have to be used, textured yarns are preferable. However, the hairs must not be too stiff. Wool fabrics in particular may be unpleasantly prickly if the diameter of the fibres is too large (Naylor and Phillips, 1997; Gehui et al., 2003). Fibre diameter is therefore the main quality criterion for wool.

The weave or knitted construction influences both thermo-physiological and skin sensory comfort. The thickness and porosity of a textile strongly affects its water-vapour diffusion or 'breathability'. The greater the number of pores included in a textile structure, the easier it will be for water-vapour molecules to escape from the micro-climate. The average path length of water-vapour molecule diffusion through a textile is directly related to the thickness of the textile. Consequently, thinner textiles exhibit a higher water-vapour transport ability than comparable, but thicker constructions. However, it is not always possible to reduce textile thickness, particularly if thermal insulation is required. This important physiological parameter is linearly dependent on thickness, which includes any protruding hairs (Goldman, 2007): the thicker the textile, the higher its thermal insulation. As a certain thermal insulation is necessary for many applications (e.g. in winter), the thickness of the textile also has to be adapted.

High water-vapour permeability is preferable in thicker materials too and materials of the same thickness may exhibit quite different properties of water-vapour transport. To find the construction with the best 'breathability', water-vapour permeability indices are defined for textiles and clothing (Mecheels, 1998; Goldman, 2007). These indices are mainly based on the ratio of thermal insulation over the water-vapour resistance, e.g. in ISO 11092 the water-vapour permeability index is given by the equation

$$i_{mt} = 60\ (\mathrm{Pa/K})\ R_{ct}/R_{et}$$

with R_{ct} being the thermal insulation and R_{et} the water-vapour resistance of the textile, respectively. Here, a higher i_{mt} value indicates a higher water-vapour permeability and thus, improved comfort.

As has already been mentioned, the design of ready-made garments has a significant influence upon comfort. Loosely fitting clothing improves both thermo-physiological and ergonomic comfort. Thermo-physiological comfort is enhanced due to a higher rate of convection and ventilation; ergonomic comfort by a greater freedom of movement. Consequently, leisure wear such as bathrobes or loungwear are often designed to be loose fitting. In principle, a loose fit is also preferable for everyday clothing, although fashion or social convention may sometimes be of greater importance.

Elastic textiles are frequently used for maximum ergonomic comfort (Voyce et al., 2005). Knitted fabrics in particular offer quite a high elasticity. The elasticity of woven fabrics may be improved by the utilisation of elastane yarns. Highly elastic textiles also enhance psychological comfort. In particular, elastic knitted textiles, usually with a higher content of elastane, are used for tight fitting or even body-shaping garments. Wearers feel more comfortable, because they feel themselves to have a more attractive appearance.

15.7.2 Textile constructions for sportswear

Many of the constructional parameters which enhance the comfort of sportswear are shared by everyday and leisure wear. Differences arise from the much higher rates of perspiration which sportswear must have the capacity to manage. Perhaps the biggest of these differences is that fibres used in sportswear should not be hygroscopic. The ability of a hygroscopic fibre to soak up moisture is limited, as is its rate of absorption. As moisture is bound into the molecular structure via hydrogen bonds, there is a high requirement of both energy and time for drying a wetted textile made of hygroscopic fibres. This may lead to post-exercise chill which is experienced as an unpleasant sensation. Its prevention is one of the most important aims of sportswear construction.

Another important comfort parameter for sportswear is the transport of liquid sweat. Liquid sweat transport is totally different from vaporous sweat transport, which takes place mainly by diffusion. As described in Mecheels (1998), liquid perspiration is transported via capillary effects, adsorption at the surface of fibres and migration along them. Thus a high rate of liquid sweat transport is only achievable with a hydrophilic fibre surface. The transport of liquid sweat can also be improved by fibres having a larger surface area, as in non-circular fibre diameters like Coolmax (Advansa, 2009) or, alternatively, micro-fibres (Miraftab, 2000; Umbach, 1993).

Synthetic fibres such as polyester, polyamide and polypropylene are employed for modern sportswear (see Fig. 15.6). These may be micro-fibres or fibres with

15.6 A typical modern sport shirt made of functional synthetic materials, being elastic and fashionable (photo: Schiesser).

non-circular diameters. Hydrophilic finishes are quite often used, especially for polypropylene, but are seldom labelled as such.

The thermal insulation of sports textiles has to be much lower than that of everyday and leisure wear as they deal with a much higher rate of metabolic heat production. Therefore, thin and open textile constructions are preferable for sportswear for the efficient dissipation of both heat and moisture.

15.8 Application examples

15.8.1 Underwear

The general principles for the differing construction of sports, everyday and leisure wear given in the previous section should be taken into consideration. As underwear is always worn next to the skin, particular attention must be given to skin comfort. This is usually achieved by textiles made of fleece, loop plush, honeycomb structures, or Piquet. Comfort may be further improved by the use of spun yarns and hydrophilic surfaces.

Fibres with a hydrophilic surface such as polyester (PES) or polyamide offer a higher liquid sweat transport than polypropylene (PP), which has a hydrophobic surface which cannot support the two main transport processes of (a) capillary effects and (b) absorption and migration. The result is shown in Fig. 15.7. Two similar sports textiles are compared, one made of PES and the other of PP. It can be seen that only the PES underwear offers good liquid sweat transport $F > 765$ g/m^{2}h

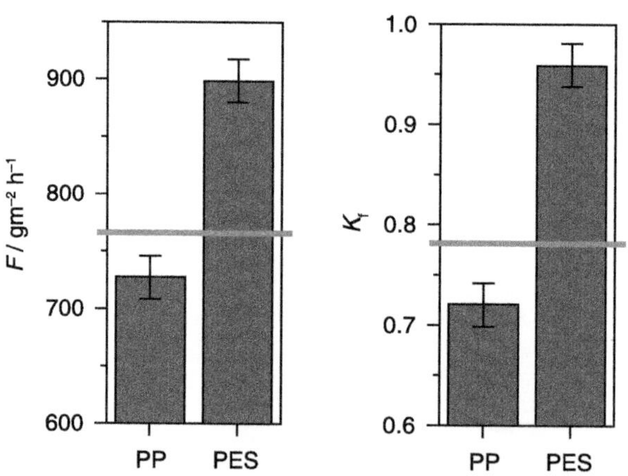

15.7 Liquid sweat transport *F* (left) and buffering capacity against liquid sweat impulses K_f (right) of two sports textiles made of polypropylene (PP) and polyester (PES), respectively. Both textiles are comparable, knitted constructions made of filament yarns, showing a honeycomb structure at the inside. It can be seen that only the polyester specimen is able to fulfil the demands of $F \geqslant 765$ g/m^{2}h and $K_f \geqslant 0.78$ (from Bartels, 2006c).

and a sufficient buffering capacity against liquid sweat impulses $K_f > 0.78$ (Bartels, 2006c). Hence, from the comfort point of view, PES performs better than PP, although a hydrophilic finish applied to PP textiles will improve their performance.

Whereas for everyday and leisure wear, loosely fitting underwear is widely accepted as being comfortable, there is ongoing discussion on the design for sports underwear. A large group of manufacturers and institutes believe that tight fitting garments should be more comfortable, as they enhance the uptake and transport of liquid sweat. However, this concentration on liquid sweat transport alone is not optimal. It would be better if liquid perspiration did not occur at all. As mentioned above, up to 30% of heat and moisture may be dissipated from the body by means of convection and ventilation, but this can only happen if the underwear is loose fitting. Tight fitting underwear impedes these important processes.

15.8.2 Fleece

Fleece became popular in the 1990s as a middle-layer in functional clothing systems designed mainly for outdoor sports. Today, fleece not only serves as a sports textile, but is an integral part of everyday and leisure wear. This success is due to the superior comfort offered by fleece. Some decisive advantages of fleece are:

- Light weight thermal insulation.
- Comparatively good water-vapour permeability. The fibres of the pile are oriented parallel to the water-vapour transport direction, enabling water-vapour molecules to escape from the textile in the shortest and most direct way.
- Very good skin comfort.
- Elasticicity, which improves the ergonomic comfort.

Some fleeces are treated with a hydrophobic finish to improve foul weather performance. However, this impairs comfort, as liquid perspiration remains on the skin. Laminated fleece textiles including a highly breathable membrane offer a better alternative, both for comfort and protection. The majority of fleece textiles are made of polyester. However, a few constructions made from wool or cotton are available. These hygroscopic fibres have a slight advantage for everyday wear, but are not suitable for sports applications.

Most of the fleece textiles on the market (even those made of wool or cotton) have a tendency to take an electrostatic charge: a consequence of their open construction which encloses air. The application of an anti-static finish solves this problem and also improves liquid sweat management. A survey on the comfort of fleece textiles with these and other findings can be found in (Bartels, 1998).

15.8.3 Breathable foul weather protective jackets

Since the Gore-Tex membrane was first introduced in commercially available textiles during the 1970s, breathable foul weather protective jackets have become a synonym for comfortable textiles. Although even the best membrane or coating is still a water-vapour barrier, consumers could readily perceive the difference to non-breathable, water-tight clothes. Consequently, non-breathable constructions largely disappeared from many markets.

Breathable foul weather protective textiles are used in sports, everyday and leisure wear as well as in other applications. Due to their high market share, much work has been done in improving the comfort properties of foul weather protective gear. Many findings relevant to comfort have been published, and we may therefore concentrate on a few less well known highlights.

The first point to be discussed is whether water-vapour permeable textiles offer a benefit, even at cold temperatures. In answer to this question, it has been shown that the breathability of foul weather protective jackets is maintained at sub-zero ambient temperatures down to –20 °C (Bartels and Umbach, 2002a). Even at such low temperatures, a breathable membrane reduces not only the micro-climate humidity, but also the condensation of sweat inside the clothing (see Table 15.1). The condensation is important if thermal insulation is to be maintained at an adequate level for protection in cool surroundings.

A problem occurs with the liquid sweat transport, which is particularly important in sportswear such as cycling jackets or running wear. A membrane or coating which repels liquid water will not permit the transfer of liquid perspiration. The problem is solved by the employment of hydrophilic linings (Böhringer, 2000; Bartels, 2006b, 2007). Even then, no direct liquid sweat transport is possible. But, as shown in Fig. 15.8, by soaking up drops and dissipating them within the lining, the moisture is spread out and a larger area of the membrane can be used for the

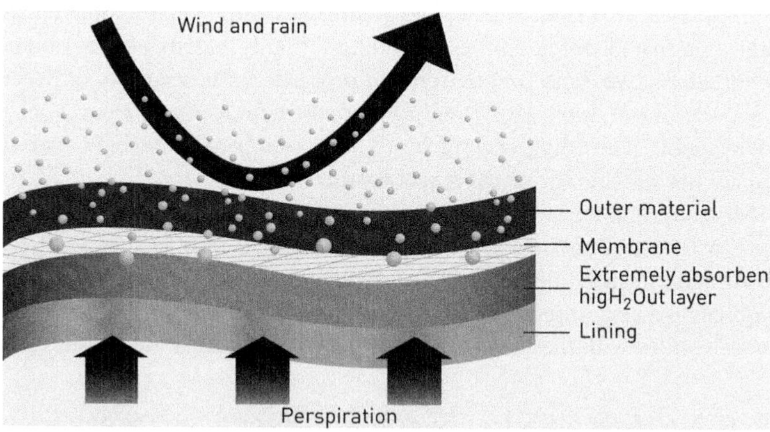

15.8 A foul weather protective textile combined with a hydrophilic lining improves the (indirect) transport of liquid sweat. Source: Sympatex Technologies.

Table 15.1 Amount of condensation in clothing systems with foul weather protective suits at temperatures of +20, 0 and −20 °C

No.	Description of foul weather protective textile	Thermal insulation R_{ct} according to ISO 11092 in 10^{-3} m^2K/W	Water-vapour resistance R_{et} according to ISO 11092 in m^2Pa/W	Condensation in clothing system in g at temperature		
				+20 °C	0 °C	−20 °C
2	2-layer-laminate with microporous PTFE-membrane and hydrophilic surface layer	5.4 ± 3.0	2.58 ± 0.30	38 ± 6	51 ± 7	101 ± 26
5	Water-vapour impermeable PU-coating	10.1 ± 3.0	164.7 ± 7.9	239 ± 60	163 ± 48	227 ± 91
7	Micro-fibre woven (identical to No. 8 but without membrane)	2.0 ± 3.0	1.45 ± 0.30	31 ± 18	48 ± 16	55 ± 15
8	2-layer-laminate with hydrophilic PET-membrane (identical to No. 7 but with membrane)	6.6 ± 3.0	5.12 ± 0.30	46 ± 12	49 ± 22	39 ± 9

Data were obtained from wearer trials with human test subjects in a climatic chamber. Additional clothing layers were used to obtain the necessary thermal insulation for each temperature. It can be seen that the water-vapour impermeable suit No. 5 leads to by far the highest rate of condensation, whereas the water-vapour permeable suits exhibit a low amount of moisture accumulation. This holds true particularly at −20 °C, hence, the breathability of the laminates is still given at these low temperatures (data from Bartels and Umbach, 2002a).

(vaporous) sweat transport. In addition, the moisture moves more quickly to the membrane. The liquid sweat transport from the skin to the atmosphere is improved by these indirect processes. Where the jacket is in direct contact with the skin, (e.g. at the forearm), the hydrophilic linings also enhance skin sensory comfort.

15.8.4 Elastic textiles

As discussed in previous sections, the use of elastic textiles (e.g. by including elastane in the fibre composition) can significantly increase the ergonomic and sometimes the psychological comfort (see also Voyce et al., 2005). However, there may be also disadvantages for the thermo-physiological and skin comfort:

- Tight fitting clothes suppress ventilation.
- Elastane is hydrophobic and non-hygroscopic, thus it cannot soak up or transport sweat.
- Many elastic constructions, especially in combination with micro-fibres, are extremely smooth (flat). These 'foil-like' constructions cause a clinging sensation on the skin.

The conclusion, therefore, is that the use of elastane alone does not automatically result in a comfortable textile. The construction guidelines given in the previous paragraphs should be taken into consideration, particularly for elastic textiles. A more detailed discussion of the comfort properties of elastic textile constructions can be found in (Bartels, 2005a).

15.8.5 Bio-functional clothing

Although not recognised as such by most consumers, there are many bio-functional textiles on the market. For example, sports textiles make wide use of the bio-functional properties of silver particles to prevent undesired body odours. In everyday clothing bio-functional textiles are mainly used in underwear for their hygienic properties. A survey on the comfort properties of bio-functional clothing is published in (Bartels, 2006a). A synopsis of the main findings follows:

- Bio-functional textiles can have the same comfort properties as normal, non-bio-functional constructions. With careful construction, it is possible to add bio-functionality without impairing comfort.
- However, some bio-functional treatments are hydrophobic, suppressing liquid sweat transport and diminishing comfort.
- In some constructions, bio-functional filament yarns are used instead of spun yarns, leading to poorer skin sensory comfort.

Hence, to avoid comfort impairments, bio-functional textiles should be constructed with the same care as normal textiles.

15.8.6 Clothing systems

In considering clothing systems, one should start with the underwear, i.e. the layer directly on the skin. This is the layer which contributes most to the overall comfort of a clothing system (Bartels, 2003). Underwear must be well constructed according to section 15.8.1 if it is to provide a good level of comfort.

But the other clothing layers also have to be functional. One inferior component will disturb the comfort of the whole clothing system. If for example, a poor water-vapour permeable layer is included, this single barrier will impair the water-vapour transport of the whole system (with the exception of that which takes place by ventilation). Condensation will take place at this barrier and cause the clothing system to become wet from the inside, so decreasing the thermal insulation.

Each component of a clothing system not only has to be functional on its own, but must also work effectively with the other layers. When combined with highly breathable foul weather protective jackets, purely synthetic underwear performs better than underwear with an outer cotton surface. This effect will be pronounced if the jacket contains a hydrophilic lining. Liquid sweat will be taken up by the underwear and transported directly to the jacket without being stored in the middle layer. However, if the outer garments are thick and have low breathability, functional underwear with a retentive external layer of cotton or wool is preferable. Further reading on sports clothing systems comfort can be found in Bartels (2005a) and Ruckman (2005).

15.9 Recent and future trends in sports, everyday and leisure wear

15.9.1 Thermo-regulating textiles

Thermo-physiologically comfortable clothing supports the thermo-regulation of the body and helps to maintain a comfortable temperature for the wearer. During the last few years, new textile concepts have been introduced into the market which actively regulate temperature and moisture particularly in so-called 'transient wearing' situations where the temperature or the activity is changeable.

One concept employs 'phase-change materials' (PCMs), usually in the form of micro-capsules (Outlast™; Magill, 2004). Strictly speaking, all materials exhibiting a phase change are PCMs. But only those with a phase change transition temperature in the vicinity of skin temperature are of interest for comfort applications, such as different paraffins or salt hydrates (Pause, 2006). The PCM takes up heat during activity or at a high temperature (phase change from solid to liquid state), and releases heat at rest or low temperature (phase change from liquid to solid state). As a phase change requires a lot of energy, much more heat will be stored or released than in a non-PCM. The promised consequence is that the wearer stays within a comfortable temperature range, even in transient wearing situations.

The phase-change temperature is a crucial point in PCM textiles as the effect can only take place if the temperature is reached within the textile. This is because the temperature of the clothing is strongly dependent on ambient temperature, body activity and other clothing layers (Pause, 2006).

The second crucial point is the implementation of PCM and care must be taken not to disturb a textile's breathability. The micro-capsules are impermeable to water vapour and if they are included in a coating, it will also become a water-vapour barrier. The phase change from liquid water to water vapour needs even more energy than the phase change from solid to liquid in paraffin. Hence, reducing breathability by using PCM means replacing a very effective phase change (evaporation of sweat) by a less effective one (melting paraffin). Therefore breathability has to be maintained when including PCMs.

A totally different concept for thermo-regulating textiles is the use of membranes with temperature-dependent water vapour permeability (Diaplex™, c_change™). At high ambient temperatures, the membrane is open to water-vapour diffusion so sweat can easily evaporate and the body is cooled down. At low temperatures, the membrane 'closes' on a molecular scale. Water vapour cannot then escape the clothing, but condenses at the membrane and releases its stored energy which keeps the body warm.

The crucial factor in this concept is the usefulness of low breathability at cold temperatures. As discussed above, in foul-weather protective clothing, condensation is not only felt as unpleasant, but also wets the clothes and reduces their thermal insulation. In addition, Gibson (1999, 2000) cannot find a temperature dependency of water-vapour permeability, which would suggest that the effect does not take place.

The real use of thermo-regulating textiles remains under discussion. However, if the full benefit is to be gained from these types of textiles, they must be exactly adapted to the application.

15.9.2 Body mapping

The dissipation of heat and sweat is not constant over the entire body area. In order to improve comfort, specially chosen textiles can be used for different parts of the body (Kitteringham, 2006). However, such constructions are much more complicated than normal textiles and are mainly used in higher priced sportswear. A prerequisite for this type of clothing is a knowledge of the exact distribution of heat and sweat production over the whole body. But skin temperature and sweat production are also dependent on the textile and this coupling of physiological parameters with textile construction further complicates the situation for clothing designers. However several well-known brands such as Adidas, Eschler or Gore market body area-specific textiles and clothes (Adidas, 2009; N.N., 2005, 2006a, 2006b).

15.9.3 Inclusion of electronics or micro-systems

An ongoing trend is the inclusion of electronics or micro-systems into garments which are known as smart clothing, i-wear, or intelligent clothing. In recent years, the development of these garments has focused on their 'intelligence', i.e. their additional functions due to electronics or micro-systems. However, the comfort properties of these constructions were often inferior (Bartels, 2004, 2005b) because of the use of stiff and water-vapour impermeable components. The development of softer components, known as 'soft switches', has followed.

Recent innovations incorporate the electronic or micro-system components into the textile itself. The material is breathable and soft to the touch. One of the newest inventions is shown in Linti and Horter (2009) with the example of a babygrow which is intended for the prevention of sudden infant death. The babygrow contains several textile sensors which monitor the baby's vital functions. Although this is essentially a medical textile, it crosses the usual boundaries of a medical product as it is worn and put on like a normal babygrow. The same holds true for clothes which monitor the vital functions of elderly people and are also worn daily. Another application is sportswear, which logs fitness parameters and checks the impact of training.

A lot of research is currently being carried out to further improve intelligent clothing. Although it is difficult to predict exactly how 'intelligent' clothes will look in the future, it seems clear that their market share will grow and that they will be of increasing importance in our daily life.

15.10 Future trends in testing comfort of sports, everyday and leisure wear

As textiles and clothes develop in respect of comfort, so must the devices used to test them. These trends are discussed in Part II of this book. For sports, everyday and leisure wear, two new pieces of apparatus may be of interest:

1. The so-called human-clothing-environment simulator of Kim and co-workers (Kim et al. 2003; Kim and Yoo, 2005), which is able to simulate a rapid change of ambient climatic conditions (Yoo and Kim, 2005, 2008). A schematic drawing of the apparatus is given in Fig 15.9.
2. Another device, designed by Fan and co-workers (Sarkar et al., 2007) simulates heavy perspiration, which is particularly important in sportswear. A novel measuring principle keeps the water supply constant throughout the process. Moreover, this 'transplanar water transport tester' is able to measure both the liquid absorption and evaporation from the textile surface.

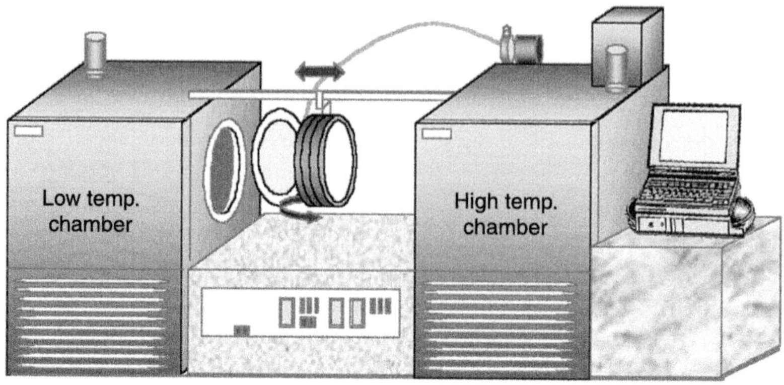

15.9 Schematic drawing of the human–comfort–environment simulator by Kim et al. The apparatus is able to simulate rapid ambient temperature changes.

15.11 Conclusions

Sports, everyday and leisure wear illustrate the importance of comfort in consumer acceptance and choice. In the long term, clothing with a high level of comfort is preferred. Thus outdoor clothing, though starting as a niche product, has become part of everyday wear and has taken a large share of the market.

Different wearer activities and perspiration levels require sportswear and everyday or leisure wear to fulfil different physiological needs. Thus good sportswear will not automatically be good everyday or leisure wear.

Laboratory apparatus exists for testing different aspects of comfort. However, different test devices measure different textile parameters. Their meaning for the human perception of comfort must be proved through mathematical correlations between laboratory trials and human beings. Different test methods differ in their ability to detect differences between textile specimens so experience is required in choosing the most suitable test procedure for each application.

15.12 Sources of further information and advice

For those who are able to read German, Mecheels (1998) is recommended as a comprehensive but easily understandable compilation on textile and clothing comfort. An overview of contemporary sportswear technology may be found in *Textiles in Sport*, edited by Shishoo (2005), which also contains surveys on sportswear comfort (Bartels, 2005a), elastic textiles (Voyce et al., 2005) and water vapour transfer (Ruckman, 2005).

15.13 References

5I3M (2003), *5th International Meeting on Thermal Manikin and Modelling*, Strasbourg, 29–30 September.

Adidas (2009). Available from http://www.adidas.com (accessed 20 May 2009).

Advansa SASA (2009), Fibres & yarns – Coolmax. Available from http://www.advansa.com/fibres-yarns-en/coolmax/coolmaxae/view?set_language=en (accessed 17 May 2009).

Anttonen H, Niskanen J, Meinander H, Bartels V, Kuklane K, Reinertsen RE, Varieras S and Sołtyński K (2004), 'Thermal manikin measurements – exact or not'. *International Journal of Occupational Safety and Ergonomics*, **10**(3), 291–300.

ASTM E96 (2005), 'Standard Test Methods for Water Vapor Transmission of Materials'.

ASTM F1868 (2009), 'Standard Test Method for Thermal and Evaporative Resistance of Clothing Materials Using a Sweating Hot Plate'.

Bartels V T (1998), 'Erweiterung der Einsatzbereiche von Fleece-Textilien durch Optimierung des Tragekomforts', Technical report No. AiF 10712, Hohenstein Institutes, Bönnigheim.

Bartels V T (1999), 'Quantitative Erfassung der taktilen Eigenschaften von Textilien in Abhängigkeit von den Konstruktionsmerkmalen der Oberflächenfasern', Technical report No. AiF 11089, Hohenstein Institutes, Bönnigheim.

Bartels V T (2004), 'The physiological function of intelligent technical textiles for protective clothing', *CINTE Techtextil China Symposium*, Shanghai, 31 August–2 September.

Bartels V T (2005a), 'Physiological comfort of sportswear', in Shishoo R, *Textiles in Sport*, Cambridge, Woodhead, pp. 177–203.

Bartels V T (2005b), 'Physiological function and wear comfort of smart textiles', *MST News*, **2**, 16–38.

Bartels V T (2006a), 'Physiological comfort of biofunctional textiles', in Burg G., *Biofunctional Textiles and the Skin*, Basel, Karger, pp. 51–66.

Bartels V T (2006b), 'Bekleidungsphysiologische Optimierung von Futterstoffen zur Verbesserung des flüssigen Schweißtransports in wasserdichter Funktionskleidung', Technical report No. AiF 13614 N, Hohenstein Institutes, Bönnigheim.

Bartels V T (2006c), 'Physiological comfort of sportswear', *Fiber 2006*, Seoul, 30 May–3 June.

Bartels V T (2007), 'Hydrophilic linings to enhance the liquid sweat transport through water tight clothing', *Avantex-Symposium*, Frankfurt, 12–14 June.

Bartels V T and Umbach K H (2002a), 'Water vapor transport through protective textiles at low temperatures', *Tex Res J*, *72*, 899–905.

Bartels V T and Umbach K H (2002b), 'Test and evaluation methods for the sensorial comfort of textiles', *Euroforum »Toucher du Textile«*, Paris, 4–5 December.

Bartels V T and Umbach K H (2003), Messverfahren zur Beurteilung der Atmungsaktivität von Textilien für Bekleidung und Bettsysteme, *Melliand Textilberichte*, **84**, 208–210.

Bartels V T and Umbach K-H (2005), 'New test apparatus to assess the stiffness of clothing textiles', *Melliand English*, **86**, 869–870.

Böhringer B (2000), 'Sympatex innovations in a mature market', *International Avantex Symposium*, Frankfurt, 27–29 November.

Breugnot C, Bueno M A, Renner M, Ribot-Ciscar E, Aimonetti J M and Roll J P (2006), 'Fabric touch: responses of mechanoreceptive afferent units and mechanical characterization', *45th Man-Made Fibers Congresss*, Dornbirn, 20–22 September.

BS 7209 (1990), 'Specification for water vapour permeable apparel fabrics'.

EN 342 (2004), 'Protective clothing – Ensembles and garments for protection against cold'.
EN 343 (2007), 'Protective clothing – Protection against rain'.
EN 471 (2008), 'High-visibility warning clothing for professional use – Test methods and requirements'.
EN 31092 (1994), 'Textiles; physiological effects; measurement of thermal and water-vapour resistance under steady-state conditions (sweating guarded-hotplate test) (ISO 11092:1993)'.
Gehui W, Postle R and Weiyuan Z (2003), 'The prickle of worsted woven wool fabrics', *Textile Asia*, **34**(3), 25–28.
Gibson P (1999), 'Effect of Temperature on Water Vapor Transport through Polymer Membrane Laminates'. Technical report No. 99/105, Natick, Massachusetts.
Gibson P W (2000), 'Effect of temperature on water vapor transport through polymer membrane laminates', *Polymer Testing*, **19**, 673–691.
Goldman R F (2007), 'Thermal manikins, their origin and role', International Society for Environmental Ergonomics. Available from: http://www.environmental-ergonomics.org (accessed 10 May 2009).
ICEE XI (2005), *Environmental Ergonomics XI*, Ystad, 22–26 May.
ISO 11092 (1993), 'Textiles; physiological effects; measurement of thermal and water-vapour resistance under steady-state conditions (sweating guarded-hotplate test)'.
Kim E A, Yoo S and Kim J (2003), 'Development of a human–clothing–environment simulator for dynamic heat and moisture transfer properties of fabrics', *Fibers and Polymers*, **4**, 215–221.
Kim E A and Yoo S (2005), 'Simulator for evaluating the performance of functional textiles', *3rd International Avantex-Symposium*, Frankfurt, 6–8 June.
Kitteringham G (2006), 'Outdoor clothing now matches insulation to specific zones', *Technical Textiles International*, **15**(3), 21–22.
Linti C and Horter H (2009), 'Baby Body Textilien retten Leben', Institut für Textil- und Verfahrenstechnik, Denkendorf. Available from: http://www.itv-denkendorf.de/images/ITV/forschung/innovationen/babybody.pdf (accessed 20 May 2009).
Magill M C (2004), 'Optimizing design with Phase Change Materials', *Innovations in Medical, Protective, and Technical Textiles*, Cary, 18–19 February, 83–95.
McCullough E (2001), 'Interlaboratory study of sweating manikins', *4th International Meeting on Thermal Manikins*, St. Gallen, 27–28 September.
McCullough E, Huang J and Kim C S (2004), 'An explanation and comparison of sweating hot plate standards', *Journal of ASTM International*, **1**(7), 1–13.
Mecheels J (1982), Zur Komfort-Wirkung von Textilien auf der Haut. *Hohensteiner Forschungsbericht*, April.
Mecheels J (1998), *Körper – Klima – Kleidung – Wie funktioniert unsere Kleidung?*, Berlin, Schiele & Schön.
Miraftab M (2000), 'Technical fibres', in Horrocks A R and Anand S C, *Handbook of Technical Textiles*, Cambridge, Woodhead, pp. 24–41.
Naylor G R S and Phillips D G (1997), 'Fabric-evoked prickle in worsted spun single jersey fabrics', *Textile Research Journal*, **67**, 413–416.
NCSU 2009, 'Vertical wicking', North Carolina State University. Available from: http://www.tx.ncsu.edu/tpacc/comfort/vertical_wicking.htm (accessed 24 April 2009).
N.N. (2005), 'Bodymapping – eine neue Stricktechnik', *Textile Network*, **3**(11), 41.
N.N. (2006a), 'Comfort mapping', *Future Materials*, **6**, 10.
N.N. (2006b), 'Body mapping develops inside and out', *World Sports Activewear*, **12**(4), 44–46.

Pause B (2006), 'Phase Change Materials – Erhöhung des thermophysiologischen Tragekomforts', in Knecht P, *Technische Textilien*, Frankfurt, Deutscher Fachverlag, pp. 399–412.

Ruckman J E (2005), 'Water resistance and water vapour transfer', in Shishoo R, *Textiles in Sport*, Cambridge, Woodhead, pp. 287–305.

Sarkar M, Fan J and Qian X (2007), 'Transplanar water transport tester for fabrics', *Meas Sci Technol*, **18**, 1465–1471.

Shishoo R (2005), *Textiles in Sport*, Cambridge, Woodhead.

Tourlonias M, Renner M, Buoeno M A and Bigué L (2006), 'Polarimetric measurements for fabrics and nonwovens related to tactile aspects', *International Fiber Conference*, Seoul, 30 May–6 June.

Tronnier H, Wiebusch M and Heinrich U (2004), 'Neues Testverfahren zur Verbesserung des Tragekomforts von Textilien', *Textilveredlung*, **39**(1/2), 24–27.

Umbach K-H (1987), 'Methods of measurement for testing physiological requirements of civilian, work and protective clothing and uniforms', *Melliand English*, **68**, E379–E383.

Umbach K.-H. (1993), 'Moisture transport and wear comfort in microfibre fabrics', *Melliand English*, **74**, E78–E80.

Umbro plc (2007), Analyst presentation, June 19. Available from http://www.umbroplc.com/doc_docs/A_I_June_2007_final_print_version.ppt (accessed 8 April 2009).

Voyce J, Dafniotis P and Towlson S (2005), 'Elastic textiles', in Shishoo R, *Textiles in Sport*, Cambridge, Woodhead, pp. 204–230.

Yoo S and Kim E A (2005), 'Effects of layer design on condensation within cold weather clothing ensembles', in Holmér I, Kuklane K and Gao C, *Environmental Ergonomics XI*, Ystad, 22–26 May, pp. 131–134.

Yoo S and Kim E (2008), 'Effects of multilayer clothing system array on water vapor transfer and condensation in cold weather clothing ensemble', *Textile Research Journal*, **78**, 189–197.

16
Cold weather clothing and comfort

I. HOLMÉR, Lund University, Sweden

Abstract: Clothing provides a means of controlling heat exchange with the environment, particularly important in cold climates as changing weather and activity require a continuous conscious adjustment of clothing to maintain thermal comfort. The most important property of cold weather clothing is its thermal insulation, which can be measured according to ISO 15831. The insulation value of a cold weather ensemble can be used for prediction of its performance in terms of providing thermal comfort using ISO 11079. Cold weather clothing builds on the layering principle: the inner layer or underwear controls moisture and air motion over the skin; the second layer is mainly insulative; and the outer layer protects against ambient factors such as wind, rain and snow. Modern materials and functional design allow clothing to be developed for quite extreme exposures.

Key words: cold stress, heat balance, physiological strain, standards, test methods.

16.1 Introduction

In many parts of the world, in particular in Northern countries and in countries with a continental climate type, the air temperature during several months of the year is low and often well below zero. People outdoors are then subject to cold stress. Native people of cold regions have adapted, mostly successfully, by a combination of physiological and behavioural responses over long time that enhances their ability to cope with the harsh environmental conditions (Young, 1988; Launay and Savouray, 2009). The successful inhabitation of colder regions of the world can be explained by the intelligent and foresighted use of technical measures, including clothing. Modern man, living in cities and in villages in temperate and cold regions spends most of his time in heated and comfortable houses at temperatures close to the thermo-neutral zone (20–25 °C). In fact, most of these people spend more than 90–95% of their time in the narrow temperature zone of thermal comfort and make only short excursions to colder climates when they travel to and from their work or shopping place. The experience of cold stress is limited and the experience gained from it often small. In contrast, people in outdoor work and outdoor leisure time activities must behave suitably and require functional clothing, selected on the basis of the specific requirements of their activity and the environmental conditions (Holmér, 2005a).

Cold is usually associated with temperatures well below the 'comfort zone'. This zone is defined as the temperature interval in which people will be in heat balance at rest or very low activity, in light clothing. This is also what we mean by indoor climate (20–25 °C). However, also at these temperatures people may be cold if exposed to strong wind and/or precipitation wets their clothing. In international standards cold is defined as temperatures below +10 °C (ISO 7730, 2006; ISO 11079, 2007).

The problems and hazards of cold environments are quite evident. The hazards are tissue cooling, slipping and falling. The problems are more discomfort and distraction by environmental and thermal stimuli, more cumbersome clothing increasing work load, deterioration in physical and manual performance and adverse health effects (Enander and Hygge, 1990; Giesbrecht, 1995; Heus et al., 1995; Hassi et al., 2002).

Cold weather is associated with low temperatures. However, the thermal effects on humans are determined also by radiant temperature, wind and humidity. These factors are dealt with in more detail in a later section.

16.2 Thermal comfort and heat balance

Comfort is dealt with in many other chapters of this volume. For many people comfort means satisfaction and pleasure. From a physical point of view, a prerequisite for thermal comfort is that the body as a whole, as well as body parts, are in good heat balance. This means that body temperature and skin temperatures must be in a certain, quite narrow range and remain constant. Accordingly, analysis of conditions for comfort in cold environments can be based on physical heat balance calculations (Parsons, 2003). The general heat balance equation of the human body (equation 16.1) applies to cold conditions as well.

$$S = M - W - RES - C - R - E - K \qquad [16.1]$$

where S is the rate of change in body heat content, M is the metabolic heat production, C, R and E are the heat exchanges by convection, radiation and evaporation, respectively and RES is the airway heat loss, all in W/m^2. External mechanical work W and conductive heat exchanges K are mostly very small and can be neglected in the analysis. RES is related to activity and may reach values of about 10–15% of M.

Activity and the associated metabolic heat production is a key factor in the heat balance equation. This heat must be lost to the environment at an equal rate as it is produced in order to avoid change in tissue temperatures. Examples of metabolic heat production are given in Table 16.1.

Equations for determination of C, R and E are given below.

$$R + C = \frac{t_{sk} - t_o}{I_T} = \frac{t_{sk} - t_{cl}}{I_{cl}} = \frac{t_{cl} - t_o}{I_a} \qquad [16.2]$$

$$E = \frac{p_{sk} - p_a}{R_{eT}} \qquad [16.3]$$

where t_{sk}, t_{cl} and t_o are temperatures of the skin, clothing surface and ambience, respectively. The operative temperature t_o integrates air temperature and radiant

Table 16.1 Examples of metabolic energy production associated with different types of work or activity

Average metabolic rate (Wm^2)	Examples
65	Resting: sitting at ease
100	Light manual work; casual walking (speed up to 3.5 km/h); spectators outdoors; winter fishing; scooter driving
165	Moderate work with hand and arm; arm and leg; walking at a speed of 3.5 km/h to 5.5 km/h; hiking; alpine skiing
230	Intense arm and trunk work; carrying heavy material; walking at a speed of 5.5 km/h to 7 km/h; intermittent activity (ball games); slow walking in deep snow
290	Very intense activity at fast pace; climbing; running or walking at a speed greater than 7 km/h; alpine skiing (well trained)
400	Sustained heavy work; long distance events in cross-country skiing

Notes: Modified from ISO 8996 (ISO-8996, 1990) and Holmér, 2005 #7129; Holmér, 2005b. Values refer to a person with 1.8 m^2 body surface area.

temperatures of the environment. Often surrounding surfaces are as cold as the air and t_o becomes equal to air temperature. The denominator contains expressions for insulation that are described in more detail below. In equation 16.3, p_{sk} and p_a are the water vapour pressure at the skin surface and in ambient air, respectively. R_{eT} is described below.

At the onset of exercise and for 30–40 minutes some heat is stored in the body, more so the heavier the exercise (Saltin and Hermansen, 1966). Body core temperature increases and stabilizes at some level determined by the relative exercise intensity. At this higher core temperature, S in equation 16.1 should approach zero to establish heat balance with the environment and steady state conditions. This can be achieved for many combinations of climate, clothing and activity, resulting in different levels of physiological strain. The values of t_{sk} and p_{sk} are related to the physiological strain. With large heat losses t_{sk} may drop and an increasing discomfort from cold is perceived. In the cold, hands, feet and extremities suffer first with excessive cooling resulting also in local cold discomfort. In particular during higher activity sweating may be provoked, raising the value of p_{sk}. The skin becomes more humid resulting in a gradual increase in discomfort from wetness. As already mentioned, a comfortable heat balance can only be preserved in a narrow range of t_{sk} and at low values of p_{sk}.

Physiological responses to cold are related to the level of cold stress and degree of tissue cooling. Thermoregulatory responses to abnormal tissue cooling comprise redistribution of blood circulation from superficial to deep vessels. In particular in the extremities blood flow may be considerably restricted, resulting in a sharp drop in skin temperature of hands and feet. A constant core temperature is long preserved, but with excessive heat losses it progressively drops. At this stage shivering is activated, producing heat to restore heat balance. Shivering may contribute as much as 300–500 W of heat production. If insufficient, hypothermia gradually develops and places the subject in a risky situation. With extremity cooling, in particular in combination with hypothermia, local cold injuries may develop such as frostbite.

Clothing is an important tool for control and regulation of heat exchange and, hence, the values of t_{sk} and p_{sk}. There are two principal thermal properties that determine clothing effects on heat exchanges by convection, radiation and evaporation.

1. *Thermal insulation* (I) defines the resistance to heat transfer by convection and radiation by clothing layers. It accounts for the resistance to heat exchange in all directions and over the whole body surface. It is an average of covered as well as uncovered body parts. This unique definition allows the introduction of clothing in the heat balance equation. The insulation of clothing and adjacent air layers is defined as the total insulation values (I_T). I_{cl} defines the intrinsic clothing insulation and I_a is the adjacent air layer insulation. They are defined in equation 16.2.

   The value of I is given in m^2°C/W or in clo-units (1 clo = 0.155 m^2°C/W). Thermal insulation is measured with a thermal manikin (ISO 15831, 2003) and the standard value is obtained with the manikin standing in calm air. Convection increases by wind and body movements (Havenith and Nilsson, 2004) and may have a significant effect on clothing heat exchange in the cold. The I_{cl} values therefore reduce with wind and walking. The air permeability of the outer layers is the critical factor but also design and softness of material.

   Table 16.2 gives an indication of the level of insulation that can be provided by different clothing ensembles. The more layers the higher will be the insulation.
2. *Evaporative resistance* (R_e) defines the resistance to heat transfer by evaporation and vapour transfer through clothing layers. As for insulation, it also refers to the whole body surface. In reality, the property is a resistance to vapour transfer. Heat transfer takes place only when sweat evaporates at the skin and is transported to the environment by diffusion or convection. The evaporative resistance of clothing layers and adjacent air layers (R_{eT}) is defined by equation 16.3. The unit is Pa m^2/W. Evaporative resistance is less important for cold protective clothing for two reasons.

   Water vapour transfer along a falling temperature gradient results in condensation and eventual freezing and this may take place in outer clothing layers. Evaporative cooling becomes inefficient. Condensation and wetting of clothing is detrimental to insulation and accelerates cooling by convection.

Most people in cold countries are familiar with wind chill, the extra cooling caused by wind at low temperatures. A new wind chill index has been introduced that provides an easily understandable interpretation of such effects (Bluestein, 1998; Url). The index applies to bare skin.

Table 16.2 Basic insulation values (I_{cl}) of typical clothing ensembles

Clothing ensemble	I_{cl}	
	m² °C/W	clo
2 layer clothing with long underwear, thin jacket and trousers	0.16	1.0
2 layer clothing with long underwear, trousers and insulated jacket	0.23	1.5
2 layers on leg: long underwear and trousers; 3 layers on torso: long underwear, shirt and insulated jacket	0.28	1.8
2 layers on leg: long underwear and insulated trousers; 3 layers on torso: long underwear, shirt and insulated jacket	0.34	2.2
2 layers on leg: long underwear and insulated trousers; 3 layers on torso: long underwear, shirt and insulated parka	0.39	2.5
2 layers on leg: long underwear and insulated trousers; 3 layers on torso: long underwear, fleece shirt and insulated parka	0.42	2.7
3 layers on leg: long underwear, fleece or fibre pile trousers and insulated trousers; 3 layers on torso: long underwear, shirt and insulated jacket	0.47	3.0
3 layers on leg: long underwear, fleece or fibre pile trousers and insulated trousers; 3 layers on torso: long underwear, shirt and insulated parka	0.54	3.5
3 layers on leg: long underwear, fleece trousers and down trousers; 3 layers on torso: long underwear, shirt and down parka	0.65	4.2
Sleeping bags	0.46–1.4	3–9

Note: Modifed from ISO 11079.

Wind effects on clothing have been studied in wind tunnels with thermal manikins and with subjects (Havenith et al., 1990; Holmér et al., 2001). A general correction formula for walking and wind speeds up to 18 m/s has been proposed (Havenith and Nilsson, 2005). Figure 16.1 shows the relative reduction in thermal insulation caused by wind and three multi-layer clothing ensembles with outer layers of different air permeability are depicted. Windproof outer layers prevent much of the cooling effect of wind.

'Breathable' membranes of various types combine the property of being windproof with some degree of water vapour permeability. This allows for evaporative cooling when water vapour escapes to ambient air. However, during strenuous work sweat production often becomes high and the vapour transmission rate of the fabrics is too low to remove all the moisture from the clothing system. In addition, the steep temperature gradient from the skin surface to the cold ambient air may result in condensation. The dew point is reached in the outer layers. Considerable moisture absorption may result and the subsequent reduction of insulation increases the risk of post exercise chilling. Best use of this kind of membranes in the cold is probably to put them in deeper layers allowing some kind of control of moisture condensation. Such applications are available for jackets and socks.

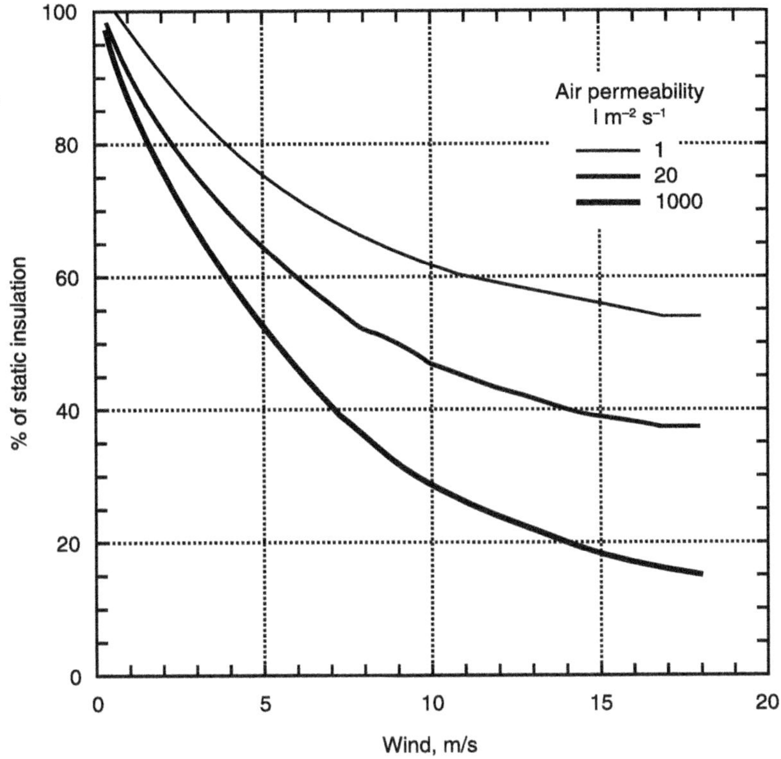

16.1 Effect of wind on three winter ensembles with outer layers of different air permeability (modified from Holmér et al., 2001).

16.3 Requirements for comfort in the cold

ISO 11079 (2007) describes a heat balance equation that can be used for analysing the conditions for thermal comfort and physiological strain in the cold. It describes two options.

1. The required clothing insulation value is calculated for a given activity level and climatic conditions. This value is called IREQ (Insulation REQuired).
2. The resulting heat balance is calculated for given activity level and climatic conditions when clothing insulation (I_{cl}) is known.

Figure 16.2 shows the required clothing insulation (IREQ) for 'comfort' values for skin temperature and sweating (low p_{sk}). With low activity the IREQ increases sharply with lower temperatures. With high activity, IREQ stays much lower and is easier to achieve with appropriate clothing.

Figure 16.3a shows the recommended exposure time to stay reasonably comfortable with various clothing ensembles. The insulation value is the intrinsic

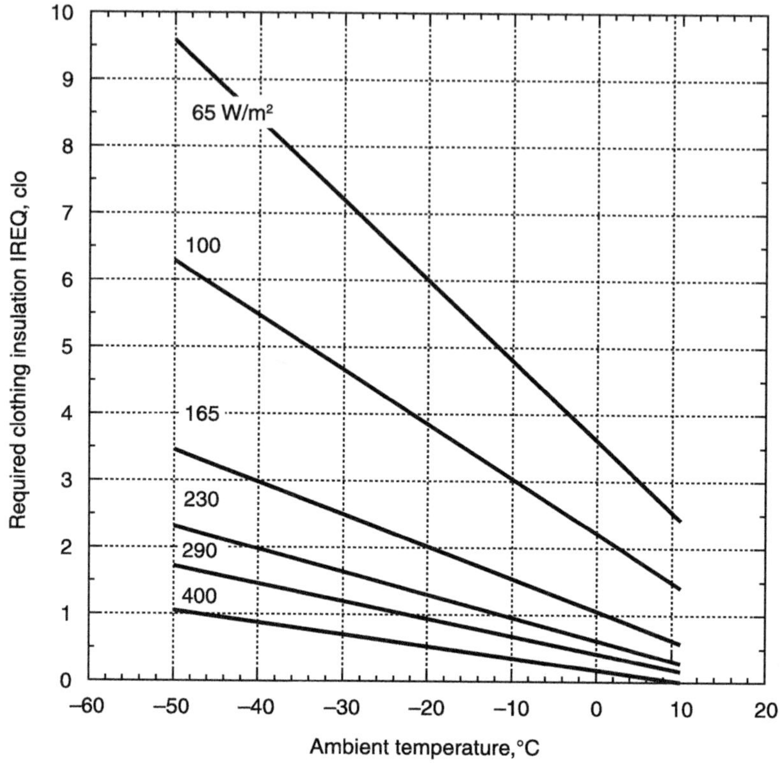

16.2 Required clothing insulation (IREQ) according to ISO 11079 for different levels of physical activity in the cold.

value (I_{cl}). It is the standard value obtained with a static, thermal manikin according to ISO 15831 (2003), and what is found in most tables about clothing, for example in ISO 9920 (2007). Figure 16.3b shows how the stay time becomes shorter and shorter at a given temperature because of the wind effect on a clothing with 3.0 clo and air permeability of 8 l/s m^2. In calm air good protection is offered for two hours or more at –20 °C. The same protection at 8 m/s can only be offered at a temperature around 0 °C.

16.4 Principles for cold weather clothing

Protective clothing against cold has been dealt with in two previous publications (Holmér, 2005a, 2005b). The main strategy for preservation of heat balance and comfort in the cold is to control heat exchange through clothing mainly by convection and radiation. The temperature gradient between the skin (about 34 °C at comfort) and the ambience may be 30–60 °C or even more in very cold environments. By donning or doffing garments or by adjustment of openings, etc., warm air is convected and heat is radiated to the environment. The processes are enhanced by body motion (microclimate ventilation). Even during heavy exercise most of the heat can be dissipated by these behavioural means. Sweat evaporation as a mechanism of heat dissipation is only required under special circumstances, for example in high intensity athletic events. Unfortunately, work that requires special protective equipment may put restrictions on the behavioural adjustments, clothing becomes too warm and causes sweating. Sweat evaporation is not as high in the cold as in the heat as the transfer of moisture through the layers is complex and much of it is absorbed or condensed in layers. There is some contribution to heat transfer but the detrimental factor is that moisture remaining in clothing dries out taking its heat from the skin (Havenith et al., 2008). This is not wanted when activity slows down or rest is taken.

Cold weather clothing for comfort must provide insulation and allow moisture control, and flexibility is required for active work. Such clothing is normally based on three layers with distinct functions and properties – inner layer, middle layer and outer layer. The *inner layer* is worn close to the skin in order to reduce

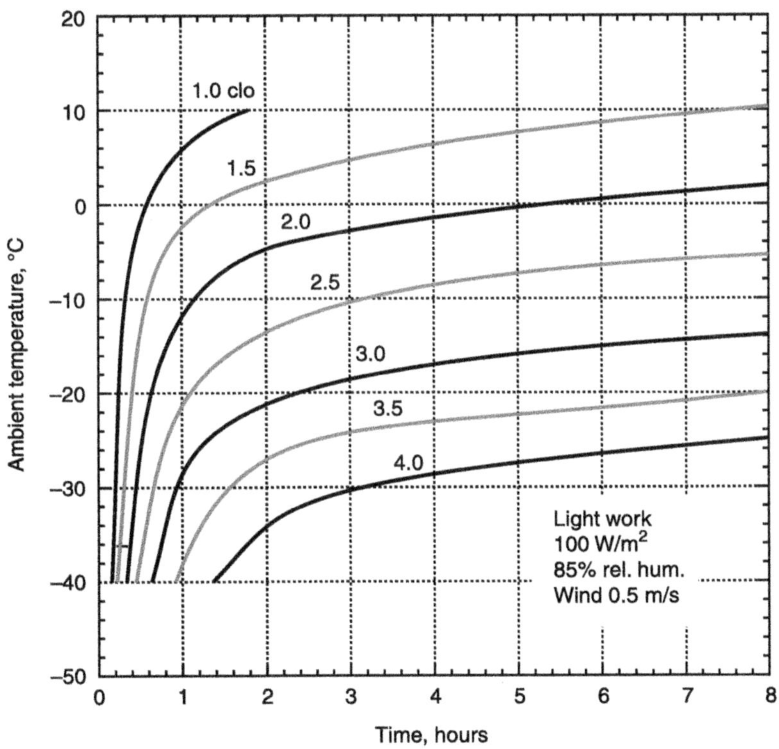

16.3 (a) Protection provided by different levels of basic insulation (I_{cl}) for light work at combinations of ambient temperature and exposure time.

(*Continued overleaf*)

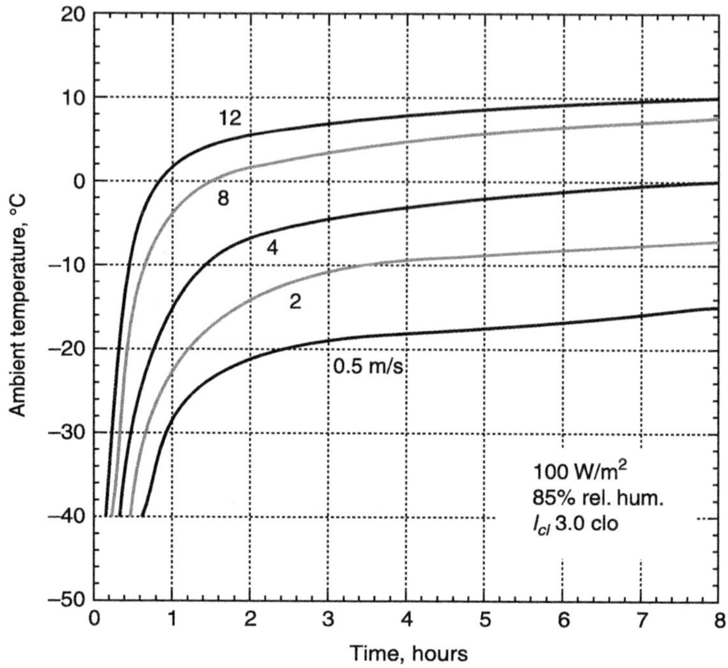

16.3 continued (b) Protection provided by clothing with a basic insulation (I_{cl}) value of 3.0 clo for light work at combinations of ambient temperature, exposure time and wind velocity.

skin convection and, eventually, control moisture. Type of fibre and fabric may depend on the exposure conditions (see below). The *middle layer* has primarily an insulating function and its thickness and design can vary accordingly. It can be composed of several pieces of garments. The *outer layer* protects against environmental factors, such as wind, precipitation, mechanical abrasion, chemicals, particles and similar hazards. In many cases this means that dense and durable fabrics are used. It may also provide additional insulation.

Design of a functional multi-layer clothing system must take account of low internal friction between layers, minimal bulk and low weight, while the clothing must not interfere with body motion. Multi-layer clothing systems add weight to the body that must be carried during walking and increase the energy cost of work. Several winter ensembles weighing 4–6 kg increased metabolic rate between 6–18% in a simulated work task (Dorman and Havenith, 2009).

16.4.1 Insulation

The thermal insulation of a clothing ensemble is determined by the properties of fibres, fabrics and air layers from skin to the outer surface. In most fabrics and clothing, fibers and membranes occupy only a small part of the total space. Much of the heat transfer is determined by free and forced convection in this space and to some extent by radiation between surfaces in the layers.

Insulation is basically provided by layers of clothing of different thickness that preserve still air layers around the body. As mentioned above, the inner layer should comprise a 'second skin' close to the skin. This layer will reduce convection at the skin surface and the cold sensation. Much of the required insulation in cold weather clothing is provided by trapped air layers in garments and by air layers between garments, so garments should be light and lofty. Wool has long been used for provision of good insulation, but modern synthetic fabrics may provide good insulation at low weight.

With exercise, the air layers are ventilated as a result of body movements and heat transfer to the environment is facilitated. Often this is not sufficient and clothing needs to be opened or even removed to balance heat production.

Very light work requires high insulation. Thicker underwear or double underwear may be required. The middle layer should be one or two garments on torso and legs. Insulation is given by shirts, sweaters, vests, or trousers made of, for example, wool blends, polyester, fleece or fibre pile. The outer garment should be a thick parka with trousers or a thick overall. The outer garment is designed with a lining of insulation, covered by fabric that protects against wind, water, etc.

With light to moderate work much of the required insulation is given by the middle layer. The outer layer may be a jacket and trousers, a parka and trousers or an overall. Depending on exposure conditions it may comprise one or more thin layers of fabric (shell) that protects against wind, water, etc.

Heavy work is associated with considerable heat production (see above), so the required clothing insulation is small. Often only thin underwear and a wind and waterproof shell (jacket and trousers) are required in order to preserve comfortable skin temperatures. In fact, in some athletic events such clothing is still too 'warm' and significant sweating and sweat evaporation is required for body heat balance. It is important that wet clothing is replaced shortly after cessation of exercise to avoid post exercise chilling. Heated shelter or thick, insulative parka is a temporary solution.

Combinations of hard work with periods of low activity, breaks or rest pauses put great demands on flexibility of clothing. Insulation requirements may vary by a factor of 3 or 4 (Fig. 16.1). One single clothing system cannot accommodate such changes. Most likely only two layers are required during work. Accordingly, the second layer must also provide some protection against wind and precipitation. The third layer comprises a thick parka or overall with lining that is donned during the low activity. This single layer may add considerable insulation to the system. Additional protection of hands and head helps to preserve heat.

16.4.2 Moisture control

As there is always a small diffusion of water vapour from the skin (not related to sweating) a strategy for moisture control must be applied. Moisture control means in practice that water in liquid or vapour form must be kept away from skin in order to preserve dryness and comfort. This is a delicate task, in particular with exercise, as significant sweating may be evoked for the preservation of heat balance. As already mentioned, sweat evaporation and the subsequent vapour transfer through clothing are aggravated by the many layers of clothing and by the lowered dew point in surface layers. The superfluous liquid or vaporized sweat needs to be handled in the best possible way within the clothing system in order to reduce discomfort, wetting of clothing and unnecessary post exposure cooling. There are several strategies that can be applied depending on the prevailing exposure conditions.

- vapour barrier
- non-absorbing
- wicking
- moisture absorption.

Vapour barrier underwear permits neither liquid water nor water vapour to pass through to the next layers. The immediate advantage is that layers on top of the underwear remain dry. This is of particular importance in very long outdoor exposures with little or no possibilities to dry clothing or change of clothing. If clothing becomes wet, insulation deteriorates and exposure may end in dangerous body cooling. This underwear does not function well with work as the build-up of moisture and skin water vapour pressure is too quick and difficult to control by ventilation alone. It may work well (with some training) in very light work, resting or standing (e.g. for hunters) and with sleeping bags during a night in the cold.

Non-absorbing underwear is usually made of synthetic fabrics with hydrophobic properties. Moisture or vapour is not absorbed, so water vapour pressure builds up driving diffusion of vapour to the next layers. Due to this effect the sensation of sweating begins quickly, stimulating behavioural action to control sweating. The skin is not wet as the underwear remains relatively dry. Wet underwear dries quickly. This kind of underwear is popular in sports and outdoor work at high activity levels.

Wicking underwear is supposed to absorb liquid water at the skin surface and transport it to the opposite side of the underwear, sometimes made of a different fibre type, or to the next layer. This results in a drier skin layer. Eventually, some of the liquid evaporates from this other layer and provides cooling of the body. The process is slow and the capacity limited. This underwear would be fine in light to moderate activity with limited sweating.

Moisture absorbing underwear is made of fibres and fabrics that absorb water or water vapour directly, although small amounts of moisture can often be retained without discomfort. However, larger amounts of moisture may cause the fabric to cling to the skin, creating both cooling and discomfort. Cotton is a typical fabric of this type. The behaviour of cotton is positive and beneficial in warm and hot environments. In the cold, however, wet clothing close to the skin is uncomfortable and cools the skin for a long time when drying out.

Wool is hygroscopic and can absorb almost one-third of its own weight without major structural changes, thus preserving much of its insulative properties. Moisture absorption helps to keep the skin dry. When oversaturated, however, wool loses insulation. Vapour absorbing underwear would be fine in light to moderate activity with limited sweating.

16.4.3 Protection against wind, water and snow

Cold weather often implies windy conditions with precipitation in the form of rain or snow. The outer layer must provide reasonable protection against these factors in order to minimize extra heat losses. Wind may penetrate the outer layer and openings in the clothing, increasing micro climate ventilation. Strong wind may also compress layers of soft fabrics. Increased convection may be beneficial with high activity levels and the wearer may adjust clothing to enhance it. However, at low activity levels and in severe cold conditions clothing should be windproof. Tests of protective clothing against cold are described in EN 342 (2004) and include measurements of thermal insulation of the full clothing ensemble and the air permeability of the fabric of the outer layer.

Rain and melting snow may wet outer layers and also penetrate into deeper layers. Wet clothing provides impaired insulation and may jeopardize heat balance, and comfort is severely impaired. When there is a risk of precipitation clothing should be worn that is waterproof or rainproof; unfortunately, waterproof clothing is often so tight that convection and vapour transmission is impeded. Clothing becomes uncomfortable due to the accumulation of condensed water vapour from the skin. The market provides several types of 'breathable' fabrics that combine water proofness with vapour permeability. Their capacity to transmit water vapour from sweating skin, however, is limited. This is due to the restriction of convection, since fabrics are windproof. Another restricting factor is the steep temperature gradient from skin to ambient air. The dewpoint temperature may fall inside clothing, resulting in condensation of water vapour on its way out from the skin (see section 16.2).

Performance evaluation of cold protective clothing is readily made with available international and European standards. This has been dealt with in previous sections. The most important property is the thermal insulation of the complete clothing ensemble. This is measured according to ISO 15831 (2003). Required insulation for work in cold environments can be determined according to ISO 11079 (2007) and EN 342 (2004). Protection against water and rain can be tested on fabrics (EN 343, 2004) or on full clothing ensembles (EN 14360, 2005). A comprehensive overview of standards relevant to cold weather apparel has been published by Maekinen (2009).

16.5 Future trends

The advent of 'smart' or 'intelligent' textiles may help to protect against cooling (Meinander, 2005). Electrically heated garments reduce 'normal' heat losses, but the heating power and capacity depend largely on size of heated area, application in the layers system and battery power. So far the latter factor appears to be limiting in many applications (Holmér et al., 2009). Phase change material (PCM) with low melting points in garments may provide additional heat. So far, however, the power and capacity of existing PCM is quite limited (Meinander, 2005). Using shape memory alloy strings embedded in a fabric allows an automatic change in fabric thickness at certain temperatures. This change in thickness improves thermal insulation (Lee et al., 2009). Reflective layers, however, appear to be less efficient. Radiation heat exchange in layers is small and the layer is often impermeable, reducing water vapour transfer.

Both industrial development and increasing tourism are likely to increase the number of people exposed to cold. The experienced worker and the outdoor specialist are usually well protected and can also use modern equipment to enhance performance and improve comfort. Less experienced persons require more information and training before exposing themselves to cold conditions. Information from clothing manufacturers are often too optimistic in terms of the performance of their products, sometimes directly misleading. Objective information about the performance of clothing is badly required by end-users.

Several international standards are available for this purpose but are mostly used only by manufacturers of protective clothing for occupational work. As mentioned, EN 342 (2004) describes methods for testing cold protective properties, such as thermal insulation and air permeability. The tested garment is marked with CE, the insulation value and a class for air permeability. It would be of great value if this standard was also used for leisure wear, for example ski clothing. The only performance test so far available for the outdoor market is a test for sleeping bags. EN 13537 (2002) describes how the thermal insulation of a sleeping bag is measured with a thermal manikin and what protection it offers in various user conditions.

Similarly, test methods for hand wear (EN 511, 2005) and footwear (Kuklane, 2004) are readily available but not yet much applied in practice.

16.6 Sources of further information and advice

www.environmental-ergonomics.org
http://www.es-pc.org/
www.eat.lth.se/english/research/thermal_environment/

16.7 References

Bluestein, M. (1998) 'An evaluation of the wind chill factor: its development and applicability', *J Biomech Eng*, **120**, 255–258.

Dorman, L. E. & Havenith, G. (2009) 'The effects of protective clothing on energy consumption during different activities', *Eur J Appl Physiol*, **105**, 463–470.
EN 342 (2004) Protective clothing – Ensembles for protection against cold.
EN 343 (2004) Protective clothing – Protection against rain.
EN 511 (2005) Protective gloves against cold.
EN 13537 (2002) Requirements for sleeping bags.
EN 14360 (2005) Protective clothing against rain – Test method for ready made garments – Impact from above with high energy droplets.
Enander, A. & Hygge, S. (1990) 'Thermal stress and human performance', *Scand J Work Environ Health*, **16**, suppl. 19, 44–50.
Giesbrecht, G. G. (1995) 'The respiratory system in a cold environment', *Aviat Space Environ Med*, **66**, 890–902.
Hassi, J., Mäkinen, T., Holmér, I., Påsche, A., Risikko, T., Toivonen, L. & Hurme, M. (2002) *Handbook for Cold Work*, Oulu Regional Institute of Occupational Health.
Havenith, G., Heus, R. & Lotens, V. (1990) 'Resultant clothing insulation: a function of body movement, posture, wind, clothing fit and ensemble thickness', *Ergonomics*, **33**, 67–84.
Havenith, G. & Nilsson, H. (2004) 'Correction of clothing insulation for movement and wind effects – a metanalysis', *European Journal of Applied Physiology*, **92**, 636–640.
Havenith, G. & Nilsson, H. (2005) 'Correction of clothing insulation for movement and wind effects – a metanalysis', Erratum, *European Journal of Applied Physiology*, **93**, 506.
Havenith, G., Richards, M., Wang, X., Bröde, P., Candas, V., Den Hartog, E., Holmér, I., Kuklane, K., Meinander, H. & Nocker, W. (2008) 'Apparent latent heat of evaporation from clothing: attenuation and "heat pipe" effects', *J Appl Physiol*, **104**, 142–149.
Heus, R., Daanen, H. A. M. & Havenith, G. (1995) 'Physiological criteria for functioning of hands in the cold', *Applied Ergonomics*, **26**, 5–13.
Holmér, I. (2005a) 'Protection against cold', in Shishoo, R. (ed.) *Textiles in Sport*. Cambridge, Woodhead Publishing.
Holmér, I. (2005b) 'Textiles for cold protection', in Scott, R. (ed.) *Textiles for Protection*. Cambridge, Woodhead Publishing.
Holmér, I., Gao, C. & Wang, F. (2009) Can a vest provide 88 clo? – serial calculation method revisited. 4th European Conference on Protective Clothing. Papendal, The Netherlands, TNO.
Holmér, I., Nilsson, H. & Anttonen, H. (2001) 'Prediction of wind effects on cold protective clothing', *Blowing hot and cold: Protecting against climatic extremes*, Brussels, NATO.
ISO 7730 (2006) Moderate thermal environments – Determination of the PMV and PPD indices and specification of the conditions for thermal comfort.
ISO 8996 (1990) Ergonomics – Determination of metabolic heat production.
ISO 9920 (2007) Ergonomics of the thermal environment – Estimation of the thermal insulation and evaporative resistane of a clothing ensemble.
ISO 11079 (2007) Ergonomics of the thermal environment. Determination and interpretation of cold stress when using required clothing insulation (IREQ) and local cooling effects.
ISO 15831, E. (2003) Thermal manikin for measuring the resultant basic thermal insulation.
Kuklane, K. (2004) 'The use of footwear insulation values measured on a thermal foot model', *International Journal of Occupational Safety and Ergonomics*, **10**, 79–86.
Launay, J.-C. & Savouray, G. (2009) 'Cold adaptations', *Industrial Health*, **47**, 26–34.

Lee, J., Park, H., Jee, K. & Kim, E. (2009) Intelligent cold protection jacket embedded with two way shape memory alloy springs. 4th European Conference on Protective Clothing. Papendal, The Netherlands, TNO.

Maekinen, H. (2009) 'Standards and legislation governing cold weather clothing', in Williams, J. T. (ed.) *Textiles for Cold Weather Apparel*, Cambridge, Woodhead Publishing.

Meinander, H. (2005) 'Smart and intelligent textiles and fibres', in Shishoo, R. (ed.) *Textiles in Sport*, Cambridge, Woodhead Publishing.

Parsons, K. C. (2003) *Human Thermal Environments*, Hampshire, UK, Taylor & Francis.

Saltin, B. & Hermansen, L. (1966) 'Esophageal, rectal and muscle temperature during exercise', *J Appl Physiol*, **21**, 1757–1762.

Url www.msc-smc.ec.gc.ca/education/windchill/index_e.cfm

Young, A. J. (1988) 'Human adaptation to cold', in Pandolf, K. B., Sawka, M. N. & Gonzales, R. R. (eds.) *Human Performance Physiology and Environmental Medicine at Terrestrial Extremes*, U.S. Army Research Institute of Environmental Medicine.

17
Achieving comfort in intimate apparel

W. YU, The Hong Kong Polytechnic University, P.R. China

Abstract: Intimate apparel is depicted as human's second skin, so is the most important clothing layer for achieving comfort. The comfort issues for intimate apparel are multidimensional and interrelated. This chapter explains the common factors in five essential aspects: (a) sensorial, (b) thermal, (c) motion, (d) aesthetical and (e) hygienic comfort. The relevant research findings are discussed and new technologies achieving the overall comfort of intimate apparel are introduced.

Key words: tactile, pressure, thermal, moisture, fit, movement, aesthetic, hygiene, underwear.

17.1 Introduction

Comfort for intimate apparel is more instant and direct than that for outerwear because it is worn next to skin (Lang, 1995; Sweeney & Zionts, 1989). Women's intimate apparel can be a functional shapewear or emotional sexy lingerie (Steele, 1996) including bra, panties, camisole, teddy, corset, girdle and stockings (Yu & So, 2001). Men's underwear is conventionally basic in style. Although it becomes more fashionable and versatile, the products are still focused on undershirts, underpants and socks only.

The fundamental purposes of wearing intimate apparel are for privacy and comfort, but nowadays customers demand more for aesthetics, sexiness and excitement. Fashion items such as laced bras and thongs could be less comfortable to wear than daily bras and basic underwear. Comfort issues for intimate apparel are multidimensional and interrelated. This chapter will explain the common factors of intimate apparel comfort in several aspects, including sensorial, thermal, motion, aesthetical and hygienic. Then appropriate materials and new technology will be introduced for achieving the overall comfort of intimate apparel.

17.2 Sensorial comfort for intimate apparel

17.2.1 Tactile sensation

Tactile comfort is often expressed as the softness, stiffness, smoothness, roughness, prickliness, dampness or clinginess of the fabrics (Ho et al., 2008). Particularly for intimate apparel fabrics, the subjective tactile comfort sensation can be evaluated by a continuous-scale chart with twelve pairs of descriptors (Table 17.1).

Table 17.1 Subjective evaluation of intimate apparel fabrics

Soft	1	10	20	30	40	50	60	70	80	90	99	Hard
Silky	1	10	20	30	40	50	60	70	80	90	99	Harsh
Firm	1	10	20	30	40	50	60	70	80	90	99	Stretchy
Floppy	1	10	20	30	40	50	60	70	80	90	99	Stiff
Crisp	1	10	20	30	40	50	60	70	80	90	99	Limp
Rough	1	10	20	30	40	50	60	70	80	90	99	Smooth
Thick	1	10	20	30	40	50	60	70	80	90	99	Thin
Scratchy	1	10	20	30	40	50	60	70	80	90	99	Slick
Tight	1	10	20	30	40	50	60	70	80	90	99	Loose
Light	1	10	20	30	40	50	60	70	80	90	99	Heavy
Dead	1	10	20	30	40	50	60	70	80	90	99	Springy
Cool	1	10	20	30	40	50	60	70	80	90	99	Warm

Inamura et al. (1995) found that the tactile comfort of girdles had a higher correlation with the surface properties of the fabric than other factors such as heat, water and air transmission. Suda and Tamura (2006) suggested that comfort could be estimated from the tactile sensations of smoothness, softness and stickiness of underwear fabrics despite the change in air temperature. However, men and women may have different subjective sensorial assessments on the hand feel of fabrics. Kweon et al. (2004) used ten descriptors (soft, thin, smooth, warm, dry, light, loose, sheer, stiff, and pleasant tactile sensation) to assess the tactile comfort of sleepwear. In that study women accepted man-made fabrics, whereas men usually preferred all cotton. It is probably because women's preferences are influenced by the sensuous qualities or style versatility of satin, but it was perceived as a stiff fabric by men.

17.2.2 Avoiding skin irritation

Women have sensitive skin especially during puberty and pregnancy. Wearing intimate apparel with irritating or allergic substances can precipitate skin reactions. Lingerie often uses lace and embroidery to enhance women's attractiveness, yet the material's rough surface or edges can directly prick the skin (He et al., 2006). Fashion fasteners, e.g. snap, buckle, zippers or wire can cause abrasion. Poor seam coverings or loose threads also bring skin irritation.

To minimize the seams and stitches, various sew-free technologies have been widely applied. The bra cup seams have been eliminated by moulded cups (Yick et al. 2010); the elastic band has been replaced by Bemis thermoplastic polyurethane (TPU) film (Covelli et al., 2007) or Lycra 2.0 polyurethaneurea (PUU) heat-activated elastic adhesive tape. Thread and stitches disappear when the hemming operation (Fig. 17.1) uses ultrasonic bonding (Herbert, 2006) with hooks and eyes integrally attached onto the inner surface of the back wing panel (Utaka, 2005) (Fig. 17.2). Brand and size labels (Fig. 17.3) are seamlessly printed

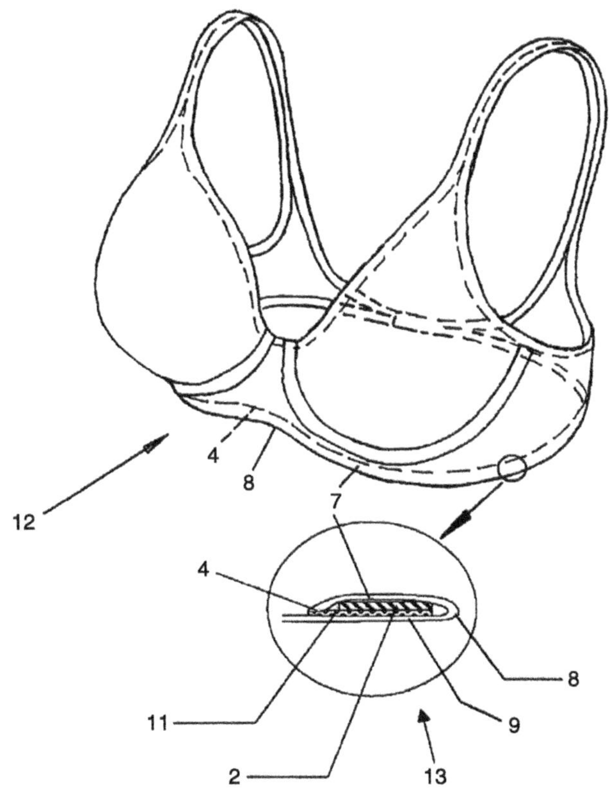

17.1 Ultrasonic bonding of bra hems. Source: US Patent App. 11/416,366.

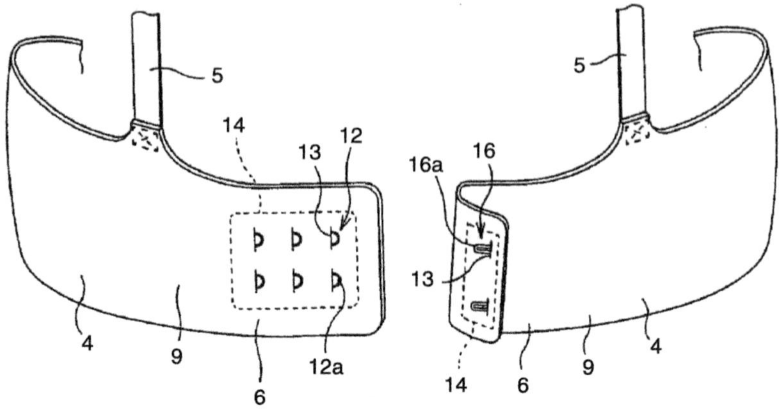

17.2 Integrally attached hooks and eyes onto the back wing of a bra. Source: US Patent App. 11/597,810.

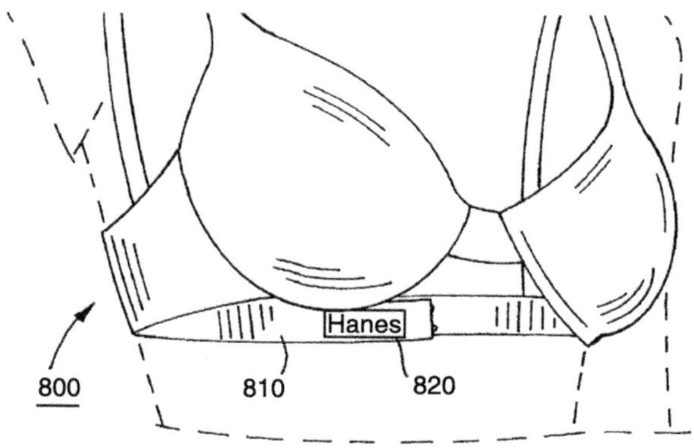

17.3 Thermal transfer printing of label pad. Source: US Patent App.10/352,536.

on the finished garment or fabric panel using thermal transfer printing (Pace & Braxton, 2003).

17.2.3 Allowing comfortable pressure

Outerwear fashion is designed to achieve the required drape and silhouette using appropriate ease allowance. In contrast, underwear requires certain fabric tension, so-called 'negative ease', to provide the wearer with a sense of support and security. Therefore, appropriate skin pressure is necessary to achieve comfort. For example, a tension of 500 gf in the waist band and leg band of panties, or in the elastic band and shoulder straps, is generally considered comfortable to be worn.

Women's intimate apparel sometimes requires larger fabric tension to create a firmer body contour and the desired shape. For example, a rigid panel is applied to exert higher pressure onto the abdomen for flattening the abdomen. Inelastic stabilizer fabric is applied at the gore (sometimes named centre piece or bridge) of a bra for bringing the breasts together and creating deeper cleavage. Firm cradle is also useful to provide a stable frame for supporting the breast weight and prevent the bra from sliding during activities.

However, excessive skin pressure may cause discomfort to the wearer. LaJean (1990) found that sport bras could effectively control the wearer's breast displacement but had lower comfort value. Ikuta (1970) reported that if the underwear was too tight, the wearer felt hotter with more sense of perspiration and tightness. Singer and Soma (1995) interviewed over 4700 women in the U.S. and found a significant correlation between the breast cancer risk factor and the number of hours per day women wore bras. Another medical study also suggested that wearing a bra could contribute to breast pain and increase the chance of breast cancer (Anon, 2000), although there has still been no creditable scientific research or clinical basis to validate these claims.

In fact, many studies have reported that tight bras or girdles could bring negative physiological effects including a decrease in cardiac output (Watanuki, 1994), skin blood flow (Nakahashi & Morooka, 1998; Tanaka et al., 1999), defecation (Lee et al., 2000b), thermoregulatory sympathetic nerve activity (Miyasuji et al., 2002) and urinary noradrenaline level (Lee et al., 2001). They also caused neck and shoulder pain (Silva, 1986), rising of rectal temperature and suppression of salivary melatonin (Lee et al., 2000a), increased core temperature (Tanaka et al., 2006), lengthened digestion time of food (Sone et al., 2000) and disturbed menstrual cycles (Kikifuji & Tokura, 2002).

Having studied the results from all the relevant literature, our earlier published monograph (Lim et al., 2006) about intimate apparel has recommended a comfortable range of skin pressure for different body parts: 4.5–9 gf/cm^2 on the waist, 8–12.5 gf/cm^2 on the abdomen, 6–11 gf/cm^2 on the hip and 6–9 gf/cm^2 on the thigh, but 17–28 gf/cm^2 on the bony regions like the side waist. The different pressure levels among different body parts are due to their different body curvature (Ng et al., 2009) and soft tissue stiffness (Lu et al., 2009).

Ito et al. (1995) found that there were weak to moderate relationships between the subjective sense of compression and the measured girdle pressure at the waist, abdomen, hip and thigh. However, Chan and Fan (2002) suggested that the tightness sensation was not only related to the pressure data, but also other factors (Fan & Chan, 2005) such as body size, fat content, resilience of soft tissues and the bone structure. Okabe and Yamana (1991) studied the skin pressure at 17 body points of 24 subjects aged between 21 and 57 wearing body suits. Their tolerable pressure level was significantly correlated with the stiffness of the body's soft tissue. Softer bodies felt less pain, particularly at the side of the waist.

Sato and Ohara (1997) found that the subjects felt the underwear comfortable but the shaping effect deteriorated when fabric elongation reached 40% or more. Nakahashi et al. (2005) analysed the factors of waist-nippers (materials, structure, size, pressure) that might affect the subjective feeling (pressure, comfort, shape) and physiological responses (heart rate, blood flow, skin temperature, oral temperature). The results showed that the clothing pressure is dependent on the tensile property of the overall garment structure, but not the individual material part. The physiological changes were smaller with larger rigid areas on the garment if the pressure was well distributed.

17.3 Thermal comfort for intimate apparel

17.3.1 Hygroscopicity of underwear fabric

Niwa (1968) suggested that the ability of fabrics to absorb sweat is more important than water vapour permeability in determining the comfort factor of fabrics. Transfer of moisture from clothing to the environment through diffusion, wicking, sorption and evaporation is affected by the thickness of fabric, tightness of fabric construction and hygroscopicity of fibre type (Mehta, 1984).

Nielsen and Edrusick (1990) evaluated five knit structures of underwear made from 100% polypropylene and found that the thickest fleece knitted fabric caused the greatest sweating and retained most moisture on the skin. Similarly, Bakkevig and Nielsen (1995) studied three kinds of underwear (polypropylene 1×-1 knit, wool 1×1 knit, and fishnet polypropylene) being worn under wool fleece garments, and concluded that more sweat accumulated in the wool underwear than in the polypropylene.

Yamaguchi et al. (2002) studied 15 women wearing different underwear with increasing humidity in a chamber. They found that the humidity sensations affected comfort more than thermal sensations did. Humidity sensitivity increased with apparel humidity regardless of apparel type, but was higher in polyester underwear than in cotton underwear.

Li et al. (1992) reported that highly hygroscopic wool fabrics were perceived as drier than polyester. Pure polyester has a moisture regain of 0.4% only, so its hydrophobic nature gave a lower level of comfort. As moisture cannot be absorbed or drawn away from the skin, the fabric will exhibit static problems such as cling during wear. However, polyester fibre can be improved with hydrophilic finish or using denier reduction and microfibre (Ferguson, 1982). Srinivasan et al. (2007) found that the wicking behaviour (water drop absorbency, drying rate and total absorbency) of the microfibre fabrics were all better than those of normal denier fabrics. However, some polyester fibres applying calcium, magnesium and manganese oxides to fabric finishing could significantly improve the moisture absorption (Jo & Kim, 1999). For example, Coolmax polyester is commonly used for designing sports bras with excellent moisture management properties (Moretz, 1993).

17.3.2 Heat transmission

According to Goldman (2005), there are six key factors involved in thermal comfort. They are air temperature, air movement, humidity, mean radiant temperature, body heat production, and clothing. The human body is normally in a state at 37 °C and the average skin temperature is around 33 °C without the presence of sweat (Yu et al., 2006). The climate also influences the preference of fabric. Rie and Teruko (2006) found that the summer-use fabrics were preferred at 34 °C than at 22 °C while the winter-use fabrics were preferred at 22 °C than at 34 °C, due to the body temperature regulation and the change in skin temperature.

Intimate apparel acts as an interface to allow perspiration and transmission of body heat from the skin through outerwear to the environment. Its thermal comfort is related to the thickness and porosity of the fabric (Yoon, 1984). For women's shapewear comprising warp knit fabric, the dense and compact structure often has poor permeability and causes thermal discomfort.

Knit structure is another factor of thermal comfort. Nielsen and Endrusick (1990) evaluated the thermoregulatory responses on five items of underwear made of 100% polypropylene fibres with different structures (1-by-1 rib, fleece, fishnet, interlock, double-layer rib). The differences in knit structure resulted in significant differences in mean skin temperature, local and average skin wetness, non-evaporated and evaporated sweat during the course of the intermittent exercise test, but no differences were observed in the core temperature.

However, the men's underwear can have important effects on the temperature of external male genitalia including penis and scrotum. Testicular function is temperature dependent and works better at 2 to 4 °C below the core body temperature (Data et al., 2003), so it should be maintained in a temperature range required for normal spermatogenesis both quantitatively and qualitatively (Mieusset and Bujan, 1995). The thermal regulation relies on the speed of heat exchange between the genitalia and the ambient environment (Shafik, 1973). If the underwear is too tight or not permeable enough for heat evaporation, it may cause male infertility. Overtight underwear may lead to higher temperature on scrotum in both walking and sitting (O'Farrell et al., 2008) and cause abnormalities of microcirculation in the epidermis and dermal vascular plexus (Sheynkin et al., 2004).

17.3.3 Moisture management fibres for intimate apparel

Moisture management of the intimate apparel is important as it helps to remove excessive wetness and carry heat away from the skin. Fibre polyethylene glycol and amino silicone in nano form has been applied to sportswear and underwear requiring perspiration absorbency (Holme, 2007). Hollow fibres have a modified cross-section which provides effective moisture management and promotes natural climatic regulation. Examples for intimate apparel include Meryl Nexten®,

Thermolite®, Aerocool®, Becool™, Coolmax®, Akwatel, Nike Sphere Cool and phase change material (PCM).

- Meryl Nexten® is the first European-made polyamide fibre (Anon, 2003) claimed to produce 25% lighter garments than conventional ones of the same thickness, and 25% greater insulation than other hollow fibres (Anon, 2002).
- Thermolite® is made from hollow core fibre claimed to dry 20% quicker than other insulating fabric and 50% quicker than cotton. It is a hollow-core hydrophobic polyester fibre that can heat up quickly, retain body heat effectively and block radiation heat loss (Dominique, 2005).
- Aerocool® is a cross-shaped polyester filament yarn that absorbs and evaporates sweat quickly and dries fast. This fibre features a 'four-leaf clover' cross-section, which helps the movement of moisture through the four capillary tubes (Hong & Kim, 2007).
- Becool™ is made of nylon and polyester that allows the skin to dry rapidly and maintain the body temperature. The four canals in its fibre cross-section drain water away by its physical advantage. The diffusion surface is two to four times larger than cotton and classical nylon with greater diffusion speed, absorption ability, absorption flow and priming pressure (Anon, 2007).
- Coolmax® fabric can transport moisture and heat to the outer surface rapidly through the four channel fibres to ensure quick drying and breathable properties (http://www.coolmaxfabric.com/g_en/webpage.aspx?id=15).
- Akwatel® polyester fabric transports moisture and assists thermoregulation using electrochemical principle (http://www.krystalartfield.com/Akwatek/Akeatek.html). It modifies the Polyethylene Terephthalate (PET) and releases hydrophilic groups at the molecular level. The modified polyester has an active surface layer with anionic end groups that transport water molecules and release them to the atmosphere before they can form into liquid water.
- Nike® Sphere Cool (http://www.nike.com.cn/product/apparel/product.htm) claims that the mesh structure accelerates the evaporation of sweat, so the wearer feels cooler and more comfortable due to the good moisture absorbency of the inner layer.
- Phase change materials (PCMs) have been used for thermal storage. It is controlled by heat absorption and release that occurs upon a change in phase. For underwear, a PCM with phase changes occurring in the range of skin temperatures should be selected (Shin et al., 2005).

17.4 Motion comfort for intimate apparel

17.4.1 Challenges of fitting

Fitting is a crucial factor of wearing comfort (Pechoux & Ghosh, 2002) particularly for the next-to-skin stretch garment. In 1950, Cain stated that fit is directly related to the anatomy of the human body and most of the fitting problems are due to the bulges of the human body (Cain, 1950). However, limited research work has investigated the assessment and influencing factors of fitting for intimate apparel.

Costantakos and Watkins (1982) suggested that a well-designed bra should prevent concentration of pressure on the sensitive areas such as veins and arteries near the skin surface. It was found that the highest pressure was at the top of the shoulders and under the front elastic band for five different bra designs; and there was a relationship between bra pressure and the wearer's comfort. Moreover, a good bra should fit the 3D complex body contours, to support the breast weight, and to provide appropriate tension by well-fitted bottom band, straps and cups (Zheng et al., 2009). Perfect fit for bras is very difficult to achieve because it involves tedious trials and errors on manipulating the shape and support provided to the soft breasts of a live model. Many pattern adaptations may be required to try different grain lines in the pattern, and tension of fabric or changing the fabric and trims for better tension recovery. Ziegert and Keil (1988) found that different stretchability of woven elastomeric fabric and different grain direction required different pattern shapes for the garments to display the same fit. Therefore, a good fitting technologist and designer must have a strong sense of 3D fitting and its relations to fabric mechanics.

As the fitting test is based on a standard size model and various sizes are graded proportionally according to the size intervals of several key measurements, ready-to-wear intimate apparel may not fit all customers. In reality, individual persons are all varied in body size, shape, proportion, symmetry, posture, fat distribution and flesh softness in different parts of the body.

Stretch clothing, like intimate apparel, presents unusual difficulties in the evaluation of fit. An appropriate stretch fit is essential to secure certain functionality, comfort and appearance. Intimate apparel is usually made up of stretch materials, so 'negative ease' is involved in pattern making (Shin, 2007). The comfort range of compression towards the body should be quantified and applied in a pattern reduction process. For easier fit, large deformation fabric (Gersak, 2005) is preferred because it has small hysteresis and good recovery of elasticity. However, elastic fabric is likely to have relaxation shrinkage and problems in dimensional stability due to the unavoidable yarn tension formed in the knitting process, so 100% recovery is not achievable.

17.4.2 Fitting checklist

The assessment of bra fit on a live model is very complicated and limited literature is available in this field. The judgement of fit is heavily reliant on the assessor's experience. In the industrial practice, pictures of fitting symptoms are taken at specific angles so as to facilitate mutual understanding between the buyer, designer and sample maker. However, the communication of fitting problems is still a problem when people are using different terminologies and expressions. Therefore, new video-based communication platforms such as 'Fast Fit' and 'Webex' are being used in the lingerie industry.

For the education of bra fitting, Yu (2009) has developed a checklist (Table 17.2) for students to assess the fitting performance of a bra in ten areas (Fig. 17.4). A good-fit daily bra should stay with the body snugly but not tightly, support and lift up the breasts with a nice smooth contour for over 12 hours a day like a second skin, and not leave red marks or irritation.

Before testing the fitting of a bra, the model should ensure a correct tension along the underband wrapping around on the ribcage, and the comfortable pressure exerted by the shoulder strap on the shoulder. Only when the bra band forms a firm frame on the body, it is reliable to check if the breasts are filling nicely inside

Table 17.2 Bra fitting checklist

(A) Gore • Gore sits against the sternum, and allows comfortable breathing down • Gore width fits for the purpose • Wire tip does not dig into the flesh	*(F) Cradle* • Keeps the breast inside the cups • Does not curl up when the wearer sits
(B) Cup • Covers the nipples • No gap inside the cup • Cup seam or lining is not itching • No irritating lace or trims • Cup peak matches the bust point • Breast is projected during motion • Projects a nice shape and curve • Cup capacity is sufficient	*(G) Wing* • Leveling around the body • Appropriate tension to hold the bra in position *(H) Strap* • Correct tension for breast support • Allows enough adjustment, but not too much turning • Strap not easy to fall off • No cutting in the shoulder • No fatigue
(C) Neckline • No gap • No bulging • Symmetric and balanced • Thin, soft and smooth	*(I) Underband* • Tension allows for comfort breathing • No riding up during motion • The bra still sits securely when the wearer raises up the arms
(D) Underarm • No gap • No extra fabric • No digging in • Not too much pressure	*(J) Fastener* • Hooks and eyes wide enough for the style and size • Front closure not touching sternum
(E) Wire • Wire matches breast root • Correct gauge • Not digging into the flesh • Correct size and width	

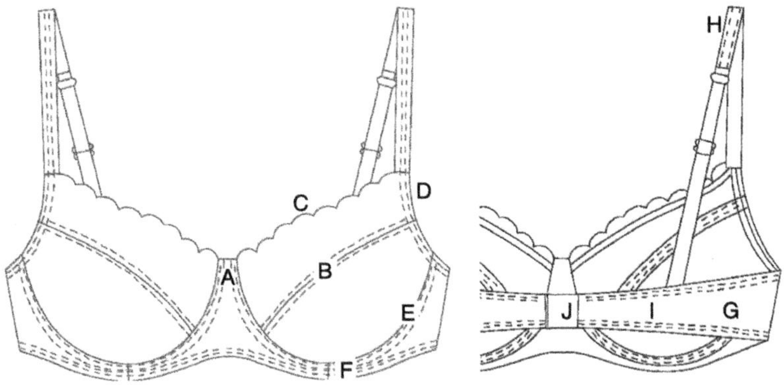

17.4 Ten areas of bra fitting assessment.

the cup or if there is any gap or bulging problem. The standard procedure of wearing a bra is outlined below.

- Bend down so that the breasts are in the best roundness and fullness.
- Put the arms through the shoulder straps and fill the breast into the bra cups snugly with the nipples matching near the peaks of the cups.
- Fasten the bra against the rib cage to ensure stability with comfortable tension, and feel that the underband and the cups remain firmly in place even when the shoulder straps are dropped.
- Adjust the shoulder straps so that the neckline edge is smooth and the tension is good enough to hold the bra in place without sliding away from shoulders or digging.
- Check if the bra stays in place when the arms rise up or the body moves.

17.4.3 Common fitting problems

Ill-fitting is sometimes due to the wrong perceptions of customers who are too conscious to slim down their sizes, so they prefer to wear tighter intimate apparel to flatten their body figures. In response to this problem, the National Association of Hosiery Manufacturers (NAMH) reminded the consumers to be honest about their sizes, for example in choosing the right size of pantyhose. Beside the customers' wrong perception of clothing size, unreliable bra sizing systems (Zheng et al., 2007) and limited choices of sizes often cause the wearing of ill-fitted intimate apparel. Since 1935, the definition of alphabetical bra cup sizes has relied on only two circumferential measurements of full bust and under bust. The existing bra sizing system is inadequate because it involves several steps of rounding-up and approximation (Pechter, 2008) and has not considered the complexity of 3D breast geometry (Zheng et al., 2006).

Nethero (2007) surveyed 1,500 women in the US and showed that over 65% of the respondents complained about discomfort when wearing their bras. The recurrent fit problems were back fat (>50%), insufficient support (>50%), straps falling down (50%), underwire hurt (35%), breast spillage over the cup (25%), bra riding up (27%), straps digging into shoulders (28%), and insufficient breast lifting (25%). Chan et al. (2001) in Hong Kong found the major areas causing discomfort in bras were the cups, underwire and straps.

Full figure women often suffer from the bulging of back fat (Fig. 17.5c) when wearing a bra, and they think it is due to the bra's tightness, so they may buy a bigger band size. This will result in insufficient support, the back riding up (Fig. 17.5a), and the straps will slide away from shoulders (Fig. 17.5i). This will result in discomfort and a sense of insecurity. The unsupported breasts will also lose strength and appear sloppy. Actually, a better solution for back fat is to use a wider and stretchy back panel with the elastic band of matched elongation.

Bra underwire is an effective tool to create and hold a clearly defined breast shape. It should be comfortable to wear if it fits the shape of the breast root. In a well-fitted bra, a woman should not feel the underwire. However, if the bra tension is incorrect (Fig. 17.5h), the cup is too shallow (Fig. 17.5m) or the wire shape is too deep, too wide or too narrow (Fig. 17.5n), the wire will hurt the breast tissue which will be harsh and uncomfortable. In fact, different bra components are inter-related in fitting. The symptom of wire digging is probably caused by the underband tension or the cup dimensions, not merely the wire itself.

Without sufficient lifting, the soft breast tissue will hang low toward the waist and spread wide toward the arms. Women try to lift up the breast weight by shortening the shoulder straps. However, this will make the bra underband ride up from the back, and the strap will dig into the shoulders (Fig. 17.5g). If the shoulder strap tension is too tight, it will pull on the neck and the spine which will affect the body posture and may cause shoulder stress and headaches. Therefore, the root cause of the ill-fitting should be systematically identified before taking remedial action. If the root cause of insufficient lifting is loose underband tension, the fastener should be adjusted. If adjustment is not available, the second step is to consider revising the wing pattern or materials.

17.4.4 Fit for movement

Human breasts are mainly composed of adipose tissue, blood vessels and lymph glands, held in place by the skin and deep fascia structures. The connective tissue strands of Cooper's ligaments are loosely attached to the underlying pectoral muscles. The breast itself contains neither muscle nor bone for support, so it is free to hang over the chest wall. Unsupported breasts will be too wide and may even be brushed with the arms while walking. Without proper support from a bra, the breasts will be bumping like a pendulum when the body moves, or suffer from

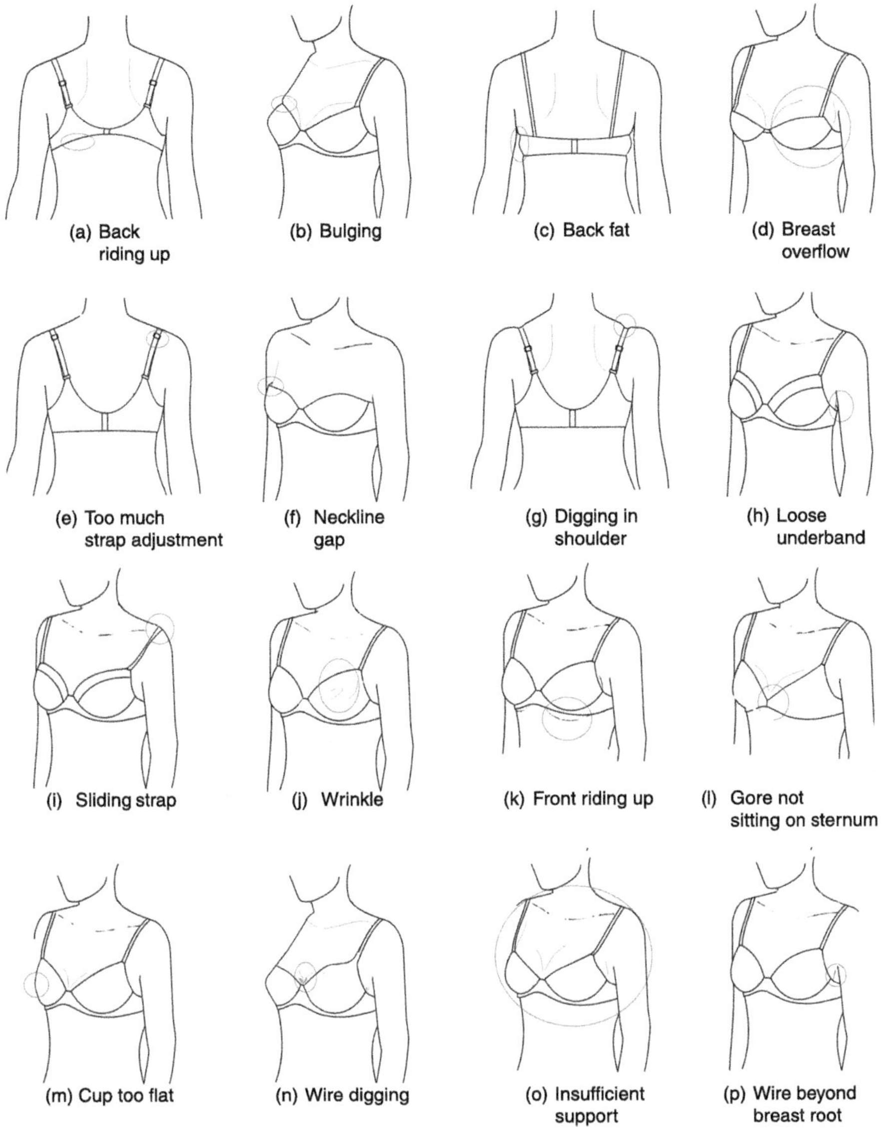

17.5 Symptoms of ill-fitted bras.

discomfort and pain during excessive rigorous movement. Breast pain after exercise is common and varies markedly in severity and clinical significance.

A sports bra is therefore expected to provide necessary breast support with a comfortably strong elastic band under bust, and wide shoulder straps that distribute the weight to the shoulder and the back torso. For effective force transmission, the straps should lie on a line with the shortest distance from the centre of breast mass and adjusted to appropriate tension. A well-designed sports bra not only avoids breast discomfort, pain or injury, but also promotes the wearer's confidence and performance in sports games.

17.4.5 Fitting for special needs

In daily life, people have basic needs, which entails putting their underwear on and off in the toilet. Therefore, the waistband of underpants should be sufficiently stretchable for passing up and down the hip without too much effort. Highly stretchy fabric and elastic is preferred. Convenience to put underwear on and off is an especially important issue for the elderly or individuals with certain disabilities, who need to undo the fastenings in the toilet, or to balance their bodies when changing clothes. Merry and Patricia (1981) developed functional panties with simplified dressing and undressing features for physically handicapped women. For visually impaired people or the elderly, a small decoration on the underpants was recommended to help them to distinguish the front and back pieces (Meinander & Varheenmaa, 2002). Elderly people have weaker arm flexibility and grip strength, so it is difficult for them to bend their arms to the back and manage the hooks and eyes of a bra. Therefore, front closure is preferred but the buckles should not be too small for them to handle.

For elderly people who have high risk of hip fracture, underpants with hip protector would be useful to prevent them from injury in case of a fall. The hip protectors contain plastic shields or foam pads fitted in the inner pockets of underwear (Parker & Gillespie, 2006).

For pregnant women suffering from low back pain, maternity under garments could provide certain abdominal support (Brady, 1991; Sour, 2007) and distribute the growing weight of fetus to the shoulder and the upper torso, which may help to relieve back pain and improve the wearer's mobility. However, some products caused skin irritation, discomfort during movement and inconvenience for toileting (Carr, 2003; Weber et al., 1972). They were complained as being too hot, ugly and not to fit (Ho et al., 2009; Johnson et al., 1994; Macintyre & Baird, 2006; O'Hare, 1997; Williams et al., 1998).

17.5 Aesthetic comfort for intimate apparel

17.5.1 Self identity

Modern lingerie is sometimes designed to be seen, so the fashion elements are as important as those of outerwear. The aesthetic values of colour, style, shape, material mix and match, fitting and finishing can all influence the clothing appearance, and physical comfort (Cheng & Cheung, 1994; Smith, 1986) as well as the psychological comfort sensation, with intensified experiences and feelings of self-identity (Jantzen et al., 2006). Through good intimate apparel, customers can enhance their self-image and self-confidence with flattering shapes and fashionable styles for their social lives. People have various desires for comfort and lifestyle such as fashion, beauty and a romantic feel. Intimate apparel should match with the necklines and surface characteristics of their outerwear, for example a plunge bra under a low-cut V-neckline to show attractive cleavage with self-confidence, and a smooth contour bra under a close-fit T-shirt.

Feeling young is most important for women who lost their youthful appearance after pregnancy, ageing or menopause. They would need a bra that makes the breasts look firm, round and natural. The bra should uplift the drooping breasts to a desirable position with necessary fullness, coverage and cleavage. However, the breast shape must not appear pointy or artificial. It requires careful use of underwire in correct shape, gauge and size, as well as a proper combination of bra components (gore, cradle, cup and wing) in correct tension under stretch.

17.5.2 Ideal body shape

Shapewear is designed to enhance the wearer's body shape toward the socially defined 'Golden proportion'. However, different cultures have various definitions of attractiveness. Western women are proud of their deep cleavages, while Asian women are shy to show the upper breast and conservative to allow nipples being visible through clothing.

The ideal breast shape also varies from country to country. Western women want a thin layer of bra cups to allow natural sexy bouncing of their breasts. Chinese women believe that a good breast shape should look firm. So the front neck point and the two bust points should form an equilateral triangle, and a well positioned bust point should be placed at the vertical, halfway between the shoulder and elbow. Average size is possibly perceived as more beautiful. Chinese women with small breasts would like to wear thick padded bras to add volume. Western women with larger breasts tend to wear minimisers to feel like a reduced size.

17.5.3 Degrading of intimate apparel

Frequent wash and wear can easily damage the surface of fabric or change the colour which results in unpleasant appearance of intimate apparel. Common surface degrading problems include pilling and snagging. Pilling mostly happens in woollen underwear and appears to have entangled fibres after washing (Ukponmwan, 1998). Snagging is due to the breaking of fibres by mechanical abrasion that appears like frosting, especially on silk fabric. Polyamide fibre is widely used in underwear because of its good abrasion resistance, ten-fold higher than cotton, wool and viscose fibres (Gruszka et al., 2005). Microfibre yarn is smooth with low lint-shedding propensity and lesser-fly during knitting, and is resistant to pilling and abrasion.

Common colour problems are fading, staining and yellowing. Explained by the 'Easy in easy out' dyeing theory, fibres with larger pore size allow easier dyeing but also easier fading and staining of colour which can be due to washing or UV exposure. Yellowing has long been a big trouble in intimate apparel. Materials such as polyurethane foam often get yellowing due to the chemical reaction under light.

Women's intimate apparel, such as a bra, is constructed by plenty of different fabrics and materials which can be made of plastics, alloys, polymers and textiles. Their abilities to be dyed and colour fastness vary, so prolonged use will easily cause colour shading among different components in a bra. Therefore, intimate apparel should be properly taken care of. A laundry bag can be used to protect the lace, elastics, and closures on a delicate cycle in cool water with mild cleaning detergent. Otherwise, the elastic materials will be damaged in hot water, cotton wire channel will shrink and the underwire may pop out. Intimate apparel is better not put in a dryer because the heat and twisting will cause degradation of the materials.

17.6 Hygienic comfort for intimate apparel

17.6.1 Environment for bacteria growth

Elegbe and Botu (1982) found that tight nylon underwear created warmth and moisture in the vaginal and cervical areas, and led to a favourable environment for the growth of candidiasis and vaginitis. Non-breathable women's panties may accumulate the secretion on the perineum, which easily cause the growth of fungus such as *monilia albican*. Tight panties can cause skin abrasion and bring fungal infection on the gluteus fold or posterior thigh (Judy, 2001).

For men, penile wetness was observed in 13% of the subjects (O'Farrell et al., 2008). Eruptions tend to occur on damp skin due to moisture of the clothing (Hatch & Maibach, 1986). Therefore, it is important for the underwear to stay dry and allow moisture evaporation. Cotton is more comfortable to wear next to skin, but it is more susceptible to the growth of bacteria than synthetic fibre. It has a porous hydrophilic structure retaining water, oxygen and nutrients. It thereby provides an environment for bacteria growth, resulting in unpleasant odour when bacteria converts human perspiration into foul-smelling substances such as carboxylic acid, aldehydes, and amines (Montazer & Afjeh, 2007).

17.6.2 Antimicrobial textiles

Antimicrobial finishing can be used to protect textiles against microbial corrosion, prevention of malodour, prophylaxis and infections (Kramer et al., 2006). Hatch and Maibach (1986) reviewed various types of chemical finish and concluded that the most significant problem of causing irritant or allergic contact dermatitis is due to formaldehyde and N-methylol. Possible allergens of dermatitis also include rubber additives, metals (nickel and chrome), glues, biocides and fire retardants (Joe, 2001). A study found that patients with skin disease pruritus who wore silk underwear experienced less itching and tended to sleep better than the patients who wore cotton underwear (Shimizu et al., 1995).

To limit the incidence of bacteria and reduce the formation of odour, anti-microbial agent can be applied directly to the spinning mass or the fibre surface. A resinogenic finishing agent (Hofer, 2006) can be coated on the underwear fabric too. Bacteriostatic fabric can be either synthetic or natural like the bio-agent 'bamboo kun' inside a bamboo fibre (Mariano, 2004).

Silver is a natural anti-bacterial and anti-microbial agent. Heide and co-workers (2006) suggested that the positive silver ions can be used to inhibit the function of bacterial enzymes as the silver ions attack the structure proteins of the germs and thereby inhibit cell division. Clothes with silver nano-particles are sterile and can be useful to prevent infection with pathogenic bacteria such as *Staphylococcus aureus* (Turek, 1998), which has influence on men's urinary-tract infection, especially for elderly men (Durán et al., 2007).

Consumers' attitude towards hygiene and active lifestyle has created an increasing market for antimicrobial textiles. Gao and Cranston (2008) reviewed the requirements for antimicrobial finishing, qualitative and quantitative evaluations of antimicrobial efficacy, the application methods of antimicrobial agents and recent developments in antimicrobial treatments of textiles using various active agents. The antimicrobial products for intimate apparel include X-Static silver fibre from Noble Fibre Technologies; Biofresh™ acrylic fibre from Sterling Fibres; Anti-microbial yarn (AMY) from Unifi; Inbue polyester yarn from KoSa; Meryl® Skinlife from Nylstar; and HealthShield™ by O'Mara.

17.7 Acknowledgement

We sincerely thank the Research Grants Council of the Hong Kong SAR for funding the GRF project (PolyU 5290/03E).

17.8 Sources of further information and advice

http://www.cancer.org/docroot/MED/content/MED_6_1x_Underwire_Bras.asp?sitearea=MED

http://www.coolmaxfabric.com/g_en/webpage.aspx?id=15

http://www.krystalartfield.com/Akwatek/Akeatek.html

http://www.nike.com.cn/product/apparel/product.htm

17.9 References

Anon (2000), 'Charities slam breast cancer claim', *Practice Nursing*, **11**, 4.

Anon (2002), 'Lingerie manufacture', *African Textiles*, **1**, 36–37.

Anon (2003), 'Microfibre first', *Textile Month*, 30.

Anon (2007), 'Stay dry, be fresh: Becool™', *TUT Textiles a Usages Techniques*, **4**, 38–40.

Bakkevig M K and Nielsen R (1995), 'Some aspects of clothing comfort', *Ergonomics*, **38**, 926–934.

Brady J A (1991), *Form fitting garment*, USPTO, U.S. 5,255,393.

Cain G (1950), *The American Way of Designing*. New York: Fairchild Publications.

Carr C A (2003), 'Use of a maternity support binder for relief of pregnancy-related back pain', *Journal of Obstetric, Gynecologic, and Neonatal Nursing*, **32**, 495–502.

Chan A P and Fan J (2002), 'Effect of clothing pressure on the tightness sensation of girdles', *International Journal of Clothing Science and Technology*, **14**, 100–110.

Chan C, Yu W, and Newton E (2001), 'Evaluation and analysis of bra design', *The Design Journal*, **4**(3), 33–40.

Cheng K and Cheung Y (1994), 'Comfort in clothing', *Textile Asia*, **25**, 48–52.

Costantakos A V and Watkins S M (1982), 'Pressure analysis as a design research technique for increasing the comfort of nursing brassieres', *Home Economics Research Journal*, **10**(3), 271.

Covelli C A, Liu H, and Farmer D K (2007), *Laminated fabric construction with heat activated polyurethaneurea compositions*, USPTO, U.S. 11/745,668.

Data R, Gupta N P and Kucheria K (2003), 'Spermatogenic arrest in men with testicular hyperthermia', *Teratogen Carcinogen Mutagen* (Suppl 1), 235–243.

Durán N, Marcato P D, De Souza G I H, Alves O L, and Esposito E (2007), 'Antibacterial effect of silver nanoparticles produced by fungal process on textile fabrics and their effluent treatment', *Journal of Biomedical Nanotechnology*, **3**, 203–208.

Elegbe I A and Botu M (1982), 'A preliminary study on dressing patterns and incidence of candidiasis', *American Journal of Public Health*, **72**, 176–177.

Fan J and Chan A P (2005), 'Prediction of girdle's pressure on human body from the pressure measurement on a dummy', *International Journal of Clothing Science and Technology*, **17**, 6–12.

Ferguson C A (1982), 'Hydrophilic finishes for polyester: durability and processing advantages', *American Dye Stuff Reporter*, **71**, 43–47.

Gao Y and Cranston R (2008), 'Recent advances in antimicrobial treatments of textiles', *Textile Research Journal*, **78**, 60–72.

Gersak J (2005), 'A study of the relaxation phenomena in the fabrics containing elastane yarns', *International Journal of Clothing Science and Technology*, **17**, 188–199.

Goldman R F (2005), 'The four "Fs" of clothing comfort', in *Environmental Ergonomics – The Ergonomics of Human Comfort, Health, and Performance in the Thermal Environment*, Tochihara Y and Ihnaka T, editors: Elsevier, pp. 315–320.

Gruszka I, Lewandowski S, Benko E, and Perzyna M (2005), 'Structure and mechanical properties of polyamid fibres', *Fibres and Textiles in Eastern Europe*, **13**, 133–136.

Hatch K and Maibach H (1986), 'Textile chemical finish dermatitis', *Contact Dermatitis*, **14**, 1–13.

He Y, Bamforth P, Jackson M R, and Rowe A (2006), 'Lace edge quality assessment', *Proceedings of the Institution of Mechanical Engineers, Part B: Journal of Engineering Manufacture*, **220**, 1863–1873.

Heide M, Mohring U, Hansel R, Stoll M, Wollina U, and Heinig B (2006), 'Antimicrobial-finished textile three-dimensional structures', *Current Problems in Dermatology*, **33**, 179–199.

Herbert A (2006), *Bonding elastic to fabric of a garment*, USPTO, U.S. 11/416,366.

Ho S S, Yu W, Lao T T, Chow D H K, Chung J W Y, and Li Y (2008), 'Comfort evaluation of maternity support garments in a wear trial', *Ergonomics*, **51**, 1376–1393.

Ho S, Yu W, Lao T T, Chow D, Chung J, and Li Y (2009), 'Garment needs of pregnant women based on content analysis of in-depth interviews', *Journal of Clinical Nursing*, **18**, 2426–2435.

Hofer D (2006), 'Antimicrobial textiles – evaluation of their effectiveness and safety', *Current Problems in Dermatology*, **33**, 42–50.

Holme I (2007), 'Innovative technologies for high performance textiles', *Coloration Technology*, **123**, 59–73.

Hong C J and Kim J B (2007), 'A study of comfort performance in cotton and polyester blended fabrics I. Vertical wicking behavior', *Family and Consumer Sciences Research Journal, Fibers and Polymers*, **2**, 218–224.

Ikuta N (1970), 'Hygienic studies of foundation garments part 1: Effects on subjective observations, clothing climate and clothing observation', *Japanese Journal of Hygiene*, **25**, 344–349.

Inamura A, Nakanishi M, and Niwa M (1995), 'Relationship between wearing comfort and physical properties of girdles', *Journal of the Japan Researh Association for Textile End-Uses*, **36**, 109.

Ito N, Inoue M, Nakanishi M, and Niwa M (1995), 'The relation among the biaxial extension properties of girdle cloths and wearing comfort and clothing pressure of girdles', *Journal of the Japanese Research Association for Textile End-Uses*, **36**, 102–108.

Jantzen C, Ostergaard P, and Vieira C M S (2006), 'Becoming a "woman to the backbone": lingerie consumption and the experience of feminine identity', *Journal of Consumer Culture*, **6**, 177.

Jo D J and Kim J B (1999), *Method for the preparation of polyester fibres of excellent water absorbency*, USPTO, U.S. 6,258,309.

Joe E K (2001), 'Allergic contact dermatitis to textile dyes', *Dermatology Online Journal*, **7**, 9.

Johnson J, Greenspan B, Gorga D, Nagler W, and Goodwin C (1994), 'Compliance with pressure garment use in burn rehabilitation', *Journal of Burn Care and Rehabilitation*, **15**, 180–188.

Judy L S (2001), 'Diagnosis and management of fungal infections of the skin', *British Journal of Community Nursing*, **6**, 186–192.

Kikifuji N and Tokura H (2002), 'Disturbance of the duration in the menstrual cycle under the influence of tight clothing', *Biological Rhythm Research*, **33**, 279–285.

LaJean L (1990), 'Selected sports bras: comparisons of comfort and support', *Clothing and Textiles Research Journal*, **8**, 55–60.

Lang S (1995), 'Clothing for safety, security and survival', *Human Ecology*, **23**, 25.

Lee E K, Hyun K J, and Tokura H (2000a), 'The effect of skin pressure by clothing on cicardian rhythms of core temperature and salivary melatonin', *Chronobiology International*, **17**, 783–793.

Lee E K, Hyun K J, and Tokura H (2001), 'Circardian rhythms of core body temperature and urinary noradrenaline secretion under the influence of skin pressure due to foundation garments worn during wakefulness', *Biological Rhythm Research*, **32**, 389–400.

Lee E K, Kikufuji N, and Tokura H (2000b), 'Field studies on inhibitory influence of skin pressure exerted by a body compensatory brassiere on the amount of feces', *Japanese Journal of Physiology Anthropology*, **19**, 191–194.

Li Y, Plante A, and Holcombe B (1992), 'The physical mechanisms of the perception of dampness in fabrics', *Annuals of Physiology Anthropology*, **11**, 631–634.

Lim N Y, Ng S P, Yu W, and Fan J (2006), 'Pressure evaluation of intimate apparel', in *Innovation and Technology of Women's Intimate Apparel*, W Yu, et al., Editors, Cambridge: Woolhead Publishing Limited, p. 168.

Lu M H, Yu W, Huang Q H, Huang Y P, Zheng Y P (2009), 'A head-held indentation system for the assessment of the mechanical properties of soft tissues in vivo', *IEEE Transactions on Instrumentation and Measurement*, **58**, 3079–3085.

Macintyre L and Baird M (2006), 'Pressure garments for use in the treatment of hypertrophic scars: a review of the problems associated with their use', *Burns*, **32**, 10–15.

Mariano T (2004), 'Clothing all needs', *Rivista delle Tecnologie Tessili*, **18**, 68–75.

Mehta P (1984), 'The requirement of moisture transport in underwear', *Wool Science Review*, **60**, 23–46.

Meinander H and Varheenmaa M (2002), 'Clothing and textiles for disabled and elderly people', in *VTT Research Notes*, 2143, VTT Information Service, Finland.

Merry J D and Patricia A W (1981), 'Panty design alternatives for women and girls with physical disablities', *Family and Consumer Sciences Research Journal*, **9**, 336–346.

Mieusset R and Bujan L (1995), 'Testicular heating and its possible contributions to male infertility: a review', *International Journal of Andrology*, **18**, 169–184.

Miyasuji A, Matsumoto T, Mitarai S, Kotabe T, Takeshima T, and Watanuki S (2002), 'Effects of clothing pressure caused by different types of brassieres on automic nervous system activity evaluated by heart rate variability power spectral analysis', *Journal of Physiological Anthropology*, **21**, 67–74.

Montazer M and Afjeh M G (2007), 'Simultaneous x-linking and antimicrobial finishing of cotton fabric', *Journal of Applied Polymer Science*, **103**, 178–185.

Moretz H L (1993), *Moisture managing brassiere* USPTO, U.S. patent no. 5269720.

Nakahashi M and Morooka H (1998), 'An effect of a compressed region on a lower leg on the peripheral skin blood flow', *Journal of the Japanese Research Association for Textile End-Uses*, **39**, 392–397.

Nakahashi M, Morooka H, Nakamura N, Yamamoto C, and Morooka H (2005), 'An analysis of waist-nipper factors that affect subjective feeling and physiological response – For the design of comfortable women's foundation garments', *Sen'I Gakkaishi*, **61**, 6–12.

Nethero S (2007), *Method for optimization of bra fit and style for all body sizes and types and for wearing with a variety of clothing styles*, USPTO, U.S. 11/829,568.

Ng S P, Yu W, and Li Y (2009), 'Photogrammetric prediction of girdle pressure', *Measurement Science and Technology*, **20**, 015804.

Nielsen R and Endrusick T L (1990), 'Thermoregulatory responses to intermittent exercise are influenced by knit structure of underwear', *European Journal of Applied Physiology and Occupational Physiology*, **60**, 15–25.

Niwa M (1968), 'Water vapor permeability of underwear', *Journal of the Japanese Research Association for Textile End-Uses*, **9**, 446–450.

O'Farrell N, Chung C K and Weiss H A (2008), 'Foreskin length in uncircumcised men is associated with subpreputial wetness', *International Journal of STD & AIDS*, **19**(12), 821–823.

O'Hare L (1997), 'School compression hosiery in the management of venous disorders', *British Journal of Nursing*, **6**, 391–394.

Okabe K and Yamana N (1991), 'Dimensional change and oppression feeling caused by wearing body suits', *Journal of the Japanese Research Association for Textile End-Uses*, **32**, 72–79.

Pace E L and Braxton G S (2003), *Pad printing on textile substrates*, USPTO, U.S. 10/352,536.

Parker M J and Gillespie L D (2006), 'Effectiveness of hip protectoes for preventing hip fractures in elderly people: systematic review', BMJ (Clinical research edn), **332**, 571–574.

Pechoux B L and Ghosh T K (2002), *Apparel Sizing and Fit*, The Textile Institute, p. 11.

Pechter E A (2008), 'Methods for determinging bra size and predicting postaugmentation breast size', in *Breast Augmentation: Principles and Practice*, M Shiffman, editor, Springer Verlag, Berlin, Heidelberg.

Rie H and Teruko T (2006), 'Effects of air temperature on the touch sensation of underwear fabrics-evaluation of the touch sensation by wearing test', *Journal of the Japan Research Association for Textile End-Uses*, **47**, 740–748.

Sato E and Ohara K (1997), 'Compliance with the extension of the skin', *Journal of the Japanese Research Association for Textile End-Uses*, **38**, 424–430.

Shafik A (1973), 'The cremasteric muscle, Role in varicocelogenesis and in thermoregulatory function of the testicle', *Investigative Urology*, **11**, 92–97.

Sheynkin Y, Jung M, Yoo P, Schulsinger D, and Komaroff E (2005), 'Increase in scrotal temperature in laptop computer users', *Human Reproduction*, **20**(2): 452–455.

Shimizu H, Sakaguchi A, Fukui J, and Shimazaki A (1995), 'The effect of silk underwear on skin', *Journal of Sericultural Science of Japan*, **64**, 23.

Shin K (2007), 'Patternmaking for the underwired bra: new directions', *Journal of the Textile Institute*, **98**, 301–317.

Shin K, Yoo D L, and Son K (2005), 'Development of thermoregulating textile materials with microencapsulated phase change materials (PCM). IV. Performance properties and hand of fabrics treated with PCM microcapsules', *Journal of Applied Polymer Science*, **97**, 910–915.

Silva M D (1986), 'The costoclavicular syndrome: a "new cause" ', *Annals of the Rheumatic Diseases*, **45**, 916–920.

Singer S R and Soma G (1995), *Dressed to Kill: The Link between Breast Cancer and Bras*, Sydney, Avery Publishing Group.

Smith J (1986), 'Comfort of clothing', *Textiles*, **15**, 23–27.

Sone Y, Kato N, Kojima Y, Takasu N, and Tokura H (2000), 'Effect of skin pressure by clothing on digestion and orocecal transit time of food', *Japanese Journal of Physiology Anthropology*, **19**, 157–163.

Sour T (2007), *Protective underwear ith abdominal support*, USPTO, U.S., 11/903,234.

Srinivasan J, Ramakrishnan G, Mukhopadhyay S, and Manoharan S (2007), 'A study of knitted fabrics from polyester microdenier fibres', *Journal of the Textile Institute*, **98**, 31–35.

Steele V (1996), *Fetish: Fashion, Sex, and Power*, New York: Oxford University Press.

Suda R and Tamura T (2006), 'Effects of air temperature on the touch sensation for underwear fabrics (part I): Evaluation of the touch sensation of the hand', *Journal of the Japan Research Association for Textile End-Uses*, **47**, 41–48.

Sweeney M and Zionts P (1989), 'The "second skin" perceptions of disturbed and non-distributed early adolescents on clothing, self-concept, and body image', *Adolescent*, **24**, 411–420.

Tanaka M, Yoshida M, and Hirata K (1999), 'Region immediately after the wear of girdle on the rate of blood flow in the skin at the bottom of feet and surface humidity on the skin', *Journal of the Japanese Research Association for Textile End-Uses*, **40**, 175–182.

Tanaka S, Midorikawa T, and Tokura H (2006), 'Effects of pressure exerted on the skin by elastic cord on the core temperature, body weight loss and salivary secretion rate at 35 °C', *European Journal of Applied Physiology*, **96**, 471–476.

Turek P J (1998), 'Editorial: Boxers and biopsies – separating fashion from fact in male infertility', *Journal of Urology*, **160**(4), 13–37.

Ukponmwan J O (1998), 'Pilling', *Textile Progress*, **28**, 1–55.

Utaka S (2005), *Clothes with back part such as brassiere*, USPTO, U.S. 11/597,810.

Watanuki S (1994), 'Improvements on a design of girdle by using cardiac output and pressure sensation', *Annuals Report of Physiology Anthropology*, **13**, 157–165.

Weber H A, Budd F W, and Curlin J P (1972), 'Iwata Obi (Japanese maternity lumbosacral support)', *Military Medicine*, **137**, 359–60.

Williams F, Knapp D, and Wallen M (1998), 'Comparison of the characteristics and features of pressure garments used in the management of burn scars', *24*, 329–335.

Yamaguchi Y, Morooka H, and Morooka H (2002), 'Effect of stimulation strength of clothing humidity on humidity sensitivity', *Sen'I Gakkaishi*, **58**, 61–69.

Yick K L, Wu L, Yip J, Ng S P, Yu W (2010), 'Study of thermal-mechanical properties of polyurethane foam and the three-dimensional shape of molded bra cups', *Journal of Materials Processing Technology*, **210**, 116–121.

Yoon H N (1984), 'Improved comfort polyester', *Textile Research Journal*, **54**, 289–298.

Yu W (2009), 'Bra fitting checklist', *Teaching Notes of Body Beauty and Fit*, The Hong Kong Polytechnic University.

Yu W and So Y K (2001), 'Special techniques for making push-up bras', *ATA Journal*, **12**, 69–70.

Zheng R, Lim N Y, and Yu W (2006), 'Breast sizing', in *Innovation and Innovation and Technology of Women's Intimate Apparel*, Yu et al., editors, Cambridge: Woodhead Publishing Limited, p. 168.

Zheng R, Yu W, and Fan J (2007), 'Development of a new Chinese bra sizing system based on breast anthropometric measurements', *International Journal of Industrial Ergonomics*, **37**, 697–705.

Zheng R, Yu W and Fan J (2009), 'Pressure evaluation of 3D seamless knitted bras and conventional wired bras', *Fibers and Polymers*, **10**(1), 124–131.

Ziegert B and Keil G (1988), 'Stretch fabric interaction with action wearables: defining a body contouring pattern system', *Clothing and Textiles Research Journal*, **6**, 54–64.

Index

CPI Antony Rowe
Eastbourne, UK
May 10, 2022